OF TIME,

PASSION,

AND

KNOWLEDGE

OF TIME,
PASSION,
AND
KNOWLEDGE

Reflections on
the Strategy of Existence

SECOND EDITION

J. T. FRASER

PRINCETON UNIVERSITY PRESS
PRINCETON, NEW JERSEY

Published by Princeton University Press,
41 William Street, Princeton, New Jersey 08540
Copyright ©1975 by J. T. Fraser; preface to the revised edition ©1990 by J. T. Fraser
All Rights Reserved

Library of Congress Cataloging-in-Publication Data

Fraser, J. T. (Julius Thomas), 1923–
Of time, passion, and knowledge : reflections on the strategy of existence/
J. T. Fraser.—Rev. ed.
p.   cm.
ISBN 0-691-08572-2 (alk. paper).
ISBN 0-691-02437-5 (pbk.: alk. paper)
1. Time. I. Title.
BD638.F67   1990
115—dc20   89-77620

This book was originally published by George Braziller, Inc., and is reprinted by
arrangement with the author. First Princeton University Press printing of the revised edition,
1990

Princeton University Press books are printed on acid-free paper, and meet the guidelines for
permanence and durability of the Committee on Production Guidelines for Book Longevity of
the Council on Library Resources

10 9 8 7 6 5 4 3 2 1
10 9 8 7 6 5 4 3 2 1, pbk.

Printed in the United States of America by
Princeton University Press, Princeton, New Jersey

TO THOSE

WHO IN FLEETING MOMENTS

REACHED WITH ME

FOR THE HORIZON

CAPTION TO
THE PAPERBACK COVER

The cover design is from a Greek bronze mirror, said to have come from Corinth, and is dated about 350–300 B.C. It shows Aphrodite and Pan playing five-stones.

Originally a magical means for divining the future, the five-stones of this engraving is a game of chance. Its throws, like those of dice, are controlled by probabilistic laws. The behavior of elementary particles is also governed by probabilistic laws. As in the game of five-stones so in the world of particles, no event can be foretold as certain to happen, but only as likely to happen with a given degree of probability. The time of such a probabilistic universe is different from the time of ordinary human experience. It misses continuity, and nothing in its nature corresponds to our notions of now versus then, or to the related distinctions among future, past, and present.

Aphrodite is seated on a stone slab next to Pan—the god of woods and pastures. Their bench is a sample of solid objects that are formed when elementary particles jell into massive matter. The universe of Newtonian physics comprises small as well as immense chunks of such matter. Events in that world are connected by deterministic rather than probabilistic laws, time is continuous, and a meaning may be assigned to the idea of well-defined instants. But it is still a universe without the kind of temporality that might be metaphorically described as flowing.

The goose in front of the bench is a traditional attribute of Aphrodite—as are doves, swans, and dolphins. Let her small entourage represent the life process itself. It is the instant by instant internal coordination necessary for the maintenance of life that defines a now in the nowless world of inanimate matter. And, it is the biological needs of organisms that distinguish between future and past with reference to those present needs, and thereby identify a direction of time.

The central figure is Aphrodite, the goddess of love, beauty, and fertility. According to Homer, "her enchantments came from this: allurement of the eyes, hunger of longing, and the touch of lips that steals all wisdom from the coolest men." Through her capacity to create designs for a long-term future and retain memories of a long-term past, the Aphrodite of this Arcadian image represents the complex time of the human mind.

The winged youth is Eros, in late Greek mythology the son of Aphrodite. Earlier, as in Hesiod, he is the son of Chaos; in Aeschylus he assumes the concrete form of rain through bringing heaven and earth into creative embrace. In Plato the principle of Eros is a perpetually dissatisfied, restless force that searches for the timeless and the eternal. In the visual metaphor of the engraving, I would like to think of Eros as representing the conflicts of the human mind between its longing for timelessness and its certainty of passage.

CONTENTS

PART TWO

Images in Heaven and on Earth

PART THREE

The Mind of the Matter

PART FOUR

Collective Greatness

THE ARGUMENTS OF TIME

Foreword to the Second Edition

The past is not a frozen country that may be discovered and described once and for all, but a chart of landmarks and paths which is continuously redrawn in terms of new aspirations, values, and understanding. A bust of Homer, considered as an expression of ideas and feelings, is not the same today as it was yesterday because we ourselves have inevitably changed. It is thus that the second edition of *Of Time, Passion, and Knowledge*, though a reprint of the 1975 edition, may nevertheless be said to be unlike its earlier self. For the study of time has come of age and in the light of its developing insights one may now return to this volume and in the words of T. S. Eliot, "know the place for the first time."

In the last quarter century but especially during the last decade, there has been a rapid increase in scientific, scholarly, and popular interest in the experience and idea of time. Scores of conferences have been held and hosts of books and papers published dealing with time from either or both of two complementary perspectives: one is the nature of time seen from the point of view of a discipline or a writer, the other is the role that time plays in the processes that are being considered. The subjects of these meetings and writings are spread across the spectrum of intellectual and practical knowledge from physics and biology to psychology and history; from political science, public and business administration to anthropology and economics; from philosophy, sociology, and religion to the arts and letters, geology, and geography.

The reasons for this upsurge of activity are numerous and complex—as is always the case with the forces that move human thought and action. But it is possible to point to a dimension of contemporary life that, while obviously not responsible for the perennial human concern with time's passage, does encourage the focusing of interest upon the nature of time.

I submit that time is of special interest to the inquiring minds of the men and women of our epoch because the socioeconomic, ecological, and ideological crises characteristic of the end of the twentieth century are, in a fundamental way, time-related. Specifically, they arise from the time-compactness of our lives and/or derive from and shape certain changes in peoples' assessments of the relative importance of future, past, and present.

The conditions that made it possible to have global problems to begin with were, until recently, all but unimaginable. In response to their unprecedented threats and promises, all received views about the position of man in the universe have become suspect and are being questioned. A result is that in the process of integrating technological and scientific progress with human needs, the second half of this century has seen a random search for guidelines rather than the continued elaboration of existing ideologies. Yet, beneath the many-sided revolution of ways and means, the most significant aspect of time for man remains its opportunities of finding ideals to die for and hence, good reasons to live. Although there is no shortage of personal heroism, there is a dearth of ideals that appear intelligible as well as praiseworthy to the majority of the heterogeneous population of the earth, nor is there anything inspiring or even vaguely satisfying in the flood of undigested data, mistaken for knowledge, that clogs the arteries of civilized discourse.

In the presence of novel challenges and untried answers, the family of people around the earth is trying to work out the checks and balances of a viable global society. Cultural, economic, military, and religious empires, while trying to retain their distinctness are creating—in spite of themselves—a single socioeconomic matrix that could accommodate a plurality of collective temperaments and ways of life. The character of that matrix will determine the fate of our species well into the twenty-first century. Simultaneously, in the world of ideas, there is a struggle to decide which of the many views of the past is most appropriate to the material and spiritual needs of a time-compact globe. The ethos most likely to conquer the minds of people will be one that can propose an interpretation of history on which a believable and desirable future of mankind could be founded.

Prompted by a sense of aloneness, men and women of all past ages have searched for and imagined other beings similar to as well as different from themselves, so that they may define their own identities with reference to these others. This search for identity has become global through the homogenizing effects of communications technology. But defining the identity of a global society is a difficult task, for all definitions of identity involve comparisons and there is no other family of man with which this one could be compared. There is no other humanity to check the excesses of this one, as one tribe would stop—because of its self-interest—the excesses of another tribe. It follows that in spite of its kaleidoscopic inner dimensions, a global society is likely to enter the set of those beings of which there is thought to be only one, such as God or the universe. The conceptual difficulties of dealing with unique beings of this kind and the practical consequences of those difficulties are well known.

Surveying the immensity of space for other humanities, looking as it were for clever siblings or rich uncles—for other technological civilizations, in current jargon—to serve as extraterrestrial saviors with answers to mankind's ills is an overly naive enterprise, even if it employs great technological sophistication. If we wish to assess the dangers and promises of our unique position in the known world, in terms of contemporary scientific and humanistic preparedness, then it is a much more practical plan to explore the nature of time, the primary domain of the life process, of the human mind, and of social transaction. *Of Time, Passion, and Knowledge* is a contribution to such an exploration.

This book is necessarily encyclopedic; that is, it offers *enkyklios paideia*, in the circle of knowledge, a comparative survey of the material minimally necessary for a study of the idea and experience of time and of its roles in the many ways of human knowledge. At the same time it also outlines a new natural philosophy—the hierarchical theory of time—which serves as a model for the interdisciplinary study of time.

The Chorus in *The Winter's Tale* is Time personified, an actor who speaks about himself (itself) in the third person:

> . . . let Time's news
> be known when 'tis brought forth. A shepherd's daughter
> And what to her adheres, which follows, after,
> Is th' argument [subject matter] of Time.

Perdita—the shepherd's daughter—is a lovely subject matter but, judging from the wealth of ideas in this book, time has quite a number of other arguments. How does one study something as familiar yet strange, as ever-present, as many-faceted as time?

Of Time, Passion, and Knowledge answers this question by recognizing a well-defined structure in the nature of time. What used to be regarded as a uniform flow which embraced equally all structures and processes, is revealed as a nested hierarchy of qualitatively different temporalities. The recognition of a structuring of time allows the development of a multidisciplinary, integrated approach to the study of time within a single vision of reality.

In our epoch of simultaneous homogenization and fragmentation of values and institutions, the interdisciplinary study of time so conceived permits a measuring up of the capacities and limits of man against the order of nature, by means that satisfy the formal demands of the sciences as well as the disciplined speculations of the humanities.

Hickory Glen
Connecticut
September 10, 1989

OF TIME,

PASSION,

AND

KNOWLEDGE

ASCENT—BY WAY OF INTRODUCTION

Some Ideas about Philosophy

ONLY A WAYFARER BORN UNDER UNRULY STARS would attempt to put into practice in our epoch of proliferating knowledge the Heraclitean dictum that "men who love wisdom must be inquirers into very many things indeed." Indeed, the classic function of all philosophies is to offer a comprehensive view of the many things that make up our one world and thereby assist and guide man in his search for meaning and order in his life.

But to philosophize today in this tradition of wide concern is a very difficult task. First of all, an overwhelming amount of information reaches us through the sciences and a very large variety of utterances come to us through the humanities. To make it worse, however, the intellectual Zeitgeist of the first six decades of this century rejected defensively any trend of thought toward universals which would include but reach beyond the confines of the sciences. The success, in the sciences, of analytical thought and of the experimental method encouraged in philosophy a flight from potentially disturbing transcendental speculation toward the security of formal argumentation, often without human significance. Though these methods and trends proved to be useful to the artisanry of science, they failed to inspire, illuminate, or unify the various ways man experiences reality in a world where the absence of inspiration, illumination, and a unity of vision are all too painfully evident.

But man's desperate search for meaning in action and order in chaos has not decreased during the twenty-six centuries since Heraclitus, or during the last twenty thousand years; if anything, it seems to have increased. Certainly, the profound spiritual malaise of our century cannot be placated and the philosopher excused by meekly pointing to the great intellectual difficulties of any program implied in the preceding paragraph. Today, as

3

in all recorded history, as man strives to create and preserve his individual and communal identity, he finds himself surrounded by problems he can neither totally understand nor conveniently reject as unintelligible, such as time, life, death, or the existence of a universe. He is also driven by aspirations whose goals he cannot hope to reach, yet cannot accept as unreachable, such as his ethical needs for justice and truth, or his aesthetic demand for consummate beauty. Beneath these existential dilemmas man still lives by and for ideas; yet, simultaneously, he remains only superficially a reasoning animal. More basically he is a desiring, suffering, death-conscious, hence time-conscious, creature. Neither the meager diet of scientific interpretation posing as philosophy, nor the obscurantism of pseudomysticism disguised as metaphysics is useful to someone stooped under the burden of daily struggle, while he is carried along into a "brave new world" with the storm of social and industrial advance.

Because of the absence of ideals inspiring as well as intelligible, in terms of contemporary scientific and humanistic preparedness, I see our epoch as essentially uninformed in spite of the spectacular results of man's control and abuse of himself and of his world. But if this be true, then there is a great need to pursue the classic task of the philosopher, which is the search for the universal.

Arrows, Thoughts, and Experiences

IN THE FAMOUS PARADOX of the flying arrow, associated with the name of Zeno of Elea, it is argued that at every instant of time an arrow occupies a length of space no longer than itself; it would follow, therefore, that the arrow does not really fly. The stationary, instantaneous arrow is a good metaphor for what we call "fact": an unchanging condition, or statement. At each epoch the body of knowledge, whether in form of scientific laws, religious beliefs, or in myths, occupies a certain volume of the intellect and no more. If the arrow is to fly so as to hit its mark, it must be permitted to behave in a way which is essentially unpredictable from its stationary condition: it must extend beyond itself; it must move. Likewise, a body of knowledge must be permitted to reach out beyond the "facts" and seek relations in ways not predictable from its present state. It is a unique and essential property of the mind that it is capable of doing so with a high probability of having been correct, when its conclusions are examined in retrospect. This function of the mind is a process of creative perception which, when it comes to intellectual matters, may be described as disciplined speculation. The tenor of this book is disciplined speculation submitted in the hope that like Zeno's arrow, and in spite of the logical difficulties implicit in its flight, it will nevertheless accomplish its purpose.

Seeking a comprehensive view of time is a task comparable to putting together a jigsaw puzzle whose pieces are alive and moving. And to continue our metaphor, the pattern we are seeking is engraved on the noetic and carnal passions which continuously emerge from the tensions that characterize the temporal existence of these living pieces. Accordingly, embedded in the formal reasoning of the book, the reader will find notes and remarks expressing the author's personal views. They are reminders that, whatever the specific area of knowledge we happen to be considering, it is never more than a selected aspect of the totality of existence. Their tenor expresses the author's revolt against the tyranny of objectivism and reconfirm his belief in the validity of conclusions reached in an earlier work. Namely, that in the study of time analytical summaries often appear meaningless, and that our thoughts and feelings should focus, instead, on attitudes "expressible only through a description of that involvement of man in life which gives rise to the problem of time in the first place." * Men are admitted into Heaven, wrote Blake in his *Vision of the Last Judgment*, "not because they have curbed & govern'd their passions or have no passions, but because they cultivated their understanding."

Leitmotivs

IN THE BRIEF concluding chapter I shall sketch the salient features of a new theory of time based on concepts which will be developed and will have been found useful in the book. According to that theory, conveniently referred to by one of its many features as the theory of time as conflict, what we ordinarily call "time" comprises hierarchically organized temporalities, each contributing something different to the temporal experience of man, and all displaying certain common qualities which do, however, take different forms. But, we do not set out following the ordinary path of axiomatization and deduction so as to prove this theory because of the profound methodological difficulties implicit in any serious interdisciplinary study of time. These difficulties fall into four major categories.

First, there is a language problem. Each field of knowing, be it clockmaking, genetics, or the comparative study of religions, has its own vocabulary which gives different meanings to otherwise identical words. Each field has a repertory of specific concepts, each uses certain stock phrases (often unanalyzed), and each has its own preferred ways of putting things. If the reader comes upon some unusual usage of words this may, of course, reflect the eccentricity of the writer but it is more likely that he has encountered a usage peculiar to a field of learning but largely unknown, in that sense, outside that field. I have tried to follow the modes of expression

* J. T. Fraser, ed., *The Voices of Time* (New York: Braziller, 1966), p. 593.

acceptable and familiar to speakers of these many languages without making the fatal assumption that speaking a specialist tongue makes one an expert or even guarantees acceptable sense. Basically, however, one has to resort to one's own ways of saying things.

Second, there are profound disagreements among professions regarding acceptable methods of reasoning. What by the standards of one profession is judged as correct and salutary argumentation often constitutes, by the standards of another intellectual discipline, a useless and perhaps even reprehensible waste of time. In this volume we consult many fields and share many types of reasoning, each for what the author believes those fields have to offer. We will try to keep in mind that the divergence among standards often stems from the practical—and unavoidable—necessity of compartmentalization.

Third, there is the problem of the personalities of knowledge. As I shall argue in the chapter on epistemology, various distinct ways of knowing display personality traits which are maintained and reinforced by the personalities of the people who create the many branches of knowledge. It follows that different methods of seeking truth are not neutral categories of inanimate tools but different ways of life sought by the disciples because of the satisfying, or reassuring, or perhaps disturbing emotional experience they offer.

Finally, these difficulties are further compounded by the fact that no one can be expected to be equally conversant with the substance of the many disciplines which must be incorporated in a study of time. This is a problem in addition to the linguistic and methodological hurdles. To alleviate the situation I have included summary reviews with each subject. The extent of these reviews was dictated by simultaneous considerations of many issues: judgment of relevance, the degree of familiarity which I could assume on behalf of the reader, and the historical or ahistorical character of the material itself. Within each topic, I found the historical approach the most efficient one, implying perhaps that only in the continuity of time can our present ideas of temporality be sufficiently understood and put in valid perspective. Decisions as to what specific material to honor and weigh, whether by endorsement or rejection, were guided by consulting peer judgements as set down in the pertinent literature. Selection still remained arbitrary, in that it had to be determined by the author, but also not arbitrary, in that it was dictated by a desire to maintain a unity of purpose. One might say by analogy that a symphony is characterized by the tones arbitrarily excluded.

So as to master or at least be able to live with these difficult conditions, the book is designed to follow certain leitmotivs that will eventually make the arguments of the theory of time as conflict appear convincing. The vocabulary and phrases of these leitmotivs will be defined as we explore the idea of time, intellectual discipline by intellectual discipline. This exploration is itself interesting as well as valuable, quite apart from the admittance or rejection of a unifying theory.

The documentation of the book ought to be regarded as a carefully prepared guide, but only as far as it goes. For more detailed data the specialist reader must consult papers and monographs which focus on arbitrarily narrowed problems so that they may provide information, rather than on large issues so that they may construct theories. But it is hoped that for the specialist reader the backgrounds as outlined might, nevertheless, be useful in orienting his preparedness toward the study of time and provide a convenient entry into that field. The general reader may find the history of the idea of time a superb intellectual whodunit, and read the book by following the instructions of W. B. Yeats:

> Because to him who ponders well,
> My rhymes more than their rhyming tell
> Of things discovered in the deep,
> Where only body's laid asleep.
> For the elemental creatures go
> About my table to and fro.
> W. B. Yeats,
> "To Ireland in the Coming Times"

Organization of the Material

I HAVE FOLLOWED, but only very roughly, the conventional boundaries of academic disciplines. In terms of these boundaries the following organization holds for the chapters of the book.

1. Philosophy
2. Time measurement
3. Preliminaries on man
4. Physical science
5. Cosmologies
6. Organic evolution
7. The mind-brain problem
8. Psychology
9. Epistemology
10. Ethics
11. Aesthetics
12. Time as conflict

The chapters do not actually bear these titles because their contents, although so focused, are not so delimited.

Through much of the work there was a need to imply certain new and useful concepts before they could be critically evaluated. Accordingly, I have had to employ certain terms in ways corresponding to their ordinary usage and then later refine or even alter their meanings. Also, I have often had to jettison the ordinary linearity of reasoning and proceed, metaphorically speaking, along several spokes of the wheel, one after the other, in the hope that we would arrive at a single hub. This was necessary because the evolution of thinking displays feats of multiple rather than single causation. Important features of body and mind have often emerged as the confluence of many and originally disparate functions. And, unlike a composer who may employ many simultaneous voices, the writer must speak only one sentence at a time.

Finally, the title deserves a note. According to the theory of time as conflict, the most comprehensive level of temporality commanded by man (something we shall call nootemporality) is associated with certain unresolvable conflicts of his faculties or capacities. One way we may describe this conflict is to regard it as one between knowledge felt and knowledge understood. These two warring projections of a single underlying tension are ordinarily recognized as passion and knowledge.

Thus, with Autolycus the Rogue in *The Winter's Tale*, we may begin:

> Jog on, jog on, the footpath way,
>   And merrily hent the stile-a:
> A merry heart goes all the day,
>   Your sad tires in a mile-a.

PART ONE

The Glass Wall

TO the distant observer the nature of time appears to be intuitively obvious, as though it were an object totally revealed to the searching eyes. As he approaches this object, however, he finds himself separated from it by a transparent wall which did not interfere with the earlier view taken from a distance.

I

THE INTELLECTUAL
QUEST

WHEN George Berkeley, eighteenth century Irish bishop and philosopher, attempted to explain to himself what he meant by the idea of time exclusive of particulars, he found himself "embrangled in inextricable difficulties." Twenty-one centuries before him the Eleatic philosophers bore witness to a similar "embranglement," though on a different basis, through their speculative argumentation on being and becoming. Fifteen millennia before the Greeks, paleolithic man already had expressed through his cave art a profound spiritual malaise stemming from his awareness of temporal passage. It seems that understanding the inevitabilities of change and permanence has been a difficult yet enticing task.

The earliest manifestations of a sense of time in man will be dealt with elsewhere: paleolithic records are discussed in the context of calendars; creation stories in the chapters on cosmologies; time and the origins of religions in the context of faiths. In this chapter we begin with the age of the Homeric epic, that is, with an epoch whose character permits us to ask questions about the conceptualization of time.

1. Representative Ideas of Time
in Western Thought

Aegean Beginnings

During the eighth century B.C. Homer, "Son of Seven Cities," recorded certain events which took place three or four hundred years before his time. Commentaries and speculations about Homer the man have been continuous from the early textual criticism of the third century B.C. to the modern classical scholarship of F. A. Wolf and others. It is difficult to identify the personality of this semimythical genius but tradition has held that he was blind. This is an important matter when we try to describe the

11

classical Greek view of temporality, for although his poetry was profoundly
influential, Homer might in subtle ways be unrepresentative as a witness
for his age. An analysis of Homeric dreams shows them to be predominantly
auditory with the visual elements obscure and shadowy, whereas, in contrast, dreams of early Greek drama are predominantly visual.[1] Since visual
descriptions of events tend to stress the spatial, whereas auditory descriptions stress the temporal, one reason for Homer's success in forming the
enduring Greek ideals might have been that he was uncommonly conscious
of mortality, and thus of heroism as a way of conquering death.

H. Fränkel among others examined the concept of time in early Greek
literature and believes to have found in Homer a complete indifference to
time, by which he means the absence of the concept of a universal, temporal flux.[2] Whitrow, while seeking reasons for the Greek failure to anticipate the scientific revolution of the seventeenth century, has observed,
following Cornford, that the Homeric epics stress the supremacy of the idea
of space and describe a world without cosmogony and creation.[3] This
attitude will reach its most articulate expression, as we shall see, in Plato's
resistance to ideas of generation.

It has been observed that the nestling and reentrance of events in
Homeric epics resemble by analogy the protogeometric art of the eleventh
century B.C.[4] Perhaps these were no more than useful "bookkeeping" devices
for the blind poet. Yet E. Auerbach, assuming close correspondence between style and world-view, argues for a Homeric indifference to time,[5]
whereas J. G. Gunnell sees in the proto- and ripe geometric forms the genius
of order created from chaos, showing a keen sense of time in those who
recited and in those who listened to the poems.[6] Homeric epic is informed
of time as duration, as before and after, life and death, as fate, youth and
aging, and as day following day but not of time as some ongoing universal
process or abstract property of the world at large. Roughly, this corresponds to the preoperational level in the cognitive development of the
child in genetic epistemology. And, just as in the language of children, in
Homer we never find "time" as the subject of a verb.[7]

Unlike Homer's praise of the sublime, the *Works and Days* of Hesiod,

....

*Homeric epic is informed of time as duration, as before and after, life and
death, as fate, youth and aging, and as day following day, but not of
time as some ongoing, universal process or abstract property of the
world at large.*

*Black figured lekythos (oil flask) ca. 500–490 B.C., attributed to the Sappho
Painter. Courtesy, The Metropolitan Museum of Art, New York, Rogers
Fund, 1941. Helios is seen rising from the sea. Nyx ("Night," daughter
of Chaos) and Eos (goddess of Dawn) are seen disappearing.*

....

his contemporary, reads rather like an early farmer's almanac of exhortations and warnings. It informs the reader of opportune and inopportune times for various agricultural activities so that want and misery may be conquered by proper judgement of natural cycles. "While it is yet midsummer command your slaves: 'it will not always be summer, build barns!' " [8] Unlike Homer, this Boeotian shepherd was a poet of metaphysical bent. His *Theogony* gives an account of the origins of the world *ex nihilo* and tells "how at first gods and earth came to be, and rivers and the boundless sea with its raging swell and the glowing stars." [9] First emerged the three primeval gods: Chaos, then "wide-bosomed earth," then Eros. His grand story of divine struggle is unsuspecting of progression in the world beyond changes in generations. Hesiod saw history as degeneration and estrangement from the divine. Man is caught in the struggle between two Erides: Eris, the Greek goddess of strife, injustice, and cruelty and her sister (invented by Hesiod), responsible for diligence and ambition. [10] Human time is then characterized by the tension between these two conflicting forces.

Whereas Hesiod remained bucolic, Heraclitus of Ephesus (540–480 B.C.), nicknamed the Sad, perceived the world in terms of ruthless conflict and change. Perhaps he projected his own character upon it: "Hard it is to fight against impulses," he wrote, "whatever it wishes it buys at the expense of the soul." [11] He saw the coexistent opposites of the world held together by Logos, an immaterial but permanent principle exemplified by the unity of God who is "day-night, winter-summer, war-peace." [12] In his view the world comprises continuous change: the opposites oscillate between their polar states and the pluralistic whole moves toward and away from its source. Particular things exist by virtue of opposites being locked in conflict. "In the same river we both step and do not step, we are and are not," [13] illustrates the unity of opposites. The Heraclitian world was a totality of processes, rather than things, existing by virtue of strife and tension between opposites.

Parmenides, a contemporary of Heraclitus and native of Elea in Southern Italy, was of a very different mind. He rejected the insecurity of restless change, even though not in those words, and sought the essence of the world in the safety of his remorseless logic, in the strength of dialectic disputation and in dogmas of permanence. Perhaps, as did Heraclitus, Parmenides also projected his own character upon the world; he belonged to the aristocratic brotherhood of the Pythagoreans or at least had a Pythagorean teacher. He focused his attention on what he regarded as the true reality of the world (hence *The Way of Truth,* part of his only extant work) in contrast to the way of the seeming and the apparent. He realized that if time is thought of as becoming, then an object would have both to be and not be—at different times. According to the representations of his views by Plato he got away from this anguishing logical difficulty by formulating three metaphysical dogmas: that which exists, is; that which exists not cannot even be named; and that which is one cannot be many. Since things do seem to come into and out of being, it then followed that the world of the sense is chimerical, change and time are virtual, and the multiplicity of the world is only an impression.[14]

As do idealistic philosophers of our own epoch, Parmenides believed that the object of knowlege must exist in and be found by the mind, not by the senses. Curiously, while his identification of the One with whatever is timeless must be regarded as correct, he did not seem to have realized that identities may be defined only in terms of nonidentities. Even as we think of the universe as One, the observer must be regarded as external to it, for an object may be defined only in terms of self-and-other relationships. Had Villon asked Parmenides, "Mais où sont les neiges d'antan?" he might have answered: the snows of yesteryear still exist, they always did and always will, for ultimate reality is timeless.

During the middle of the fifth century B.C. a medical-naturalistic image of time emerged from the writings of the Sicilian-Greek philosopher, statesman, and medical man, Empedocles (490–430 B.C.). In modern terms it may be called biological or organic. He identified love and strife as the two cosmic forces whose interaction in many forms determines the texture of existence.[15] Love and strife were dynamic fluids or hypothetical substances somewhat in the way that seventeenth century chemists regarded phlogiston. The idea of all-filling material reappeared in our epoch first as ether, then as the ideal fluid in the cosmology of general relativity theory. The Empedoclean idea of a universe oscillating between poles of unity and diversity also reappeared in contemporary clothing in the physical model of the oscillating universe. His celebrated theory, that perception is due to an encounter of an element in us with the same element from the outside, resembles the representative theory of perception. These continuities across twenty-five hundred years do not imply in any way that he anticipated modern science; they show only a continuity of patterns of thought in the history of ideas. In any case, he did try to reconcile the permanence of being, as emphasized by Parmenides, with the experience of change, as emphasized by Heraclitus. He put his theory of perception and his philosophy of organic interaction together and described nature as the mixture and the separation of things mixed.[16]

After the middle of the sixth century B.C. a protoscientific and mystical view of the universe and man came into existence through Pythagoras and the Pythagoreans. According to the witness of Philolaus of Tarentum, the Pythagoreans believed that "actually everything that can be known has a Number, for it is impossible to grasp anything with the mind or recognize it without [Number]."[17] Aristotle knew that the Pythagoreans "supposed the elements of numbers to be the elements of all things, and the whole heaven to be a musical scale and a number. . . ."[18] To early Pythagoreans the writing down of a number was a generative act in the same way that we regard the painting of a picture. Quite consistently, they identified the creation of the world with the generation of numbers from an initial unity. Within this world they saw souls as transmigrating even among species, while retaining the soul's affinity to the stars of the sky. Just as numbers could be odd or even, limited or unlimited, so the world itself consisted of the harmony of opposites, such as one-many, male-female, good-evil.[19]

This fascination with number as the essence of the world, and of time, survives into our own epoch both in the scientific and in the social domains. Our sciences regard mathematization as their final goal, while the pre-occupation of all industrialized societies with number instead of quality is an uncomfortably familiar aspect of modern existence.

The sketch of pre-Socratic thought given, implies sophisticated and complex world-views with respect to time. Though they were mostly in-direct and preconceptual, the ways these men approached reality are as fresh in our epoch as they must have been in their own age. If we were to select one salient feature for each poet and thinker we mentioned, we should say that for Homer the world was the struggle of Greek mortals and immortals. Hesiod was earthy, agricultural; Heraclitus a metaphysician; Parmenides a logician of timelessness; Empedocles was a medical-natural-istic philosopher; while the Pythagoreans were committed to a world of numerical order, albeit containing the strife of opposites.

Zeno and His Virtual Disjunctions

The conclusions of Parmenides regarding the illusory nature of time and change received support from his student, Zeno of Elea, in the form of certain exciting dilemmas that have survived to our own days. Although there is some uncertainty with regard to the original wording of the para-doxes, the thoughts and purpose of Zeno are fairly clear, mainly because Aristotle, Plato, and later commentators dealt with them in great detail. They were designed to discredit the idea of change and multiplicity as advocated by Heraclitus and the pluralists, by showing that belief in motion and in distinguishable qualities leads to self-contradictory conclusions; hence, movement and plurality are virtual. It would presumably also fol-low that reality is timeless. To gain this end, Zeno manipulated concepts and mental images of certain opposites, such as motion versus rest, the finite versus the infinite, continuity versus atomicity with the proverbial shrewd-ness of a Greek merchant. I will now consider two of the paradoxes per-taining to time and motion and imply certain solutions which contain what, at this point, must be taken as working assumptions regarding the nature of temporality and knowledge.

Perhaps the most powerful one is that of the flying arrow.[20] At each instant of its flight, so Zeno claims, the arrow occupies only a region of space equal to itself, but not more. Hence the idea of motion can not amount to more than a description of static relations, certainly not a true phenomenon in itself. It then follows that motion, and with it time, must be virtual.

Whether a moving arrow is longer than a stationary one, or perhaps shorter as special relativity theory has it, does not matter. The fundamental problem is how to compound motion from no motion, or time from no time.

Let us note first that rest (no motion) may be generated through the superposition of motions of equal magnitudes but opposing directions, whereas there is no analytical superposition possible that would generate motion out of no motion. This suggests that motion be considered as epistemologically prior to no motion. The abstraction created by the mind, in the spirit of Zeno's argument, is not motion but rest. Indeed, quantum theory and relativity theory both suggest that the world is fundamentally restless. Lawfulness, which is no change, is an abstraction which combines with our existential knowledge of motion in the totality of our experience of time. Furthermore motion subsumes rest as perhaps time subsumes timelessness, but not vice versa. But if time versus timelessness and motion versus rest do show asymmetrical exclusiveness of ranks and not mutual exclusiveness of conditions then they form only virtual disjunctions. It would follow that the basic fallacy of the paradox of the flying arrow must be a subtle category mistake. Turning around Plato's famous metaphor of time as a moving image of eternity, we might then say that the stationary arrow is a frozen image of change, timelessness is a stationary image of becoming.

Zeno's paradox of Achilles and the tortoise was stated by Aristotle as follows: if the tortoise is given a start, Achilles can never come up to him for "the quickest runner can never overtake the slowest, since the pursuer must first reach the point whence the pursued started." [21] This paradox has been successfully handled by means of converging infinite series which permit the formulation of the problem in terms of the total, finite time taken by the aggregate of an infinite number of ever decreasing mathematical intervals.[22] In reality as well as in its mathematical image, Achilles does catch up with the tortoise. But the usefulness of the mathematical simulacrum stops at the correct prediction of this rendezvous.

We know that Achilles and the tortoise will meet, just as we know that the arrow moves, we need no theoretical demonstration. The challenge of the paradox pertains instead to the question of atomicity versus infinite divisibility of time and to its partial corollary, whether or not it is possible to perform an infinite number of acts within a finite period of time. The mathematical model, though useful as far as it goes, cannot shed any light on these questions because number and the uses of number belong to primitive levels of temporality, whereas acts actually performed comprise several, more advanced integrative levels. Thus, the celebrated solution using the theory of converging infinite series only transfers the paradox from classical philosophy to the philosophy of mathematics where it becomes an exercise in distinguishing the symbol of an event from the event itself.

For over twenty-four hundred years Zeno's paradoxes have been refuted, praised, ridiculed, and solved at each epoch according to the metaphysical views prevailing and the intellectual tools favored and available at that epoch. Since about the middle of the nineteenth century the increasingly sophisticated machinery of mathematical logic and semantics in its many forms have been employed to show that the atemporal world of implications can accommodate motion and time—by refuting them formally. I believe that these approaches must lead to a dead end. The solution lies in reconciling, in a unity of hierarchy, the members of such virtual disjunctions as motion and rest, finity and infinity, atomicity and continuity. For the moment, we shall rest our case with this statement.

Plato and Aristotle

Frightened by the turbulence of the Athens of his days, shocked by the execution of Socrates, the aristocratic and wealthy Plato set his sights on the safety of being and detested the insecurity that attends philosophies of becoming. The unexpected, that Heraclitus bid us await if we are to find the truth, was relegated to the inferior aspects of the world, certainty and timelessness were judged superior, and praise of the immutable became the basis of Platonic thought.

Plato regarded time as coeval with the world; it was created when the world was, and if the world were ever to vanish, time would vanish with it. In phrases reminiscent of Genesis, Plato tells us how God rejoiced in seeing the creature he had created move and live. Now, the nature of the Creator was everlasting, but to bestow this attribute upon the created was impossible. Hence, God did the next best thing. "He resolved to have a moving image of eternity, and when he set in order the heaven, he made this image eternal but moving according to number, while eternity itself rests in unity; and this image we call time." "The sun and moon and five other stars, which are called planets, were created by him in order to distinguish and preserve the numbers of time. . . ." [23]

These beautiful noetic metaphors are integral parts of the total Platonic world-view in that they identify the ultimate and real basis of existence with the incorruptible forms of geometry and suggest the influence of the Pythagorean philosophy of number. As mentioned earlier, the Pythagoreans regarded the writing of numbers as generative acts. But such a coming-into-being would have been distasteful to Plato, who distrusted the senses and all ideas of generation. Accordingly, he preferred to see in number a necessary knowledge stemming from pure, timeless intelligence rather than a generative act.[24] In *Meno* he demonstrated to his own satisfaction how such knowledge manifests itself, by posing questions about geometry to an uneducated slave and urging him to "recollect what he does not know." [25] The correct answers given by the slave were explained by Plato as coming from a collective store of eternal and absolute knowledge, suggesting further that the soul is immortal, it is One, it does not partake in change, it is unbegotten and indestructible.[26] What today we regard as mental abilities favored by natural selection, Plato took to be proofs of the immutability, hence timelessness, of knowledge. Consistently then, he believed to have found the ultimate essence of the world in a corpus of unchanging laws or forms underlying all things. Such a substratum was then regarded as knowable and real, while particulars were real only to the extent that they partook in, or somehow reflected final reality. He softened the extreme monism of Parmenides by accommodating the world of the senses, though he degraded that world by pointing to its changeable, hence inferior, nature as compared with the timeless world of ideas. This still left certain intelligible though unobservable matters unclassified, which were then relegated to space, "the mother and receptacle of all created and visible and in any way sensible things." [27]

No doubt Plato was as much interested in his works and days and nights as was Hesiod before him or Max Weber after him. But the principles which determined his relationship to the world at large, to society, and to his own self were predicated on the timeless. Since Weltanschauungs determine what methods are permitted and judged useful in the search for knowledge, and through knowledge determine the fabric of communal life, distrust of the temporal survived in the West through Platonic and Aristotelian philosophy. It came into collision with new and emerging world-views only with the Renaissance discovery of time and the subsequent scientific and intellectual revolutions.

For Aristotle, a student of Plato, intelligibility continued to be the norm of philosophizing, exemplified in that which remains identical with itself. Yet for Aristotle time, motion, eternity, and change became concepts to be analyzed and taken seriously. His often quoted definition of time evolves along arguments as follows: we measure more or less by number; we measure more or less movement by time; hence, more or less movement is more or less by number, therefore "time is just this—number of motion in the respect to 'before' and 'after.'" [28] This definition cannot be interpreted, however, without reference to Aristotle's views of motion and of number.

In agreement with Plato, Aristotle regarded movement and change as inferior aspects of reality. Linear motion is perishable since it must end and remain therefore forever imperfect; circular motion, however, since it has no definite beginning and end is imperishable, hence real; rotary motion therefore is prior and superior to linear motion. The Pythagorean contention that all things come from number was criticized in *Metaphysics*. Aristotle held that number, whether number in general or of abstract units, cannot be the cause of things, nor can it form matter, nor can it be the final cause of anything, and "how are the attributes—white and sweet and hot—numbers?" [29] Time, then, in the definition given in *Physics* is not a number of some sort in the Pythagorean sense, but is rather *that aspect of motion* by which it is measurable. In this interpretation, the Aristotelian "number of motion" is not a definition of time but an explication of the operation of time measurement. Temporality is smuggled in through the provision of before and after.

Reflecting on the meaning of before and after, he concluded that time has cycles, hence, a rhythm, but not a direction.[30] Finally, that all things are measured by the regular circular movement of the heavens explains, he wrote, what in his epoch was regarded a common saying, "human affairs form a circle." [31] But if all things are like circles then it is legitimate to say, as he did, that we are both after and before the Trojan War. What final view one is to take with respect to the details, he does not seem to have said. From his attitudes it would follow however that, in his view, no radically new things can arise in history.

Aristotle did insist on coming-into-being, but his was a very weak sample of that principle. In *Physics* he wrote that "in time all things come into being and pass away; for that reason some called it the wisest of things, but the Pythagorean Paron called it the most stupid, because in it we also forget; and this was the truer view." [32] Although coming into being is the subject of his book *On Coming to Be and Passing Away*, his insistence on becoming sounds very much like an insistence on being. "Things which come to be do so of necessity because a cyclical series of changes is absolutely of necessity." [33] He concluded by observing that even destruction takes place only incidentally in time, time being only the measure of the change represented by perishing. The impression is unavoidable that he was more interested in the various static conditions connecting different states than in processes in time.

The Aristotelian view of time comprising, as it does, mainly instructions for time measurements, referenced to the imperishable rotary motion of the heavens, was characterized by Piero Ariotti as a "celestial reductionism of time." [34] By reductionism he means, following Carnap, an expression of ontological priority; in this case the priority of the circular motion of the heavens to our ideas and sense of time. This is a fortunate phrase. It expresses the historical desire of philosophers to identify some features of the world as more basic than time, and of which time is only a manifestation. In Plato, and in many thinkers after him, this reductionism has led to a radical elimination of time and its replacement by the timeless, in the tradition of Parmenides. Others, from Heraclitus to Bergson, reduced time to becoming; again others reduced it to the functioning of the mind.

Although no brief summary can do justice to the many nuances and varying opinions about time implied on the preceding pages, some generalizations are in order. The Greek ideal was intelligibility, embodied and represented best by whatever remains identical to itself, to wit, lawfulness. The most evident of such identities is the circular motion of the heavens, unending and incorruptible. Insofar as the singular, the contingent or the unexpected cannot be fitted in lawfulness, the unique and unpredictable were judged to be inferior aspects of reality. Accordingly, the modern idea of a philosophy of history, that is, a body of principles which can accommodate the regular as well as the unique and progressive events of a past, could have had no place in Greek thought. They admired what is permanent, rational, beautiful, or grandiose and suffused this admiration with a melancholy desire for an absolute order. It is in this rich intellectual soil that the seeds of many later ideas about time in the West may be found.

The Eastern Mediterranean

While the intelligent and quarrelsome Greek world lived its centuries of wars and ambition, searched for rational order, and questioned the reality of time, a view of time of an entirely different texture emerged around the eastern Mediterranean. This view of time was distinguished by its concern with what its proponents believed to have been historical facts pertaining to a select group of people, the Hebrews.[35]

The difference between the temporal views of the Hebrews and of the Greeks has been widely studied. Keeping our sight mainly on philosophy and intellectual issues, we remark here that for the Greeks the past was the tradition of heroes. Great individuals interacted with each other and with a society of divinities in an essentially cyclic world. In contrast, the Hebrew tradition was preoccupied with a divine purpose and with time as a straight line. History reflected the tension of a drama derived from a symbiosis between Israel and her God, Yahweh; a type of mutual aid arrangement known in Greek terms as amphictyonic, with God bearing to his people a relation best described as that of the father to his son. Jewish philosophy coeval with pre-Socratic philosophers and with the Schools of Athens does not exist; philosophizing was very much a Greek invention. Greek philosophical ideas about time and the Hebrew idea of time as history merged some time between the first century B.C. and, perhaps, the second century A.D. Before this merger the two ideas cannot even be discussed in terms of similar *Fragestellung*.

A brief psychological aside is appropriate. Freud noted that the Jewish people were "the declared favorite of the dreaded father." [36] For various reasons he believed that this relationship was correlative with emphasis on the intellect, renunciation of the instinct, and the return of the repressed. These conditions constitute a sufficient source of anxiety, and anxiety seems to be coemergent with our sense of time.

Reverting back to the history of religions, from Hebrew texts of the Old Testament (believed to have originated between the eighth and fifth century B.C.), S. G. F. Brandon concluded that:

> The fact, that, throughout all these writings, history is presented essentially as a drama concerning Israel's faithfulness to its god, is symptomatic of that original tension which stemmed from the amphictyonic relationship. . . . This tension is not apparent in any other ethnic religion, and the cause surely lies in the peculiar origins of Yahwism. . . . history for the zealous Yahwist was essentially *Heilsgeschichte* (salvation history), the record of Yahweh's

original deliverance of Israel and of his continuing providence according to the nation's deserts.[37]

The relationship was manifest as a continuous reminder of a covenant with God and of the mutual expectations, thus Heilsgeschichte is obviously a process and not a proposition. From the sixth century B.C. until the destruction of Jerusalem in 70 A.D. the disappointments and miseries of life did not change the Heilsgeschichtlich interpretation of history, but rather forced its extension to the largest stage conceivable: the world of man.

M. Burrows points to the poignant attitude of the author of the Book of Daniel. He rose above the immediate concerns of his epoch (possibly the second century B.C.), and perceived in the vicissitudes of Israel's fortunes a divine plan involving the totality of the known world and directed to the sole purpose of the final salvation of God's selected people.[38] Thus, the concept of history was born through the claim of an all-embracing purpose governing the community of man. The plan manifested itself not in the destiny of heroes as Homer would have had it, or in conflagrations consuming the world as the Sumerian epics would have had it, but in producing on earth an eventual haven of righteousness and peace.

If there is any single quality that distinguishes the Greek and the Hebrew interest in the past, it is the vastly different type of inspiration that may be derived from each. The Yahwist eschatology tried to alleviate the pains of existence by promising a world of justice and goodness, if not for the individual, at least for his progeny, right here on earth. Whereas the philosophers of Athens erected their teaching of timeless forms as bulwarks against the ravages of passing, the Yahwist sought escape, including resurrection and postmortem life for the soul, in the temporal category of future.[39] They came to regard time as a forward motion with an identifiable beginning in the divine act of creation, and leading to the fulfillment of the divine purpose.[40]

Christianity and Patristic Philosophy

It was in the background of expectations implied in the Hebrew Heilsgeschichte and applied to the conditions of the Jews in the land of Israel that Jesus of Nazareth conducted his ministry during the first decades of Christian chronology. The teachings of Jesus, from his very first utterances, carry a note of urgency which, in one form or another, seems to have remained with Christianity through the heaven-directed medieval times to its current secular forms.

The first Christian theology of Paul changed the Hebrew Heilsgeschichte from a linear, uninterrupted development to a plan of two phases, separated by the unique events of Christ's birth, death, and resurrection. (In a curious opposition to St. Paul, the religious philosopher Karl Jaspers would regard the period between the eighth and second century B.C.

as the axial period of history "which gave birth to everything which, since then, man has been able to be, the point most overwhelmingly fruitful in fashioning humanity.") [41] The important point here is that to the one-way progress of Hebrew history Christianity added the idea of a unique event or, philosophically, the idea of contingency. Thus, we have these two elements: lawfulness or being, in form of the divine design; contingency or becoming in form of the idea of the unique. Time now could be regarded as progressive and nonrepetitive; and a philosophy of history, including the idea of progress, became thinkable.

The emphasis of Christianity on the unique role of the Savior was acquired as early as the first century. In the view of Brandon, it arose largely from the need of the followers of Christ to prove his messiahship.[42] During the first four-hundred years following the crucifixion of Christ, the ancient lands around the northeastern Mediterranean had experienced an influx and mixing of ideas resulting in many and often contradictory views of time among the leading lights of thought and faith. Thus, classes of views, which in cautious retrospect could be called philosophies of history, did not achieve any articulate, definitive patterns until about the fifth century. We will consider only a few major trends of thought during this span of time leading eventually to the medieval synthesis.

The philosophy of Plotinus (205–270), founder of the Neoplatonist school and first great thinker of the Christian era, may be described as an attempt to reconcile the lawful and eternal with the free will of man. Perhaps his adventurous youth and mystical feelings were the influences which made him seek the reconciliation of the strains and burdens of everyday with the idealistic beauty of Platonic thought. His attempts to create a working unity between timeless and temporal concepts influenced Christianity in its formative years, and through St. Augustine, Meister Eckhart, Nicholas of Cusa, and others such a unity became incorporated in Christian intellectualism.

As did Plato, Plotinus also regarded the generation of the ordered universe as timeless and perceived in it a hierarchy of suffused beings. He protested against the Aristotelian notion of time as the measure of motion for he found it a circular argument: we measure time by motion and motion by time. He noted that since motion can cease but time cannot, the sources of time must be in the soul of man. To state what time really is, "Something thus the story must run":

Time, before it was time, lay "in the Authentic Existent together with the Cosmos itself. . . ." There also existed an active principle of nature set to realize and govern itself "and it chose to aim at something more than its present: it stirred from its rest and the Cosmos stirred with it. . . . And we (the active principle and the Cosmos), stirring to a ceaseless succession, to a next, to the discrimination of identity and the establishment of ever new difference, traversed a portion of the outgoing path and produced an image of Eternity, produced time." Even more deeply, however, "the Soul contained an unquiet faculty, always desirous of translating elsewhere what it saw in the Authentic Realm, and it could not bear to retain within itself all the dense fullness of its possession." [43]

For Plotinus the soul had two levels. The higher one is that of the intellect, the lower one is that of experience and desire. The higher level is forgetful because it seeks unity and tries to resemble timeless eternity. Time then is the relationship of the soul to eternity and constitutes the life of the soul which creates time through a progressive derivation:

> Time, then, is contained in the differentiation of Life; the ceaseless forward motion of life brings with it unending time; and life as it achieves its stages, constitutes past time.

He registered a continued, almost painful awareness of time:

> But the movement within the Soul—to what are you to refer that? Let your choice fall where it may, from this point there is nothing but the unextended: and this is the primary existent, the container to all else, having itself no container, brooking none.

He found the apex of human existence in the culmination of mystical experience, in a nondiscursive grasp of nonspatial and self-contained ultimate reality. In the philosophy of Plotinus the unquiet principle of self-realization coexists with the living organic whole of the material universe. He attempted to express this through the synthesis of the lawful and the unexpected in the nature of time. His poetic grasp of the Cosmos may, though it ought not, sound farfetched to modern ears. For it is precisely this combination of the lawful and the contingent which found explicit, formal expression in the mathematized physics of the sixteenth century.

A comparison of Plotinism with the *Meditations* of Marcus Aurelius, written less than a century before him, offers a rather dramatic contrast:

> The rational soul . . . wanders round the whole world and through the encompassing void, and gazes into infinite time, and considers the periodic destructions and rebirths of the universe, and reflects that our posterity will see nothing new and that our ancestors saw nothing greater than we have seen. A man of forty years, possessing the most moderate intelligence, may be said to have seen all that is past and all that is to come; so uniform is the world. [44]

A century and a half after Plotinus the idea of directed, purposeful, and linear time received passionate support from St. Augustine (354–430), who argued against the cyclical theories prevailing in his days by pointing to the necessity of novelty. If one assumes, as Christians must, that the labor of life be rewarded by blessedness of which the soul will forever be assured, there must be unique and unrepeatable events. Otherwise the soul after one single life could not remain forever blessed, and Christ himself would have to die and rise an infinite number of times. But "once Christ died for our sins, and, rising from the dead, He dieth no more." [45]

As Plato, so St. Augustine held that the world and time were created together, neither anticipating the other, before the beginning there was only timeless eternity.[46] The "infinite ages of time before the world" or "the infinite realms of space" cannot be conceived of in our world of time, neither can one comprehend infinity as a number, though one can understand that it is comprehensible to God.[47] Thus, to the question "What did God before He made Heaven and Earth?" he would simply answer "I know not," rather than eluding the question with the merry reply "He was preparing hell . . . for pryers into mysteries." [48] Within the humanly understandable existence of time, God's eternal plan is being worked out according to the design contained in seed in the act of creation. The beauty of God's plans for this temporal world would be appreciated by man only if it were spread out before him like a complete sentence.[49] But this cannot be the case, for secular history is a hopeless flux of generation and destruction. The final meaning of history, then, is the potentiality of transcending time into blessed eternity where man may find rest from the terror of time.

As compared with Plotinus, St. Augustine can be accused of insensitivity to the existential predicament of man. He did write that "our heart is restless until it reposes in thee," [50] and his *City of God* finishes with the description of "the eternal felicity of the city of God" and of the perpetual Sabbath where bodies will be fulfilled as the spirit wants them to be. But he did not seem to have sensed that time, anxiety, and the fear of death are a tight syndrome. This is the more curious for his greatest contribution to an understanding of time is his sensitivity to what in our epoch would be called the psychology of time.

Instead of contemplating the motion of things and thereby joining the Aristotle-Plotinus debate, St. Augustine preferred to think about time in strictly temporal and introspective terms. Consider, for example, reciting the measured syllables of a hymn. Its syllables are not attached to a moving body; rather, they are the voices and fleeting images of time. Accordingly, he concludes:

> It is in thee, my mind, that I measure time. . . . The impressions, which things as they pass by cause in thee, remain even when they are gone; this is which, still present, I measure, not the things which pass by to make the impression. This I measure, when I measure times. Either then this is time, or I do not measure times.[51]

St. Augustine's famous question and answer: "What then is time? If no one asks me, I know; if I wish to explain it to one that asketh, I know not," [52] is a particularly sharp insight into the psychology of time. I would describe it as an indeterminacy principle between the reciprocal qualities of time felt and time understood. Time is both felt and understood in ordinary experience, yet these two ways of knowing become mutually exclusive at their extremes: a purely noetic analysis becomes timeless, as does the experience of ecstasy. We shall return later to this particular issue.

J. F. Callahan juxtaposed four views of time in ancient philosophy. Although we have considered more than four views, we may close this subsection by his summary:

Thus time which was treated metaphorically by Plato as the moving image of eternity, physically by Aristotle as the number or measure of motion, and metaphysically by Plotinus as the productive life of the soul, receives at the hands of St. Augustine a new facet, psychological, and emerges from ancient philosophy a well favored but still provocative problem.[53]

Islam

Extant archeological data of Ancient Arabian culture suggest a naive appreciation of the universe prevalent during the first millennium B.C. It displays a concern with material prosperity and a belief in generally benevolent supernatural beings who could be bribed to extend privileges to man. According to the interpretation of S. G. F. Brandon, a simple belief in postmortem existence implied no more than the inability to conceive of complete annihilation upon death.[54] By the time of Muhammad (570–632) the Arab evaluation of the human predicament became, in Brandon's phrase, that of sophisticated fatalism. Unlike simple fatalism which is the belief that fate is essentially unpredictable, the stronger fatalism discerned by Brandon holds time to be the supreme arbiter of human destiny. Whereas Europeans could afford to seek the grand design in the nature of time, the ruthless demands which physical environment imposed on Arabian life may have predisposed the Arabs to fatalism.

Medieval Moslem thinkers were educated in Greek thought, which they naturalized to the Islamic condition. Early Moslem thought does not admit continuous duration, but only instances of time. Similarly, Arabic grammar does not admit tenses as states of things, but only as verbal aspects of things.[55] The Islamic scholar, Louis Massignon, suggests that if we are to know the essential Moslem idea of time we should not inquire from the theoreticians who follow Plato and Aristotle but turn to the practitioners, such as the psychologists of ecstasy. They will tell us that time is made up of instants without duration as exemplified by the continuity of experience in joy or pain, and that duration obtains only when man turns to the external world.

The Islamic Weltanschauung seems to have remained unchanged to our own days, the same as that which originated in ancient Arabia. Though transformed by Muhammad in terms of the central issue of his own calling, time still remained the final arbiter of events. By the age of Arab and Spanish-Arab thinkers at the turn of the first millennium A.D., its form was an amalgam of Arab, Greek, Hellenistic, Jewish, and Christian thought.

The Schoolmen and the Late Middle Ages

Between the death of St. Augustine and the birth in 1126 of the next seminal and important thinker, Averroës of Cordoba, some seven centuries passed in a homogeneous tradition of thought with the Platonic doctrines dominant. The beginning of medieval scholasticism is conveniently associated with the name of Averroës, its end with that of Nicholas of Cusa, who died in 1464.

Although the dominating influence through the scholastic period was the writings of Aristotle, his dialectics did not become known until the twelfth century. Ideas concerning time consisted mainly in the restatement of Platonic, Neoplatonic, and patristic concepts; it is to this stagnant scholarship that Averroës brought a sense of freshness. He regarded the sacred texts as simplistic; rational theology as muddling: he attempted to reconcile reason and faith in a philosophy freer than one founded on faith alone. He was trained in medicine and astronomy, thus related to his contemporaries somewhat as Empedocles related to the other Eleatic philosophers.

Averroës thought of the world as an organic unit, a dynamic structure with a flow of creativity downward and noetic discovery upward, comprising a single intellectual soul for humanity. According to this doctrine of monopsychism, or oneness of the intellectual soul, the human intellect at large is imperishable, the individual soul is only an agent receiving its content from a store of intellect that belongs to the whole of humanity. He held that scientific knowledge is eternal, ungenerated, and incorruptible—in other words, timeless. He postulated a universal time suffused through man and the cosmos, of which the celestial motion is only one example. Time, then, is a cosmic property of everything included in the universe: heaven, earth, and man.[56] Ultimate religious reality is reached in timelessness, as for example in the ecstasy of the dance—a view which is, it will be recalled, sympathetic to the Islam. Such doctrines of eternal and universal human soul and knowledge are sometimes described by philosophers as eternalism; they are sympathetic to twentieth century pantheism and are current in such concepts as the Jungian idea of the collective unconscious.

A contemporary of Averroës, the Christian mystical philosopher Joachim of Fiore of Calabria (1135–1202) held an apocalyptic view of time and history differing from that of Averroës, both in content and in form. The division of history into numbered epochs was central to his thought. He perceived God's dispensation of good and evil in three ages: that of the Father or the Law, that of the Son or Grace, and the age of the Spirit or spiritual understanding. He forecast and his followers proclaimed the Third Age as that of communism, and preached the community of goods and women. While the influence of the Averroëan stance on time worked mainly through the channels of learning, the apocalyptic view of Joachim became a direct source of social forces whose effects may be traced ahead to the German Peasant Wars of the sixteenth century and the antinomian views of William Blake six centuries after Joachim.[57]

Opposing eternalism we find the doctrine of creationism, advocated by Saint Thomas Aquinas (1225–1274). Creationism arose from the scholastic debate regarding the individuality of the soul, an issue resembling the polarization of views in the mind and body debate of modern philosophy.

Aquinas held that material as well as formal causes are produced ex nihilo. In the case of man, the formal cause is the soul, created independently of the material cause, which is the body. Thus, the soul is not transmitted from parent to child but is created by God at the birth of the infant. Otherwise it would have to be thought of as a natural process and subject even to cyclic changes, which Aquinas assumed it is not. An opposite doctrine, namely that the soul and the body are created simultaneously, was promulgated by Tertullian, a Latin church father who died circa 220. This is the view held by most Lutheran confessions of our days.

The issue between the creationists and Tertullian on the one hand and of the Averroëan stance of collective, indestructible soul on the other hand is, clearly, that of becoming versus being. These abstract and, in a way, pedantic concerns, have a fascinating ancestry in the ancient question of when life enters the embryo which, in its turn, reflects the puzzlement regarding the origins of life. Ancient Egypt is the starting point of a train of thought traced by Needham in his superb history of embryology.[58] As early as 1400 B.C. a hymn to the sun god, Aton, praises the creator as the one who gives life to sons in the bodies of their mothers, and to fledglings in the eggs. Pre-Socratic antiquity showed many signs of appreciation of the mystery of embryonic growth, often coupled with cosmogonic doctrines of the world-egg. For Hippocrates, but especially for Aristotle, the origins of life became of systematic and scientific interest coupled with problems regarding the ethical status of abortion. The naturalistic inquiry may be traced to our own days, merging with and separating from theology as the latter rose in importance, and remains embedded in Roman Catholic theology.

The explicit contribution of Aquinas to the corpus of opinions about time is unoriginal and uninteresting even though he deals with time at many places in his extensive writings.[59] Essentially he only elaborates and occasionally slightly modifies the Aristotelian views of *Physics*.

In 1277, that is only three years after the death of Aquinas, Etienne Tempier, Bishop of Paris, condemned as erroneous 219 philosophical propositions, most of them Aristotelian, and some twenty of them corresponding to the teachings of Aquinas.[60] Eleven days later Thomas Kilwardby, Archbishop of Canterbury, condemned a shorter list at Oxford. Theirs was a reaction of conservative theologians to what they regarded as unwarranted deviations of philosophers from the theological truth. Propositions 83–92 dealt with the eternity of the world and attacked, as untenable, such ideas as the Great Year of Plato and the unreality and infinity of time. In these and in some of the other propositions the dependence of time on motion was questioned. Since Tempier exceeded his mandate of inquiry by the severity of his edict, under the pressure of Dominicans and others the decree was revoked by the Curia in 1325.[61] But by that time the genie was out of the bottle; the Philosopher had been openly attacked. Tempier certainly did not envisage it that way, but his attack on the authority of Aristotle did not return philosophers to the fold of theology but supplied them, instead, with a paradigm of skepticism which (eventually) came to characterize scientific knowledge. The separation of time from motion and from events in time was an essential step, as seen in retrospect, for the construction three and four centuries later of the Galilean and Newtonian kinematics wherein time and space must be regarded as mutually independent variables.

If not a large-scale revolt, the fourteenth century did produce a lively discussion about the nature of time. William Ockham (1280–1349) agreed in principle with the naturalistic-cosmological propositions of Averroës. Man, he said, notices the motion of the heavens while also aware of changes in himself. Therefore it is possible for him to perceive time through his own coexistence with the world.[62] Nicholas Oresme (d. 1382) rearticulated the Plotinian stance when he held that, "even if all things were at rest, still time would exist; or if all things in motion were moved more quickly than they are now moved, still the time would not be quickened."[63] The Scottish theologian Duns Scotus (1265–1308) distinguished between the material and formal essence of time; the former was the time of the world, the latter that aspect of time by virtue of which it may be identified as the number of motion.[64] His follower, John Marbres, argued that time is a quality that varies in intensity and not a quantity that changes, also that it is independent of motion, continuous and indivisible.

An opposing view was maintained by the Franciscan Nicholas Bonet (d. 1360) who believed that time is atomistic, with each atom itself being indivisible. The idea was not new with him. Isidore of Seville (d. 636) proposed it in his *Etymologiae* as did the Venerable Bede (d. 735) in his work significantly titled *De Divisionibus Temporarum*.[65] The twelfth century Jewish-Arab philosopher, Maimonides, wrote in his *Guide for the Perplexed* that "time is composed of time atoms, that is of many parts which because of their short duration cannot be divided."[66] Bonet distinguished between natural time as "that which time possesses in sensible matter" and mathematical time as that separated by abstraction from sensible matter. He put forth the stunningly modern view that mathematical time is something abstracted from the totality of experience and detached from the multiplicity of things; hence it may be considered infinitely divisible, whereas real time is atomistic.

The end of the Middle Ages and the revival of letters may be conveniently, if somewhat arbitrarily, associated with the mystically minded Neoplatonic cardinal, Nicholas of Cusa (1401–64). He combined the theocentric outlook of scholasticism with a philosophy that asserts the dignity of man and his capacity for self-realization. He broke with Aristotelian cosmology when he described the universe as "a sphere with its center everywhere and its circumference nowhere,"[67] verbally anticipating the perfect cosmological principle of contemporary scientific cosmology.

Cusanus analyzed the traditional doctrine of noncontradiction (that "A is B" and "A is not B" cannot both be true) and found that it applies neither to God nor to things. As he saw it, God himself is the synthesis of opposites, both the greatest and the smallest (without size), both eternal and temporal. In geometry the arc of a circle coincides with a cord when they are both minimum (an intuitive anticipation of differential geometry); in mathematics, infinite series can have final limits. Although reason can declare some things mutually exclusive, intelligence sees a unity in them. The title *Learned Ignorance* signifies the status of the mind in his philosophy.[68] Namely, the mind tries to avoid knowledge which would lead it toward the irrational and approaches instead, intelligence, that would help it find unity in the world. By claiming that the curve coincides with a straight line and the state of rest with motion, he was attacking some basic but usually unstated assumptions that supported Aristotelian physics.

St. Augustine, Averroës, Aquinas, the lesser known scholastics, and Cusanus represent the intellectual and spiritual labor of medieval Christianity. In some ways theirs was a majestic mire whose minutiae tend to alienate the modern reader. Yet, it was an operational synthesis of the eternal law of God with the unpredictable fate of man. By the turn of the fifteenth century the signs of a new intellectual temper in Europe were numerous, exciting, and complex. While no one announced that the Middle Ages (a term first used in the early 1600s) had come to an end, there was clearly a feeling afoot that something had passed and that a new era had begun.

The Renaissance

Between the demise of the Middle Ages and the birth of Kantian critical philosophy one may observe two simultaneous processes in the intellectual history of time: the secularization of the idea of historical continuity and change, and the birth of its formal corollary, the mathematization of natural phenomena progressive in time.

Theophrastus Bombastus von Hohenheim, better known as Paracelsus, was born one year after Columbus landed in America and thirty years after the death of Cusanus. As was Cusanus, Paracelsus was also a Neoplatonist and a mystic seeking the links between the macrocosmos of the world and the microcosmos of man. But unlike Cusanus, who was still theologically oriented, Paracelsus tended toward the protoscientific, with considerable interest in alchemy and magic. He opposed the quantitative-numerical determination of time in Peripatetic philosophy and preferred to interpret time qualitatively.[69] Stars, he wrote, are luminus indicators of time but do not generate time or govern man's actions; on the contrary, those who are wise will dominate the stars. While thus calling for a control of fate by man, he believed that an individual's life is guided by the stars through an internal knowledge which each organism uses to direct its activities towards fulfillment. He called such fulfillments "monarchies" and conceived of time as a class of such monarchies, and as a qualitative connecting principle sub-suming fulfillments, and not something to be counted by number. He observed that not only life but the course of illnesses have their tempo, hence, he insisted that the physician must enter into a relation with the monarchies of illnesses—that is with time. The physician "should act against time. For physic has to overcome time." [70] He anticipated the idea of bio-logical clocks when he held that living processes are clocks in themselves: "thyme blooms all year round, whereas the crocus has its time in autumn," a rose cycle occupies less than half a summer, while a juniper cycle extends over three astronomical years.

This biologically oriented outlook on time, represented earlier by Empedocles in a more metaphysical cloak, reached its full Renaissance development in the work of Jan Baptista van Helmont (b. 1577), chemist, physiologist, and physician. As with many other scholars of his epoch, he was equally interested in alchemy, the supernatural, the experimental method of Francis Bacon, the findings of William Harvey, and the works of Galileo.

His treatise, *De Tempore* (1648), is divided roughly into a philosophical-biological and a protoscientific-medical portion. [71] In Part I, he delivered a rough criticism of Aristotelian and scholastic ideas of time. He argued against time as being inconceivable without motion, against the divisibility and the continuity of time, against time as number, and against time as a real object of nature. His own concept of time was that of a divine principle represented by the speed of biological processes, or velocity of life, something immanent in objects and not of reason. He summed up his views by stating that time has no parts, hence it cannot be divided, hence it is not of the nature of succession; it is independent of motion and repre-sents instead the relation between the creator and the created. It is to be gauged by life processes and not the latter by some abstract units. In Part II, Van Helmont held forth against the doctrine of critical days central to Greek and medieval medicine. (It consisted of the belief that health and illness depend on calendrical and astronomical configurations.) He suggested that each disease had its specific rhythm and duration and it was these biologi-cal time parameters which should be studied by the physician, instead of assuming general regularities associated with the motion of celestial bodies.

In 1548, three decades before the birth of Van Helmont in Belgium, Giordano Bruno was born near Naples, Italy. Known to his contemporaries as the Nolan, after his birthplace, he has been described as a man who was "despised and miserable during almost the whole of his tempestuous life course . . . yet played a crucial part in the reshaping of European thought that began in the sixteenth . . . century." [72] His interest in the stars was metaphysical and cosmological rather than alchemical and medical. Best known for his work *On the Infinite Universe and Worlds*, he made the infin-ity of the worlds within one infinite universe a basic tenet of his philosophy. To reconcile such an infinity with our understanding, which is limited to the finite, he explored, as did Cusanus, the coincidence of the contraries. He perceived in the subject-object relationship a process of admixture culminating in identity extending to the infinity of the universe, which was envisaged by him as also bound up with the identity of the contraries.[73] As though taking to task modern theorists of transfinite numbers, he criticized those who believed that "the eternity of time involveth the inconvenience of as many infinites, one greater than another, as there are species of numbers." [74] He believed, however, that this objection did not hold for his thesis because, although there are an infinity of worlds, there is only one infinite universe.

> In conclusion, he who wants to know the greatest secrets of nature should regard and contemplate the minima and maxima of contraries and opposites. It is a profound magic to know how to draw out the contrary after having found the point of union.[75]

In 1600, this seeker of worlds larger than that accommodated by Roman theology was burned at the stake in a "terrible silence," as "his tongue [was] imprisoned on account of his wicked words." [76]

Galileo and William Shakespeare were both 36 and Sir Francis Bacon was 39 years old when Bruno died in the Piazza del Fiori in Rome. But Bacon and Shakespeare lived and worked in the balance and relative safety of Elizabethan England where Bacon's new philosophy turned man away from the books of tradition to the light of nature, from searching the past to questioning the present, and from an inward-directed search of the self to inquiries aimed at knowledge of the nonself. Galileo was yet to face a punishment, rather milder than the stake, forty years later.

Most historians of science would agree that it was Galileo's mathematical rationalism against the logical and verbal methods of Aristotelian physics that laid the foundations of modern science. In his *Dialogue on Two New Sciences* (1638), Galileo switches back and forth between strictly verbal arguments concerning uniform and accelerated motion and visual (spatial) and mathematical representations of the same arguments, somewhat in the fashion of modern high school texts on geometry. His great success in dynamics derives from the exploitation of this type of symbolic transformation between verbalized concepts and geometric parameters. We may say that the geometric metaphor is an intellectually pleasing static image that conjures up the experience of watching moving bodies.

The Galilean description combines the lawful aspects of motion, as expressed in mathematical relations, with the becominglike aspects of motion in the necessary contingencies, sometimes called boundary conditions. The use of graphs would have pleased Plato, for such abstractions correspond to the incorruptible forms of his geometry and point, in a Platonic stance, to the ultimate, real bases of existence. The quantification made possible through mathematization would have pleased Aristotle for it gives instructions on how to measure time by the number of motion. But the Aristotelian "before and after" with respect to which motion may be counted had to be supplied by the experimenter who was not part of the nomothetic (lawful) portion of Galilean kinematics but of the contingencies.

The total separation of time both from the equations of motion and from the contingencies, including the observer himself, is the essence of Newton's dictum that "absolute, true and mathematical time, of itself, and from its nature, flows equably without relation to anything else."[77] From the distance of 250 years one can detect in this pronouncement the fortunate confluence of many trends working through the genius of Sir Isaac.

Physically, he regarded absolute time as an obvious correlative of absolute space, as he believed to have demonstrated in his famous rotating bucket experiment.[79] In the view of Whitrow,[80] he might have been attracted to the idea of an absolute rate measurer, something I would call a disembodied clock, because of its implicit mathematical simplicity. Since, as Ariotti has noted, all attempts to reduce time to number of motion had failed, adopting the idea of time as totally independent of events in time offered a way of bypassing the requirement for reduction altogether.[81] The idea of an absolute substratum whose God-given temple constitutes the scaffolding of all events might also have pleased him. While he admitted that possibly there is no such thing as "equable" motion, time as Sensorium Dei is an idea with religious appeal. Yet, this absoluteness of Newtonian time does not necessarily signify a Jasperian "axenzeit," or a quality of being anchored in history. The concepts of velocity and acceleration demand only separation of instants, that is, conditions of before/after but not any reference to a present, such as an absolute date.

The idea of absolute time permitted the clear separation, and subsequent reunification, of the lawful and the contingent in the nature of time. Before it became possible to formulate the dynamics suitable for the determination of future behavior of moving bodies, it was necessary to distinguish with precision the invariant and contingent components of motion. Eugene Wigner has stressed that

> The surprising discovery of Newton's age is just the clear separation of laws of nature, on the one hand, and initial conditions, on the other. The former are precise beyond anything reasonable, we know virtually nothing about the latter.[82]

The revolutionary contribution of the Newtonian synthesis to our understanding of temporality resides, therefore, not primarily in the idea of absolute time but rather in the separation of being and becoming in the nature of physical time, permitted by that idea.

The Newtonian idea of absolute time was opposed by Leibniz on logical grounds. Absolute time, being independent of events and things in time, would also exist even before the creation of the world. But such a totally empty time can have no identifiable instants and thus, in such a time, God could have created the world not only when he did, but perhaps even a year earlier. However, taking that the succession of events and positions of bodies would be the same in any such two (or presumably more) created worlds, the worlds could not possibly be told apart because, "to suppose two things indiscernible is to suppose the same thing under two names." [83] Thus, it is meaningless to ask when (in absolute time) God created the world. It follows that "instants, considered without the things, are nothing at all"; [84] time cannot exist independent of bodies. Instead of absolute time, Leibniz proposed to understand time as the order of succession, or relation of events which take place in a world of bodies coexisting in space.

Newton's empirical argument as derived from the rotating pail experiment came under criticism by George Berkeley (1685–1753), who realized that it included a reference frame vaster than that implied by Newton. He sensed that there is no such phenomenon as strictly local motion, but only motion with respect to other bodies. If we try to imagine absolute (that is totally empty) space we must think of our own body as removed from that universe. But there is no other physical universe into which we might move. Hence there is no absolute space. Furthermore, "since absolute space in no way affects the sense, it must necessarily be quite useless for the distinguishing of motion." [85] It follows that its corollary, absolute time, is also useless. The important point here is that when speaking of time, the observer cannot be eliminated.[86]

The importance of Berkeley's reference to the universe was explicitly stated two-hundred years later by Ernst Mach in considering motion in general. "When we say," he wrote, "that the body preserves unchanged, its direction and velocity in space, our assertion is nothing more or less than an abbreviated reference to the entire universe." [87] Clearly, the only equivalent alternative to Newton's rotating pail experiment would be to stop the pail and rotate the universe. The implications of the fact that what we are talking about is a relative motion was fully realized only in Einstein's general relativity theory in 1916.

While Newton, Leibniz, and Berkeley debated the philosophical and scientific ideas of absolute time, Gianbattista Vico (1668–1744), usually thought of as the first modern historian, tried to make sense of time as a pattern in the collective affairs of man. He took to task those who believed that the course of human events is "a deaf chain of cause and effect" and proposed, instead, a demonstration of historical lawfulness. "For though this world has been created in time and in particular, the institutions established therein by providence are universal and eternal." To find the rules of these institutions, their origins, and to identify the eternal structure or order in history, he proposed a new science of "a rational civil theology of divine providence." [88]

Vico's rational civil theology sees man as guided toward a final goal by the powers of providence, even against his own will. His "providence" belongs to a class of social laws perceived as directing the community of man. The class has many members, such as Bacon's laws of the marketplace; "the invisible hand" of Adam Smith; the "cunning of reason" of Hegel;

the directedness of evolution of Kant; the dialectical materialism of Marx; and, in modern thought, entropy.[89] We shall return to the question of whether or not history has laws in the broader context of historicity and epistemology. We are closing this subsection with a brief mention of Vico because he symbolizes—though perhaps no more nor less than Van Helmont, Galileo, or Newton—the Renaissance drive for the formulation of statements about nature and man in forms which combine the nomothetic and the generative (becominglike) aspects of time in various principles of unity.

Kant and Critical Philosophy

The desire to identify qualities and actions which regularly recur in nature and man—a trend which characterized the developing thought of the Renaissance—was a continuation, in changed form, of the scholastic debate about universals. A universal is any quality which belongs to several members of a class of objects or actions, and which is not expected to change with time. Thus, Plato's forms may be considered in this category. For Aquinas, universals were functions of the mind because, as he saw it, even self-evident knowledge must be derived by the mind from experience. For the thirteenth century scholastic, Albertus Magnus, universals existed as a matter of faith; for Duns Scotus, they existed with some qualifications. Medieval realists generally held that universals were entities whose being was independent of mental apprehension. Opposing them were the nominalists who held that universals were general notions, or concepts which had no reality of their own. The Newtonian idea of absolute time may be described as realist for it is ontologically prior to events; the relational time of Leibniz may be described as nominalist for in it events are ontologically prior to time. The views of Immanuel Kant (1724–1804) represent a third position, incorporating some aspects of both.

Kant was influenced by the argument of the mathematician Leonhard Euler who held that the ideas of space and time could not be obtained through the senses, hence they could not be abstracted, but must somehow be assumed, such as in the Newtonian idea of absolute time. In his inaugural dissertation at the University of Königsberg (1770), Kant suggested that the idea of time is not something abstracted from sense experience but rather something presupposed by it. This may be seen from the fact that all attempted definitions of time in terms of before and after or past-present-future already presuppose time. He concluded that the idea of time is pure intuition, meaning that it constitutes direct experience of sensory content, but not the content itself. Time is not something objective and real but something resting "on an internal law of the mind," which permits both "A is B" and "A is not B" to be true, provided they represent conditions successive in time. "Hence the possibility of changes is thinkable only in time; time is not thinkable through changes, but only *vice versa*." [90]

During the twenty years following his inauguration he developed what he called the critical method, the essence of which is an inquiry into the limitations and boundaries of knowledge and the knowable. "Criticism" here is understood to mean a survey of the potentialities of the intellect, something which, in his view, should precede the process of philosophical speculation itself. Thus, through critical philosophy, any opinion about time also implies a theory of knowledge. Kant understood that certain concepts, such as time, freedom, causality, or one and the many contain certain logically unresolvable contradictions. He called them antinomies. He formulated certain theses about these concepts and proceeded to show that both such theses and their antitheses are equally plausible. Since he did not hold with Cusanus and Bruno that the law of contradiction is invalid, except as just qualified, he came to regard his seemingly faultless but contradictory arguments about time as examples of the limits of pure reason, hence also the limits of the knowable.

To accommodate these limitations he postulated two coexistent worlds: the phenomenal and the noumenal. The phenomenal is that of pure reason; it is one of lawfulness, necessity, causality, and determinism. The noumenal world is that of final truths, and unlike the phenomenal world, it is free or "unconditioned" and forever unknowable. The noumenal world must be postulated by practical (rather than pure) reason and it is the task of science to study that world as it appears even if the final, underlying truth, the *Ding an sich*, can never be known. Thus, he did not reconcile the beinglike (lawful) and becominglike (contingent) worlds but provided instead a static structure to contain both.

Kant's dictum that "the intellect does not derive its laws (a priori) from nature, but prescribes them to nature" poses some profound epistemological questions which Kant himself clearly saw.[91] How does it happen, for example, that the potentialities and principles of our experience, presumably limited and determined by the mind, are in such remarkable agreement with what nature exhibits to us? Kant perceived two possible explanations: either the laws are borrowed from nature, or our perception of nature as a receptacle of events and things is derived from the laws of possible experience.[92] He would probably say either that time is something we learn from nature, or our idea of time-in-nature is a law, that is, a possible form of sensitivity. It is this second solution which he would tend to favor.

As did Vico before him, Kant also believed that history consists of a working out of a plan; for Vico it was that of Providence, for Kant a plan of nature. Nature and providence were not incompatible. Kant believed that one should seek the universal, natural law first, for only in the totality of history can one hope to find the purpose of providence. He used the well-known feature of probabilistic laws, today widely employed in the sciences, according to which, although the fate of a member of a set making up a universe may be causally unpredictable, the history of the assemblage can be probabilistically lawful. What is "chaotic in a single individual may be seen from the standpoint of the human race as a whole to be a study of progressive though slow evolution of its original endowment." [93] Thus, mankind is controlled by lawfulness, wherein the connection between the individual and the aggregate is supplied by God. It is the task of the philosopher "to try to see if he can discover a natural purpose in this idiotic course of things human." He insisted that it is possible and desirable to discover such a natural plan, without remarking that the coexistence of individual free will and causal lawfulness of society constitute, together, an example par excellence of the thesis and antithesis of his Third Antinomy on causality and freedom.

Kant was not the only thinker who saw in history the operation of natural law. His contemporary, Johann Gottfried von Herder (1744–1803), believed that the laws of history are as binding for mankind as are the laws of nature for the earth, plants, and animals. In terminology that resembles that of the natural scientist, Herder described history as the progression toward an equilibrium of national forces, leading to a final structure embracing all of mankind. The human race is destined to proceed through various degrees of civilization, and the laws of nature will, in the course of time, produce a "more durable" humanity.[94]

Hegel and the Dialectics of History

The complete philosophical transition to modern views of history, man, and time is the work of Georg Wilhelm Friedrich Hegel (1770–1831). His philosophy is that of idealism in the service of the reconciliation of the contraries. In his early life of Sturm und Drang he found in love a practical union of the opposing elements of human desire. In his later years he found the communal life of man more interesting, and identified history as the true arena for the struggle of opposing forces of man. As did Faust after having left Gretchen, Hegel also transferred his concern from the individual to history. The early Hegel identified man with life and love; the older Hegel identified history, man, and time with death, as summed up in his famous dictum that time is what man makes out of death.

The idea of nature, according to Hegel, is that of reality bifurcated into two external aspects: the spatial and the temporal. By "external" he meant that objects in space and events in time exclude one another. Opposite to nature is man's *Geist*, which translates both as spirit and mind and is, therefore, a broader concept than its usual translation as "spirit." Hegel speculates about the development of the spirit from simple consciousness to self-awareness, to absolute knowledge, and sees its communal manifestations in the creation of institutions. History itself is identified with the "process of development and realization of the spirit." [95] "History in general is therefore the development of spirit in time, and nature is the development of the idea in space." [96] The technique of history, if we are permitted to use a very non-Hegelian word in descriptive retrospect, is dialectical motion. A positive thesis, whether thought or action, produces its reaction, a negative antithesis; the two together interact and produce a synthesis which, in its turn, forms a thesis that brings forth its own antithesis. This way the dialectical motion, or process, advances through categories of increasing complexity reaching its pinnacle in the absolute idea, while constantly following the plans of God.

Unlike Kant, who keenly felt the limits of man's intellectual faculties, Hegel believed that "reason is sovereign in the world, that the world therefore presents us with a rational process," [97] thus he formulated one of the metaphysical assumptions of modern science. The force that drives the dialectical motion of history is the "cunning of reason" that sets human passion to work.[98]

In this process, "Time is the negative element of the sensuous world. Thought is the same negativity but it is the best, the infinite form of it, in which therefore all existence generally is dissolved." [99] In this heavy and obscure passage "negativity" means something of an innate opposition, such as between objectivity and subjectivity. Finally, "Time is just the notion definitely existent, and presented to consciousness in the form of empty intuition." [100]

Hegel emphatically rejected the idea that pure mathematics, or reason or logic may be somehow considered the science of time. Time deals with the sheer restlessness of life and with its inherent differentiation, while mathematics deals with quantity and with whatever one can do with quantity, such as form differences, sums or equalities in abstract, lifeless, structural unity. Mathematics, at best, can only image the arrested, paralyzed, quantitative aspects of the totality of time. Knowing time through mathematics "degrades what is self-moving to the level of mere matter, in order thus to get an indifferent, external, lifeless content." [101] In this warning one can almost hear Plotinus saying that time is the life of the soul. Yet, Hegelian time exists before events; hence it is relational, such as that of Newton or of special relativity theory. Hegel's philosophy makes no sense in an empty universe; it is a drama unfolding among relations determined by nature and the mind.

We witnessed an about-face regarding man, time, and history: from awareness of time as a source of deistic religion in Heilsgeschichte we went to awareness of time as a source of secular religion. Hegel's philosophy, rich enough to inspire Nazism, Fascism, and Communism, attempted to give to the modern world a comprehensive account of the universe and thus remains the last great philosophy in search of universal wisdom.

The Fragmentation of Philosophy

During the last two-hundred years the philosophical evaluation of time in Western thought has been made increasingly difficult for three main reasons.

(1) There is a demand upon philosophy to accommodate scientific truth. Speculation has thereby become an intellectually dangerous exercise. The fear of failure has sent many academic philosophers scurrying to the safety of logic, methodology, and the philosophy of science, away from the classic functions of philosophy.

(2) Profound social upheavals have altered the context in which the sources of temporality, such as the self and the other, must be considered. A new and satisfactory context is not yet at hand.

(3) A good portion of modern thought has left the traditional confines of philosophy and established itself through other images of reality such as the arts and letters.

The result is a mass of impressive but incoherent speculation centered on many and different guiding principles such as phenomenology, existentialism, positivism, and others. Because of this incoherence it would be impractical to survey modern views of time along historical and chronological lines. We shall encounter nineteenth and twentieth century philosophers when we consider subjects and questions that correspond to the major thrust of their inquiries. Meanwhile we must keep in mind that, for the study of time, philosophy itself is only one of the many sources of knowledge.

2. Oriental Concepts of Time

THE MUTUAL geographic accessibility of the regions around the Mediterranean and in Western Europe did not guarantee a uniform view of time in those regions. We have seen ample evidence to the contrary. It did permit, however, their interaction for at least the last two millennia. But until the industrial revolution, detailed interaction of Oriental and Western ideas were scant and, with some notable and important exceptions, quite uninfluential. Accordingly, the spectrum of Oriental views of time retained its unique Asian colors even if a sufficiently thorough search would reveal, no doubt, that each major Oriental view had at least one independent advocate in the West. A foremost characteristic of Oriental philosophies is that they consist almost entirely of the pronouncements of sages and not of analysis or criticism. Consequently, they are rich in the appreciation of human emotions and in the understanding of values, rich also in cosmological imagery and even in instructions for behavior, but poor in the type of argumentation that the West regards as empirical.

Time in China's Past

The perennial philosophy of Chinese culture, in the words of Joseph Needham, was an organic naturalism.

> Europeans suffered from a schizophrenia of the soul, oscillating forever unhappily between the heavenly host on the one side and the "atoms and

the void" on the other, while the Chinese, wise before their time, worked out an organic theory of the universe which included nature and man, church and state, and all things past, present and to come.[102]

With a tradition of seeing the world as an organic whole and with metaphysical idealism generally occupying only a subsidiary place in Chinese philosophy, the idea of time was rarely considered to be independent of timepieces, astronomy, religion and the mores and folkways of society.

The affairs of world and man comprised classes of carefully compartmentalized cycles of changing hierarchical position and power. As Granet concluded in a classic study of the character of Chinese thought, neither time nor space were conceived of as receptacles of abstract properties,[103] nor entities apart from concrete actions.[104] Succession or duration were seldom emphasized and achieved the status of reality only where they were useful and necessary for the life of society.[105] The linear time of clocks was counted off for utilitarian purposes and checked against astronomical cyclicity.[106] The ambiguity of linear counting versus cycles did not seem to have bothered the Chinese thinkers and in spite of the regular (numerical) progression of clocks and calendars, the world cycle was believed to restart every 23,639 years, these being the smallest common multiples of planetary periods. In this understanding of Chinese thought, based by Granet not only on philosophies but also on mythology, folklore, and the world outlook of ancient writings, man appeared as a living, cyclic portion of a universe which comprised cycles upon cycles.

In an essay of superb scholarship entitled "Time and Knowledge in China and the West," Joseph Needham examined Chinese philosophy, natural philosophy, historiography, techniques of time measurements, and attitudes to biological, social and scientific changes in time.[107] He noted that although some Buddhist schools in China held, as part of their general doctrine, that the world is an illusion and that there was something unreal about time, indigenous Chinese philosophers invariably accepted the reality and importance of time. In agreement with Granet and others, Needham acknowledges the Chinese preoccupation with cycles but adduces evidence from many facets of Chinese Weltanschauungs and daily practices attesting to an interest in linear time. Taoism, for instance, sees history as "a continuous process ever renewed," making the Taoists' emphasis on history comparable, in Needham's judgement, to the importance of history for Hegel and Marx. "It would really be true to say that in Chinese culture, history was the 'queen of the sciences,' not theology or metaphysics of any kind, never physics and mathematics." [108]

One related issue needs to be mentioned. Since Aristotle, Western thought has held causality and its corollary, lawfulness, almost sacrosanct. This was not the case for Chinese philosophers who stressed organism and pattern, and permitted principles of connectedness between simultaneous events, corresponding to something like a resonance.[109] This idea resembles the Jungian concept of synchronicity and permits an effect to precede its cause.

All the aforesaid pertain to Chinese ideas of time as held in the past. Chinese philosophy has entered a new phase. To what extent the profound social changes will influence, if at all, the traditional attitudes to time, is yet to be seen. We shall leave the debate on this tentative note and will return to the Chinese genius in other contexts.

India and the Eternal Present

Indian philosophy is roughly coeval with Chinese, both dating to the Vedic period circa 1500–600 B.C. The multiplicity of Indian branches of speculation seems to dwarf the multiplicity of philosophies in the West, with most speculations blending with myths more intimately than they have in European thought since Greek antiquity.[110] In spite of this multiplicity one may say without reservation that the essence and character of time in traditional Indian philosophy is that of cyclicity, or the idea of perennial and eternal return. History, or more accurately cosmic duration, is a creation-destruction-creation process, made up of vaster and vaster cyclic processes contained in the doctrine of the *yuga*.[111] This constitutes the belief in a staggering hierarchy of cycles with the smallest unit, the yuga, an "age," measuring perhaps 360 human years. Upon the innocuous yuga rest a score of superior cycles reaching up to the life of the gods, which is of the order of 311,000 billion human years. But the gods themselves are not eternal either, and thus the cosmic cycle of creation and destruction goes on forever.

Cyclical time represented by the metaphor of the "sorrowful weary wheel" is accepted by Buddhism and Jainism. In the Bhagavad-Gita the divinity is "Time, that makes the worlds to perish, when ripe, and bring on them destruction" [112] thereby creating an existence based on unpredictables. While Parmenides and Zeno sought support for the illusory nature of time through arguments about motion, rest, continuity, and other analytic and local qualities, in Indian idealism time is illusory because what exists today will not exist, say, in a thousand million years, hence it could not have been real in the first place.[113]

The identification of time with cyclicity and with unreality appears to be responsible for such hallmarks of Indian thought as the depreciation of any metaphysical content to history, emphasis on perfect beginnings and subsequent deterioration, and, of course, on the eternal repetition of a fundamental cosmic rhythm. Value judgements derived from such views, combined with the insecurities that have been rampant on the Indian continent perhaps for millennia, necessarily suggest suffering as the only reality—and destroy the hope for future improvements as unrealistic. Man, stooped under the weight of being, has only one avenue if he wishes to escape from his miseries, and that is to escape from temporality by transcending the human predicament of existing in time. True wisdom, therefore, resides in running away from suffering by means of grasping the favorable, timeless moment of the eternal present. The working out of the techniques of how to escape from the servitude of time into the timelessness of the eternal present seems to have been the major motivating power of Indian religious practices and communal structures. We will return to them, later, in different contexts.

Japan and the Unity of Opposites

Japanese philosophies of life rest on three major and interlocking moral and behavioral systems. In a very loose way they may be associated with respective emphases on the categories of future, past, and present. Buddhism, introduced in its Chinese form via Korea in the sixth century, comprises a large and varied structure of metaphysics derived from the enlightened teachings of the Buddha and addressed to "him who has given up the world." It is generally more concerned with the future and with eternity than with the past. The archaic and indigenous religious cult of Japan is the Shinto, literally the teaching of the way of the gods. It is a syndrome of beliefs which, though unorganized, has preserved an ethos of respect for the past. Teachings regarding praiseworthy behavior and concern with the moral order in the present are stressed by the Japanese version of Confucianism.

The structures of opinions and norms embodied in these three great systems are adaptations of ideas originally foreign to the Japanese soul, mind, and landscape. Treatises on specifically Japanese ideas about time (as distinguished, for example, from the Chinese) are rare, and such writings as may exist are not easily available.[114] Yet, even the occasional but interested traveler to the Orient will sense certain attitudes which appear to be characteristic of Nippon. For these reasons, we will limit our glimpse to twentieth century thought only and even there, to the work of Nishida Kitaro (1870–1945), leading representative of modern Zen.

Zen Buddhism pertains to pure experience and not to discursive cognition, holding that it is not possible to render other than indirect and suggestive material, verbal projections as they were, of inexpressible mental and spiritual happenings. This feature of Zen introduces at the very basis of its practice the possibility of bypassing the limitations of Western logic, while also relinquishing its advantages. The source of Zen is the experience of sudden enlightenment called *satori*. While mystical experiences are certainly widely known in the West and are incorporated in the history and theology of Christianity, mystical experience as such had not been made the source of any major Western philosophy.

For Nishida Kitaro, committed to Zen and educated in Western thought, ultimate reality is the Nothingness of Buddhist tradition. In Western existential thought "nothingness" is a basic datum of living experience, a mental state of complete indeterminacy. Buddhist Nothingness is thought to comprise a framework within which the form and the formless coexist and rejects any explicit reference to existence. The resistance of anything formless to being given a verbal form is well known to Gestalt psychologists who try to deal with unconscious perception. The difficulties of putting into philosophical form the formless content of Zen meditation is certainly not easier, as judged from the uncompromising difficulty of Nishida's text.

One of the translators of Nishida found it useful to incorporate in his introduction to Nishida's works the representation of one of the philosopher's poems both in its visual *kanji* form (created by the philosopher through the stroke of the brush as the Pythagoreans created numbers) and in translation:

> The bottom of my soul has such depth
> Neither joy nor the waves of sorrow can reach it.[115]

For Nishida the ultimate Nothingness when embodied in the form of reality is an example of the unity of contraries. "The world of reality is essentially the one as well as the many, it is essentially a world of the mutual determination of single beings. . . . That is why I call the world of reality 'absolute contradictory self-identity.' "[116] In the concrete present Nishida sees the coexistence of innumerable moments in the experience of one, hence the present is an example of the unity of opposites. His last work bears as its title *The Unity of Opposites*. As was true for Hegel, his concerns had led him to a philosophy of history. In his formulation he sees the element of *poiesis*, or creation in the world, as the joint action of the form and the formless, somewhat as a distant echo of Hegel's *Geist*.

The insistence of Zen upon the primacy of experience makes Zen Buddhism sympathetic to the mysticism of Meister Eckhart and to the Hegelian and Marxian dialectics. Conversely, it makes Zen-based philosophies attractive to those Western thinkers who have "found their way back from intellectual virtuosity to existential philosophy."[117] The type of philosophy proposed by Nishida attempts to see the form of the formless and hear the voice of the voiceless. That his methods are those of Japanese traditions and that his ideas may sound strange should not prevent us from recognizing an identical struggle in Western natural science where, through the geometrization of matter and motion, the practitioners of relativity theory claim to have given form to formless space, and to the silence of passing, the voices of time.

3. Being, Becoming, and Existential Tension

OUR REVIEW of the history of the philosophy of time, though brief, covers, nevertheless, too many views to permit clear-cut conclusions. It is quite legitimate, however, to make some assumptions in the light of the survey. Then the value of any new insight so claimed will have to be judged not so much by how closely it describes the past which gave rise to it, but how useful it eventually may become in solving some of the difficult problems raised by the past.

The history of the philosophy of time, outlined in the preceding sections, is characterized by the interplay of two distinct and contradictory views of the nature of time. When understood in their polarized extremes, one of these holds that beneath our experience of time there is an ultimate reality of permanence, the other one that the ultimate reality of the world is that of pure change. World views which derive from or relate to the former embody the principle of being; they tend to regard substance and space as ontologically prior to function. World views affined to the latter embody the principle of becoming; they tend to consider process and function more fundamental than matter and space. The notion that such a polarization is detectable, albeit in many cloaks, throughout the history of thought about time, and that it is fundamental to the problem of time, is itself a philosophical stance.

But being and becoming are mutually exclusive concepts. In a (theoretical) world of pure permanence everything is predictable, at least in principle, and coming into being cannot be accommodated. Likewise in an (imaginary) world of pure becoming, lawfulness has no place. Similarly, mutual exclusiveness may be found in many other pairs of concepts. Consider, for instance, the diadic form of before/after and the triadic form of future/past/present. Our experiential world is triadic; it is the world of passing and change. Before/after does not suggest a static world, yet it cannot contain becoming, for it does not have a present in which a condition or thing may emerge or vanish; it may be called a world of pure succession. The diadic form of time suggests being, the triadic form becoming. (The reader may recognize in them McTaggart's "extensive" or "B-series" and "transitory" or "A-series.") [118] A philosopher of German romanticism might have designated the same ideas as nunc stans (eternal present) and nunc fluens (changing present). Determinism and strict causation are beinglike ideas, free will and indeterminism becominglike. Or, consider the "one" and the "many." One signifies whatever remains identical with itself; only a world of many can contain becoming. Again, necessity concerns happenings which must come about, contingency concerns happenings which may or may not come about. Finity implies a sudden coming into (or going out of) being, whereas infinity implies a continuity which remains unchanged.

It seems we are dealing here with a polarization of some sort which manifests itself through a family of the pairs of time-related concepts. This polarization reflects two aspects of temporality which, though not separated by hard and fast rules, are nevertheless distinct: they are the nomothetic (lawlike, stationary, or permanent) and the generative (creative) aspects of time. The history of the philosophy of time, as sketched, revealed no compelling reasons why one or the other group of concepts ought to be identified exclusively with temporality to the detriment of the other group. If anything, it suggests that time subsumes both. But if we hold that in temporality both the nomothetic and generative features of nature coexist, that would make the idea of time a paradox par excellence, for it would assign to a presumed real entity two mutually contradictory properties, thereby making the idea of time self-contradictory. This sounds Kantian, but it is not, for the problem is deeper. Namely, our experience of time includes both being and becoming in an unproblematic unity. What, then, are the sources of the difficulties in giving the idea of time a noncontradictory status so that it may correspond to our existential awareness of it?

First, let us note that the existence of purely contingent and purely necessary events is a metaphysical assumption justified only by its usefulness in analysis. Each obtains only after some mental stripping away of phenomena regarded as nonessential in an application of Husserl's eidetic reduction. In one case the reduction removes everything that is not lawful or, in the limit, mathematically expressible; in the other case everything that is predictable, even in principle. It seems to be the nature of the world that such reductions can never be complete. Consequently, no thoroughly convincing example of being or becoming can be given. A shrewd and well-versed intellect can make any example look somewhat suspect as to its being purely necessary or purely contingent.

Second, let us think about those canons of reasoning which tell us that being and becoming or, more generally, the nomothetic and the creative are incompatible. These are the rules of logic which have been regarded in the West as suitable for dealing with propositions, including paradoxical ones.[119] Although logic has grown from its Aristotelian form to the advanced modalities of formal logic, it retains, unlike the empirical sciences, three fundamental rules of thought. They are (1) the law of contradiction (sometimes called the law of noncontradiction), which postulates that a proposition and its negation cannot be simultaneously true; (2) the law of the excluded middle, the famous tertia non datur, which states that anything is either a proposition or is not that proposition; and (3) the law of identity, which states that anything is identical with itself.[120]

Those who have held to the doctrine of the coincidence of the opposites have sometimes violated all of these rules—but certainly the first one. Leibniz believed that the law of contradiction (if in it we include the law of identity) is sufficient to demonstrate all truths independent of experience, such as mathematical or logical truths are assumed to be.[121] If Leibniz is correct, as I believe he is, and if the idea of time refuses to conform to the law of contradiction, then we must conclude that time is not independent of experience. In that case, however, the perennial debate of Heraclitus and Parmenides cannot be concluded, or even properly considered, without an understanding of those deeper-lying conditions in the nature of man, or perhaps in the world at large, which gave rise to the experience of time in the first place.

Assuming now that such deeper-lying conditions are identifiable, I wish to postulate that the sources of temporality are neither in permanence nor in change alone (neither in the nomothetic nor the generative aspects of nature), nor in the coincidence of appropriate contraries, but rather in the conflicting separateness of certain opposites. Specifically, I shall assume that temporality comprises a multitude of conflicts between the nomothetic and generative aspects of the world at large, of which our experience of time is a part. I shall attribute these conflicts to a universal condition which I shall describe by the purposefully broad concept of existential tension. Furthermore, I shall assign ontic status only to existential tension but not to the polarized opposites which we recognize in the nature of time.

As far as man is concerned, tension and conflict have been the staple of philosophies concerned with the dialectic of selfhood, such as existentialism. The first to single out the experience of tension, or more poignantly that of dread, as a hallmark of selfhood was Sören Kierkegaard (1813–1855) who noted its uniqueness in the nature of man. He distinguished dread from fear (which is usually identified with a specific cause) and regarded it as an unavoidable accompaniment of selfhood. By selfhood is meant the establishment of unchanging identity in the presence of change. For Kierkegaard the origins of selfhood must come from an external source, namely, from God. "The self cannot itself attain and remain in equilibrium and rest by itself, but only by relating itself to that Power which constituted the whole relation." [122]

Kierkegaard saw man as the final crowning product of Divine Creation. If, instead, we think of him as a product of organic evolution and one member (as a species) of an undetermined and open chain of beings, then the necessity of demanding from him equilibrium and rest is removed, for there is no final destination or reference with respect to which that rest could be had. Dread, then, will take its place in a scheme of more generalized tensions as that specific one which corresponds to man's peculiar level of organic complexity. Confirmation of the existence of stresses and conflicts to fit such a scheme, and the character of the conflicts, may then be sought through natural science, and through any and all means of rational knowledge. As a first step in such an enterprise, we shall examine one of the earliest known activities by which man has tried to identify the nomothetic aspects of the world, and an activity in which he must be both a sage and a tinker. That activity is clockmaking.

II

THE EMPIRICAL
SEARCH

WHILE Homo Sapiens tried to understand time in the multitudes of its meanings, Homo Faber learned to organize his activities with the assistance of processes which, according to the changing lights of various epochs, he judged to be predictable regularities. Such processes constitute the basis of timekeepers such as calendars and clocks. The function of timekeepers is said to be that of measuring time. In this chapter we will deal with time measurement primarily as a search for order and only secondarily as a technical virtuosity.

1. Selected Regularities:

Predictable Futures

ACCORDING TO an Eddic poem recorded in the thirteenth century

> woe's in the world,   much wantonness;
> axe-age, sword-age—sundered are shields—
> wind-age, wolf-age,   ere the world crumbles;
> will the spear of no man   spare the other.[1]

The unknown authors of this Norse apocalypse described sternly and beautifully the existential pathos of their lives: the times of ax, sword, wind, and wolf come and go with deadly regularity, but just exactly when the next ax will fall, the nearest sword will cut, or the next windstorm blow there is no way of telling. Although predictions such as these are based on certainties of past events, they leave even the "predictable" happenings uncertain and indeterminate to different degrees. It is precisely the simultaneous predictability and unpredictability of ax, sword, wind, and wolf that makes the sigh so true: "woe's in the world." I shall try to show that the major methods of timekeeping embody the paradox of coexistent opposites in the nature of time, mixed with a healthy sampling of ethical, aesthetic, and value judgements made by the clockwatcher and calendar maker. All these amount to saying that the empirical quest for regularity is a program in applied metaphysics.

The Skies, Seasons, and Epochs

The earth had been spinning on its axis for some 1,000 million years before living matter appeared, and for another 3,000 million years before man emerged. There is no period in the story of life when the cyclic process of day and night and that of the seasons did not operate. The unquestioned rank of the rotating heavens and of the changing of the seasons above all other cyclicities, is probably responsible for the fact that we measure all human activities in terms of suns, moons, and seasons (days, months, and years) and not vice versa. The origins of archaic timekeeping are indistinguishable from those of interpreting and measuring the motions of the stars, planets, and the moon. Therefore any review of early timekeeping must consider timekeepers and the heavens together.

(1) The sun, the moon, and some "hours"

Microscopic analysis of certain upper Paleolithic bone and stone artifacts revealed that what used to be regarded as the scratches on these artifacts were in fact toolmarks.[2] But the existence of identifiable toolmarks signifies the work of a maker, which suggested to Alexander Marshack to seek some purpose expressed, perhaps, in the patterns of the marks. Marshack succeeded in establishing sequential grouping with an underlying rhythm of lunations. If his interpretation corresponds to fact,[3] we must then recognize that as early as thirty to thirty-five thousand years ago our ancestors were able to transform their experience of regular processes in time into a set of static symbols representing lawfulness in nature. G. C. Hawkins coordinated the results of a statistical analysis of the relative positions of boulders at Stonehenge in Wiltshire, England, with astronomical data extrapolated to English neolithic times (1800–1400 B.C.) He concluded that Stonehenge, a place of Druid worship, was also a highly sophisticated lunisolar calendar.[4] These two examples together suggest that the traces of archaic man are those of a creature ready for and capable of searching for order in the motion of the heavens and changes of the seasons.

The earliest known device employing the diurnal rotation of the sun for the measurement of time is an Egyptian shadow clock of green schist, dating to the tenth to the eighth century B.C.[5] The sun stick, still used in Tibet, tells the time of day. That is, certain events in the lives of the clockmakers coincide with the shadow's attainment of a given length. A sunstick placed parallel with the axis of the earth comprises a sundial. When so placed, the stick is called a "gnomon." Its shadow indicates the time of the day not by changing length but by changing direction. The earliest known version of a sundial is a Roman "hemispherium," which is the inside surface of a quarter sphere. The motion of the shadow of its pointer upon the inner surface is used to tell time. Its invention is attributed to the Babylonian astronomer Berossus (fl. c. 300 B.C.).

In the sixteenth century the idea of projecting the shadow of a gnomon on the inner, curved surface of a body was revived in Germany and in Italy. The hemispherium was replaced by a chalice, cup, or goblet, or by two or more rings of various configurations. Such structures were believed to give more precise indications of the time of the day, according to increased knowledge of the motion of the sun. From the fourteenth century on, other types of sundials followed: vertical columns with horizontal gnomons, dials made of two rectangular tablets, the popular flat sundial, and single ring dials. Their variety was limited only by the ingenuity and interest of the dialer; their artistic beauty by his skill and taste.[6]

It will be recalled that the geometrical representation of time suggested quantification because lines may be divided into segments. Likewise, sun-clocks of all types, when engraved with static gridworks, suggest that order is implied by the motion of the shadow. The sun does the moving according to the judgments of the heavens, while the dial reflects the judgement and life-style of its maker. Thus, a sundial by its very existence demands some regularity in the division of time, and by its geometrical nature suggests numbering and naming. The link between heavenly and earthly events is reflected in the language of the divisions of time. Most modern and ancient tongues, for example, describe the period of twenty-four (modern) hours as a single unit by means of a word which also denotes that period during which the sun is over the horizon. A day, as we know but do not question, comprises a day and a night. Nilsson had called this practice pars-pro-toto reckoning: a portion of a duration and the total duration are known by the same name.[7] This practice makes for easy description and quantification: one may travel for seven moons or be away for six sabbaths.

In most civilizations days are counted from some part of the night; counting from dawns is rare.[8] The Sumerians, about 3500 B.C., divided their days into 12 hours, starting at sunset; the Babylonians into 24 hours starting at midnight.[9] In the Chinese horary system three methods of sub-division existed since ancient times: a day (midnight-to-midnight) could be divided into 12 "double hours" or 100 "quarters"; a night (sunset to sunrise) into five "night watches" of variable lengths.[10] Similar divisions of the night into watches were also used by the Romans and remain in use in most of the armed forces of the world. Medieval, monastic time divided the day into 12 equal periods of daylight, and 12 equal periods of darkness hours, each such division, therefore, varying with the seasons. Japanese division of the natural day after about 660 A.D. consisted of the segmentation of an astronomical day into 50 kokus and each koku into 6 bus.[11] There are many variations to these divisions both in the Orient and in Europe.[12] In the Chinese horary system various parts of the day were given names of animals: 5-7 A.M. "hare," 7-9 A.M. "dragon." [13] In the Western horary system of modern times various parts of the day are given names of numbers: 3-4 A.M. "three," or 11-12 P.M. "eleven." Both of these are pars-pro-toto practices.[14]

(2) The moon, calendars, and more "hours"

The sun is not the only astronomical body whose motion has been judged sufficiently regular to be used as a basis of timekeeping. The Egyptian *merkhet* was a device for telling the hour of the night by the culmination of stars, attesting to the Egyptian familiarity with the elements of astronomy in remote prehistory.[15] Nocturnal dials which told hours by means of the fixed stars were first made in Europe in about 1520 and were commonly used during the seventeenth and eighteenth centuries. Such night dials are simplified embodiments of the oldest scientific instrument, the astrolabe, which was the chief astronomical instrument of both Arab and Latin astronomers of the Middle Ages and was known as "the mathematical jewel." [16] It is believed to have been invented by the Greek Hipparchus during the second century B.C. In its simplest form, that of the planispheric astrolabe, it may be described as a flat model of the heavens. It may be put to one of three uses: (1) to compute the position of heavenly bodies at any given time; (2) to determine the time of the day from the altitudes of the stars or the sun, provided the position of the observer is known; and (3) if the position of sun or stars as well as a reference time are known, to determine the position of the user. Thus, an astrolabe may be described as a memory device upon which the laws of planetary motion and the motion of the earth are engraved as functions of the observer's position. In simplified ways, the same holds for a nocturnal dial—or for a sundial.

The cyclic regularity of the moon played an immense part in the elaboration of cyclical concepts in archaic cultures.[17] The moon represents more obviously than other heavenly bodies the changing and permanent aspects of time. It is dark, then increases, it is full, then decreases, then vanishes and reappears, yet it remains identical with itself. The average length of the month is one synodical lunation, suggesting the latter as the origin of the former. One may guess why lunar periods must have been regarded as very important by early man. All sea creatures have periods that coincide with the tides which, in their turn, contain a fundamental cycle of a period that corresponds to one half that of the moon's apparent revolution around the earth, as well as many other astronomical cycles, such as those corresponding to the conjunction and opposition of the moon with the sun. Menstrual periods of women average 29 days thus, even in our epoch they coincide, on the average, with some phases of the moon. (Pregnant women of the Tiv tribe in Nigeria count lunations to determine the status of their pregnancies.)[18] Land animals also display many lunar periodicities. Perhaps the moon was the most obvious and reliable natural clock, with a period appropriate for man's memory.

All ancient calendars are primarily lunar. From among them the

Muhammadan calendar remained strictly lunar. It is based on twenty-eight lunar mansions which make up a solar year of 354 days, calling for occasional intercalation. The lunar months retrogress with respect to the seasons making one complete cycle every thirty-two and a half years.[19] The day begins at nightfall at the instant when the constellation containing the new moon rises.

Much older than the Islamic calendar (whose era begins with the Hegira, that is, Muhammad's flight from Mecca to Medina on July 16, 622 A.D.) are the classical calendars of India which date to about 1500 B.C. In them the daily positions of the moon and the monthly positions of the sun are referred to twenty-seven or twenty-eight zodiacal constellations, corresponding to two estimates of the length of lunation. A lunisolar accommodation is achieved by intercalating months as well as days according to a complex, empirical tradition. Six seasons: spring, hot season, rains, autumn, winter, and dewy season were spinning their weary wheels simultaneously with at least three other divisions of the civil year. Civil days in the Vedic tradition were divided into 30 units of *muhurta*, or 60 units of *qhatika;* each qhatika contained 30 *kala*, 60 *pala*, 360 *prana*, 3,600 *vipala*, and 216,000 *prativipala*. This last unit corresponds to 6⅔ milliseconds, of doubtful practical value to the ancient Hindus. Going toward longer periods of time, we have already discussed the system of yugas wherein the Brahma's life is about 3×10^{14} years. In the chronology of the Vayu Purana each order of magnitude has a name: 10^1, *Dasa;* 10^2, *Satam;* . . . 10^{12}, *Padman*.[20]

Time reckoning in Babylonia and Assyria was based on astronomical as well as seasonal periodicities. Those who worshiped the heavenly bodies preferred the former; those who followed agricultural deities preferred to arrange their lives following the periodicities of grains and fruits. The month was lunar and the years lunisolar, with intercalary months inserted to keep the months in proper phase with the seasons. Looking at long periods, Assyrian calendrical units of the second millenia B.C. included one of 350 years, multiples of which were judged propitious for the reconstruction of sacred buildings.

(3) Calendars and chronologies

The traditional Chinese calendar is based on the use of two interlocking cyclical characters, ten celestial names and twelve terrestrial names resulting in a sexagenary cycle. In the primarily agrarian civilization of China, the promulgation of a lunisolar calendar was "the numinous cosmic duty of the imperial ruler." For a concise summary we turn to Needham:

> Since celestial magnitudes are incommensurable and subject to slow secular change, continual work on the calendar was necessary through the ages, and few were the mathematicians and astronomers in Chinese history who did not work upon it. Between 370 B.C. and 1742 A.D., no less then 100 "calendars" or sets of astronomical tables were produced, embodying constants of

ever greater accuracy, and dealing with the determination of solstices, day- , month- and year-lengths, the motions of sun and moon, planetary revolution periods, and the like.[21]

We find the word "year" also meaning "harvest" since the Chou dynasty (eleventh century B.C.), an example of Nilsson's pars-pro-toto counting. The tenth year of a king's reign was designated as his tenth harvest. Traditional reckoning of years since the Han dynasty (second and first century B.C.) was from events proclaimed by each new emperor as his *nienhao,* or reign-year base.

A highly sophisticated astronomical calendar was developed by the lowland Maya civilization of pre-Colombian America. In its simplest form it consisted of 20 names cycled in units of 13. Since the least common multiple of 13 and 20 is 260, each day in the 260 day cycle could and did have a unique designation made up of a numeral and a name.[22] The seasonal year was 365 days divided into 18 units of 20 days each, with 5 "evil days" intercalated. It was a calendar of forbidding complexity, of cycles upon cycles of classes of years, extending to many millions of years into the past. It permitted the writing of glyphs, symbols representing elements of the numerical calendrical cycles, to represent millions of years in the future. The calendar was closely tied to the experience of passing. Time was personified as a burden carried by a procession of divine bearers of different units of time, with the various divinities changing places at the end of each day's journey according to a rule of permutation. The numerical representation encouraged mathematical extension of the calendar to immense eons, but the Maya concern wtih periodicity prevented the rise of a conception of progressive time.[23] Quite clearly, the Mayan calendar was much more than a method of keeping track of days and arranging daily activities with its assistance. It was a unique form of mathematized religion and mythology of time.[24]

Compared to other calendrical systems the Western one tends to appear simple, especially for those raised in it. In the Greek system a year of 12 lunar months and the solar year of 365.25 days were reconciled in calendrical units of eight years called octaeteris, with three lunar months intercalated. Days began at sunset, but were reckoned from the following dawns, with no regular division into hours. Periods were defined by "cock-crow" and the like. In the classical period of Greece, years were not numbered but identified by the names of the ruling magistrates. From the third century B.C. on, counting in terms of the four-year units of the Olympiads was popular. During the lives of Aristotle and Plato, civil years were identified by the name of the standing committee of councilors who reigned that year.

The early Roman republican calendar contained twelve months. Seven of them constituted lunations of 29 days, four of them 31 days and Februarius 28 days. There were three nodal days each month: *Calendae* (the first), *nonae* (the fifth or seventh), and *idus* (thirteenth or fifteenth days). The other days were counted backward from the nodal days. New moons and the appropriate *nonae* were officially announced; as in China and Egypt, the supervision of the calendar was the duty of a high public official. The various corrections which modulated this basic scheme [25] made the calendar useful yet hopelessly out of date toward the end of the republic. This prompted Julius Caesar to initiate a total revision. The lunar calendar was abolished by decree and replaced with an entirely solar calendar implying a lessened overt dependence of man on the changing phases of the moon. The Julian calendar year with one intercalated day once every four years differed from the tropical year by only about a day in 130 years. Thus, it was successful in bringing the months in phase with the changing seasons and offered a reliable scaffolding for any religious or civilian scheme of preferred days desired by church or state. With the passage of time the vernal equinox and its calendrical date still drifted apart, but no reform was forthcoming until the German Jesuit mathematician Clavius, under the reign of Pope Gregory XIII, redesigned the calendar (1582) by inserting a second order correction: a century year is a leap year only if it is divisible by 400.

Western chronology, as any other chronology, must count years from an arbitrary event. Its strength is the universality it achieved around the world through the industrial and scientific revolution. Counting years from the birth of Christ was first introduced in 525 by the Roman Abbot Dionysus Exiguus. Counting backward from the birth of Christ originated as late as the eighteenth century and is of uncertain authorship.[26]

It is evident from what we have seen that the regularities of the skies above and the seasons on earth are the primordial sources of timekeeping.* The ubiquity of these cyclicities may explain why they became the generating rhythms of all natural processes, extending eventually to the temporal organization of the individual and collective functions of society. In the heavens all but the North Star move in eternal cycles. In Chinese symbolism the emperor, who represented the steady light of the North Star, was seated on his throne facing southward as does that star (and as did Chinese magnetic needles) and he was expected to oversee everything but theoretically do nothing himself, as did the North Star. For everyone else on earth, one must assume, Shakespeare's admonition remained valid:

> Thou by thy dial's shady stealth mayst know
> Time's thievish progress to eternity

Sonnet 77

An inscription on a 1772 sundial in Yorkshire, England, called it by the name *Certa Ratio*, or sure reckoning.[27] This phrase has its perennial charm,

---

* The only universal division of time which does not seem to have an evident astronomical reference or seasonal origin is that of the week. Perhaps it is favored by some obscure biological rhythm; perhaps it relates to the sacredness of the number seven; its relation to the seven planets of antiquity is quite clear, but we do not know what came first: the week, or the seven names of the days. See F. H. Colson, *The Week* (Cambridge: The University Press, 1926).

for, in each age whatever device was used for the interpretation of the
motion of heavenly bodies in the service of man's life must have been
regarded as sure reckoning. Sundials and star dials were relied upon
because the motion of the planets and the cyclical recurrence of the seasons
were trusted to constitute unchanging laws.

But what for Clavius was a year of 365.2422 days is, for the twentieth
century astronomer a year of $(365.25636556 + 0.00000011 \ T)$ ephemeris days,
where T is a correction factor for epoch. Timekeeping in terms of the
rotation of the skies and variations of the seasons has left the domain of
unaided observation of the sun, moon, and stars and come to depend on
sophisticated instrumentation and on complex theoretical arguments.

The Flow of Water and Sand

Waterclocks are the oldest known nonastronomical devices used for
the measurement of time; they employ the regularity in the rate of flow of
water through an orifice, into or out of a container. Bucket-shaped, outflow-
type water clocks survive from the Temple of Karnak (fifteenth century
B.C.),[28] and there is evidence for their use by the Assyrian King, Tukulti-
ninurta from the thirteenth century B.C.[29] They are known today by the
name "clepsydra," or "water-thief" given to the water clocks introduced
from Egypt into classical Greece.

The outflow type was marked on the inside of the bucket; as the water
escaped through an orifice near the bottom, the level of the remaining
water indicated the hour of the day. Calibration involved two problems:
how to get the flow uniform as judged by the clockmaker's sense of time,
and how to correlate the readings with the time determined from astronomical sources. The Roman architect and historian of science Vitruvius, writing
in 27 B.C. suggested that to assure uniform flow, the waterhead be kept
constant by continuously replenishing the container from another source,
such as another clepsydra. To indicate the hours as they varied according
to the season, either the rate of flow or the calibration were to be adjusted.
He describes an automaton comprising a very slowly rotating drum driven
by a sinking float, with the rate of water flow controlled by orifices.[30] A figure
which itself sinks as the day passes points out the hours on the drum. The
drum carries an engraved gridwork corresponding to the hours changing
with the seasons. From our point of view, this amounts to the transformation of a linear temporal process (the fall of the water, judged uniform)
into circular motion (of the drum) in which the yearly cycle of the sun is
implicit. The transformation from linear motion (again, a sinking float
counterbalanced by a weight) into circular motion, which is that of the
heavens, becomes explicit in clocks which Vitruvius called *Anaphorica*
(from the Greek ἀναφορικόν, implying relation to star ascension), clocks
whose rotating dials carry the signs of stars and of the zodiac. Anaphoric
clocks are the ancestors of the clock dial in that both are fixed images of the

zodiac.[31] Vitruvius also described a monumental Greek water clock, the Tower of the Winds, so called for a set of decorative reliefs personifying the winds. This complex structure, reconstructed between 1962 and 1966, was the combination of wind-vanes, sundials and water clocks. It may also have employed an anaphoric disc as a display device showing the seasonal hours.[32]

Chinese clepsydra techniques date from the sixth century B.C. They were of the outflow type, possibly imported from the Fertile Crescent.[33] From about 200 B.C. on, the inflow type became popular, equipped with a float and indicator rod resembling the one described by Vitruvius two centuries later, in Rome. One or more compensating tanks were used to keep the pressure head in the penultimate outflow tank constant and thereby the motion of the float uniform. In later centuries sophisticated cumulative regulation was achieved by using as many as five successive tanks and providing the last one with an overflow orifice. In some arrangements a steelyard balance held the lowest compensating tank as a continuous check on the weight of the water and the possibility of seasonal adjustments for changing hours. Instead of water, which would freeze in the winter, sand was sometimes used, or mercury, with all pipes and reservoirs made of chemically resistant materials such as jade. The oldest printed illustration of a water clock in any culture is from a Chinese manuscript of about 1155 A.D., showing an inflow type device with a torch near it to ward off the frost.

There was a resurgence of interest in clepsydrae in seventeenth century Europe as part of the Renaissance interest in principles of nature and the desire to harness natural forces.[34] We find compartmentalized metal cylinders whose rotation by a weight was checked by the controlled flow of mercury [35] and many other devices, often cute, ornate, and gaudy. Sand-glasses first appeared in the fourteenth century [36] and are still broadly used. In the earliest example, two separate glass bulbs were bound together at their necks, having a metal plate with a pierced hole clamped in between. One may look at such a device as the combination of an outflow and inflow clepsydra, with fine sand replacing water or mercury.

The dripping faucet of contemporary America, reminding the sleepless of the unpredictable regularity of time through its drip-drip, expected and yet each time unexpected, is a water clock of long ancestry.

Burning Rates

Matteo Rizzi, founder of the Jesuit missions in China, observed around the year 1600: "As for their clocks, there are some which use water, and others the fire of certain perfumed fibers." [37] Rizzi might not have realized that the use of controlled rates of burning for timekeeping has an extensive history in the Orient, going back to the sixth century A.D.

The simplest fire clocks of China and Japan were match cords which burned like a wick or fuse, dropping weights as they were consumed, informing people of the passage of time. Graduated candles were used, also incense in the form of hardened paste sticks, straight or spiral, marked off into suitable intervals. An incense-seal clock consisted of a hard wooden disk into which was carved a continuous groove. "The Greatly Elaborated Incense Seal" is such a timepiece, featured in the *Hsian Ch'eng*, a work on aromatic incense popular in Medieval China.[38] The incense made from a variety of aromatic powders according to prescribed recipes was placed into its grooves, lit in the center, and burned for approximately twelve hours. The length of the path is estimated to be about twenty feet. Incense seals served in Taoist and Buddhist temples to indicate the time for the striking of the bell and to remind the villagers that, once again, it was time for prayer.[39] That water clocks played identical roles in the monasteries of the West, although the two cultures were not in communication, is a reminder, if one is needed, of the psychic unity of man.

Although Plutarch knew that "When the candles are out all women are fair," the use of sweetly smelling incense sticks as timers in Geisha houses was reserved for the artistic Japanese. Their Japanese name, *senko*, retains the meaning of "flower girl incense stick," [40] to which the broad use of incense among the flower children of contemporary America is pleasingly akin. The use of burning rates of candles was known in Europe. Alfonso X, King of Castile, (mid-thirteenth century) owned a candle clock attributed to "Samuel el Levi, the Jew of Toledo." [41] In it the weight of a new candle was counterbalanced by a pulley system; as the candle was consumed and became lighter, the system raised a tablet which indicated the time of the day with reference to an index mark. Geronimo Cardano, a contemporary of Copernicus and inventor of the universal joint, also invented the portable time lamp now known by his name. It employed the amount of unburned oil as a measure of time.

In incense clocks, oil lamps and candle clocks, burning is accompanied by flame and smoke. But this is not the only way heat may be transported or the lawfulness of such transport employed as a clock. A device called an entropy clock was suggested by Eddington.[42] Take a hot and a cold body, such as two iron cubes. Place them into contact and keep them in a container that isolates them from all outside sources of heat. Let the two blocks have imbedded in them thermoelectric couples used to measure their temperature difference. It is almost certain that the warmer cube will cool off while the colder one will warm up. During this heat exchange process a galvanometer attached to the thermocouples will indicate a flow of current decreasing with time (as experienced by the observer), and a suitable ammeter may be calibrated in units of time. Generally, smaller temperature differences will correspond to later instants, just as the shorter candles in the clock of King Alfonso X corresponded to later hours.

Eddington's purpose in suggesting this somewhat odd scheme was not clock design as such but to reveal the correlation between the physical concept of entropy and our sense of time. The irreversible burning of

Giordano Bruno's body would have been an equally valid example, but it would have missed the soothing abstractness of physical science. In any case, the correlation among the experience of the passage of time, body heat, and fire clocks is summed up by Yeats:

> But a coarse old man am I,
> I choose the second best,
> I forget it all awhile
> Upon a woman's breast.
> > Daybreak and candle-end.
>
> "The Wild Old Wicked Man"

Controlled Oscillations of Large Bodies

While some of the Greek philosophers worked on reducing time and the perennial motion of the stars to permanent intellectual notions, some of their fellow Athenians attempted to reduce these motions to those of mechanical models. Thus, a bronze planetarium is said to have been made by Archimedes and described by several authors. This is from Cicero:

> When Archimedes put together in a globe the movements of the moon, sun, and five wandering [planets], he brought about the same effect as that which the god of Plato did in the Timaeus when he made the world, so that one revolution produced dissimilar movements of delay and acceleration.[43]

Actual fragments of a geared planetary device of Hellenistic origin, dating circa 65 B.C., were recently recovered and identified as a planetarium or astronomical clock, the oldest extant member of its species.[44]

I wish to draw attention to the fact that all our dial clocks, such as wrist watches, are planetaria which show the motion of the sun and do so at a rotation rate which is two (or twenty-four) times faster than that of the sun.[45] The making of such planetaria was not feasible with candle clocks or sand glasses, they became practical only with the appearance of mechanical clockworks. Before a uniformly rotating pointer became possible, two technical problems had to be solved: the slowing down and the making uniform of the rotation of a wheel driven by a weight or any other source of power. Solutions to these joint problems are many, but they all involve devices which control the rate at which the driving force can act upon the wheel through some process which the clockmaker judges to be regular. I shall try to illustrate some of these principles.

What Needham called "the grand ancestor of all clock-drives," [46] is the escapement of "The Water Powered Armillary (Sphere) and (Celestial) Globe Tower" described by the Chinese astronomer Su Sung in 1090 and designed and built by him and his associates in the years preceding that date. This awkward yet sophisticated escapement controlled the rotation of a primary wheel according to a nonuniform but recurrent cyclic program. Power was obtained from the torque of a water wheel whose scoops were filled and then emptied.

The escapement which checked the forward motion of this wheel was a device of weight bridges and linkwork which remained stationary while each scoop was filling, but then operated instantaneously so as to open a gate and release one spoke, the next scoop being brought into position under the constant flow water-jet. Steady motion was thus secured by intersecting the progress of a powered machine into intervals of equal duration. . . .[47]

Su Sung's escapement provided uniformity of rate. The rate itself was determined by the time needed to fill each scoop with water; that in turn, was controlled by a clepsydra-type arrangement of water tanks, with the ultimate tank being of a constant-level type. In all mechanical clocks we find components whose functions are similar to those in Su Sung's clock. They may be identified as elements which (1) assure uniformity, (2) control the rate, and (3) supply power.

The earliest known European device that combines these dynamic elements is the verge and foliot escapement invented during the thirteenth century. Its oldest extant and still operating example is in the cathedral tower at Salisbury, England.[48] Built in 1386, the oscillating component is a foliot (which is a weighted bar rotatable in a horizontal plane) and a verge (that is a vertical bar attached to the center of the foliot) with two pallets on the verge. A saw-toothed crown wheel rotated by a weight alternately engages the upper and the lower pallet, imparting an oscillating motion on the verge-and-foliot unit. The uniformity of motion is secured by the fixed moment of inertia of the oscillating foliot; its rate is determined by the torque imparted on the verge by the crown wheel, itself a function of the driving weight. The power that moves the clock comes from the work needed to raise periodically the weight that drives the crown wheel.

The genesis of the mechanical clocks, such as the Salisbury clock just mentioned, the famous astronomical clock of Giovanni Dondi (1364), and the First Strasbourg Clock (1354) [49] is also the epoch just after Aquinas, that of Ockham and Nicholas of Oresme, leading up to Cusanus and the Renaissance. Somewhat earlier, some time during the thirteenth century, the regularity represented by, and built into the clock began to be returned to society in an amplified manner. First this took the form of improving coordination of activities, first within, then without the monastery walls. Later, as we shall see, the increasing coordination tended to grow into an increasing subjugation of man to those of his ideals which could be expressed in schedules.

The verge and foliot clock was accurate only to within about two minutes per day; the reduction of the heavenly clock to that of an earthly clock was unsatisfactory. In the history of successive approximations, or improved reductions, a radically novel means was born with the invention of the principle of pendular timekeeping by Galileo. According to an apocryphal story, the young Galileo discovered the isochronism of the pendulum by checking the constancy of the period of a swinging candelabrum against the constancy of his pulse beat.[50] Later he constructed a pendulum of adjustable length to measure the pulse rates of patients; this

device was called a pulsilogium. In 1602 he claimed that the period of
the pendulum is independent of the amplitude of its oscillations and of the
weight of the bob. Finally, in his *Dialogues* (1632) he wrote:

> As to the times of vibrations of bodies suspended by threads of different
> lengths, they bear to each other the same proportion as the square roots of
> the lengths of the threads; or one might say the lengths are to each other as
> the squares of the times; so that if one wishes to make the vibration-time of
> one pendulum twice that of another, he must make its suspension four times
> as long.[51]

Here was a process that combined into one device two of the three necessary elements of a mechanical timekeeper: uniformity of motion and control
of rate. It needed only power to be complete. Although Galileo invented an
escapement using a pinwheel instead of sawtooth wheel, he never actually
built a working model.[52] It is believed that one such model was completed
by the end of the seventeenth century for the Duke of Tuscany by a clockmaker from Augsburg, Germany.

Comparison of the performance of pendular clocks with astronomical
references and with each other began no sooner than the idea of using a
pendulum for timekeeping had emerged. Christian Huygens concluded
that a truly isochronous pendulum should have its bob swing along a
cycloid rather than the arc of a circle and constructed a pendular bar in
1673 approximating this condition. The story of pendular clocks thereafter
is that of the technology of second, third, and fourth order corrections to
the theory of isochronism. Temperature and barometric pressure compensa-
tions were introduced and methods of suspensions and drive improved. Yet,
these and later improvements had to remain imperfect. The relationship
between the period and the displacement even of an ideal (mathematical)
pendulum is valid only as an approximation for small angles. But mathematical pendulums cannot be built, and a complete theory of nonidealized
physical pendulums has not been worked out. In the absence of an accurate
theory, our belief in pendulums as dividing time into equal intervals rests
only on the mapping of their signals into some other more trustworthy
clocks.

A timekeeper that was small enough to be carried on a person was
made during the first decade of the sixteenth century by Peter Henlein of
Nurenberg. The use of a falling weight as the source of power was replaced
in it by the potential energy of a bent spring. In 1675 Huygens published
the details of a balance wheel which was to replace the pendulum in
watches. The pendulum is displaced from its position of gravitational
equilibrium to which it tends to return; the balance wheel is displaced
from its equilibrium position as determined by a coil spring. Technical
advance did not stop there, however. Quite recently the use of miniature
tuning forks became practical. These are displaced by magnetic attraction
from an equilibrium position determined by the elastic properties of steel.
In these three analogous oscillators (the pendulum, the balance wheel,

and the tuning fork) uniformity of operation obtains through what the
clock watcher believes to be a lawful, cyclic exchange of the driving energy
between its potential and kinetic forms. In all three, continuous improvements have been necessary for improved reduction of the heavenly clockwork to earthly ones.

Mechanical timepieces, such as the ones mentioned, may be described
as Newtonian, even if constructed long before the seventeenth century. In
the Newtonian universe inertial and gravitational forces kept the planets on
their appointed rounds; forces, energy, and Laplacian determinism controlled the motion of the heavenly bodies, and the same forces, energy and
determinism, moved and controlled the clock. Alexander Pope gave an
intuitive eighteenth century account of these conditions in his *Essay on
Man* as he sent Homo sapiens off in pursuit of his destiny.

> Go, wondrous creature, mount where Science guides,
> Go, measure earth, weigh air, and state the tides;
> Instruct the planets in what orbs to run,
> Correct old Time, and regulate the Sun.

This Newtonian world, however, has been irrevocably superseded by relativity and quantum theories. And our clocks have changed accordingly.

Controlled Vibrations of Small Bodies

Elements which assure uniformity of motion, control rates, and supply
power to drive the clock are also present in devices that work by absorption
or emission of particles or electromagnetic waves, or by microscopic oscillations.

An electromechanical device, the quartz clock, was developed in the
United States early this century. The physical phenomenon which forms
the basis of its operation is the piezoelectric effect, or pressure electricity,
identified in the year 1880 by the Curies. The application of mechanical
stress produces in certain dielectric crystals an electric polarization and,
conversely, a voltage applied between certain faces of the crystal produces
mechanical distortion. Certain crystals, such as quartz, will have a natural
mechanical frequency determined by their size and shape. It is at this
resonant frequency that the crystal is most responsive to cyclic voltage
variations, and it will develop alternating voltage at this frequency if its
mechanical equilibrium is disturbed. A quartz crystal of appropriate size
and weight may therefore be made part of an electric circuit; frequency
thus controlled is stable compared with the aggregates of other clocks and
astronomical sources. For quartz clocks, this stability is about 1 millisecond
per day (or of the order of 1 prativipala in the Vedic tradition). The rate
of the clock is controlled by the physical dimensions of the crystal and by
holding constant such environmental factors as temperature and humidity.

Power is supplied from an electric source. Characteristic oscillation frequencies are of the order of 100 KHertz which, by the electric equivalent of a gear train, may be stepped down to lower frequencies. A display of the time, if called for [53] may be a dial or in the form of numerals. Unbeknown to the designers of digital display clocks, their practice is a metaphysical feat. They replaced Plato's "time as a moving image of eternity," as represented by the planetariumlike dial of Jacopo Dondi, by Aristotle's "time as the number of motion with respect to before and after." The clocks display the numbers of motion, the clockwatcher supplies the before and after.

The most accurate reference clocks are quantum mechanical instruments, their uniformity in measuring time is assured by the constancy of atomic structures; their rates are determined through the selection of specific atoms for resonance and through the control of environmental conditions. The frequencies at which atoms and molecules absorb and emit electromagnetic waves most efficiently are characteristic functions of physical substances. Thus it was possible for the Thirteenth General Conference of Weights and Measures in 1967 to define a second of time as follows:

> The second is the duration of 9,192,631,770 periods of the radiation corresponding to the transition between two hyperfine levels $F=4$, $m_f=0$ and $F=3$, $m_f=0$ of the ground state $^2s_{1/2}$ of the atom of cesium-133 undisturbed by external fields.[54]

The notation is the spectroscopic designation of certain atomic states. The environmental conditions to be controlled are temperature, vapor pressure, the magnitude and direction of magnetic fields, and the like.[55] But one second of duration is also defined as $1/31,558,149.747$ part of a tropical year. Do the two definitions of the second agree? Generally not.[56] From the point of view of the astronomer the situation was cogently described by G. M. Clemence.

> We are . . . in a position to see what is meant by an invariable measure of time. It is a measure of time that leads to no contradiction between observations of celestial bodies and the rigorous theories of their motions, yet more precisely, it is the measure of time defined by the accepted laws of motion.[57]

An identical claim for minimum discrepancy between theory and observation can also be made for the spectroscopist. In fact, even nonuniform processes such as the random emission in radioactive decay have been considered as clocks.[58] The hidden feature that makes radioactive decay a possible means of time measurement is the fact that we know the laws of that decay, albeit in probabilistic-statistical form.[59]

Thus, the radioactive clock and Su Sun's Heavenly Clockwork have this in common: they are clocks by virtue of their builders' belief in the timelessness of natural laws upon which their functions are predicated.

Three important points may now be made. (1) The decision as to which clock is more reliable depends on judgement about permanent regularities. For instance, in the case of the atomic resonance versus tropical second, there are known and suspected physical reasons to which the unevenness in the rate of the earth's rotation may be attributed; there are no known reasons why the cesium clock should exhibit cyclic or secular variations. Should such reasons be discovered, the responsibility for the disagreement would be partly shifted to the cesium clock. (2) No clock can be said to be accurate in itself: measurement of time always involves at least two clocks. Often one of them is hidden in unstated assumptions, such as when I claim that my watch is accurate within two seconds per day. (3) Whether or not two clock readings may be reconciled depends on whether or not an acceptable theory exists to connect the two.

With the increasing sophistication of the processes employed to measure time, increasingly sophisticated natural laws (nomothetic features of time) are called for and, conversely, may be tested. Thus, the accuracy of Christian Huygens' pendulum could be tested against the laws of Kepler which were regarded as exact. Kepler's laws of planetary motion could have become suspect as incomplete, or at least as puzzling, when in 1845 Leverrier found an unexplained variation in the motion of the perihelion of Mercury, using the improved pendular clocks of his day. The reliability or regularity of current atomic clocks is being tested in connection with certain first and second order effects predicted by Einstein's relativity theory.[60]

What we see here are successive reductions of ideas about natural laws involving time into working models which confirm these ideas to a sufficient approximation. Whereas Plato perceived in faultless forms the true and final home of all and thus projected man upward, the clockmaker tried to reduce those forms into approximate models here on earth. Plato's cosmos worked in harmony but the world of all clocks never did. This was noted by Charles V, Holy Roman Emperor, who in 1550 retired to the solitude of a monastery and busied himself with clocks. "To think that I attempted to force reason and conscience of thousands of men into one mold" sighed the Emperor, according to hearsay, "and I cannot make two clocks agree." [61]

The history of timekeepers demonstrates the reduction of successively more refined ideas about natural processes to working devices which confirm those ideas.

The pendulum and escapement of Huygens's clock. From Christian Huygens, Horologium Oscillatorium, *Paris, 1673. Courtesy, The Burndy Library, Norwalk, Connecticut. Figure II shows the cycloidal cheeks used to improve the isochrony of simple pendulums.*

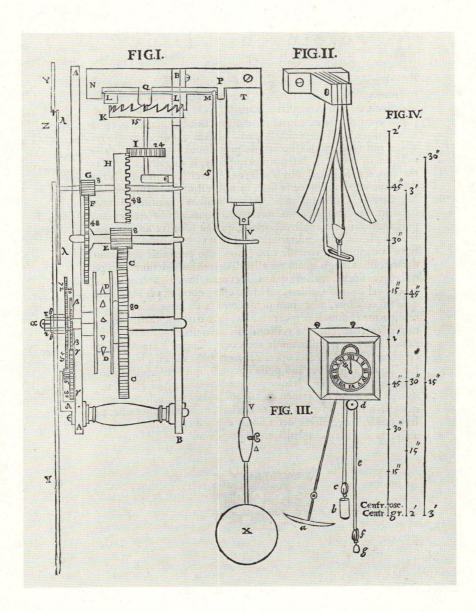

FIG. I.   FIG. II.

FIG. IV.

FIG. III.

2. Clocks and
Clockwatchers

ALTHOUGH clocks and calendars bear some undisputed affinity to the concept of time, it is difficult to define exactly what that relationship is. That they "measure time" assures us only that they employ predictable regularities. I would rather think that the devices used for reckoning time bear a "sympathy" to temporality, that is, they communicate something about time. But communication assumes a source, a link, and a sink, and for this reason it is the system of clocks and clockwatchers together whose nature we must consider if we are to place into perspective the relationship between timekeepers and time. We may begin this by examining the double role of the clock as a device and as a metaphor.

The clock in the Strasbourg cathedral was built in 1354, and rebuilt in 1574. The early model had a calendar, an astrolabe, and an automaton. The stroke of the hour brought the three Magi before the Virgin Mary who was carrying the child Jesus; she bowed toward the Magi and walked on while the carillon played religious songs. Another automaton, a cock, flapped its wings, lowered its tail and crowed, reminding the faithful of Mark 14:72: "And immediately the cock crew again. And Peter remembered the word that Jesus had said unto him: Before the cock crows twice, thou shalt thrice deny me."

We have seen that the ancestors of the mechanical clock (such as the Hellenistic gear train of Price) were primarily astronomical devices, not timekeepers. Neither were the Strasbourg clock specifically, or medieval monumental clocks in general, timekeepers. They were religious and astronomical reminders.[62] In the words of Lynn Whyte:

> Suddenly, towards the middle of the fourteenth century, the mechanical clock seized the imagination of our ancestors. Something of a civic pride which earlier had expended itself in cathedral building, was now diverted to the construction of astronomical clocks of outstanding intricacy and elaboration. No European community felt able to hold up its head unless in its midst the planets wheeled in cycles and epicycles, while angels trumpeted, cocks crew, and apostles, kings and prophets marched and countermarched at the booming of the hours.[63]

A fourteenth century astronomical clock studied by John North was described in a contemporary manuscript as an "Opus quorundam rotarum mirabilium. . . ." "A device of certain remarkable wheels by which the true places of all the planets are known, and also the hours of the day and night."[64] Remarkable it was, as was the astrarium of Giovanni Dondi, son of the inventor of the dial, constructed between 1348 and 1364. The latter was described as "a perfect machine, where the intricate mass of the orbs and planets is clearly and distinctly known to be moved in an orderly manner. . . ."[65]

With one eye on the heavens and one on earth, the earlier Chinese clockworks were equally ambitious in their purpose. Beyond guidance in the timing of crops, they were employed for ritualistic purposes, for assistance in maintaining a cosmic unity and harmony between the earth (the Emperor) and the universe, and in the planning of genetic lines in China's governing families. From among the concubines "... the women of highest rank approached the emperor at times nearest to the full moon, when the Yin influence would be at its height, and, matching the powerful Yang force of the Son of Heaven, would give the highest virtues to children so conceived." [66]

Thomas Aquinas observed in his widely ranging theology that all things moved by (man's) reason act as though they had reason, even if they lack reason, and illustrated his point by reference to a moving arrow and a clock.[67] Arrows and clocks retained their privileged positions as images of space and time, respectively, with the clock slowly rising into the position of a peculiar metaphysical device. Nicholas Oresme likened the universe to a mechanical clock that was created and set moving by God, who also saw to the proper proportions of the celestial wheels.[68] The clockwork metaphor was taken seriously by Kepler when he attempted to understand the world as a unit interconnected by actions at a distance. In 1605 he wrote:

> I am much occupied with the investigation of the physical causes. My aim is to show that the celestial machine is to be likened not to a divine organism . . . but rather to a clockwork. (Whoever thinks that the clock is an organism attributes to it the glory due its maker). Insofar as nearly all the manifold movements are carried out by means of single, quite simple magnetic force, as in the case of a clockwork all motions [are caused] by a simple weight.[69]

Giovanni Alfonso Borelli (1608-1679), an Italian physiologist, sought to explain the motion of animals on mechanical principles in his *De motu animalium*. He also attempted to account for the fertilization of the ovum by the male sperm by an odd comparison of the process with the workings of a timepiece made of cogwheels and moved by a weight.[70]

But the clock in the West grew to be more than a cosmological or biological metaphor. Medieval Christianity with its otherwordly directedness prepared the believer for timeless eternity. To assist him to escape from time by making better use of time on earth, it used the clock to give religious activities a collective beat. First, clocks reminded him of his tasks; then, by representing organization and order, they began to mark "a perfection towards which other machines aspire." [71] When projected outward on the world, clocks suggested an awful degree of lawfulness in life, society, and in nature at large; they reinforced the heilsgeschichtlich world view that the essence of the world is its temporal trend.

Simultaneously, from the early mechanical clocks to our atomic devices, timekeepers underwent a profound change in what may be described as their form of intelligibility and, perhaps "color of feeling." Shakespeare could still attribute to King Henry VI the wish

> To be no better than a homely swain;
> To sit upon the hill, as I do now,
> To carve out dials quaintly, point to point,
> Thereby to see the minutes how they run
> How many make the hour full complete . . .
> How many years a mortal man may live.
> When this is known, then to divide the times:
> So many hours must I tend my flock. . . .
>
> *Henry VI* Part 3 (III, ii, 22)

By inspecting a dial, or a simple mechanical clock one is usually able to read function directly from structure. Thus, their clockness may be confirmed by inspection.[72] This does not hold for non-Newtonian clocks. In them the transfer from structure to function must proceed through the acceptance of sophisticated physical theories. Hardly anyone would find a cesium clock quaint or even suggestive of the time of life. But clocks without man, as Dickens saw them, are hours without passion. In *Hard Times* he described the dull engagement of Mr. Bounderby with Miss Gradgrind:

> Mr. Bounderby went every evening to Stone Lodge, as an accepted wooer. Love was made on these occasions in the form of bracelets. . . . The Hours did not go through any of those rosy performances, which foolish poets have ascribed to them at such times; neither did the clock go any faster, or any slower, than at other seasons. The deadly statistical recorder in the Gradgrind observatory knocked every second on the head as it was born, and buried it with his accustomed regularity.[73]

The Marxian evaluation of the clock did not reinsert a relationship between the individual and the clock but only between society and the clock. "The clock is the first automatic machine applied to practical purposes;" wrote Marx to Engels in 1863, "the whole theory of production of regular motion was developed through it." [74]

There is a profound interdependence here not only between clocks and clockwatchers, but among these two and nature at large. This important, many-sided connection escaped Reichenbach when he maintained that pendular clocks are not clocks because they are open systems, that is, they depend on their environment. In his view, a spring balance clock may be regarded as a true clock for it employs only internal devices, hence it is a closed and self-contained system.[75] On the contrary, I believe that no timekeeper can be a closed system; there is certainly none discussed on the prior pages. Sundials, calendars, star clocks and the like are obviously astronomical. The flow of water and sand as used in clocks keep time under the influence of the earth's gravitational field; in free fall they would not work. Balance wheels, vibrating forks, any and all rotating gadgets are inertial devices and neither inertia nor gravitation can be made sense of in a universe empty of everything but the clock, or even the clock and a local chunk of matter. Burning rates are thermodynamic processes inconceivable without a universe in which entropy may freely decrease as well as increase. It is precisely connections such as these between clocks and the world that doom to failure any totally physical theory of time. Physical theories of processes are never completely accurate and it is their almost-correctness that leads to the dialectical motion between ideas about time and principles of time measurement.

We recall that the Tempier and Kilwardby declarations of 1277 proclaimed the independence of time from motion, thereby acknowledging the failure of at least one type of reductionism of time. Likewise, each successive improvement in the operation of a timekeeper is a declaration of the failure of prior principles of time measurement. The operation of clocks and calendars tests theories which claim to be able to predict the future accurately, but actually never do so. This is the failure of the mensurational reductionism of time that is, time cannot be completely reduced to its measurement process. New theories of time are necessary guides for the construction of new timepieces, and new timepieces are accepted only in the context of new world views and needs.

Processes suitable for timekeeping are everywhere: pine cones in my study which open when ripe, geese migrating every spring and fall in response to some ancient call, the sun rising with great probability every morning, and the hum of my Bulova Accutron doing its regular $1.136,003,-398,424 \times 10^{10}$ cycles per sidereal year. Each time one speaks of the reliability of such regularities one equates them with a degree of certainty about the future. But what seems to characterize the future is unpredictability, hence, in the human context, uncertainty is also part and parcel of all clocks:

> It seems that the Clepsydra
>   Has been filled up with the Sea
> To make the long, long nights appear
>   An endless time to me.
> The incense-stick is burnt to ash
>   The water clock is stilled
> The midnight breeze blows sharply by
>   And all around is chilled.[76]

Here we encounter once again the incompatible aspects of being and becoming in the nature of time. At this occasion they are not expressed in the abstract reasoning of philosophers but communicated through the practical concerns of the clockmakers. Since the measurement of time always involves the comparison of at least two clocks, we may regard time measurement as an intellectual and practical scaffolding that conjoins in rational unity the clocks, the clockwatchers, and whatever events are unexpected by the clockwatcher. For, although the ticks of the clock follow with trusted regularity as do the ax-time, sword-time, wind-time and wolf-time of the Vikings, there is no way of knowing whether the next strike of Big Ben or the next oscillation of the quartz crystal will or will not occur.

3. The Stuff that Clocks
are Made of

WHAT we have learned about timekeepers and time measurement is useful in elucidating some concepts common to the vocabulary of the philosopher and the clockmaker.

Time Scales

I have stressed already that time measurement involves the reading of two or more clocks. We may now formalize this finding and define a *time scale* as any set of specifications of how the readings of one clock are to be transformed into readings of another clock. Time scales so defined are implicit in all time measurement.

It is not difficult to identify time scales. Sidereal months transform to mean solar days through relations partly empirical, partly theoretical, with a residual inaccuracy due to unknown sources remaining. A plot of the circadian rhythm of the spontaneous muscular activity of a female albino rat constitutes instructions on how to change rat clock readings to laboratory clock readings or vice versa.[77] A rat clock is very inaccurate. By this we mean that connections among events selected as rat-ticks are controlled not only by regularities we know but also by processes as yet unknown. Hence, the temporal positions of the tick-events are to a great degree unpredictable, as compared with those of the laboratory clock. Cesium clocks are judged to be of very great accuracy. Yet, as we have seen, the time scale that connects their readings with astronomical readings still contains some unknown, hence unpredictable, processes. We cannot even tell how the differences should be split between changes in the earth's rotation and shifts in cesium frequency. That John F. Kennedy was assassinated on November 22, 1963 is also an instruction. It tells us how to connect calendar readings with events in history, which is a temporal process made up almost entirely of unknown constituent processes.

Time scales may thus be seen as forming the network or substratum of the temporal universe. They are the operational equivalents of the Principle of the Unity of Time,* for mutual interchangeability of clock readings

---

* Two working assumptions which are necessary, though not necessarily sufficient for interdisciplinary argumentation about time, I have called "the Principle of the Unity of Time." (J. T. Fraser, *The Voices of Time,* New York: Braziller, 1966, p. xxi.) They are: 1. when specialists speak of time they speak of various aspects of the same entity; 2. this entity is amenable to study by the methods of the sciences; it can be made a meaningful subject of contemplation by the reflective mind, and it can be used as proper material for intuitive interpretation by the creative artist. This "unity of time" is to be distinguished from that of Aristotle, meaning the identity of dramatic and real duration, and that of Heidegger, meaning the coexistence of permanence and change.

can derive only from some genetic identity among all clocks. Time scales do not imply the subordination of one class of events to another but assume, instead, that temporality unites all conceivable timekeepers.

Simultaneities

The reading of clocks demands judgements of simultaneities or temporal coincidences. We regard two signals as *simultaneous* if, in terms of the empirical givens or logical limitations of the system receiving the signals, the temporal separation of the signals proves impossible.

For instance, two shots heard within a sufficiently brief period of time will be judged as one single shot by a listener, even if distinguished as two successive shots by an instrument. For reasons that derive from the physiology of hearing, the man will perceive the two shots as one because of his inability to identify an audible signal structure with such features as he may recognize as conditions of before and after. Putting it by way of time scales, we must say that he does not possess an inner clock in terms of which the separation of the external signals could be measured. It follows that certain periods which the instrument recognizes as temporal must be regarded from the point of view of the human listener as essentially atemporal.

By atemporal in this context I mean that there is no empirical way in which the listener can recognize this period in terms of some hallmarks of temporality such as before and after, or future, past, and present. In some other cases man may take the role of the "instrument" and separate two events which, in the experience of another organism or in the self-consistent operational scheme of a device or theorem, must be judged as atemporal. An important feature already implied and common to all states of simultaneity is that they presume temporality. Simultaneity, recognized as an atemporal condition by one system or organism or accommodated by a logical structure, can be understood only if there is an associated system, organism, or logical structure for which the phenomenon is recognized to be temporal.

Events and Processes

I wish to define as an *event* anything that remains self-consistent and identical with itself through a period of time. Events so defined may be recognized and given names. For instance, "one shot in the dark" as perceived by a listener is an event; so is "one tick of a clock" as defined by a physicist, "one day" or "one night" as defined in the Book of Genesis ("and he divided the light from the darkness. And he called the light Day, and the darkness Night"); "one explosion of a supernova" measured by the constancy of light intensity on a photographic plate; "one ball" (as judged by a child, as long as the ball is fuzzy) or "one life" as declared by a gynecologist or an executioner. Events so understood are specific simultaneities, hence they are atemporal phenomena. Consider, for instance, that "one day" is that condition when the sun is over the horizon; "one ball" is the condition of something being round as well as fuzzy. Neither of these identities (as far as their definitions go) have a differentiated temporal structure. They are comparable to "one shot in the dark" for a listener who cannot tell that, for a microphone the one shot is two shots.

What to mean by an "event" has been of concern to modern philosophers. In the process philosophy of A. N. Whitehead, event is defined as that portion of time through which a specific character of place is discerned.[78] Thus, event remains temporal and tied to space. This idea is reminiscent of the principle of minimum time put forth by R. G. Collingwood; this principle states that the way nature appears depends on the length of time taken to observe it. For instance, as he sees it, there exists a shortest period of time in which a substance may exist, "because the specific function or process whose occurrence is what we mean when we speak of the specific substance as existing, cannot occur in a shorter time." [79] For molecules this may be the minimal time in which the atoms can be said to have formed them; for Zeno's arrow there might be a minimum time such that in shorter time the arrow could not be regarded as moving and, for Aristotle, happiness demands a full lifetime and cannot exist in less. Both Whitehead's and Collingwood's ideas resemble the one I proposed. But they differ from it, because neither of them makes the crucial point, or implies as essential that an event, although spoken of in the context of temporality, must in itself be regarded as intrinsically atemporal.

In his pioneer work on linguistics, B. L. Whorf noted that English suggests a polarization of nature: nouns are static, verbs are dynamic.[80] For instance, "man" and "house" are nouns and they do present lasting objects, while "strike," "turn," and "run" are verbs and they do describe transitory and brief events. But what about nouns like "lightning," "storm," or "noise"? "It will be found," says Whorf, "that an 'event' to us means 'what our language classes as a verb' or something analogized therefrom. And it will be found that it is not possible to define 'event, thing, object, relationship' and so on, from nature, but that to define them always involves a circuitous return to the grammatical categories of the definer's language." [81] Insofar as languages perceive reality differently, there remains an arbitrariness in my definition of an event. But this only demonstrates that an event cannot be thought of as a happening unto itself but only as something against the background of other identities, even if the identities are sliced out of the totality of our experiences in many different ways.

The primary reference simultaneity in man, that is, his basic reference event, is his life. A person's life is one single unit: "my absence of life" cannot be followed by "my life" and again another "my absence of life." "My birth" and "my death" from the point of view of the life event cannot be separated any more than two brief audio stimuli, offset by less than about

2 milliseconds, can be separated by hearing. This formulation of the life-event has overtones of Heidegger's being-unto-death, Kierkegaard's sickness-unto-death, and the whole problem of personal cosmology known under the philosophical-psychological category of "my death." I believe that the identity of the self serves as an archetype in the formulation of knowledge about all other events regardless whether they come directly from sense experience or are mediated by instruments and theories. For this reason it would seem that the beginning or end of an event, that is, the creation and cessation of an identity, are not in themselves separate events from the point of view of the identity.

Let us now define as a *process* any set of events connectable by rational arguments, such as the phases of the moon or the stages in Lady Macbeth's madness. If the connection comprises regularities expressible, at least in principle, by mathematical relationships, the events may be said to form a *stationary process*. Examples are the motion of an (idealized) planet around an (idealized) sun, or subsequent emissions of alpha particles known to be regulated by strict probabilistic laws. If the connections consist of value judgements relative to the clockwatcher we may speak of *creative processes*. Examples are the stages of a love affair leading up to pregnancy, or stages in the deterioration of a Venetian statue because of air pollution. Because of their usually abstract nature stationary processes are often judged as objective conditions; creative processes usually involve selective judgements normally influenced by communal values and tend to constitute privileged disclosures.

This analysis reminds one of Whitehead's division of time into the "perpetual transition of nature into novelty" and "the single time series we naturally employ for measurement." [82] But whereas for Whitehead both the creative advance and the measured series are temporal, thus both having the same genidentity as their sum, in the philosophy of time as conflict only the tension between the two processes is given ontic status and is associated with temporality. Moreover, the separation between the two types of processes is considered as determined by the nature of the formulator of the ideas.

We may return now to the multitudes of devices which we encountered in our survey of timepieces. In terms of the mild axiomatization of this section, clocks may be seen as employing stationary processes, so judged by their makers. Clocks represent the ways in which man the toolmaker identifies and puts to use what he regards as the lawful content of his experience of time. It is appropriate, therefore, to consider man himself.

III

THE SEEKER

THE FORMULATOR of philosophies of time and the fabricator of time-keepers are one and the same organism. In this chapter we have our first look at man who is a clock, a clockwatcher, and a clockmaker in one person and explore some features which these conditions imply.

1. The Perception and Conception of Children and Ideas

THE RELATION of the clockwatcher to the flow of time is often described as that of perception; with blissful simplicity we say that man perceives time. For a discussion of the historical status of perceived external reality, a matter of concern to philosophers since Parmenides, Democritus, and Aristotle the reader must be referred to the literature. Presently I shall single out two difficulties associated with the interpretation of the meaning of time perception.

The Error of Misplaced Precision

Those studying perception seem to share an intuitive certainty that the term refers to mental processes which manifest a sufficient degree of unity and coherence to justify their being grouped as a class of functions, namely, perception. The growing emphasis on quantitative methods that characterizes experimental psychology demands increasing precision in the analysis and interpretation of perceptual processes and tends to direct attention away from epistemological and ontological questions related to perception. For the study of time this trend has a distinct disadvantage and has led to a curious situation. Whereas, on the one hand, we find repeated emphasis on the fact that time perception has no obvious sense organ through which time may enter as light enters the eyes, on the other hand there is an ever

72

increasing body of experimental work on an assumed perception of time. Consequently, much of the psychology of time has been directed to fragmentary issues which do not fit together as would pieces of a jigsaw puzzle,[1] nor do we have any assurance that they are parts of the same single and coherent picture.

In a paper entitled "The Fallacy of Conjunctive Analysis," R. Ackerman criticized the empiricist analysis of perception on the basis that although analysis of perception may produce sense data valid in its own right, when the aggregate of such data is reassembled, it need not yield the percepta. He argued that empirical analysis will remain useful only if accompanied by a phenomenological study of time, because sense-data analysis loses the temporal integrity of the molar object.[2] I believe that the active part the perceiver plays in time perception makes an understanding of that process impossible when considered solely through the experimental methods of science, for those methods are designed to be blind to the creative character of nature at large. It is both interesting and significant that these difficulties are implied, hence acknowledged, by our language.

In ordinary use "perception" describes sense-oriented awareness of change and no-change in the self or in the environment; also the internal events of the sensory system believed to be responsible for that awareness.[3] This ambiguity is probably not accidental but intrinsic in the processes so designated and is so reflected in the words we use to describe them. It is a generally accepted view of cultural anthropology that language as the vehicle of cultural continuities displays certain features analogous to the biological inheritance process. Thus, there are, for instance, deep layers in the language which the individual did not create, just as he did not create the morphology of his skeleton; the lore of our inherited verbal expressions is collective and archaic.[4] Words, just as individual thoughts, are shaped and may be repressed by communal usage; in such cases their meanings may be partly or totally forgotten, yet they remain unconscious latent powers beneath dictionary definitions. Whatever the mechanism or reason, the forms of words in spoken language often reveal unconscious beliefs about phenomena they describe and carry something of a primitive tacit knowledge.[5]

Both *perceive* and *conceive* denote mental activities related to *grasping* (understanding) as well as to creation. The second part of the Latin *percipere* contains the idea of "taking hold" from the Latin *capere*. The German *begreifen* originally meant grasping with the hands, then changed its meaning to the generalized concept we now understand as "grasping an idea." For fifteen million years before the relationship between conic sections on the one hand, and planetary orbits on the other, were grasped and celestial mechanics was thus conceived, women were grasped so as to make them conceive new life—even if the lofty goal of preserving the race was no more uppermost in the mind of the copulating male than the benefit of humanity in the mind of a creative scientist. If we do assume that the evolution of language is not a totally arbitrary and random process, then the clear biological priority in the meaning of perception cannot be neglected. Thus, I believe, talking about "time perception" (or for that matter, about any other perception) implies the awareness that perception includes a creative stress not unlike that of carnal knowledge.

This example is a hint: it is not evidence. But it is sufficiently cogent to make me feel that any attempt to deal with the sense of time via the minutiae of experimental psychology alone is bound to remain fruitless because of the error of misplaced precision.

The Difficulty of Regressive Sharing

There is no obvious point of entry through the biological sciences into studies of time experience or of the experience of timelessness. More promising is an entry through depth psychology, provided that one admits with Freud, as I wish to do, that the mental faculties of man contain a display of phylogenetically recent as well as phylogenetically archaic evolutionary functions. Here I want to draw attention to one particular difficulty which follows from this evolutionary organization of man and something which we shall often encounter. It is the blindness of a creature with very complex perceptual machinery for the world as perceived by simpler organisms.

As the clockwatcher tries to contemplate the structure of his integrated behavior by noting those constituents in it which represent, in various and modified forms, aspects of earlier and simpler selves (both phylogenetically and ontogenetically), he finds himself hampered by the difficulty of regressive sharing. Infants, horses, and plants all live in the time of a human adult observer, just as animals without eyes live in space. But the imaginative sharing by man of their primitive temporal experience is either impossible or prohibitively difficult for reasons which derive, I believe, from the directionality of evolution.

Marcel Proust, master of the psychology of the twilight zone between the mental states of time and timelessness, described the childhood experience of struggle associated with rising from time-ignorance to time-knowledge:

> . . . in my own bed, my sleep was so heavy as completely to relax my consciousness; for then I lost all sense of the place in which I had gone to sleep, and when I awoke at midnight, not knowing where I was, I could not be sure at first who I was; I had only the most rudimentary sense of existence, such as may lurk and flicker in the depths of an animal's consciousness; but then . . . memory . . . would come like a rope let down from heaven to draw me out of the abyss of not-being. . . . in a flash I would traverse and surmount centuries of civilization, and out of a half-visualized succession of oil lamps, followed by shirts with turned down collars, would put together by degrees the component parts of my ego.[6]

If we regard time and timelessness as mutually exclusive conditions of the same rank we cannot make sense of the efforts of young Proust to rise "by degrees" from the atemporal to the temporal. His struggle becomes intelligible, however, if it is understood as the rising from a simpler to a more complex perceptual environment.

Universes of Perception

These observations strongly suggest that we distinguish perceptual universes and thereby acknowledge the active contribution of the perceiver to the nature of the world as known to him. The opposite of this view, that the organism is a passive observer of an external world which acts upon it, was appropriately christened by B. Kaplan as "the dogma of immaculate perception."[7] I would rather endorse a stance of "passionate perception" and engage the idea of active, species-specific *Umwelts* as suggested by the philosophical anthropologist Jacob von Uexküll.[8] According to Uexküll, for each animal the world-as-perceived is determined by the potential functions of the totality of its receptors and effectors. Its receptors determine the world of all possible stimuli that the animal may experience; he calls this *Merkwelt*, that is, the animal's universe of signals. The sum of all possible responses as determined by the effectors of the animal form its *Wirkwelt*, or universe of possible actions. The dynamic combination of the *Merkwelt* and *Wirkwelt* makes up the animal's *Umwelt*, best rendered into English as the animal's "specific universe."

The idea of Umwelts, or specific universes may be further generalized. Consider, for instance, the large body of psychological literature concerning the accuracy of time estimates. The findings are fragmentary and contradictory,[9] but there is clear evidence that the accuracy of short time-estimates is a function of a host of physical, physiological, and psychological conditions. These estimates vary with motion, temperature, motivation, age, health, cultural conditioning, and, in a very important and fundamental way, with sense modality. The suggestion thus emerges that we include with each sense organ its associated nervous and brain structures and maintain that each of them has its Umwelt. We might even seek the temporal features of such sense-specific Umwelts.

Beyond accommodating the many universes of perception, Uexküll's *Umweltlehre* (Umwelt principles) is rich in potentialities. By taking it seriously, creatures of different psychobiological organization (and in the same creature, different senses) might eventually be understood as "grasping" reality differently. A description of the world by a living organism, whether expressed through language or through any other integrated behavior, might then be best conceived of as a dynamic model created by the knower and the known together. External reality then will not appear as an immense store of well-defined information from which an individual may select some, resembling the student of a language who selects words from a dictionary. Instead, it will comprise acts of creation whose uniformity for each species is guaranteed by the psychobiological uniformity of members of that species.

Thus, the world as perceived becomes an ontological statement and an epistemological demonstration. For Kant the Ding-an-sich, though not knowable, existed nevertheless, independently of the perceiver. In the view held here final reality is relative to the perceiver. Yet, for the reader and this writer it is not ambiguous because the reader and this writer share the psychobiological unity of their species.

2. Three Modules of the
Short Term Present

IN ANTICIPATION of some later views which incorporate a hierarchical relation of time and the timeless, I shall now distinguish two major psychological modes in man's dealing with time. Under *time perception* I will class man's functions as a timekeeper, such as his perceptual and motor skills related to brief time intervals. Under *time sense* I will subsume the relevant aspects of those behavioral functions in which the symbolic transformation of experience is preeminent, such as long-term expectation and memory, personal identity, language, and the communal structuring of time. These two modes often blend one into the other, and a sharp line between them cannot be drawn. Yet, as do Shakespeare's "seven ages of man," this division also makes good practical sense. What is loosely known as the time sense of animals we shall call by a different name and will regard it as related to time perception rather than to the sense of time.

....

A description of the world by a living organism, whether expressed through language or through any other integrated behavior, might best be conceived of as a dynamic model of the external world created by the knower and the known, together. The external world, then, is not an immense store of well-defined information from which the individual may select some, as the student of a language selects words from a dictionary. Instead, it comprises acts of creation whose uniformity for each species is guaranteed by the psychobiological uniformity of the members of that species. For each animal the world-as-perceived is determined by the functions of its receptors and effectors. A species-specific world so determined is called the animal's Umwelt.

Paul Klee (1879–1940) Erkenntnis eines Tieres (Knowledge of an Animal). *Courtesy, the Busch-Reisinger Museum, Harvard University, Cambridge, Massachusetts. Museum Association Fund.*

....

The Physiological Present in Man, and Its Structure

Much of the experimental work in the psychology of time deals with problems of judgements of sequence and duration as functions of experimental variables, and with the attendant concern of constructing mathematical models to accommodate these findings. The extensive literature on this subject employs a number of loosely defined (or undefined) and ill-coordinated terms, such as physiological, specious, psychological, and mental presents. I will attempt to develop specific meanings, from empirical considerations, for some of these terms.

Two sounds separated by less than about 2 milliseconds would tend to appear to all human subjects as one single sound.[10] Sounds separated by more than 2 milliseconds are recognized by most subjects as separate, but the separation does not yet enable them to tell which signal came first. Similar tests performed for visual and tactile inputs reveal that each sense modality has its distinctive powers of discrimination and displays a certain degree of dependence on the other senses.[11] A period of about 20–50 milliseconds normally enables subjects to tell which signal came earlier. Concentrating now on the duration of a stimulus rather than the separation of two stimuli, one can ask whether there is a lower limit to the perception of duration. Robert Efron found that there is, and summed up his findings in two general statements:

1. The duration of the perception of the stimulus is equal to the duration of the stimulus, provided the stimulus is longer than some critical value.
2. For stimuli shorter in duration than the critical value, the perception is of constant duration.[12]

For visual excitation this threshold is about 130 milliseconds, depending on the parameters of the physical stimuli. This, then, is the lower limit of stimulus identity, for stimuli are judged to be either this long, or longer. Although the neurological processes underlying this limit are obscure, it is likely to represent the shortest physiological processing time.[13] If this is the case, we could then not expect any conscious action to take place in periods shorter than this limit, not even in theory. I will call this limit the physiological present in man.

Thus, at the lower limits of time perception we discover in temporality a hierarchy of levels. Periods of less than about 2 milliseconds must be regarded (for the conditions specified) as *atemporal*. How else would one describe an Umwelt wherein temporal separation of events is impossible? Periods that last to 20–50 milliseconds (again for the conditions specified) may comprise events which are countable but which cannot be placed in any temporal order. I will describe these events as *prototemporal* and the corresponding world as a prototemporal Umwelt. Periods which last to about 130 milliseconds (in man) are characterized by a temporality of before and after but not one of future, past, and present, for the physiological present is not sufficiently long for the establishment of a conscious present. This condition has some rather curious logical properties. Namely, insofar as it is not possible to place the "after" event in the future and the "before" event in the past, all one can say is that they relate as pure succession but without a preferred direction regarding future and past. Pure succession is a purely diadic form of temporality, one of pure asymmetry, and not a triadic one. I shall describe this temporality and its appropriate Umwelt as *eotemporal,* implying the dawn of time. We note that from the point of view of our mature, noetic temporality all these lower temporalities appear to be incomplete. This, I think, is an example of the difficulty of regressive sharing. They should be considered, instead, as quite complete as far as they go; as worlds sufficient unto themselves.

Throughout this book we shall encounter many examples of atemporality, prototemporality, and the pure succession or asymmetry of eotemporality. Though it should be unnecessary, let this warning be voiced: the significance of these temporalities and the Umwelts they determine resides in the quality of the levels just described and not in the specific numbers which happened to be associated with them in the perceptual apparatus of man.

The Creature Present of Animals

The generalized definition of simultaneity introduced above was suggested by the inability of human subjects to distinguish between bursts of sounds separated by less than about 2 milliseconds, and by the subsequent realization that this then signifies atemporal conditions. There are examples in animal behavior which may be interpreted as simultaneities experienced by the animals.

The oilbird of tropical America uses echolocation for its nocturnal navigation. The bird sends out short pulses separated by as little as 2 or 3 milliseconds and identifies the position of the echo between two outgoing signals.[14] Thus the bird must be assumed to be able to discriminate conditions of before and after within periods shorter than 2 or 3 milliseconds. We must also assume that its ability to do so ceases at some shorter interval which, then, is the oilbird's simultaneity. This is the only bird known to use this method of navigation; but bats, well known for their use of "sonar," have auditory acuities comparable to that of the oilbird.

It was first reported in 1932 that four or more tactile stimuli applied to a snail's belly within one second make it extend itself as it attempts to crawl on a solid surface.[15] This behavior may be interpreted as a perception of simultaneity: what we judge as four events separate in time, appear to be judged by the snail as simultaneous impressions characteristic of a continuous, solid surface. According to Haldane the present of the bees may extend to as long as five or ten minutes. He interpreted the rhythmic and patterned motion of the bees upon their return from a foray as a ritual dance serving as a propositional function expressed in motion. Whereas a man may write down his experience, the bee acts it out through dancing and thereby conveys information, such as the location and value of food found, through parameters of motion. Haldane reasons that this is possible only if the flight and the dance are copresent in the bee's "mind." [16] There is little doubt that with properly designed experiments the experience of simultaneity in many animal species could be identified. I will call such simultaneities (presumably) experienced by animals their "creature present." This must be distinguished from the present of conscious experience in man, to be discussed later; for this I shall adopt the term "mental present," following Whitrow.[17]

In man we have identified an experience of simultaneity, and above it, prototemporal and eotemporal Umwelts. Do animals, as a class of beings, experience temporality above that of the atemporal creature present? The answer is in the affirmative if we are willing to ignore lower-order organisms. First, we note that anticipating and expectancy, that is, a general future-orientedness is evident throughout the animal kingdom.[18] Surely, the raison d'être of the bee's report is that of action in the future; it implies an organic belief that the future will copy the past. Several species of birds, otters, lions, wolves, chimpanzees, and monkeys have been studied for reward expectancy and found to be capable of anticipation. Rats can sustain a delay of about four minutes, cats seventeen hours, chimpanzees forty-eight hours.[19] For man, as we have seen, the delay can extend to the rewards of a postmortem world; divinities are said to be willing to await the end of the world before getting revenge or handing out rewards.

Back on earth we note that L. W. Doob, in his summary survey of psychology and time, drew attention to the important relationship between the experience of a present and elements of delayed gratification.[20] As he sees it, operating in a present amounts not simply to a capacity to select among alternatives, but, more importantly, to be able to select delayed gratification in favor of present gratification, or else, to escape from present nongratification. The capacity in many animals for delayed gratification is prima facie evidence for the existence of temporal levels above their creature present, although the accurate delineation of the temporalities so involved may not be easy. Furthermore, I think that Doob's insight about the necessary relationship between the experience of a "present" (unanalyzed) and delayed gratification, or escape from nongratification, implies the existential stress I have associated with temporality.

The Psychological Present

Ernst Mach, in his pioneering work on experimental psychology, extended the idea of sensation to include that of time and space without the necessity of specific receptors. He connected the sense of time with "the growth of organic consumption," and saw it as related to but not deriving from rhythmic biological processes.[21] I shall argue in chapter 7 that we have an anatomic structure, not itself a receptor but rather the coordinator of all receptors, namely, the central nervous system, which is sensitive to the dynamics of the organism and which may be regarded as the organ of our time sense. On our way to establishing the validity of this claim we will have to learn not to dissect the human body in search for receptors and effectors but to search in the phenomenological totality of behavior and consciousness for the sources of these existential tensions which manifest themselves in our knowledge of time. This enterprise is made possible by the remarkable unity-in-multiplicity of the psychobiological organization of man.

There are reasons to believe that the development of space perception is an adaptive process [22] aided by visual perception, which itself is the evolutionary result of kinesthetic exploration of the environment.[23] Whether by sight, scent, or touch, to "know space" the organism must reach out. But reaching out into space is also a prerequisite for "knowing time," because some of the experiential elements of the future are distant events and conditions yet to influence the organism. The development of time perception, insofar as it assists the organism in surviving the conditions seen to challenge him, is quite clearly also an adaptive process. The "reaching out" is an integrated function of all the senses of an organism, quite a remarkable feature, for each sense may be said to have its sense-specific Umwelt and each Umwelt its evolutionary stages at least partly surviving. There is no competition among them that would lead to a Darwinian preference for a "preservation of favored senses" (even if some do change in relative importance), for evolution favors the total organism. The strength of the unity that links the senses was recently illustrated in a study of intermodal phenomena by Cohen and Christensen. They even suspect a basic isomorphism among the senses, suggested by the similarity of verbal analogies we can give to our visual, auditory, tactile, and olfactory experience.[24]

I would like to call "psychological present" that most complex form of creature present which is appropriate to the sophisticated nervous and brain structure of man. Thus psychological present is the experiential corollary of time perception (with "time perception" as defined earlier), a type of simultaneity but, so to say, somewhat loose at the edges. It is the cooperative venture among all of man's receptors and effectors, but it need not imply conscious experience. Once conscious experience is implied, we must speak of mental present, to which we shall now turn.

3. The Mental Present

MENTAL PRESENT obtains when we enrich the psychological present by the addition of such capacities as long-term memory and expectation, personal identity, communication by language, and, generally, the symbolic transformation of experience. To emphasize these intellectual aspects of the mental present, I shall call the temporality which corresponds to the mental present *nootemporality* and the world determined by it the nootemporal Umwelt.

The various constituent elements of the mental present are so tightly intertwined that it is impractical to deal with any of them apart from the others, yet the linear nature of thought demands that we nevertheless try to focus on one issue at a time. To indicate and emphasize this intricate inter-dependence, I have placed three dots in front of the following four sub-section headings, such as ". . . Language." It is to be read as meaning, "Mainly about language but also about other related issues." Furthermore, I have placed a • sign in front of each group of thoughts which we must consider here because it relates to the structure of the mental present but which does not necessarily form a continuation of the arguments preceding it. Most of the problems touched upon in this section will be dealt with in detail later in the book, though the reasoning given here will not be re-peated. This section may be regarded as a glossary of some ideas neces-sary for the appreciation of chapters 4, 5, and 6.

. . . *Memory and Recall*

• The adaptive advantages of purposeful behavior, as exemplified by delayed gratification, are obviously great. In the case of the bee, if Haldane's interpretation is admitted, the dance is preparation for communal action leading to cooperative control of the future of the hive; a similar command of fate is not open to a bee acting alone. In its turn, the communication necessary for the future-oriented communal control of the life of the hive—or of any similar community—is made possible by the corollary manifesta-tions of a creature present and a mastery of continuities.

That animal rituals, such as the dance of the bees, might be possible forerunners of human language was pointed out by Haldane,[25] who was quite aware of the anthropomorphic danger of his thought. Bypassing, how-ever, the difficulty of regressive sharing, and justifying his method by the consistency of his conclusions, he translated the message of the dance into English as "I shall fly," a statement of intention, for example, in connection with the finding of a new home before swarming. Other bees observe the report of the dancer, imitate the dance by sympathetic induction, and through this imitation, "I shall fly" becomes a first person plural of the future, or even the imperative, such as "let us fly." The danced debate may be interpreted as a ritual for reaching unanimity.[26] Engels held that human language also originated as a necessity of labor.[27] Independently, and on purely logical grounds, Wittgenstein had held that private language is not possible, though he also held that something of a private language, neces-sary for privileged knowledge, nevertheless, is possible.[28] These and similar interpretations remain incomplete, however, unless the idea of time is explicitly brought in. For social communication to make any sense, one must presume a temporal present of some sort and some awareness of continuity between past and future.

• Let it be assumed that, for reasons unknown, the expectations of our distant ancestors began to extend further and further into the future. Somewhere in this growth process the creature must have come upon the fact that his own death was inevitable. In the immediate face of death all creatures show mortal terror. This is not new; what was added was a more subtle concern with death. This new awareness was likely to have urged our ancestors to explore the past for cues to future expectations; total fulfillment of their desires in the present was forever disturbed by a type of knowledge better to be forgotten. With these thoughts in mind, let me now distinguish between "recall" and "memory." By recall I mean a response to external stimuli in a manner that indicates that the stimuli have been encountered before. In contrast, by memory I mean a re-presentation of past experience in response to stimuli which at the time of representation are only contingencies. Thus, one can have memories of memories and expectations, but one cannot recall an earlier recall. I would like to postulate further that the evolutionary emergence of memory over and above that of recall was somehow connected to the discovery of the inevitability of an unpredictable event, namely death. Since the past does not contain similarly paradoxical events, memory could become characterized by certainty—or almost so. Perhaps it is for this reason that Proust held the past to be the only reality; the true meaning and significance of experience can be grasped only in retrospect. For Goethe, what he possessed appeared far away and what was gone became reality:

> Was ich besitze, seh ich im Weiten
> Und was verschwand, wird mir zu Wirklichkeiten.*
>
> *Faust,* "Dedication"

If my reasoning corresponds to fact, then, unlike simpler presents whose limits were more or less expressible in numerical readings of clocks, the mental present of man has no definite limits because it comprises, in addition to the archaic creature present, the totality of his private and communal expectations and memories.

• Gauging the depth and searching for the structure of the mental present is difficult; most of us would be hard put to imagine what is entailed in living in a purely atemporal, prototemporal, or eotemporal Umwelt. But we do have occasional geniuses capable of regressing into regions of archaic experience. We turn again to Marcel Proust.

> Suddenly, I was asleep, I had fallen into that deep slumber in which are opened to us a return to childhood, the recapture of past years of lost

---

* What I possess is but in the distance,
  All that vanished appears reality.

feelings, the disincarnation, the transmigration of the soul, the evoking of
the dead, the illusions of madness, retrogression towards the most elementary
natural kingdoms (for we say that we often see animals in our dreams, but
forget almost always that we are ourselves then an animal deprived of that
reasoning power which projects upon things the light of certainty; we present
on the contrary to the spectacle of life only a dubious vision, destroyed
afresh every moment by oblivion, the former reality fading before that which
follows it as one projection of magic lantern fades before the next as we
change slides), all those mysteries which we imagine ourselves not to know
and into which we are in reality initiated almost every night, as we are into
the other great mystery of annihilation and resurrection.[29]

Into this world of continuously fading and reappearing reality the feeling of identity is introduced. The Umwelt of the infant, both regarding
objects as well as persons, is one without identities; it is an Umwelt of
undifferentiated continual change. With the learning of distinguishable
identities, which is the recognition of permanences, atomization of this
continuity obtains. States of simpler temporalities of unproblematic change
are replaced by conflicts between earlier levels on the one hand and the
newly discovered multiplicity of permanences on the other hand. In this
development the individual changes from a nonproblematic being to an
organism which is both a thing and a process; its thing-ness is its identity,
its process-ness is the continuous change of archaic experience. These two
now define each other. Proust's morning discovery of the self and the other
changeless identities maintained through time, recapitulate, even if loosely,
the ontogeny of the sense of time.

> . . . out of a half-visualized succession of oil lamps, followed by shirts
> with turned-down collars, [I] would put together by degrees the component
> parts of my ego.
> Perhaps the immobility of things that surround us is forced upon them
> by our conviction that they are themselves, and not anything else, and by
> the immobility of our conception of them.[30]

As the sleeper emerges from his primitive present he tries to grasp the
certainty of permanence in things, in the self, and eventually, in the laws
of God, nature, and man. In this process permanence is added to an organic
sense of proto- and eotemporal change. Some sleep is without existential
tension and, as one may hope, a state of bliss. Upon awakening, however,
slowly but surely there arises, as Hamlet well knew,

> The heart-ache and thousand natural shocks
> That flesh is heir to . . .

and with it a deeper conflict: that between the psychic reality of "natural
shocks" and that of the "dubious vision, destroyed afresh every moment by
oblivion." The ability to distinguish between the primitive present of sleep
and that of the waking state is something one must learn as the capacity
to distinguish between dream and nondream develops.[31] I would think that
one meaning of the physiological present is that minimal time in which this

conflict between the Umwelts of undifferentiated selflessness and the world of self/nonself may come about.

. . . Language

• If the early beginnings of social communication are identified with the reaction of one organism to the presence of another, one may say that social communication exists for bacteria, unicellular organisms, plants, colonial organisms, and invertebrates.[32] Under the more restrictive terms of information-transfer through the senses from one individual to another, social communication has been studied in zoosemiotics.[33] That chimpanzees can be taught language is interpreted by some as evidence for an essential continuity of cognitive processes between subhuman and human primates, a conclusion which must be regarded with caution when inquiring into the phylogeny of language.[34]

Much thought has been expended on determining the origins of human language, a subject of intellectual fascination and a touch of mystery, because the ability to communicate in language bestows fearful powers upon those who can do so. That human speech arose out of animal communication is a pleasing assumption whose main strength is its consistency with organic evolution and the fact that any alternative would tend to be dependent on divine intervention. But great is the step between the utterance of a phonetic form and the capacity to associate with such an utterance a symbolic transformation of experience. In his study of the early growth of language, Leonard Charmichael remarks that

> no one who considers with care the sensory and behavioral development sequences in the growing human infant during its first months can fail to recognize the descriptive value of the concept of *emergence* in noting the steps that are always antecedent to the uttering of the first meaningful word by the individual. Each developmental stage of vocalization is in some respects unique in its characteristics, and no developmental level is ever a mere sum of previously described and pre-existing antecedent capacities or processes. . . . Each such stage thus marks a step, as it were, in the progress from the mere emission of sound to true, meaningful human speech.[35] [italics his]

Some scholars, such as Noam Chomsky, welcome the emphasis on emergence in language growth for it seems to support the idea that children bring to the language certain linguistic universals which, at a certain phase of development, become available to them with great relative rapidity. Eric Lenneberg suspects that language is made possible by an as yet unidentified species-specific biological capacity in man.[36] He holds that the human brain is a biochemical machine that computes the relations expressed in sentences, following certain rules which resemble the rules of generative grammar studied by Chomsky.

Though the peculiar biological organization of man will surely be proved a prerequisite of speech, and we shall come back to this problem several times, we may find some more immediate information about language and time in the morphology of speech. Specifically, I am thinking of the importance of timing for intelligibility as brought out recently by work done on the psychology of speech in a noisy environment.[37] A sample of speech was prepared containing only sounds of high energy vowels and occasional consonants. The result was a stacatto sequence with very low intelligibility. If, however, white noise was added in the gaps of the signals, the intelligibility increased drastically, such as from 20 percent to 70 percent. The result was expected on the hypothesis that speech consists essentially of accurately timed patterns of sound, though the degree of success was unanticipated. Clearly, then, an appreciation of temporal patterning of short intervals is demanded, both from the speaker and the listener, if speech is to be useful.[38]

•   Robert Efron's work on the effect of handedness on the perception of simultaneity and temporal order seems to support the hypothesis that judgements of simultaneity and sequence can be performed only by the speech-dominant hemisphere of the brain.[39] His results suggest a possible phylogenetic link between speech and the perception of temporal sequence. Some hints as to what might have happened come from recent studies on verbal learning which reveal that for short periods, up to thirty seconds, words are stored in forms approximating their acoustical characteristics. Confusion during short periods is most likely to occur among words of similar sound.[40] For longer memory, however, words are "filed" according to meaning. Short-term memory is believed to be functional in holding acoustic sequences (resembling the messages of Haldane's bees held in the creature present); while long-term memory is organized according to meaning (and is used in the complex machinery of human language in the mental present). If we evoke a recapitulation theory we may detect in these findings two steps in the phylogeny of language and also accommodate the fact that in contemporary man verbal information does not have to pass through short-term memory to reach the source of long-term memory. In any case, a division between short-term and long-term memory is indicated, paralleling my earlier distinction between recall and memory.

Going now from the study of words to the global question of speech and music, we find in Anton Ehrenzweig's work on the gestalt of artistic vision and hearing that

> the conscious ear picks out the articulate ("substantive") tone steps, rhythmical beats, articulate chords . . . at the same time it represses "transitive" pitch inflections, free rhythm, and inarticulate chords which are sandwiched between the articulate tone events.[41]

By "substantive" and "transitive" he distinguishes between sounds which are consciously recognized and those of which we are normally not aware in intelligible speech. He believes that in music it is possible to become aware of transitive sounds, such as in the performance of a violinist who would intensify his vibrato into a wobble. In contrast, the repression of inarticulate speech sounds is almost impenetrable.[42] The success of this repression, that is, the unconscious behavior which makes it almost impossible for us to hear "inarticulate transitive sounds" in speech, depends on the capacity of the hearer to establish temporal order. When the temporal order is destroyed, such as when listening to speech played backward on a tape recorder, the total acoustic content is heard. The transistive sounds become audible because part of the repression is lost, and smears, grunts, and squeaks obtrude on our attention. In reversed-order speech, articulation is lost and with it the fundamental content of speech, meaning.

It seems, then, that language capacity is intimately bound to time perception as well as to the sense of time. For, on the lower end it calls upon our ability of timing, that is, upon our functioning as a clock. On the upper end it involves the conscious/unconscious structuring of the mind which, as we shall see, is one feature of our sense of time.

. . . Permanence and Change

• Earlier I warned against regarding any specific separation of time into its aspects of permanence and change as a statement about final reality. Such separations are conditioned by a variety of physical, physiological, and psychological factors and are weighted by cultural filtering. I will illustrate my point with reference to perception.

Certain motional phenomena, popularly known as optical illusions, depend on the ambiguity of separating motion into permanence and change. For instance, under suitable experimental conditions a changing image may be alternately apprehended as (a) a rotating loop of steel wire, or (b) a rubber loop compressed from a circle to an ellipse, to a straight line and relaxed back to a circle via an ellipse. Another changing image may be apprehended as (a) a steel sphere disappearing in the distance or, (b) a balloon being deflated. In both cases, earlier decisions must sometimes be reversed because of information extraneous to immediate perception. "Illusion" as used above implies the metaphysical belief that a final, nonillusory, determinable truth does exist. The real thing, we would say is either (a) a permanent steel band changing position, or (b) a permanent rubber band changing shape and length; also, either (a) a permanent sphere moving away, or (b) a permanent balloon changing size.

Turning our attention for a moment to the physics of motion, we note that the separation of the permanent (lawful) from the contingent (boundary conditions) in the Newtonian equations of motion is equally arbitrary. It is subject to change if a different bifurcation of temporality appears to be more satisfactory for such reasons as accuracy of predicting future events. This is what was done by relativity theory. If the separation of permanence from change is difficult in physics, it becomes prohibitively difficult if we wish to separate the lawful from the contingent in the actions of man or in the known course of history. We return now to perception.

J. J. Gibson studied the perception of moving things and processes as functions of the state of locomotion of the observer. In his analysis of tests on sequential perception, the experiments began to be intelligible only when the motion of an image on the retina was distinguished from the relative motion of subject and object. He found that

> continuous optical transformations can yield quite simple perceptions, but they yield two kinds of perception at the same time, one of change and one of nonchange. The perspective transformation of a rectangle, for example, was always perceived as both something rotating and something rectangular. This suggests that the transformation, as such, is one kind of stimulus information, for motion, and that the invariants under transformation are another kind of stimulus information, for the constant properties of the object. The constancy of the phenomenal object in such momentary sensation, may be enhanced by it.[43]

Gibson also worked out what he calls the theory of information pickup and made it the backbone of a novel view of perception.[44] He asserted that the conventional view, that the brain constructs or computes objective information about the outside world from happenstance sense impression, is inadequate. Instead, he maintained that the aggregates of sense organs are responsible for actively isolating external invariants through actions coordinated by the neural system. Whatever *is* they prove to be invariant we tend to regard as objective information. But this brings into the action of perception many functions not normally regarded as directly contributing to the perceptive process, such as language, or the values of objects.

• Gibson's arguments turn our attention to cultural filtering as one element that conditions our habits of separating permanence and change. That language is one bridge between culture and life follows from its biological origins, as has been extensively argued by E. H. Lenneberg.[45] Kalmus went as far as to note an isomorphism between language as a communication system and the transfer of biological information from cell to cell.[46] As we put these seemingly unrelated findings together, the suspicion grows that what we separate as permanence and change has its roots in the biology of man and is modulated by social values as communicated, for instance, through language. Regarding this latter, we may turn to Chomsky who built his theory of syntax on the reasoned belief that there exists a set of rules by which a speaker (most impressively, a child learning to talk) can generate an infinite number of meaningful and new utterances. That in some ways each utterance be new is essential for language; to be communally useful, it must contain unpredictable information. The seemingly permanent deep structure of language (the permanent rules which the child brings to language) pertain to those aspects of language which have been judged by its speakers as necessarily or desirably permanent; they are continuities learned in the past. The unpredictable elements of language communication, by their very character of unexpectedness, pertain to the future. These combined representations of the world of objects and feelings by sound, function within the confines of the mental present. If these conditions were not so, we could not expect language to be able to convey meaningful messages (or engender meaningful though preformed reactions) about future, past, and present.

Instead of lofty syntactics, consider temporality as embedded in Sanskrit, the classic language of the Hindu inhabitants of India. In Sanskrit change itself is preferably comprehended as permanence. In utterances pertaining to change ("change," as comprehended, for example, in English) nouns are more likely to be employed than verbs. Nouns, as we noted with Whorf, designate the unchanging aspects of things, as judged by the speaker. For Heraclitus, $\pi\alpha\nu\tau\alpha$ $\rho\epsilon\iota$, that is, "everything flows"; here the verb conveys a sense of motion. The equivalent Sanskrit phrase is *sabbe saṅkhārā aniccā*, or "all things are impermanent." [47] This phrase exudes a sense of permanence. A different differentiation of temporality may be found in the language of the Hopi Indians, which reflects the world in terms of two cosmic forms. The manifested comprises the historical universe with no attempt to distinguish between past and present; everything that is or has been manifested to the senses belongs here. The subjective or unmanifested comprises all that we call future or mental.[48] But the division is, again, between what is judged permanent (sensible, or real) and what is judged unpredictable. Clyde Kluckhohn suggested that different cultures see the world as divided into "determinate" and "indeterminate" elements according to their general judgement of the limits of lawfulness and caprice.[49]

• The separation of continuities from the unpredictable in sense impression is an important part of the child's growth. Piaget's work revealed some of the developmental stages of learning to recognize the permanence of objects.[50] At seven or eight months the identity of objects is not yet established: if a ball disappears, the infant acts as though it vanished from existence. Later through a process of adaptation the infant learns to recognize identities external to himself: what is out of sight and grasp is not necessarily out of mind any more. Between eighteen and twenty-four months he accomplishes what Piaget calls a "miniature Copernican revolution." The undifferentiated world becomes differentiated into identifiable continuities and indeterminacies. His Umwelt now comprises permanent objects (much later, permanent substances and conserved numbers) and all other features which are judged as unknown or unpredictable. We need no detailed arguments to realize that the process of differentiation is very strongly conditioned by social guidance.

. . . Personal Identity

• According to the developmental psychology of Piaget the child learns to see himself as one continuity among many continuities of his perceptual world;[51] by learning to define permanences external to himself he comes to define one permanence which is his personal identity. Let us now recall our assumption that through the operation of natural selection the creature present of man's immediate ancestor expanded to include some continuous reminders of the inexorability of the individual's death. Then, let us speculate further. It is likely that the constant awareness of his demise now urged him on to seek an increased mastery of future contingencies through better use of his ability to recall past conditions and events. Perhaps, as did the morning thoughts of Proust, the waking consciousness of early man spoke to him about the uniqueness of his memories and expectations and impressed upon him the specificity of his body, thus defining his self against the nonself. Personal identity amounted to a loneliness unknown to his fellow creatures and probably acted as another force driving him in search of protection against the unwelcome certainty of passing. The unity of the self is likely to have been the archetype of one-ness; with one-ness came the potentiality of the many, and with the many, the potentiality of the quantification of experience.

It would seem that the capacities for identifying permanences and quantifying experiences were necessary prerequisites for the transformation of sense impressions and feelings into symbols. The external world could now be conceived of as a stage upon which identities, including that of the self, play their fatal games; in addition, it became conceivable to abstract invariances from feelings and other sensations whose external source could not be identified. With a private world of meaningful experience we could expect to see metaphor-making emerging and speculate that, as the death-less world of symbols (as permanent beings) was now pitted against the knowledge of inevitable passing (such as death as becoming), our early ancestors might have experienced a deep sense of unresolvable conflict. But the conflict need not have been a contradiction.

In a work on the natural philosophy of perception, R. L. Gregory called attention to the logical categories of deductive and inductive reasoning in their relation to the functions of the brain.[52] He stressed that in ordinary use deductive reasoning is not permitted to contain self-contradictions and cannot be refuted by sense impressions. If a finding does not correspond to a deductive law, we discard it as inaccurate but retain deductive reasoning. ("Facts" are selected or deduced from contingencies; otherwise we would not have to perform experiments to find them.) Inductive reasoning can, however, and often does contain self-contradictions. New discoveries can and do modify inductive inferences. From considerations of the possible major modes of the operation of the brain, Gregory concluded that problem solving by means of abstract thinking or by language are deductive processes; problem solving by perceptual processes are, in contrast, inductive.

Let us assume that Gregory's reasoning is correct. It would follow that the paradoxes of the philosophies of time belong to the world of abstract language-thinking; they would be features of the noetic Umwelt. There need be no similar paradoxes of time perception, however, because of the inductive character of perceptual processes. Let us recall St. Augustine: "If I think of time I don't know what it is; if I do not think of it, I do know." Here might be the sources of the dichotomy between knowledge felt and knowledge understood. There are no existential paradoxes of time: in the lives of horses, dogs, savages, children, and saints, life and death, for example, form an unquestioned unity. The paradoxes seem to be introduced by whatever processes are responsible for speech-thinking. This does not mean that life itself cannot or does not have unresolvable conflicts; it does, and we shall discuss them at length. But those conflicts are, of course, not intellectual.

• Gregory also holds that the brain, which is essentially an analogue device (for it can produce without analysis a topological equivalent of the external landscape), learned to operate as a digital device with the invention of language (for it must make digital distinctions, that is, manipulate identities). Considering the fundamental role played by controlled rhythm in the genesis of intelligible utterances, this conjecture seems reasonable. But if this is truly the case, we have an impressive coincidence of co-emergent qualities: the definition of the self; invention of language; memory added to recall; conflict between the world as felt and as understood. These qualities would tend to mutually reinforce one another: improved language skill increases the potential content of the memory store; this hardens the definition of the self; which improves the chances of survival, etc. From the many closed loops of interconnected developments we could surely expect an explosive rather than gradual advance. The all but miraculously rapid appearance of the humanoid features of the brain—perhaps tens of thousands of years out of millions of years—indicates such an explosive rather than a gradual emergence of man's identity.

4. Resolutions of
Perceptual Conflicts

GENERALLY we perceive far more than actually sensed. For instance, visual information can lead to associations appropriate to touch, such as when we perceive the hardness of a table by watching a heavy object rest on it; or auditory information can lead to visual imagery such as when we hear a window broken. The associations appropriate to "hardness" or "broken window" are hypotheses based on prior explorations and amount to inferences from sensory data. Perception, in the words of R. L. Gregory, is a "gamble of hypotheses." [53] They are based on expectations derived from memory and operate in the complex experience of the mental present.

Differences between the hypotheses and sense impressions may subsequently be resolved by identifying what we call a "fact." Yet conflicting information is essential for our perceptual world.

For instance, the eyes often get two simultaneous but partially contradictory images. Such conflicts are resolved through the integrative activity of the nervous system in a process described by Georges Schaltenbrand as the elimination of apparent empirical contradictions.[54] A new existent is introduced which is of higher order in an appropriate way than are the sense impressions. Specifically, the two partly contradictory two-dimensional images are combined in a three-dimensional unity known as space. The new higher-order unity may now display some noncontradictory (that is, invariant) features, such as objects.

Both phylogenetically and ontogenetically the perception of motion is prior to perception of space or time, as amply demonstrated by the work of Piaget.[55] The dog's reaction of getting out of the way of a flying object, or the infant's resistance to crawling over a visual cliff far precede any ideas about space or time. Clearly, it is not the experience of motion which is being constructed from elements of distance and interval, but it is the removal of certain contradictions perceived in the sensation produced by motion which is expressed in terms of space and time. Now, except for unusual conditions, motion is clearly assigned to and satisfactorily comprehended as that of the subject with respect to a stationary environment (such as when we walk), or that of the environment with respect to the subject (such as when someone pulls the carpet from beneath us), or as relative motion.

The question of why the moving observer perceives himself as being in motion and does not instinctively postulate the motion of the environment has been studied in animals and man and is satisfactorily explained by the reafference principle.[56] It is postulated that higher-order stimulus variables, such as the patterning of visual, olfactory, or auditory space, enable the subject to compare afferent, that is, inflowing stimulation with stimulation which should have occurred according to commands given by its higher control centers and derived from an "internal map." The difference between the expected and the actually encountered is then chalked up to the movement of the environment. In the interpretation of Mackay, the stabilization of the environment is something taken for granted until there is sufficient evidence to the contrary.[57] The voluntary motion of the eyes, or kinesthetic exploration, does not evoke a perception of environmental motion, because the changes on the retinal image or of the motor system are expected stimuli. They are confirmations of predictions based on environmental regularities judged as permanent. Thus, at the rudiments of the perception of motion and change we can recognize the instinctual separation of the world into the expected and the unexpected or, philosophically, the necessary and the contingent. Enduring identities seem to make up our "internal map" and form the store of knowledge on which actions may be based.

Our senses inform us that differences between what is expected and what is encountered and tensions that accompany such incongruences are chronic conditions of existence. But then temporality, if identified with these tensions, will itself become a fundamental invariant, a framework provided by our desire to identify the permanent in our experience. In visual perception it is possible to introduce a higher-order unity into certain incongruencies, such as by fusing two different two-dimensional pictures into a three-dimensional one. An analogous resolution of contradictions in time, such as between the necessary and the contingent, can manifest itself only in changing the qualities, levels, or intensities of temporality, for temporality itself remains, as it were, a basic invariant. Much of this book deals with the delineation of the levels of temporality appropriate to various integrative levels of nature.

"It is not in the nature of reason," wrote Spinoza in his *Ethics,* "to consider things as contingent but as necessary." Or, in a corollary form, "It is of the nature of reason to perceive things under a certain form of eternity." [58] The subjective experience attendant to this function of the mind is recognized in Spinoza's *Improvement of Understanding.* It is only "love toward a thing eternal and infinite [which] feeds the mind wholly with joy, and is itself unmingled with any sadness." [59] I would like to interpret this as a reference to that quality of the mind which tends to extract from the coordinated experience of the senses whatever—subject to linguistic, communal, and other influences—it judges as nomothetic and thereby useful for controlling future conditions. Such an insight, I believe, was already implicit in the Greek image of man even though concern with temporality did not become explicit.

In *Protagoras,* Plato recounts how Prometheus, one of the Titans inhabiting the earth before the creation of man, had grieved at the Olympian neglect of man. He tells us that the capacities necessary for survival were all heedlessly squandered on brutes and, even though the day was at hand when man was to emerge from the earth to light, he had not yet been suitably provided for. In his perplexity Prometheus "stole from Hephaestus [the god of fire and metal working] and Athena [the goddess of wisdom and feminine crafts] wisdom in the arts together with fire." [60] The attributes stolen by Prometheus are described by Plato as *sophia* which means both knowledge and skill. After Prometheus, man came to partake in knowledge originally apportioned to gods alone and "he soon was able by his skill to articulate speech and words and to invent dwellings. . . ."

Prometheus was something of a mischief maker but superior to his peers, mainly because he was gifted with prophetic foresight. This is already implied in the meaning of his name, literally "Forethinker." Aeschylus sensed the profound implication of forethought and made

Prometheus not only the bringer of fire and civilization but also the pre-
server of man. I believe that the Promethean gifts of man—which are just
about everything that distinguishes him from the higher animals—all
derive from his specific skill and knowledge (*sophia*), pertaining to time
(*chronos*). It appears reasonable, therefore, to coin the word *chronosophia*,
or *chronosophy* as a name of this constellation of skills, and of the discipline
studying the content and the ramifications of man's knowledge of time.[61]

PART TWO

Images in Heaven and on Earth

WE set out to learn more about the presumed object behind the glass wall by examining its reflections in the world around us. We conclude that it is not a single object at all but a vast, changing landscape, and that we are part of the countryside. The glass wall vanishes.

IV

THE ROOTS OF TIME
IN THE
PHYSICAL WORLD

AN EXAMINATION of some of the characteristic teachings of physics suggests that they pertain to surviving elements of a primordial world in which we may discern the roots of time. It is argued that understanding the physical world as a vast aggregate of particles (sec. 1) helps identify the origins and elucidate the nature of the primitive levels of temporality; understanding the physical world as a single, coherent unit (sec. 2) helps identify the boundaries of these levels. Both paths of inquiry reveal certain intractable formal almost-symmetries in the ways that physics deals with temporality (sec. 3).

1. Aspects of Time
and the Many

OUR MOST comprehensive view of matter and the void is Quantum Theory, the science of the "many."

Probability

When Abraham Lincoln declared that you can fool all of the people some of the time, you can even fool some of the people all of the time, but you can't fool all of the people all of the time, he illustrated some of the curious features of probabilistic statements and implied a relationship between probability and time. To see what these features and relationships are we have to reformulate Lincoln's words for a world of inanimate models.

97

We may select a quantity pertaining to a single dynamical system and measure this quantity by an ensemble average and a time average. An ensemble average refers to simultaneous actions of many identical systems; a time average is obtained from consecutive actions of one system. According to a scientific theorem introduced by Boltzmann in 1871 and known today as the ergodic hypothesis (also called "the assumption of the continuity of paths," by Maxwell), these two averages should agree.[1] [*] The conditions which must be fulfilled for the ergodic hypothesis to hold are (a) a complete identity of the dynamical systems, and (b) a complete causal disconnectedness of events observed.[†] The agreement between ensemble and time averages predicted by the theories has been so excellent for a wide variety of systems that the ergodic theorem is considered valid in spite of the absence of rigorous justification from first principles. Reflection on this state of things does lead, however, to at least three puzzles. (1) How can simultaneous, unconnected events in an aggregate of indiscernibles give predictably integrated results? Whence this esprit de corps? (2) How can consecutive unconnected events in the life of an individual give predictably integrated results? Whence the historic consciousness? And (3) what is the source of agreement between the two? That is, how can we exchange spatial and temporal distributions? [‡]

Questions such as these are ordinarily dealt with through the idea of mathematical probability related to the experience of chance; the intellectual ancestry of probability may be traced from the pagan notion of fortune to modern ideas of game theory. The first treatise on chance is probably Geronimo Cardano's *Liber de Ludo Aleae* (*Book on Games of Chance*), published posthumously in 1663. Pascal still had gambling in mind when he developed his *aleae geometrie* or the "geometry of dice." In its modern form, we encounter the concept of probability in Bernoulli's *Ars Conjectandi* (1713). But whether in archaic divinatory practices or in the faith of game theories, probability has always been handled as a type of law-

---

[*] A simple example will elucidate these statements. Consider 10,000 identical dice, each slightly (but equally) "doctored." On 10,000 identical dice-throwing machines we throw all of them at once and find that the numeral 5 turned up 5,236 times. We assume that had the dice been fair, the numeral 5 would have turned up only about 1,666 times. We conclude, therefore, that the "doctoring" favors 5. Now, we take any one of the dice and throw it with any of the machines 10,000 times. According to the ergodic hypothesis we should again find that 5 turned up about 5,236 times.

[†] The requirement (a) is self-explanatory even though in practice it can only be approached but never achieved. Condition (b) demands (1) that there be no communication among the 10,000 machines as they throw their dice all at once; and (2) that the single machine-dice system which does the 10,000 throwings be completely unprejudiced by past results or future expectations of its own operations.

[‡] In terms of our examples: (1) if 10,000 simultaneous events must be taken as unconnected so as to produce a probabilistic outcome, in what way do they operate to yield outcomes that hold for their aggregate? (2) If 10,000 consecutive events must be taken as unconnected to produce a probabilistic result, in what way do they operate to yield outcomes that hold for a process in time? And (3) what is the source of the interchangeability of the two actions: one spatial, the other temporal?

fulness. From among modern theories of probability, the frequentists hold that it is a measure of relative frequencies of occurrence and they choose to ignore any deeper significance; for subjectivists it is a measure of belief; for objectivists probability implies (and covers up) ignorance. But all agree that for probabilistic laws to be valid they must concern universes of identical systems because the laws of large numbers apply only to aggregates of indiscernibles, be they molecules, mice, or men.

Leibniz held that there could be no such things that differ by number only. We will recall his opinion that indiscernibles are identical: if two objects are exactly identical then we have given two names to the same object. His own monadology postulates teleological atoms called monads, but they are not alike since each reflects the totality of the universe from its own vantage point; cooperative functioning among them is assured through preestablished harmony: a condition comparable to several clocks which remain in phase because they were set in phase at the time of their creation. Connections among apparently unconnected members of aggregates would not in themselves have been a problem to Leibniz or to the natural philosophers of Newton's era, for they took it for granted that the universe is the work of a Supreme Author and we are only trying to piece together His work. We might be ignorant about the workings of His laws, but there are no disconnections in the Mind of God. This type of belief is still present in Einstein's famous critique of the probabilistic nature of quantum theory, namely, that God does not play dice; it also survives in the debate about possible hidden variables whose discovery would change quantum theory from a probabilistic to a deterministic discipline.

In modern applications of the theory of large numbers the Leibnizian critique holds only partly. Namely, we buy the privilege of saying something, rather than nothing, about the expected future conduct of aggregates of identicals by giving up some of our classical ideas of connectedness. When the National Safety Council predicts 1,250 fatalities over a Labor Day weekend, they take each unit ("one driver") as completely identical with all others. Should one insist on sufficient differentiation until the statement is narrowed to a specific individual, the statistical prediction implicit in the forecast would become useless for all practical purposes.

Whereas the assumed identity of all drivers is a necessary artifact imposed by the observer, the indistinguishability of elementary particles seems to be an important and fundamental property of nature. It is one of the working assumptions of particle physics and is necessary for the formulation of quantum statistics.[2] The indistinguishability among particles, for example, that all electrons are exactly alike, leads to a freedom of interchangeability which was unimaginable in Leibniz's time and also has profound consequences for the physics of microscopic as well as macroscopic systems.[3] Indistinguishable items may be counted but cannot be placed in numerical order.[4] This is of no consequence to their sum, for that is independent of the sequence in which the items are counted, but it does make their ordering, including temporal ordering, impossible. But happenings which may be counted but may not be placed in any temporal order define a prototemporal Umwelt. Thus, a world made up of identical particles only, must be regarded as prototemporal.

Quantum theory, the most comprehensive theory of matter we have, is intrinsically probabilistic whereas classical mechanics is deterministic. The two are ordinarily reconciled through the correspondence principle which states that systems of high quantum numbers behave classically. A corollary phrasing is that deterministic laws are also probabilistic but with such high degrees of probability that they appear as certainties. This still leaves open the questions about connectedness. But the correspondence principle, when taken together with the prototemporal character of the particle world, does point to a possible evolutionary answer.

When the world of primordial radiation began to change into ponderable mass the first products were some of the elementary particles. Although in due course the particulate world became differentiated and matter more complex, this ancient world is still with us just beneath the macroscopic structure of matter. I regard the intrinsically probabilistic nature of quantum theory as the surviving remnant of the primordial world, coexisting with the advanced levels of the evolutionary hierarchy and manifesting its nature beneath the more complex properties of matter.

At the archaic level of prototemporality the distinction between space and time ought to be taken as a rather weak one. Pure succession characteristic of the eotemporal world and nootemporality characteristic of our ordinary world of "time" do not yet exist. Thus, individual histories and simultaneous group actions are interchangeable, strange as this may sound, because time and space are not yet differentiated. The interconnectedness among the identicals of the prototemporal world, rather than something more sophisticated than causation, is, I believe, something more primitive. In that world individual fates cannot yet exist.

How do contingencies fare in the prototemporal Umwelt? Necessities, it will be recalled, are beinglike events which must come about. Contingencies are becominglike: they may or may not come about. In probability theory, Bayes's Theorem states that if the relative frequency of certain events is determined by past tests as n/m, the n/m is also the probability of future occurrences. For a finite number of samples any probability may be made to look like a contingency, because we have no way of telling whether or not a probable event will occur. But, whereas by Bernoulli's Theorem, as the number of samples approaches infinity probability approaches certainty, contingencies have no percentage figures which they can approach, for they cannot be described by ratios of relative frequencies of their own kinds. In the world of nootemporality we could point to Napoleon, or to the Newtonian theory of gravitation as unique entities inexpressible in terms of their past histories or future occurrences. In the prototemporal world, becoming must take simpler forms, such as the complexification of indistinguishable particles into distinguishable aggregates. But in the prototemporal world as elsewhere, coming-into-being cannot be described by lawfulness, even the lawfulness of probabilistic necessity.

On an English heath, once upon a time, two men encouraged three witches to tell the future of individual grains, out of an aggregate of identical grains.

> BANQUO: If you can look into the seeds of time
> And say which grain will grow and which will not,
> Speak then to me . . .
> FIRST WITCH: Hail!
> SECOND WITCH: Hail!
> THIRD WITCH: Hail! . . .
> MACBETH: Stay, you imperfect speakers, tell me more. . . .
> *Macbeth* (I, iii)

As were the witches, so is the probabilistic world of quantum theory; it is an imperfect speaker and cannot yet say things more specifically than in statistical, depersonalized forms.

Controlled Randomness

The concept of randomness implies total unpredictability about individual occurrences. Thus, for the outcome of a process to be truly random we must be totally ignorant of any underlying regularity: we can produce a well-shuffled pack of cards only if we do not know how we did it. But, as I have argued, probabilities are types of necessities, hence we do know something about them. To express this ambiguity of probabilistic events I shall describe their regularity as "controlled randomness."

A quantitative measure of "controlled randomness" may be given through the use of the concept of entropy invented by Rudolf Clausis (1822–1888) and named by him from the Greek τροπη, meaning "transformation." In its remarkable career, the concept of entropy came to be a useful measure of variables as different as the flow of heat, the flow of energy in general, the rate of information transfer, degrees of indeterminism, orderliness, and structural complexity. Common to all proper uses of entropy is that it describes complex conditions of aggregates of elements which undergo change. Origins of the need for a concept such as entropy may be traced to the recognition by the early nineteenth century founders of thermodynamics that nature displays certain large-scale trends. For example, Sadi Carnot in his *Réflexions sur la puissance motrice du feu et sur les machines propres à développer cette puissance* (1824), drew attention to the intrinsic limitations of the efficiency of a heat engine which employs regular successions of cycles. He observed that although energy may be conserved (following what later became known as the first law of thermodynamics), some of the energy during each cycle may be rendered unavailable for more work. His findings were reformulated in mathematical

terms by Clausius who enunciated a second law of thermodynamics in the
simple statement that heat cannot of itself pass from a colder to a hotter
body. In 1854 he introduced the entropy concept as the ratio of the
quantity of heat gained (or lost) by a body to the absolute temperature at
which the heat transfer occurs.

Because of the principle embodied in the second law, entropy became
a measure of that energy which becomes unavailable during any physical
process.[5] The entropy concept also made possible the restatement of the
second law in an elegant and concise form: all physical processes within an
isolated system lead to an increase of the entropy of the system. An
isolated, or closed system is defined as one that does not exchange energy
or matter with its environment. The need to so delimit the validity of the
law is an example of idealization often found in physics, usually demanded
by the need for conceptual simplicity. Since the creation of closed systems
did not appear problematic, or not more so than similar simplifications
invoked in other theories (for example, neglecting three-body problems in
the formulation of the laws of planetary motion), and since the description
of energy transfer through entropy change was taken to apply to all conceivable processes, the second law assumed an impressive power of universal
validity. Philosophically, it implied a universal and large-scale one-wayness
in nature.

In a brief scholarly survey Whitrow drew attention to three distinct
phases in the history of the concept of entropy.[6] The first one he described
as phenomenological, represented here by Sadi Carnot. The second one is a
statistical reformulation that began with Clausius and continued with
Boltzmann. The third one is the use of quantized information for the
generation of entropylike variables, a method which permitted the extension of entropy concepts to fields outside physics.

In a series of papers in the 1870s Boltzmann showed that the second
law can be reinterpreted as the combination of laws of classical mechanics
and the theory of probability. The second law was shown to be essentially
statistical in nature and the entropy of a system a measure of the probability
of the state of system. His proof, known as Boltzmann's H-theorem, permitted again a rephrasing of the second law of thermodynamics: in the
course of time, all closed systems evolve towards more probable states.[7]
Objections to the validity of the second law were voiced by Loschmidt in
1876 and Zermello in 1896.[8] A hybrid statement of these two objections
would say that, since each process that makes up the system is completely
reversible and/or because the finite number of particles making up the
assembly sooner or later will revert to some earlier configuration, it follows
that sooner or later the whole assembly will return to an earlier condition.
Hence, the entropy of a system cannot be thought of as a monotonically
increasing quantity. These challenges were answered by Paul and Tatiana
Ehrenfest by pointing to the statistical character of the H-theorem and
stressing that it concerned only average variations and did not exclude

fluctuations. The "recurrence" and "reversibility" objections were later re-applied to problems of time and the universe. Namely, since the second law holds only for isolated systems, it does not necessarily apply to the Universe, which may be open. This challenge has been answered by two main arguments. (1) By imagining that whenever an isolated closed system of concern gets close to equilibrium conditions, it is extended to include new regions of lower entropy (which happen to be available).[9] (2) By assuming that our part of the universe, or even the universe at large, is in a low entropy state of small probability, on the upswing towards equilibrium.

Early in this century, the second law of thermodynamics stood, if not unchallenged, certainly undisturbed as one of the powerful generalizations about physical processes, and one that exhibited a one-way trend in the world, namely, that of an increase of entropy—on the average. The power of this generalization did not escape the attention of Eddington who perceived in it the physical basis of time. He coined the phrase "the arrow of time" to express that one-way property of time which is "vividly recognized by consciousness."[10] He held that "nothing in the statistics of an assemblage can distinguish a direction of time when entropy fails to distinguish it,"[11] that "consciousness with its insistence on time's arrow and its rather erratic ideas of time measurement may be guided by entropy clocks in some portion of the brain."[12]

All of this would not be quite so interesting were it not for the fact that the H-theorem, expressing a temporal trend of physical systems toward more probable states, belongs to a very limited class of physical phenomena. That class consists of only two members: the H-theorem and the expansion of the universe. All other laws of physics are either independent of time or invariant under time reversal.[*] It is as though, except for the expansion of the universe and the H-theorem there is a conspiracy in nature to hide the direction of time.

We shall leave the expanding universe alone for the moment, but note that our reflections reveal something very important about this intriguing state of affairs. First of all, the world described by equations which do not respond to sign changes in t is not timeless but eotemporal. The world of controlled randomness is not timeless either but, as we have argued, it is prototemporal. The H-theorem pertains to the interface of these two worlds (if such a spatial metaphor as "interface" is permitted).

---

[*] Time invariance is a formal property of certain equations of physics; the idea pertains to the manner in which time is included as a variable. For instance, time might appear in a quadratic form such as in $s = \frac{1}{2}gt^2$, the expression for displacement in free fall. Whether we think of time as "positive" or "negative" (whatever these might mean) the sign of displacement is always positive. Or, consider Maxwell's theory of electromagnetism expressed in partial differential equations of the first order with respect to time. Because of the first-order quality of the equations, a change in the sign of time would appear to make a difference. However, the reversal of time also entails the reversal of current flow and that of the magnetic field intensity. When all these are taken together, the sense of changes turns out to be independent of the sign of t qua the direction of time.

In its probabilistic character it is appropriate to the prototemporal world; in its macroscopic character to the eotemporal Umwelt of pure succession, to the pure before/after, or pure asymmetry of the macroscopic laws of motion.

Levels of Causation

In a most general sense a cause is anything that can be held responsible for an event. The relationship between causes and their events, or effects, is called causality. Causality has been acknowledged in all civilizations, but opinions about its possible and permissible structure have varied widely.

In pre-Aristotelian Western thought causality may be found in attitudes toward chance and fate, as expressed and explained in the teachings and practices of magic and religion. The Aristotelian idea of material, formal, efficient and final causes signaled, among other ideas, the birth of speculative philosophy and remained up to our days the substance of thought about causality. The suggestion that time order itself be reduced to causal order is due to Leibniz [13] but was developed in detail by Kant who admitted causality as the basis of time order though not of the lapse of time.[14] Hume held that causal connections are never actually observed but only recognized by the temporal proximity of causes and effects and by repeated observations, thus he put time before cause. So did J. S. Mill who defined cause as the antecedent, or concurrence of antecedents, upon which an effect is invariably and unconditionally consequent.[15] He suggested that complex events be analyzed into their constituent factors, identified by observation and experimentation as invariable connections. In modern science causality is taken as an intrinsic and essential feature of a world wherein causal laws must be applicable to man himself.

But the position of causality in science is not one without problems. On the one hand, deterministic causation is an inviolable assumption of scientific thought; its alternatives would be magic or religion. On the other hand, however, our most complete theory of matter, quantum theory, cannot admit events which *must* follow an antecedent, as Mill would have it, but only events which *might* follow. Consider, for example, the theory of radioactive processes. Here the lawful behavior of the system is represented by a smooth curve, a state function which depicts the average behavior of the radioactive radiation. When a particle is emitted, there is no change in the lawful description of the aggregates. As David Park put it, "Thus there is nothing in the physics pertaining to the nucleus itself that reflects the most characteristic quality of the event: *something suddenly happens,* and it is this quality of events that the advancing front of time expresses." [16] Clearly, this is an example of controlled randomness implicit in the probabilistic nature of quantum theory. But then, does probabilistic causality hide ignorance, or is the strict macroscopic causality, so essential to our daily lives and to our methods of thought, no more than a mirage? The question is isomorphic with the one we asked about connectivity among indiscernibles, but it is extended to discernibles. Some useful hints may be found in the development of the idea of causality in the child, set forth in the work of Piaget.[17]

The child's idea of causality begins with a dim feeling that his actions somehow relate to happenings, even though he does not yet separate himself from his actions. This leads to convictions that certain actions consistently produce certain consequences. There develops simultaneously a feeling that temporal contiguity of two events means some correlation between the events. (Here contiguity does not assume a sense of time but only the existence of the physiological or psychological present). This feeling eventually grows to a belief that certain external conditions consistently bring about certain other external conditions. Continued visual exploration of his environment and the feedback between actions and impressions begin to define the infant's identity. The first step in his explorations, according to Piaget, is one of curiosity (such as the desire to move and observe his hand) leading to a sense of power and sensation (such as pulling a string that pulls a rattle). Then emerges an interest in external events (which the infant tries to get repeated by magical gestures). Then there appears a realization that an external object can be causally related to another external object which is of interest to him (such as a person's hand holding the object). Then follows a recognition of causes independent of himself. Finally, in the "preoperational" level he learns to infer a cause from its effects and can foresee an effect, given its cause.

Let us now integrate Piaget's findings into our philosophy of time. At the earliest stages there is a manifest need to reach out for the environment, resulting in conflicts between need satisfaction and the restraints of the environment. The outlines of the self are then defined through the exploration of the spatial boundaries of the infant's body and the temporal boundaries of his present. Early behavior resembling belief in magic causation suggests itself as the source of later beliefs in connectedness among events such as is often expressed in religion, art, and science. An increasing intensity of intentionality stemming from the child's needs becomes associated with the self as a distinct entity. Learning about causal connections among external objects reveals to the child a visual-spatial type of causality (such as the realization that a stick must touch a stone if it is to roll the stone). The ability to infer a cause from its effects, or vice versa, is a nonvisual, temporal understanding of causality. It signals the discovery of correlations between before/after and cause/effect. At this level, spatial connectedness among events is not required any more for the postulation of causation.

We are witnessing here the development of two distinct though mutually dependent components in the structure of causality: (1) the realization that it is possible to account for change by means of connectedness among events and (2) that causal chains may display intentionality. Connectedness by necessity becomes distinguished from any intentionality of the connections. Whereas eventually intentionality is likely to become associated with ideas and processes such as the self, volition, nature, other people, or God, necessity tends to be associated with things. Our present words for "cause" reflect, however, both aspects of the concept of causality. Thus its source, the Latin *causa* is continued in the French *chose* and Italian *cosa*. While the Latin translates into our "cause," the French and Italian translate into "thing." Both of the latter are used as filling words for anything for which one does not find the right word, very much like "cause" is employed as an absolute, all-filling feature of nature. The German *Ursprung* denoted "origin" and by implication, "original cause." Translated literally it means "primordial jump" where "jump" implies "leap" or "spring" of the infant in the womb, as "when Elisabeth heard the salutation of Mary, the babe leaped in her womb" (Luke 1:41). This meaning survives in the English "offspring." In scientific and philosophical context, to denote "cause" in German *Ursache* came to be used, meaning "primordial thing," a static image (rather than *Ursprung*, "primordial jump").[18]

As in the origins of words describing causality, so in the cognitive faculties of infants, in ahistorical cultures and in the primordial physical world, connectedness and intentionality are not yet differentiated. The early Umwelt of the child, as far as causality goes, bears a striking resemblance to the probabilistic world of indistinguishable particles. (In that proto-temporal world things are sure to happen, but we do not know where or when and would not know how to influence them.) Later, when Piaget's "preoperational" level is reached but before complete command of future/past/present obtains, the child's Umwelt corresponds to the reversible, pure succession of the eotemporal world. I believe that it is no mere coincidence that the levels of our early cognitive development and those of concept formation correspond so closely to successive levels useful in the description of the development of the physical universe. Perhaps the growth of the child's idea of causality recapitulates in a loose and approximate fashion the growth of degrees of causation in the physical universe. This would imply that causality itself is of an evolutionary nature, that it began with the probabilistic causation of the prototemporal world, continued with the deterministic causation of the eotemporal world (to be discussed) and eventually reached far beyond physics, to the modalities of multiple causation in biology and to historic causation. Since these various levels manifest distinct and different lawfulness (as well as different degrees of indeterminism and undeterminism, as we shall see), we must postulate the emergence of new first principles in the course of evolution. I will call such a hypothetical process nomogenesis, or the birth of lawfulness, and return to it subsequently.* To remain consistent with our earlier views, we must

---

* An early warning is appropriate, however. The postulate of nomogenesis is *not* to be confused with the occasional suggestion that the laws of nature may be changing.

accommodate the several types of causation simultaneously, and take them as surviving and operating within the hierarchy of creative conflicts.

On the level of sophistication that corresponds to the noetic powers of homo sapiens, connectedness is often expressed in mathematical form, while intentionality is thought of as a manifestation of the mind or of some mind-like agency. Uncritical extrapolation of this dichotomy to the beginning of time can be easily challenged if nomogenesis is taken seriously. Indeed, as we glance back along the history of the universe, deterministic causation itself might have to be judged as inapplicable earlier than some cosmic state. There is evidence that this may be the case. In the very early universe, the cosmology of general relativity theory breaks down and quantum effects become dominant.[19] Above a certain critical temperature, the primordial fire might have behaved according to probabilistic rather than macroscopic laws. Thus, at the origins of time, connectivity and intentionality may have been undifferentiated and only their unity, a creative tension, might have existed in an atemporal world. Looking toward the future, it would then also follow that even the most sophisticated form of causation known to us today is not necessarily the last stage of possible forms which the creative stress may take.

Atomicity, Continuity, and Uncertainty

The idea of atomism relates closely to the logic of explanation: what do we explain by what? Lucretius could envisage without mental reservations a multiplicity of substances made up as various configurations of atoms which were, in themselves, exactly alike. Isidore of Seville, the Venerable Bede, and other atomists could understand time as made up of temporal atoms because they left the nature of the atoms themselves unquestioned. We have encountered various entities which might be called atoms of time, such as the physiological, creature, psychological, and mental presents. In fact, anything that qualifies as an event (by our earlier definition) is also a time atom for it is a unit of unchanging identity. Our concern in this subsection is whether there is anything in the physical world which may so qualify.

Salecker and Wigner considered the requirements one must put to an idealized clock in its most general use of measuring timelike proper time (this chapter, section 2) and sought the limitations imposed on the measurement by the principles of quantum mechanics.[20] They wanted their clock to be as accurate as possible but sufficiently light so that gravitational effects among its parts could be neglected. They found that these two requirements were working against one another. As the parameters of the clock varied there always remained a minimum uncertainty of $\sim 10^{-24}$ seconds in the measurement. If their findings are taken to be of universal validity we may then say that there are no physical processes which could be used to identify the temporal separation of two events to closer than about 10^{-24} seconds.

A number of the same order ($\sim 10^{-23}$) obtains if one divides the shortest experimentally meaningful distance, the radius of the classical electron, by the greatest experimentally meaningful speed, that of light. Thus, the 10^{-23} seconds may be interpreted as the shortest period in which causal interaction may occur. H. T. Flint, also starting from quantum theory, concluded that there can be no physical significance to identifying a particle-event in time to greater accuracy than about 10^{-21} seconds.[21] While we are looking at small intervals, I may add that the highest known electromagnetic frequency, that of cosmic photons, is about 10^{23} Hz. The suggestion emerges that 10^{-24} seconds represents some type of limitation of temporality.

It will be recalled that every time measurement consists of the comparison of two readings. We should have had no difficulty with the claim that, for audio discrimination in man, "periods of less than 2 milliseconds must be regarded as atemporal" because (1) "How else could one describe an Umwelt wherein temporal separation of happenings is impossible?" and (2) we could always think of measuring 2 milliseconds by another clock. What the preceding paragraphs suggest is that 10^{-24} seconds must be regarded as atemporal as far as the physical world goes, for "How else could one describe an Umwelt. . . ." The 10^{-24} will again have to be determined by another clock, albeit a complicated one, involving a series of measurements and calculations. Whether or not such limits exist is a matter of experimental testing. Once such limits are made believable, as I think they have been (they even have a name, "chronon"), the question is that of Zeno: How can we put time together from atemporal elements?

Those who have held out for the infinite divisibility of physical time in spite of the evidence suggesting its atomic nature, see their cause supported by a belief in isomorphism between time and its mathematical metaphors.[22] First it is assumed that time is an extensive quantity such as length, that is, that there exists some physical process of addition by which from two smaller quantities of time a greater quantity of time may be produced. (An intensive quantity is one for which processes of addition and subtraction do not hold. For instance, unlike an extensive quantity, two people each with an IQ of 80 cannot be combined to give one of 160; two bodies at temperatures of 38°K and 42°K put in contact, will not combine into one body with a temperature of 80°K.) If extensiveness of time is assumed, then magnitudes must be taken as capable of describing everything we ever wanted to know about time and, of course, magnitudes can be identified with real numbers. Since the set of real numbers in pure mathematics is a continuum, it follows that time itself must also be mathematically dense.

Taking the variable t in the equations of physics as a continuous quantity is a useful assumption: it is easier to solve differential equations than difference equations. Beyond that, however, the assumption of infinite divisibility runs up against the Eleatic problem of how to constitute time from timeless instants (even apart from the experimental evidence of chronons). From among the many attempts designed to show how mathematically continuous time may be derived from abstract, timeless points, Whitehead's method of "extensive abstraction" is probably the most sensitive and sophisticated one.

Whitehead defined his method as "the law of convergence to simplify by diminution of extent." [23] His method is analogous in the domain of logic to a type of mathematical ordering of real numbers, known as the Dedekind Cut. In the procedure of Dedekind a dimensionless point is constructed as a set of natural numbers, rather than postulated as was done by Euclid. The Dedekind Cut amounts to a replacement of the One by the Many, which is probably why it suggested itself to Whitehead as a means of reaching a single, assumed physical quantity from a multitude of its elements. He asked us to think of duration as forming a sum of a series of descending temporal extensions, where each member progressively contracts and "the successive durations are packed one within the other like a nest of boxes of a Chinese toy." [24] He called this an abstractive set and noted that "as we pass along it, it converges to the ideal of all nature with no temporal extension, namely, to the ideal of nature at an instant." His method is to guide our thought "to the consideration of the progressive simplicity of natural relations as we progressively diminish the temporal extension of the duration considered."

I wish to illustrate the incompleteness of Whitehead's beautiful thoughts by comparing them to the spermist doctrines of the mid-seventeenth century. According to the spermists, the human sperm contained a homunculus, a little man capable of growing into a mature man. Carried to its logical consequence this is an encasement theory: the sperm contains a miniature adult with its miniature sperms, which contain their homunculi, ad infinitum. Experimental biology and genetics proved the case for epigenesis, however, and not preformation. Organization obtains through dynamic interactions, and the morphology of the adult needs to bear no resemblance to the morphology of the sperm. The debate whether time is, or is not, continuous is strikingly similar to the preformation-epigenetic debate. Those who insist on infinite divisibility may be likened to the spermists; the theory of time as conflict is akin to the stance of the epigeneticists. The noetic form of time need bear no resemblance to its more primitive forms, hence those forms cannot be combined through the method of extensive abstractions or described as a dense set. As we "progressively diminish the temporal extension of the duration considered" the nature of temporality changes.

Somewhere around 10^{-24} seconds there is a natural limit in the physical world. Shorter intervals have no physical meaning; hence that period must be regarded as atemporal. Perhaps this is also a parameter of the primordial universe, below which its history cannot be probed,[25] and the chronon a heritage from the creation of time. The creation of elementary particles came to define a prototemporal world, commencing with periods longer than 10^{-24} seconds. As matter further complexified and formed distinguishable aggregates, the physical laws of motion were born with their reversible temporalities of before/after. These laws belong to those of an eotemporal world: one with pure succession but no preferred direction.

It is the transitions between the atemporal, prototemporal, and eotemporal worlds which give rise to Heisenberg's uncertainty principle, as they gave rise to the H-theorem. The uncertainty principle states that it is impossible to specify simultaneously *and* to any desired accuracy certain canonically conjugate physical quantities, such as momentum and position, or energy and time. The principle is usually stated in difference equations or, more precisely, inequalities, such as $\Delta E \, \Delta t \geqq h/2\pi$ or $\Delta x \, \Delta p_x \geqq h/2\pi$ where E is energy, t is time and h is Planck's constant. Because of the smallness of Planck's constant, the quantity which pivots the mutually restrictive character of energy and time, or position and momentum, the uncertainties (that is, intractable errors in specifications) have no practical consequence to distinguishable aggregates of particles. Macroscopic bodies follow their Newtonian (or relativistic) laws of motion appropriate to an eotemporal world. However, as the world changes from that of distinguishable emsembles to that of indistinguishable particles, the relationship expressed by the equations becomes increasingly important. This change is a gradual one and is not to be imagined as a sharp limit comparable, for example, to the similar step in the auditory Umwelt of man. The intractable error in specifying the temporal position of particle-events increases until all we can say about such an event is that within a specified period it *may* happen, according to probabilistic prescriptions. Thus, from the eotemporality of the macroscopic world we shift to a prototemporal world. Finally, through the curious mutuality among the parameters of quantum mechanical equations we obtain the chronon of the Salecker-Wigner analysis, an entity which in our philosophy must be regarded as atemporal.

It seems, then, that even though Heisenberg's famous equation uses the same symbol t as does, for instance, Kepler's equally famous Third Law (simply stated as $t^2/R^3 =$ constant, where t is the planetary period, R the average distance to the sun), the t's are denizens of two, or perhaps three, different temporal Umwelts.

Thermodynamic Arrows of Time

In evaluating the reasons why in all branches of physics, except in thermodynamics, time invariance prevails, Mehlberg concluded that

> There would be neither a miracle nor an unbelievable coincidence in the concealment of time's arrow from us only if there were nothing to conceal—that is, if time had no arrow. On presently available evidence time's arrow is therefore a gratuitous assumption.[26]

Before and after Mehlberg many other keen minds have examined this problem, though not all were equally taken by the idea of timeless time. Although details of the arguments have varied, their rudimentary form remained a categorical syllogism: (A) All physical processes obey the H-theorem, (B) our sense of time derives from physical processes, albeit from very complex ones, hence, (C) our ideas and our sense of time must also derive from the law of increasing entropy. I believe that both the major and the minor premises of this syllogism are sufficiently incomplete to make the conclusion untenable. It is not the validity of the second law of thermodynamics that is doubted, as far as it applies, it is rather that the proto- and eotemporal Umwelts in which that law operates form only the lower reaches of our sense of time.

Classical thermodynamics, which gave rise to the second law, deals with closed systems; with processes which, on the microscopic scale, must be taken as completely reversible and which sooner or later will reach an equilibrium condition. All of these features are details of the idealization employed to simplify the conditions for which thermodynamics is called to account. It is easy to conceive, but difficult to handle formally, systems which have boundary conditions complementary to the classical system. Such a complementary system would be open, in that it would be permitted to exchange matter and energy with its environment; it would subsume irreversible processes and would avoid operating near equilibrium conditions. In a treatise on open systems first published in 1955, Ilya Prigogine suggested that mature living organisms constitute a thermodynamically open, nonequilibrium, stationary state. He showed that such a system would require that the rate of its entropy increase due to its internal, irreversible processes be at a mathematically local minimum when the system is in steady state.[27] In a later work he associated the law of increasing entropy with progressive disorganization. He concluded that "a theory of 'destruction of structure' must in some ways be complemented by a theory of 'creation of structure,' a theory lacking in classical thermodynamics." [28] Presently, let us revert to the critique of the physical arrow of time.

It has been a virtually unchallenged opinion among those who have considered this problem that our sense of time parallels the trend of the H-theorem. Admittedly, entropy increase of closed systems is not monotonic; statistical reversals are permissible. But the general thrust of increasing entropy was held to point from past to future. It seems now that living and self-organizing things in general are characterized by a trend of entropy decrease. One could consider this condition an uncharacteristic deviation from the general "rule," and identify life with the statistical, random possibility clearly permitted by the H-theroem. Such a stance would be ill-advised, however, because entropy decrease (on the average, and with minor statistically permissible deviations of entropy increase) is a necessary, unalienable, and perhaps thermodynamically sufficient feature of life and of the creation of structure. If living and self-organizing material ceases to decrease entropy * with any significant consistency, it also ceases living and self-organizing; thus, operation in a decreasing entropy mode is not an incidental and occasional but a permanent and essential feature of life. We have, therefore, not one but two thermodynamic arrows of time: that of inanimate matter and that of living matter. It follows that the identification of past-present-future with the direction of either arrow is arbitrary. We will recall that the eotemporal t of kinematics is that of pure succession: it belongs in an Umwelt where events may be ordered according to sequence but no preferred direction may be attached to the sequence. The existence of two thermodynamic arrows lifts us above the kinematical level. Namely, now we may at least focus on one or on the other arrow and speak about a "preferred direction," with respect to the other arrow.

Historically there has been a universal preference for associating time's arrow with the increasing entropy mode of matter. There are three main sources from which this preference stems. The first one is the fact that our acquaintance with closed systems preceded by a century our familiarity with open systems. Had we learned about the decreasing entropy mode of self-organizing systems first, we might now be insisting that our sense of time parallels some sort of an "inverse H-theorem." The second reason is that we are surrounded by an overwhelming preponderance of systems which are not self-organizing, and scientific judgements have been generally informed and inspired by this fact rather than by the character of life and creation. Finally, since modern western civilization is emphatically physical and mechanistic in its outlook, our clocks are exclusively inanimate. Were our timekeepers organic, we would probably have favored a parallel between time and decreasing entropy, and observed that the inanimate universe "runs the other way."

But if both the increasing and the decreasing modes of entropy production parallel our sense of time, and if our preference depends only on cultural conditioning, then neither may really be trusted. We can see from

---

* The awkwardness of this phrasing, or its even more awkward equivalent of "producing negentropy" stems from Clausius's unfortunate nomenclature which implied that it is the loss of usable energy (rather than its production) which corresponds to an increase in entropy.

....

We have not one but at least two thermodynamic arrows of time: that of inanimate matter and that of living matter, and they point in opposite directions. It follows that the identification of past-present-future with the direction of either arrow is arbitrary, hence neither may be trusted. When discussing temporality, we must consider the coexistence of the two.

M. C. Escher, Ascending and Descending (1960). *Courtesy, Escher Foundation, Haags Gemeentemuseum, The Hague.*

....

prior arguments that either of them taken alone corresponds to proto- and eotemporal worlds. But, their coexistence has something interesting to say. Prigogine's finding, that an open system, to be in steady-state nonequilibrium condition, must produce through its internal functions a minimal rate of entropy increase, implies that such systems strive to maximize their opposition to the increasing entropy thrust of their environment. If a certain closed system produces entropy at a given rate, the same closed system containing an open system will produce entropy at a lesser rate.

Now, the quantity of biomass known to us as compared with the inanimate masses of the universe is minute; the biomass of man is almost nonexistent.[29] But from the point of view of the coexistence of entropy increasing and decreasing processes the biomass on this earth, and even the ephemeral biomass of man, must be taken on equal footing with all the inanimate matter of the universe. Although the observer is indeed small compared with the universe, it is he who holds the idea of the universe in his mind and asks questions about time. But if he is taken as an essential and equal partner in a combined system made up of closed and open thermodynamic systems, then we may perceive a substantial conflict between the two eotemporal arrows. We may describe this conflict as an existential tension in which self-organizing matter opposes the trend of inanimate matter.

2. Aspects of Time
and the One

OUR MOST comprehensive scientific view of the universe derives from the theory of relativity, the science of one single world.

From Absolute Rest to Absolute Motion

In the world of Newtonian kinematics, which is the study of motion in terms of velocities, answers to the inquiries, "When is then?" and, "Where is there?" are easily found. Universal frameworks of absolute rest and absolute time are assumed and to them all positions in space and all events in time may be referred. This comfortable belief was rejected when it became clear that if we are to obtain self-consistent physical laws to account for observed physical phenomena we must specify, among other matters, what we mean by time at a distance.

The revolutionary new system of kinematics, dynamics, and cosmology, which does provide physical laws of the necessary self-consistency, directs our attention away from time and space, for it constitutes physics built on the primacy of motion. Einstein's special relativity theory of 1905 offers a metaphysical receptacle, called space-time, in which the physical quantities of length, time, and mass are intricately dependent on relative velocities and on the speed of light. His general relativity theory (1916) deals with the geometrical properties of this substratum in the presence of energy.

Accounts of relativity theory often begin by pointing to the negative outcome of the A. A. Michelson and E. W. Morley experiment in 1887 designed to measure the velocity of the earth with respect to a hypothetical substance called ether. That some substance exists which is capable of supporting and transmitting electromagnetic oscillations would follow from the electromagnetic wave character of light. Since Maxwell's celebrated demonstration of the existence of electromagnetic waves, this wave character had not been questioned; but wave motion was held to require a substance of which the waves are the motion, hence the belief in an ether filling all of space. Had such a substance been found, it could have represented a frame of absolute rest. The failure to detect the motion of the earth with respect to an ether, by a clear-cut and conceptually simple experiment, remained uninterpreted at the end of the nineteenth century. The velocity of light, as determined by dividing known distances by known times, was proved to have been independent of the relative motion of source of light.[30]

Recent inquiry suggests that contrary to earlier beliefs Einstein's thoughts were not triggered by the negative outcome of the Michelson-Morley experiment.[31] From our point of view this is irrelevant; we shall leave this act of creation alone and let its logical sequence, if there was one, be discovered by historians. It is convenient to approach the physical principles of relativity theory by juxtaposing the search for a physical reference of absolute rest with Einstein's realization that such a frame does not exist. For, "the phenomena of electrodynamics as well as of mechanics possess no properties corresponding to the idea of absolute rest." [32] This is the fundamental and revolutionary insight of relativity theory.

In our mental imagery "absolute rest" is a timeless, motionless existence; it is the eternal rest of the soul from self-definition and strife, even if such ideas are far from the physicist's conscious mind in the context of physical theories of motion. Before the time of Copernicus, the idea of "motion" was ordinarily interpreted in terms of change of position with reference to an absolutely resting earth; horses, ships, and stars all moved with respect to the solid countryside. The Copernican revolution confused this issue or, more precisely, extended the meaning of the word "rest." Absolute rest was now attributed to the aggregate of matter in the solar system and motion was assigned to the earth with respect to the new frame of rest. The Einsteinian revolution is more radical. We are asked to dispense with all references to absolute rest and rebuild our ideas of motion with reference to constant unrest. The physical process identified as the new reference is the propagation of electromagnetic waves in a vacuum, a process usually abbreviated as the "propagation of light," using the name of its visible region pars pro toto.

It is appropriate to call the motion of light "absolute," for its role in relativistic kinematics is analogous to that of absolute rest in Newtonian kinematics. Consider that in Newtonian physics the speed of absolute rest is invariant at $v = 0$ cm/sec. This appears to us as something of a self-evident definition that reflects the logic of experienced reality. That such frames exist may be said to be a fact of nature; they may certainly be identified by simple experiments. The statement that no meaning could be attached to speeds less than that of absolute rest, such as to motion slower than that of an ether (had that been found a true feature of the world) appears to be so self-evident on a logical basis that it sounds patently banal, almost as though it were a careless joke. But consider further that in relativity physics the speed of light is invariant at $c = 3 \times 10^{10}$ cm/sec. That such frames exist may be said to be a fact of nature; they may certainly be identified experimentally. The statement that no meaning can be attached to speeds more than that of light sounds rather surprising. The symmetry between these two limits is not immediately evident, however, because we have no experience of very high speeds. Had we, as a species, evolved with such experiences, then the existence of an upper limit to relative velocities would probably be quite self-evident. All the experiential consequences of such features of nature would have already been built into our language and our thinking and anchored in them so deeply that any assertion that "speeds larger than that of light cannot be found" would probably appear to be self-evident on a logical basis and would sound patently banal, almost as though it were a careless joke. It is partly because of the symmetry between absolute rest and the motion of light that the motion of light may be quite appropriately called absolute motion. While stressing this symmetry, we must, however, remember an important asymmetry. Whereas a universal frame that could correspond to the idea of absolute rest both as regards mechanics and electrodynamics could not be found, a universal frame that does correspond to the idea of absolute motion both as regards mechanics and electrodynamics can be found and it was so identified. Thus the motion of light deserves to be called "absolute" more so than the absolute rest of prerelativity physics. For the consistent physical descriptions of the motion of a multitude of translating bodies, Newtonian physics would select one arbitrary frame of motion and would refer all velocities to that frame; the speed of light plays an identical role in relativity theory. The security of no change in a Sensorium Dei was replaced by the less reassuring absoluteness of continuous change. It is important to understand, however, that the propagation of electromagnetic waves is not to be interpreted as absolute motion with respect to anything specific at rest, just as the idea of absolute rest was not to have been interpreted as rest with respect to anything specific in motion. The motion of light must be comprehended as an absolute motion for and by its own self, an absolute Parmenedian One of the physical world.

The New Invariant

Special relativity theory is built on two fundamental principles and one set of clock calibration instructions. (1) The principle of relativity of motion replaces the privilege of absolute rest by a democracy of the physical equivalences of all inertial frames; it demands that all laws of nature are to have identical forms in all of these frames. (2) The principle of the constancy of light velocity introduces the propagation of electromagnetic waves as a referent of absolute motion in the sense just discussed. (3) The calibration instruction is an operational definition of how two clocks at relative rest are to be synchronized by means of light signals.[33] From these three items follow the rules of transformations which, unlike the Newtonian transformations of similar scope, are capable of yielding self-consistent readings of lengths of moving rods and of times of moving clocks. These rules, in their turn, give rise to the peculiar answers which the theory gives to such questions as "When is then?" and "Where is there?" when the distant event happens to be moving with respect to the questioner.

In classical physics the most characteristic feature of motion is its velocity, constructed as the ratio of lengths and times, thus presuming the concepts of distance and time. That this is not at all an instinctively self-evident way of describing motion has been shown by Piaget's work on the perceptual growth of the child. The child learns to form ideas of space and time from perception of relative motion, and only much later can he construct the idea of velocity from perception of distance and time. The practical value of velocity = distance/time resides, thus, not in any simple obviousness but in something more subtle: it is a quantity invariant for all observers stationary with respect to each other.[34] In relativity theory motion is represented by a combination of distances and times no more obvious to the uninitiated than velocity as distance/time is to the perceptual faculties of the child. It is through the use of this new combination that the separate absolutes of distances and times may be replaced by a new quantity which relates all motion to that of light. The validity of the representation so obtained is demonstrated by the satisfactory account which the complete theory can give for a variety of physical phenomena, and by its success in predicting others, all unexplainable or even unintelligible in terms of Newtonian concepts.

Motion in relativity theory is described in, or inscribed into a four-dimensional space called four-space or space-time. The coordinate axes of this four-space are mutually orthogonal; its geometry is obtained by extending the rules of ordinary mathematics to multidimensional spaces. Each of the four axes is taken to be extension calibrated in units of distance. But whereas the first three are calibrated as are ordinary, spatial Cartesian coordinates, the fourth one has a calibration constant of the inverse velocity of light, which is about 3×10^{-11} sec/cm. Points and lines in four-space are called world-points and world-lines, to distinguish them from their three-dimensional relatives.

A spatial model of space-time cannot be constructed because four-space is not space in the ordinary perceptual sense. The visual model most often used for the demonstration of relativistic kinematics, the Minkowski diagram, is a model of plane-time rather than space-time.[35] It is sufficient, however, if one understands without being able to visualize a geometry which is capable of identifying four mutually orthogonal coordinates of a world point, usually designated by x, y, z and t. The history of a moving point, then, is described as a four-distance whose consecutive points carry information on consecutive locations of that point (in the observer's time external to the manifold). Three-space in itself is as timeless as was St. Augustine's universe before God created it; in such a space motion cannot be given either a logical or an operational status. The four-space of relativity theory is not different. It is not a magical extension of our kinesthetic experience into a world beyond, but a mathematical construct; neither is it, however, completely arbitrary. Rather, it is a suitable scaffolding for keeping order in our thought about the world, akin to Plato's immutable forms or the timetables of airlines.

In ordinary analytical geometry, points in space are represented by ordered sets of numbers called coordinates. Beginning with points one then builds up the formalism for representing lines, surfaces, and spaces. Identical formalism may be used for the identification of world-points and world-lines in space-time. Components of a world-line along the three space axes are called spacelike, along the time axis, called timelike. Spacelike and timelike components of a given world-line may be calculated for any new four-coordinate system, provided the system's orientation and location with respect to the first coordinate system is known. The relativistic properties of time and limits of temporality do not arise, however, from the method of four-dimensional representation; Newtonian physics could also be written in four dimensions. The relativistic features arise when the principle of relativity and that of the finitude and constancy of the velocity of light are applied as restraining conditions to motion in space-time, together with Einstein's instructions on the calibration of a distant, stationary clock.

Let an inertial observer record the motion of an inertial particle from position (x,y,z) at time t to a position $(x+dx,\ y+dy,\ z+dz)$ at time $(t+dt)$. It can be shown[36] that the most general expression for the four-distance ds covered by the moving object, in terms of ordinary distances and times, is

$$ds^2 = c^2 dt^2 - (dx^2 + dy^2 + dz^2) \qquad (1)$$

Restated in a nondifferential form we have

$$s^2 = c^2 t^2 - r^2 \qquad (2)$$

where t is ordinary time between two events, and r is the distance* between the two points where the events happen. Whereas in general t and r will be different for observers in relative translation, the calculated quantity s will be the same for all of them. Equation (1) for ds is the new combination of time and distance measurements used for the description of motion in relativity theory. The integrated quantity s describes motion with the same specificity with respect to a four-coordinate system as does velocity in Newtonian physics with respect to absolute space and time.

Now, velocity as defined by distance/time has the simplest possible invariance; it is the same for all observers in relative rest. The new quantity s represents a higher order invariance: it holds for all observers in relative translation. This new invariant is called "proper time." [37] Since, as we shall presently see, proper time itself may be timelike or spacelike, the term is confusing. To avoid this, I shall refer to the invariant s as "space-time distance" or "four-distance," and reserve the term "proper time" to the quantity s only when it is timelike.

Inspection of equation (2) suggests the way that space-time distance is tied to the absolute motion of light. As the quantities r and t vary among inertial observers, the four-distance may be real, zero, or imaginary. These alternatives are intricately tied to the space-time structure of the world as revealed by physical measurements. Let us imagine a mouse translating between two extragalactic nebula. Let E_1 and E_2 represent the events of conception and delivery in the life of a mouse; this is a fairly constant nineteen days for mice, determined by endogenous processes. For the mouse E_1 and E_2 happen in the same place, meaning, within its own body. The quantity of r in (2) is zero and s as measured by the regular processes of its body is its proper time, that is, nineteen days. All other inertial observers would calculate this period of time as nineteen days even if their own calendars show more than nineteen days, provided they know relativity theory. They will also all agree that conception preceded birth.

Because of the nonlinear and very rapid increase of energy with high relative local velocities, relative speeds of c would require an infinite amount of energy; hence they do not obtain. [38] It follows, therefore, that for all conceivable observers, even if their local (coordinate) measurements of r and t separating E_1 and E_2 may differ among themselves, E_1 will always be judged ex post facto as having preceded E_2 (as determined by the sense of time of the people making the measurements). It also follows that for all macroscopic bodies the separation s between any two events in their own lives must be temporal. Thus, the new invariant of relativity theory for the motion of ponderable matter, is time. The level of temporality of proper

---

* I am following a notation which yields real numbers for timelike intervals, for this formalism is the best for our purpose. Many writers prefer to define ds^2 as $dx^2 + dy^2 + dz^2 - c^2dt^2$ which is consistent with the imaginary nature of the relativistic time axis but confusing for the idea of time.

time is determined by the nature of the processes we are trying to describe. If E_1 and E_2 are the birth and death of a man, proper time may stand for nootemporality. If the events are in the kinematical world, such as those in the history of an orbiting planet, proper time is eotemporal. If the events are probabilistic, proper time must be prototemporal. If the two events are in the life of a photon then $r = tc$ and $s = o$; the two events are simultaneous. Furthermore, since s is invariant with respect to translation, all events in the life of a photon will be recognized as simultaneous by all inertial observers. Clearly, traveling photons (and they always travel) determine an atemporal Umwelt. Since at the beginning of time the universe was that of pure electromagnetic radiation,[39] and much of the original energy is still in form of electromagnetic waves, it follows that the electromagnetic universe is one, single, atemporal event.

An event E^* may also be so located that even the "first signal" quality of light (a descriptive phrase coined by Reichenbach) is not sufficient to connect it with a here-now. For such a case the second term in equation (2) will become larger than the first term and the four-distance becomes imaginary. The separation of a here-now from E^* for such a case is said to be spacelike. (Sometimes a new real quantity is introduced by rewriting (2) as $a^2 = R^2 - c^2 T^2$ to make the handling of spacelike four-distances more conventional). Since the four-distance is invariant, all inertial observers who obtained information ex post facto about the coordinates of that here-now and those of E^* will judge the separation to have been spacelike, and will agree on its magnitude. Since there exist no means of interacting faster than c, there will be universal agreement that E^* (for me, let us say, here and now) belonged to a world which Eddington called "absolute elsewhere." [40] The same region may, of course, contain any number of other events, all characterized by the fact that physical interaction between them and my here-now cannot exist. But that means that I have no way of identifying them as events at all, hence they belong, with respect to my here-now, in an atemporal world.

We have already identified the chronon as an example of physical atemporality. It seems, then, that as we contemplate larger and larger regions of the universe the chronon expands, as it were, and we are forced to regard larger and larger spans of time as necessarily atemporal. For the smallest physically meaningful object, the classical electron, the chronon was found to be about 10^{-24} seconds. For a sphere with the lunar distance at its radius the atemporal world (in the sense specified) is about 1.28 seconds; when the radius is one astronomical unit the "chronon" is about 8 minutes; when the radius is equated to the "radius of the universe" the length of this atemporal, single event is about 1.77×10^{17} seconds, a quantity known as the age of the universe.

Some events in the absolute elsewhere, say E^* and E°, may themselves be separated by a spacelike four-distance. The application of relativistic transformations to such events shows that different inertial observers may encounter them as happening in opposing sequences. For one observer, E^* precedes E°, for another observer E° precedes E^*. Since we must be talking about countable but temporarily unorderable events, they must be judged as belonging to a vast prototemporal world. But whereas our earlier account tied prototemporality to the world of indiscernibles, here we encounter countable, yet unorderable events. This is one of the purely relativistic contributions to our understanding of temporality. Other events in the "absolute elsewhere" may lie on a timelike world line. They will be so recognized (ex post facto) by all inertial observers, and all will agree as to the temporality represented by the events: perhaps successive emissions of alpha particles in a prototemporal Umwelt, or a body in free fall in an eotemporal Umwelt, or stages in the pregnancy of our mouse controlled by its biological clock. The universe of the absolute elsewhere may have a complete hierarchical structuring of its own. For my here-now it is atemporal, for reasons stated. In the here-future, however, I may find reasons to regard some (then past) events as having displayed prototemporal, eotemporal, or nootemporal characteristics.

Be these as they may, the relativistic invariant of four-distance which replaced the classical idea of velocity is a much richer concept than its predecessor.[41] Furthermore, since it is invariant with respect to all inertial observers and not only with respect to all observers at relative rest, its use is consistent with the historical thrust of scientific knowledge toward mathematical descriptions of nature with increasingly universal validity.

Clocks and Proper Time

In his "Notes" to Minkowski's paper on *Space and Time*, Arthur Sommerfeld remarked that all conclusions drawn from the proper-time concept of relativity theory regarding indications of moving clocks are based "as Einstein has pointed out, on the unprovable assumption that the clock in motion actually indicates its own proper time, i.e., that it always gives the time corresponding to the state of velocity regarded as constant, at any instant."[42] A remark of identical substance was made by Einstein in the context of his general theory of relativity which successfully predicted that clocks of any type, when placed in the gravitational fields of massive objects, will tick slower than identical clocks in gravity free environments. He cautioned that "Everything finally depends upon the question: Can a spectral line [of atoms-as-clocks oscillating in distant stars] be considered as a measure of 'proper time' (if one takes into consideration regions of cosmic dimensions)?"[43] These warnings which hold for translating and rotating clocks and for clocks in a gravitational field, also hold for stationary clocks (by the principle of relativity) and for clocks away from a gravitational field (by the principle of equivalence).[44] Thus, all motional conditions are covered.

When Sommerfeld and Einstein spoke of proper time, they must have meant the four-distance s, more specifically, a timelike s, employed in the equations of relativity theory to specify that property of the cosmological substratum which we unproblematically call "time." When they spoke of clocks they must have meant processes in time employed as clocks. If we admit our earlier conclusions that clocks are taken to be reliable if they act in accordance with theoretical constructs judged by the clockmaker as predictable regularities, then as an approach to a possible answer to Einstein's question, one ought to begin with a convincing theory (relativity theory) and seek processes which conform to the theory. Thus, the clock in motion will be said to indicate its own proper time if its readings are consistent with theoretical expectations, and the same holds for spectral lines. A close look at the clocks versus proper-time problem reveals two distinct questions. (1) Are relativistic predictions valid for all types of clocks, and (2) what epistemological and/or ontological assumptions are hidden in, or must be made for, any claim that a clock shows its proper time?

The answer to the first question must be a simple affirmative. Let us imagine that the opposite is the case, that is, thermodynamical, mechanical, and biological clocks follow different transformation laws under translation. Since the earth has been traveling very rapidly with respect to many objects, circadian rhythms should have gotten systematically out of phase, with respect to the rotation of the earth, long ago. But this is not the case. On the contrary, the correlated unity of all temporal processes has been a tacit assumption in all human knowledge, a necessary prerequisite of science— and an inspiration of much thought. A different phrasing of the same question, however, is less easily answerable. What assures us that clocks which employ different temporalities, proto- and eotemporal clocks, will remain in phase with biological clocks? The answer to the question so phrased must come from ideas yet to be developed, but it amounts to this: In the hierarchical structure of temporality each integrative level is restrained by the nomothetic features of the lower levels from which it has arisen. While each new type of clock can extend the set of those features which we can subsume under "clockness," it must abide by the earlier lawfulness.

The second problem is the physicist's way of questioning the identity of time with processes in time, a matter which, as we have seen, has been important in the history of ideas. In Platonic thought the link between time and processes in time is the absolute rotation of the sky; in relativistic thought the link is the absolute motion of light. It is through the phenomenon of the propagation of electromagnetic waves that times and distances become mutually transformable in the space-time continuum, and it is to the quality of absolute motion that all other motions are referred. Plato could not have asked for a more satisfying system. Yet these remain instructions for measurements which beg the problem of clock versus time. Einstein already recognized this gap in 1922 when he wrote that "we are still far from possessing such certain knowledge of theoretical principles as to be able to give exact theoretical construction of solid bodies and clocks." [45]

The theoretical construction of a clock comprises, as I see it, the opinions of its maker regarding the relationship between temporality and clocks, that is, between time and processes in time. There are many such opinions; perhaps too many. The intuitive idea of time suggests that temporality is independent of man or of any other clockwatcher imaginable; time even "feels" to be independent of matter itself. This type of temporality has been called *constitutive*. Yet it is impossible to imagine or make sense of a totally empty and unbounded volume where no events take place yet time passes. Time seems to demand the existence of change in the states of matter.* When temporality is thought of as following from physical happenings in the world, it is often called *relational*. Both propositions stem from the same introspective notion of time, yet they are logical opposites. From the constitutive aspect, time is ontologically prior to events (or clocks), whereas from the relational aspect events, or clocks, are ontologically prior to time. *Absolute* time means two distinct ideas. It may mean *Newtonian absolute* time which "from its own nature, flows equably without relation to anything else"—hence it is constitutive. It may mean *nonrelativistic* time signifying that the temporal separation of two events is measured by the same number of ticks by all identical clocks regardless of relative motion or gravitational environments. Time may be said to be *special relativistic* if identical clocks in various states of translation give different readings for the temporal separation of two events, corresponding to the instructions of the Lorentz transformations. Time is recognized as *general relativistic* if clock readings also show the influence of gravitational environments according to various specifications derived from Einstein's field equations.

Reverting now to clocks and proper time, we note that the variables of the Lorentz transformations are lengths and times. It is not possible to include among them any effects of energy either in the form of gravitational potential or in the form of ponderable mass. The mere relative motion of two disembodied mental constructs in an empty universe is sufficient to call for the transformation of the time metric. Hence special relativistic time is constitutive. In the general theory of relativity, matter and inertia are introduced through the quantity of energy-momentum and make themselves manifest in various physically identifiable ways. Time transformations are called for between regions of different gravitational potentials associated with matter or, by the equivalence principle, between accelerating frames. Thus the time of general relativity suggests relational time because it derives from a preexisting physical world. But the world of the general theory is that of galaxies which form a more restrictive class of elements than do the observers contemplated in the special theory, in that they keep a cosmic time.[46] This time admits simultaneity among fundamental particles (galaxies) in relative motion though it insists on special relativistic times within galaxies. It may be called *cosmological-common* time which is thus absolute in an almost Newtonian sense. We must add to these considerations our

---

* The reader will recognize in some of these concepts the rephrasing and regrouping of views already encountered in philosophy under different headings.

earlier conclusions that the level of temporality of proper time depends on the Umwelt in which we operate.

One might hold that diligent search may reveal which physical idea of time is closest to reality and assist in an eventual decision (even if in a scientifically tentative manner) as to which type of clock is the most reliable one (it will be that one whose operation corresponds most closely to the physical theory judged as most convincing). The substance of this subsection suggests, however, that the issue is too intricate to be settled through such a simple-minded program. Or, if the reader wishes, he might prefer to say that the program is not as simple as its physicalistic phrasing suggests, for it seems to call for nothing less than a complete theory of time.

The Astral Geometry of Gauss

Questions about time, timekeepers, and the world were certainly among those of concern to early relativists, but they were of very low priority. In keeping with the Platonic emphasis on the immutable, and dislike of the generative which characterizes physical science, the creators of relativity theory preferred the static and the geometrical aspects of the world to the dynamic and the temporal. As in medieval times, when all good minds went to work for the church of Rome, hence theology flourished, so all good relativists applied themselves to the exploration of geometries, hence relativity theory became overwhelmingly geometrical. To be able to understand what relativity theory teaches about time, in addition to the details already discussed, we must make a detour to problems of measuring distances and, instead of asking what happens to clocks when they are thought of as operating in the distant past or future, we must find out what happens to geometrical lines in distant space.

The origins of geometry, according to Herodotus, reach back to the ancient Egyptians who invented it for the purpose of measuring land,[47] and it is safe to assume that some need for mathematical organization of spatial forms has arisen in all early civilizations. The early Christian intellectual, Clement of Alexandria, wrote of Egyptians who were *harpedonaptai* or "rope stretchers": rope stretching was an activity employed for astronomical determinations as well as for land measurement. From the harpedonaptai to Immanuel Kant, the tendency has been not to combine but rather to separate the logical and the experimental or experiential aspects of geometry. This trend was radically reversed by Carl Friedrich Gauss (1777–1855), who applied his genius to the possibility of constructing a geometry based on axioms partly at variance with those of Euclid, and who judged it appropriate to subject his theories to experimental tests. To describe his new geometry Gauss adopted the term "non-Euclidean," after having first called it by the prophetic name of astral geometry.[48] Presently we must concentrate not on this intellectual revolution at large, but only on a very few of its pertinent details

Euclid's axiom of parallels is Postulate V in the first book of his *Elements*, written about 300 B.C. It states that

> if a straight line falling on two straight lines makes the angles, internal and on the same side, less than two right angles, the two straight lines, being produced indefinitely, meet on the side on which are the angles less than two right angles.[49]

By default, if the two angles, internal and on the same side, are exactly two right angles, the two straight lines being produced indefinitely will never meet. Such lines in Euclidean geometry are said to be parallel, and the axiom is often stated in the following form: given a line l and a point p not on that line, there exists in the plane of p and l one, and only one line m which does not meet the given line l.

Gauss was not the first one to feel uncomfortable with this postulate; so did Euclid and Ptolemy. There were several attempts, at least since those of the sixteenth century mathematician Clavius, to derive the postulate of parallels from one of the other axioms. But Gauss was probably the first one to recognize that, for the resolution of what appears to be an abstract problem of the mind, an appeal might be made to other than the reflective faculties of man. Specifically, he felt that the problem of whether or not two straight lines at exactly right angles to a third line do or do not meet is a matter to be decided in the depth of space. Accordingly, in 1817, when he wrote that geometry is not an a priori science [50] the validity of the parallel postulate of Euclid became a matter which could be experimentally tested.

A degree of epistemological prudence is appropriate here. No physical meaning can be attached to a demand that we ride along two parallels and observe the region where they cross. The proposal is, instead, that a geometry which can self-consistently accommodate certain non-Euclidean postulates should lead to locally testable predictions. Just as we have no experience as a species, hence no intuition, about the character of the world of very high speeds or very long times but must depend on logical extrapolation from what we know of the here and now, so we have no intuitive sense of geometries that involve the astral distances of Gauss. It is one hallmark of the human mind that it can arrive at valid statements pertaining to things and events not directly within our perceptual world or evolutionary history. Thus, when Gauss had three observers standing on three mountain peaks, measuring the angles of a large triangle to see whether their sum differs from 180°, he began making the geometry of the universe a part of the physics of here and now.[51]

In studying the properties of surfaces Gauss made use of the notion of a geodesic, which is a line lying within a surface, being the shortest distance between two points on that surface. Thus, in a plane a geodesic is a straight line; on a sphere it is part of a great circle; on a doughnut, a variety of curves. Gauss also showed that the nature of a geodesic depended on a property of the surface which he named its curvature. For a plane, the curvature is zero; for a sphere, a constant number; for a doughnut, a family of numbers. Einstein once gave a verbal description of Gauss's idea of curvature by talking about minute rods connecting infinitely near points in a surface. The curvature, then "is the quantity that expresses to what extent and in what way the laws regulating the positions of the minute rods in the immediate vicinity of the point under consideration deviate from those of the geometry of the plane." [52]

It was the German mathematician Riemann (1826–1866), a student of Gauss's, who first generalized the concept of curvature to manifolds of any arbitrary dimensions, including, therefore, the three-dimensional continuity known as a space. The curvature of a space is the quantity that expresses to what extent and in what way the laws regulating the geometry of that particular space differ from those of Euclidean space, which is taken as having zero curvature. Empirical tests for Euclidean space, then, comprise tests designed to determine whether the geometry supported by a space is Euclidean. One may seek, for instance, confirmation of the Pythagorean theorem regardless of the size of the triangle. Thus, since the four-distance s of special relativity theory is itself an extension of the Pythagorean theorem, a physical space for which equation (1) (of section 2, this chapter) is valid, is flat, or Euclidean. Since the Lorentz transformations of special relativity derive from the expression for the elementary four-distance (completed with the principles of relativity, constancy of light velocity, etc.) and since they were found valid for the electromagnetic world, we can say that the electromagnetic universe has zero curvature, or, that it is geometrically "flat." The expression, as given by equation (1) holds anywhere in that space.

The Pythagorean expression for s can assume, however, a more generalized form such as

$$ds^2 = Ac^2 dt^2 - (B dx^2 + C dy^2 + D dz^2) \tag{3}$$

A, B, C, and D are themselves functions of x, y, z and t. Elements ds determined from equation (1) may be linearly added. In mathematical practice, differential distances obtained from (1) may be replaced by difference distances. Lines constructed from such difference distances will be straight. The lines obtained by integrating ds of equation (3), however, will, in general not be straight because of the functional interdependence of the variables. If the resulting lines are identified with geodesics of a space, the space will not be Euclidean.

We will recall the first signal quality of light; this suggests that light beams be identified with geodesics of physical space. This indeed is the case in relativity theory where light beams, when so considered, are called null geodesics. Thus, studying the geometry of light beams amounts to studying the geodesics of four-space. In ancient Egypt the harpedonaptai, or "rope stretchers," were employed to determine the metric of the globe; we now have "light beamers" seeking to determine the geometry of the universe. If the world of light beams turns out to be non-Euclidean, we should be able to describe that world by the measure of its non-Euclidean character, that is, by the quantity of its curvature. It is this determination which is intrinsic in the general relativistic description of motion, in general, and motion of light beams specifically.

Theoretical studies of curved, multidimensional spaces were undertaken by mathematicians many decades before Einstein came to identify the geometry of the physical world as a curved space. The Hungarian, János Bolyai (1802–60), and the Russian, Nikolai Lobaschewski (1793–1856), independently developed ideas of a non-Euclidean space described as hyperbolic. It involves a conception of parallelism wherein parallel lines approach each other asymptotically at one end and diverge infinitely at the other. Through one point external to a line an infinite number of lines may be drawn (hyperbolically) parallel to the original, though they are delimited by two boundary lines which separate these so-called "nonintersecting" lines from all the "intersecting lines." The space-time structure of the universe was subsequently found to conform closely to this type of non-Euclidean geometry.

As the idea of geodesics replaced straight lines and as curvatures came to specify non-Euclidean spaces, our ideas of how the elements of geometry behave underwent a radical change. For instance, Riemann challenged the validity of Euclid's definition of a straight line as extensible in either direction indefinitely. He admitted that our experience does contain examples of endlessness or "unboundedness" but not of infinite length. Arclike inhabitants of a circle may travel endlessly around and around and judge their circle as unbounded. But to those who are able to leave the one-dimensional confines of the line, the circle is finite in its area, and the area is measurable by its radius. If the arclike creatures were intelligent, they might even find a way to use inductive reasoning and learn about the finiteness of their world. The surface of a sphere for two-dimensional "spherelanders" living in it is also unbounded; yet for "spacelanders" it is clearly finite. Again, if we are willing to continue this science fiction, the crawling spherelanders could not learn of the finiteness of their sphere through kinesthetic exploration, though they might discover it by studying spherical trigonometry. They would then conclude that the number of degrees in a triangle depended on a quantity not itself part of their world, namely, the curvature of the sphere. Einstein once speculated about such imaginary creatures and observed that what the creatures would consider their straight lines (shortest distance between points), a three-dimensional creature would consider curved. "The great charm resulting from this [and similar] consideration[s] lies in the recognition of the fact that the universe of these beings is finite and yet has no limit." [53] The important point is that the existence of a dimension external to their world cannot be directly experienced by such creatures but they must infer it from the geometry of their world. Yet, for a three-dimensional observer the coexistence of unboundedness (in two dimensions) and finiteness (in three dimensions) can be a matter of sense experience, and need not be an inductively obtained quality.

If we permit ourselves, as did Riemann, to speculate about higher-order spaces, we may continue from one-dimensional arc-creatures, to two-dimensional sphere creatures, to three-dimensional space-creatures and leave the possibility open that the world of the latter may also be unbounded and finite. By exact analogy to the one- and two-dimensional worlds, a spatial creature might also be able to proceed endlessly in his space, hence judge it unbounded, and might also be able to infer from his geometry yet another dimension of his world not directly available to him through kinesthetic exploration of his space. But, to someone more advanced than the spatial creature, the coexistence of unbounded space with a finite measure of that space which, however, is itself not part of space, might be directly evident.

Matter and Inertia

We noted that descriptions of motion in terms of velocities are invariant with respect to all frames in relative rest; also, that special relativistic descriptions of motion are invariant for all frames at relative rest or in translation. In 1916 Einstein extended his postulate of relativity to accelerating frames by demanding that "the laws of physics must be of such nature that they apply to systems of reference in any kind of motion." [54] His general relativity theory, which derives from this demand, provides descriptions of motion which are equally valid for frames at rest, in relative translation, or in acceleration. Then through the principle of equivalence of inertial and gravitational masses, the equations of motions valid for accelerating frames become equations of universal gravitation. Thus, wrote Einstein, "pursuing the general theory of relativity we shall be led to a theory of gravitation." [55]

Through the general theory of relativity it became possible to attend to phenomena involving acceleration and to that property of matter which resists acceleration: inertia. The mathematical formalism was an extension of the identification of world-lines with geodesics. But the geodesics were now not straight world lines appropriate to the flat space of the electromagnetic universe but curved lines obtained by integrating equation (3): $ds^2 = Ac^2dt^2 - (Bdx^2 + Cdy^2 + Ddz^2)$. It will be recalled that the A, B, C, and D are themselves functions of the position of ds and embody instructions on how each line element depends on that position (in ordinary space and time). Einstein constructed a set of field equations and required that the instructions incorporated in A, B, C, and D satisfy those field equations. The new expression for ds, hence its integrated form of continuous world-lines, remained, of course, geometrical objects. They specified lines characteristic of non-Euclidean geometries. In Einstein's field equations the differential geometry of the four-space (a geometrical object), is equated with the energy-momentum tensor of matter and radiation (a physical object). Free motion was restricted to motion along the geodesics of a four-space whose curvature became a function of the distribution of matter and radiation in the vicinity of the geodesic. Consequently, one could say either that the description of matter and energy and motion was made to incorporate geometry, or that the geometrical description of the world was made to incorporate the matter, energy, and motion. Although the two statements are equivalent, the second phrasing is the popularly preferred one because of the flight from time that characterizes physics.

The history of the meaning of geometrical forms shows an increasing sophistication in assigning to physical objects properties first only familiar and contemplated for mathematical objects.[56] For the synthetic geometry of the Greeks, the objects of concern were points, curves, and surfaces— reflected in the designs of their objects of art. The Renaissance mathematization of science came to extend the precision of geometry to that of the physical world, such as to the lawfulness of curves of missile trajectories. The work of Descartes and Fermat gave birth to our knowledge of how to transform formal functions into lines, a transformation known in its modern version as the plotting of equations. W. K. Clifford, an English mathematician, suspected a century ago that matter and motion might be expressible in terms of extension [57] and indeed, in general relativity theory, space and time are handled as the perceptual equivalents of geometrical objects. Thus, the formalism easily leads to geometrical terminology when describing physical phenomena conjugate with geometrical notions.

For instance, the geometry of space-time around a massive body has a curvature determined by the mass of that body. The coefficients of coordinate elements in equation (3) are so weighted that the geodesic of a small body, given certain initial conditions of velocity, is a Kepler orbit with very slight modifications.[58] Given different initial conditions the geodesic may be any conic section, corresponding to classical orbits of freely moving bodies again with slight modifications. Light itself moves along zero geodesics implying that proper time vanishes along its path and retaining for the motion of light a quality of absolute uniqueness. The local curvature of space-time (its degree of deviation from Euclidean space!) is proportional to the local energy content, whether in the form of ponderable mass, gravitational potential or electrical energy. It is only one step from here to the identification not only of gravitational and inertial forces and energy, but mass itself with geometry. As Eddington put it, "When we perceive that a region contains matter, we are recognizing the intrinsic curvature of the world." [59] Of course, we see in things what we know of them; a physicist hit by a hard stick might recognize the strength of intermolecular forces; others may see only a furious mule driver. In the four-space of relativity theory the energy and the momentum of a particle are represented by a single four-vector whose components are formed the same way one forms the components of a world-line. The timelike component of the energy-momentum corresponds to the energy of the particle; the three spacelike components are those of its ordinary moments of inertia. If we permit ourselves to identify the curvature of space-time with matter, we can also say that the same curvature expresses itself as the momentum and inertia of matter. Interestingly, however, the properties of the electromagnetic world do not seem to derive from those properties of space-time which can be measured by Riemann curvature. This is another way of saying that the geometry of the electromagnetic world is Euclidean.[60] Although charges and masses coexist, charges populate a flat, masses a curved space-time (at least to low orders of approximation); the magnitude of charges does not depend on motion while the magnitude of masses does. Curved four-space is that of matter and inertia; flat four-space is without inertia.

There is a cosmological asymmetry which gives rise to the different geometries. Consider Mach's Principle (so named by Einstein), which interprets the inertial properties of local matter in terms of its interaction with all other masses of the universe. This amounts to an assertion that the local inertial frame is determined by some average of the motion of the distant astronomical objects. If inertia is assumed to be due to long-range gravitational-type interaction, it then can be shown that inertia depends almost entirely on the most distant masses of the universe.[61] For the argument to hold, it must further be assumed that the universe consists almost exclusively of matter with conventional mass and not matter with negative mass. (Since the physical properties of negative masses are fraught with ambiguity, one can make a good case for this cause). Thus, whereas the distant reaches of the universe are gravitationally "charged," they are likely to be electrically neutral (partly because this is consistent with local experience, partly because it would be difficult to give a reason for the opposite condition). Assuming now that long-range electrostatic and gravitational interactions follow formally identical laws, it is then not surprising that the world of electric charge is without inertia. What we see here is an interpenetration of two primitive universes: the atemporal electromagnetic world without inertia being on the cosmic scale electrically neutral, and the proto- and the eotemporal world of ponderable mass, with inertia, being on the cosmic scale gravitationally "charged."

The behavior of electromagnetic fields is described by solutions to Maxwell's famous differential equations. These are two solutions, mathematically, equally valid: one describes the propagation of electromagnetic waves from a source in the present toward sinks in the future; the other one the propagation of the waves from a source in the present toward sinks in the past. Since the atemporal electromagnetic world has no preference, the choice between the two solutions as regards to which one corresponds to reality must be made on the basis of information that originates in higher Umwelts.[60a] We perceive here, at the roots of time in the physical world, a suggestion that temporality is structured: above the atemporal level (neglecting for the moment the prototemporal world) we find a world appropriate for the inertial manifestations of matter. This latter is the kinematic Umwelt of pure succession.

There is another way to approach the problem of inertia and time. We note that motion along a short distance r in three-space may imply two different questions: possible times of transfer between the terminals of r, say P_1 and P_2, and possible times of transfer between events E_1 and E_2 occurring at P_1 and P_2. The time of transfer along r has a minimal value as seen from any coordinate attached to ponderable mass: namely, the time taken by light to travel that distance. It also has a zero value, that for the traveling photon. Transfer between E_1 and E_2, however, is more restrictive because reaching point P_2 before E_2 happens would not help. Curiously, this separation has a well-defined maximum value in time, namely, that measured by a clock for which P_1 and P_2 occupy identical spatial positions. The relativistic specifications that free bodies move along geodesics is equivalent to the requirement that clocks carried by such bodies indicate the maximum possible temporal separation between two events. The separation of the two point-events E_1 and E_2 also has a minimum value, namely, zero. But for a photon to realize this minimal time it would have to go on a distant journey and get back just at the right instant.

Now we have two limiting but asymmetrical conditions. For a clock in a universe of ponderable mass attached to a chunk of matter, there can exist a minimum spatial separation between events E_1 and E_2 (zero distance) coupled with a maximum temporal separation. For the same clock (with an intelligent observer attached) there also exists a maximum separation between the two events (the distance traveled by the light beam connecting the events) as well as a zero temporal separation (inferred on behalf of the photon, from relativity theory). From the photon's point of view, the two events are simultaneous and happen at the same place. (Thus, an atemporal world is also something of an a-spatial world). Whereas the maximum distance and zero time of the light beam (inferred by the macroscopic observer with his clocks and theories) demands no more than the atemporal Umwelt of the electromagnetic world, the maximum time and zero distance separation requires the coexistence of the atemporal electromagnetic world with the kinematical, eotemporal world of ponderable matter.

Let us assume now that we are ignorant of the concept of time but know relativity theory, have clocks, and will travel. We would certainly note that the number of ticks between events E_1 and E_2 may have a maximum value and that this maximum is a privileged condition somehow tied to ponderable mass in free motion. We would also know that fewer ticks may obtain between the two events if we do certain things to overcome what we recognize as inertia of matter. Under these time-ignorant conditions the possibility of zero ticks between two events might not appear to us as interesting as the existence of a well-defined quantity, a maximum

possible number of ticks connecting the two events under some well-de-
finable conditions. We might postulate then a new maximal principle of
physics (not a principle of maximum time, however, for that would be
circular). That capacity of matter that makes any clock tied to it signal a
maximum number of ticks between two events would deserve to be called
by a separate name, for it would appear as an important feature of nature;
we might call it "time." Inertia, then, would not be associated with a
resistance to changes in states of motion but with a universal trend of
change in nature directed away from the atemporality of the absolute
motion of light.

The Largest Set of Objects

When modern physical theories began to formulate answers to such
questions as "What is the largest set of physically significant objects?"
and "What are the limits of applicability of physical laws?" views about
temporality became functions of cosmological models. A cosmological
model is a body of interpreted equations judged to predict correctly the
behavior of matter at large. The beginnings of scientific cosmology,
dedicated to the formulation of models, arose from the long history of
narrative cosmologies. First there was a successive shift of the center of
the world from such an object as a local mountain to the whole earth, then
from the earth to the sun; then from the importance of objects to the importance of universal laws.

The application of the Newtonian law of gravity to the largest possible
dynamical system assumed, though it did not formulate, what later became
known as the perfect cosmological principle. This is a postulate of simplicity which states that, on the average, the universe appears to be the same
in all aspects, for all observers, everywhere and at all times. Employing
his universal law of gravitation, Newton speculated about matter evenly
distributed in an infinite space which

> could never convene into one mass; but some of it would convene into one
> mass and some into another so as to make an infinite number of great masses
> scattered at great distances from one another throughout all the infinite
> space.[62]

Newton's speculations were followed by a line of scientific cosmogonies
from Thomas Wright, Immanuel Kant, and Pierre Simon de Laplace to our
own days.

It was, I believe, the stunning change in scale that forced the thoughts
of scientific cosmologists toward increasingly abstract models. Three generations of Cassinis, astronomers of Paris (1667–1784), provided working
observers with a meticulous survey of the heavens. Sir William Herschel
(1738–1822) mapped some of the stellar regions beyond the solar system,
showed that the sun has a motion of its own, and pioneered statistical

astronomy by determining the distances of some 3,400 stars. He also made reliable estimates of the size of our Milky Way and both he and his son Sir John speculated about island universes. Our present astronomical image of the universe is that of over 10^9 galaxies, with distances up to 10^{10} light years. Modern cosmology has to accommodate the local astronomical world as well as this largest set of objects.

The substratum of the Newtonian universe was an ideal, static fluid,[63] whose particles were the island universes or galaxies. The main feature of Newtonian cosmology, first developed by the mathematician Carl Neuman, a contemporary of Mach, is that by considering the universe as an object, its whole behavior could be characterized by a single function between distance and time, $R(t)$, satisfying the differential equations which describe the total gravitational force acting on a particle.[64] Distance and time were the only variables of this function; which thus introduced into scientific cosmology the concepts of cosmological distance and that of cosmological-common time. But Newtonian cosmology was not able to answer satisfactorily a number of experimental facts which emerged during the nineteenth and early twentieth centuries. Foremost among them were (1) the darkness of the night sky, (2) the cosmological red shift, and (3) the absolute motion of electromagnetic radiation.[65] Eventual explanation of these puzzles, as well as considerable new insight into related matters, was derived from scientific cosmology when in 1917 Einstein showed how his theory of gravitation may be employed for the description of a universe which is spatially finite yet unbounded, and in which Mach's principle strictly holds.[66]

In relativistic cosmology the particles of the substratum do not make up an ideal liquid; they are described, instead, as bundles of geodesics in space-time. They are bundles because within the confines of each island universe the general distribution of matter determines an average direction of geodesics in four-space, including that of the null-geodesics of light. Therefore the motion of light as well as the inertial motion of matter within the confines of galaxies follows a pencil of preferred direction, named by Gödel the "compass of inertia." [67] Mach's principle can now be restated for space-time: for an observer sharing the average motion of matter of a galaxy, the local light compass and the local inertial compass coincide. This is a generalized equivalent to such Newtonian-like statements that the light rays arriving from very distant objects determine a framework which, in Newtonian physics, is absolutely nonrotating. Although the precise meaning of Mach's principle in general relativity theory does sometimes appear to be elusive,[68] the cosmological character of inertia cannot be seriously doubted.[69] Two conclusions immediately follow. One is that any timekeeper which employs inertial manifestations of matter (and we have argued that all of them do either explicitly or implicitly) is a cosmological device. The second one pertains to our reasoning that the coexistence of the electromagnetic and the inertial worlds are necessary for a definition of the lowest levels of temporality. This issue may now be seen to possess a cosmic reference.

This cosmic world may be said to be represented by the field equations of world gravitation that were put forth by Einstein subsequent to his earlier work on the general theory of relativity. Whereas the cosmological equations describe the universe at large, the general relativistic field equations may be seen in retrospect as pertaining to local gravitational fields superimposed on those of the world model. Further superimposed on both are the local anomalies of the special relativistic Lorentz transformations. Now, if one wishes to find a world model in scientific cosmology, this is the way to proceed: Find a solution to Einstein's differential equations of world gravitation, identify its variables with physical parameters, then determine whether the solutions fit available astronomical data or else can reliably predict such data. Physical variables which must be accounted for include the observationally determined average density of the visible universe, the universal constant of gravitation, and the distance-dependence of red shift in the nebular universe. This last item is of special interest from the point of view of time.

A linear relationship between distance and the red shift was discovered in the third decade of this century; this relation is now known as Hubble's law.[70] As an increasing number of extragalactic objects were observed and their distances and red shifts measured, Hubble's law became further confirmed. The simplest and therefore the favored interpretation identifies the changes in frequency with Doppler shifts caused by the radial motion of these objects away from us. The further the object, the faster it moves away from us. Assuming that these radial velocities have remained essentially constant through the past history of the universe, the Doppler shift interpretation points to an extraordinary state of affairs. At the presently accepted rate of separation,[71] all the nebulae were in one single cluster about eleven billion years ago and since then they have been flying away in all directions for all observers associated with the particles of the cosmos.

One embarassing feature of the cosmological field equations is that they permit many mathematically valid solutions (which do accommodate the empirical parameters of the universe) but their predictions about the long-range temporal behavior of the universe are contradictory, and there is no known intrinsic criteria that would help us to select from among them. In this regard Herman Bondi distinguishes five classes of acceptable solutions.[72] They are models which (1) expand monotonically from $t=$ finite and $R=0$; (2) expand from $R=$ finite and $t=-\infty$; (3) start at $t=$ finite and $R=0$ and expand with decreasing speed; (4) oscillate between $R=0$ and $R=$ finite and (5) one model which contracts from $R=\infty$ to $R=$ finite and again expands to infinity. One model which differs from all of these is that of the steady state theory which assumes the continual creation of matter ex nihilo: as matter appears from nowhere, it pushes the already existing galaxies into the infinite reaches of space. The most revealing way in which these solutions may be classed is by their attitudes to temporality and it is some of these attitudes which we shall now examine. Though some of them have lost their scientific significance, all of them remain interesting because of the world-views which they often only thinly hide.

Bondi saw the advantages of the steady state theory and its associated idea, the continual creation of matter, in that "the problem of creation is brought within the scope of physical inquiry and is examined in detail instead of, as in other theories, being handed over to metaphysics." [73] "Although this process is not directly observable it is of great interest to discuss the physics of the creation mechanism." [74] But such a discussion, that is, a description of how matter may come about from nothing, does not seem to exist anywhere in the literature of scientific cosmology. The physics of creation ex nihilo is interpreted to mean the theoretical determination of such quantities as the amount of newly created matter and the type of motion of such matter that satisfy certain equations. The underlying metaphysical assumption goes unrecognized and unmentioned.[75]

The philosophical preference for continuous creation rather than the acceptance of a singularity in time (especially in the absence of any overwhelming evidence for the former) would be puzzling were it not for the witness of its distinguished Parmenidean ancestry. When observation and evidence made the continuous creation theory untenable, the emphasis was shifted to models of pulsating universes; of course, a pulsating universe, in the long run, is again a steady state one. Any model which does not demand a $t=o$ condition relieves us "of the necessity of understanding the origin of matter at a finite time in the past." [76]

To be able to account for the universe without the concept of becoming, it was found permissible to allow for its coming about as the continuation of an old one. One inquiry into the origins of radiation associated with an early state of the universe proposed a model in which "it is essential to suppose that at the time of maximum collapse the temperature of the universe would exceed $10^{10}°K$, in order that the ashes of the previous cycle would have been reprocessed back to the hydrogen required for the stars in the next cycle." [77] Since there is no conceivable information-carrying structure that can survive this cataclysm, the prior universe is a handy metaphysical assumption, though it is not so labeled. In the moving image of eternity the unceasing rotation of the heavens was replaced by the unending oscillations of an unbounded universe.

Cosmological models which start with a $t=o$ condition are known as Big Bang theories. As observational support for a Big Bang theory increased, new attention was given to possible physical meanings of a unique event of creation.[78] Rather than rejecting a beginning of time, Charles Misner concluded that an absolute zero of time must be regarded as an essential element of scientific cosmology.[79] The mathematics of the situation suggested to him that at a finite proper time in the past all matter (in whatever form) had experienced infinite density. One must work one's way back to $t=0$ imagining a series of plausible clocks, permitted by the increasing temperature of the universe. First the heavier elements, later even ammonia molecules disappear. At $10^9°K$ nuclear transitions vanish; near $10^{12}°K$ we can use only muon decay clocks. At $10^{16}°K$ thermal muons vanish and we would have to seek for some other standard of time-like proper time. Although the pre-Big Bang situation is ill understood, there seems to be a general agreement that at about $10^{18}°K$ the universe was filled with radiation and with such particles of zero rest mass that could move with the speed of light. Matter began filling the universe only after it began to expand and cool. To accommodate these early happenings Misner formulated a cosmic time scale upon which epochs are labeled by means of their relation to volume of space (at that epoch). Both he and Milne, in earlier attempts, postulated logarithmic relations between two types of variables (volume versus age; gravitational versus atomic processes) because a logarithmic relation permits some interesting mathematical transformations between certain time readings. For instance, what in the measure of volumes (or gravitational processes) corresponds to a finite time in the past, on Misner's cosmic scale (or Milne's atomic processes) corresponds to a past infinity of time.

But what we are describing here cannot be expressed through quantitative transformations of any sort, because it constitutes a creative and not a stationary process: namely, the emergence of the early levels of temporality. Certainly, the primordial universe of pure radiation is atemporal for special relativistic reasons already discussed. We may add a general relativistic reason: in a universe of infinite density a clock, even if it were thinkable, would slow to a stop through an extreme red shift.[80] In the primordial fireball, connectedness and intentionality (which make up our idea of causation) were not yet differentiated, thus causation as ordinarily understood could not have existed. Only a primordial existential tension can be given any meaning and perhaps even operationally recognized in the probabilistic behavior of the prototemporal world. As I have reasoned earlier, the atemporality of the radiation-filled universe survives in the atemporality of the electromagnetic world and in the simultaneity known as the chronon. Subsequent appearance of matter came to define a prototemporal Umwelt whose remnants we encounter in the world of indiscernible particles. The formation of discernible aggregates signaled the emergence of eotemporality represented by the reversible laws of physics.

Regarding the post-Big Bang evolution of the universe, physical cosmology accepts the notion that the world is an object which changes according to unchanging laws. This type of formal separation of the lawful from boundary conditions has been very successful in describing the physical behavior of local matter. But, if the universe is by definition the most inclusive set of objects, from where do the boundary conditions come? The program of physical cosmology could thus lead to an epistemological cul de sac were it not for the fact that in that discipline the universe is contemplated by an observer who, by being an observer, is external to it. The boundary conditions thus derive from that part of the world which is not included in the purview of physical world models: the clockwatcher and seeker of order.

3. The Living Symmetries
of Physics

GEORGE SARTON remarked in his history of science that "each language evidences not a perfect symmetry, like that of a geometric drawing, but one that is imperfect in many ways, like that of a tree or a beautiful body—a living symmetry." [81] A number of such almost-symmetries arise in modern physics in connection with its handling of time.

•   The invariant differential element of space-time (our equation (1)) may be rewritten in the form $ds^2 = dx^2 + dy^2 + dz^2 + d\tau^2$ if its last term is multiplied by $i^2 = -1$, and length and time units are selected so that the velocity of light is unity. Then we could exclaim with Minkowski that the equation is perfectly symmetrical and space and time are now equal, and that this equality will communicate itself to any law derived from the postulate of world-lines. He believed that "the essence of this postulate may be clothed mathematically in a very pregnant manner in the mystic formula $3 \times 10^5 \mathrm{km} = \sqrt{-1}$ sec." [82] It is difficult to share this enthusiasm and this distrust of generation, for the formal symmetry on paper will no more prevent anyone's demise than does a belief in afterlife. Muses has shown that in a multidimensional geometry whose coordinate axes are real, an imaginary axis "is by its very nature perpendicular to all ordinary numbers, positive or negative. Hence the diagonal of a nega-dimensional [n-dimensional] hypercube is independent of all spatial dimensions, however high." Thus, "time is ever other than space, however high the spatial dimension. Conversely, higher dimensional forms in no way imply time as one of their dimensions." [83] What to Minkowski appeared as a mathematical necessity dictated by a desire to spatialize time, implies, upon more detailed examination, an incommensurability of time and distance.

•   All too much has been written about time dilation and the attendant problem known as the clock paradox.[84] Many minds might be put at ease if they were to reflect on Whitrow's simple but profound words:

> The anomalies and discrepancies of time-ordering that arise in connection with the special theory of relativity are due not to the nature of events themselves but to the introduction of observers moving through the universe relative to the fundamental observers in their neighbourhood [85]

or, if they were to weigh the relative importances of Einstein's cosmological equations, his teachings of general relativity theory, and the epistemological significance of special relativity theory as put forth earlier. Long after the twin paradox is settled, however, there remains an asymmetry between the peculiarly relativistic properties of space and time: the Lorentz-Fitzgerald contraction and time dilation. Because of the uncritically assumed equality of space and time in relativity theory, the phenomena of motional contraction of rods (distances) and the motional dilation of time are usually taken as complementary phenomena. But the Lorentz-Fitzgerald contraction is only apparent; a traveling meter rod upon its return home will again be known in its original length. But a traveling clock upon its return home will be, and will remain, different from its stay-home twin; that is, it is and will remain younger. To say that this simply follows from the formal properties of the Lorentz transformations is true but does not settle the problem, for those transformations are said not to distinguish between distances and times.

Some light may be thrown on this asymmetry if we remember that the "physicist's t" stands for various temporalities which should not be carelessly mixed. In an atemporal universe there would, of course, be no rod contractions or time dilations—for the atemporal world has only radiation. In a prototemporal universe we must employ processes appropriate to that Umwelt, that is, processes involving countable but unorderable particles. We would have to measure the variation of their aggregates by statistical means, such as by entropy changes. But entropy, as also the speed of light, is a privileged quantity in relativity theory; the entropy of a moving system equals the entropy of the system measured by a co-moving observer. Thus a prototemporal clock cannot undergo motional time dilation. As Landsberg reasoned, the temperature of a body is a statistical concept which derives from the relative motion of the molecules and thus remains unaffected by any translational motion of the center of gravity of the aggregates.[86] Instead of an entropy clock, the traveler may carry a macroscopic inertial clock whose operation belongs to the eotemporal world of pure succession. Such a clock, if its trip deviated substantially from the local compass of inertia, should, according to special relativity theory, return home "younger" than its stay-home twin. A nonlinear clock employing particle decay should also return home indicating a shorter period of elapsed time. Although the mechanism of particle emission is probabilistic, hence prototemporal, the lawfulness of decay, as lawfulness, belongs to the world where events are countable as well as arrangeable. The same argument holds for a cesium-beam clock whose governing mechanism is the flipping of the magnetic moments of particles. Here, again, although the process is probabilistic, hence prototemporal, and even if relaxation times may be Lorentz invariant, the lawfulness of relaxation as lawfulness suggests eotemporality.

It seems, then, that those levels of temporality which share the undirected feature of space (the atemporal and the prototemporal) do not, for they cannot, display an asymmetry between traveling rods and clocks. Only for clocks that belong in higher Umwelts can we distinguish between the behavior of returning rods and returning timekeepers.

- Events in the absolute elsewhere cannot influence the here-now because of the limited speed at which causal connection can propagate; some such events, however, can influence the here-future. There are, therefore, some future events whose occurrence is completely unknowable in the here-now even in a totally deterministic world. No similar arguments hold for the here-past. This strictly physical phenomenon introduces a weak but clear asymmetry between future and past into a local physical world which may otherwise be locally prototemporal or eotemporal. In the primordial, atemporal fireball causality spanned the whole world instantaneously (more precisely, it was atemporally present); no asymmetry between past and future, not even the primitive type here suggested could have existed. This asymmetry was born simultaneously with the coming about of the indistinguishable particles and their prototemporal Umwelt. As the universe further cooled and expanded, so did the four-volume of the absolute elsewhere, and with it the store of potentially unknowable future events in the here-now. This store of events became again enlarged as particles slowed into larger aggregates of matter. The enlarged store of potentialities could then contain causes of future events whose coming about we could predict in principle, but not in practice. Thus, even in a strictly deterministic proto- and eotemporal world, the finite speed of propagation of causation and the expansion of the universe introduced first a new, then a strengthening asymmetry between future and past in the form of intractable elements of indeterminism in the here-future.

Writing by the Yalu River in 1895, the French philosopher and poet, Paul Valéry, recorded an imaginary dialogue with a sage. The scholar reflected on Western ways of doing things:

> You have neither the patience that weaves long lines nor a feeling for the irregular, nor a sense of the fittest place for a thing. . . . For you intelligence is not one thing among many. You . . . worship it as if it were an omnipotent beast. . . . A man intoxicated on it believes his own thoughts are legal decision, or facts themselves born of the crowd and time. He confuses his quick changes of heart with the imperceptible variation of real forms and enduring Beings. . . . You are in love with intelligence, until it frightens you. For your ideas are terrifying and your hearts are faint. Your acts of pity and cruelty are absurd, committed with no calm, as if they were irresistible. Finally, you fear blood more and more. Blood and time.[87]

These remarks apply uncomfortably well to the opinion often expressed in the philosophy of science, namely, that the timelessness which a superficial examination of physical laws suggests does, in fact, assign a status of unreality to time. The great but parochial appeal of such views would remain unimportant were it not for the fact that philosophical antipathies and sympathies held by the learned, at various places and in various epochs, bias the set of prevailing values, hence also the fabric of communal life. And the flight from the conflicts of asymmetries that underlie temporality inhibits the enterprise of creative imagination.

V

TIME CONTAINED: COSMOLOGIES

COSMOLOGIES deal with the world as an all encompassing unit, called the universe. But whether or not it makes sense to talk about a universe is not self-evident; and what we might mean by its beginning (or no beginning) or by its end (or no end) are puzzling questions. Scientific cosmologies are mainly concerned with the "long present," that is, with the continued existence of all there is; narrative cosmologies dramatize the beginning and the end. Scientific cosmologies are deadly silent on the history and aspirations of man; narrative cosmologies closely relate his destiny to the fate of the world. All cosmologies, however, express value judgements through the particular ways in which they interpret the strategy of existence.

1. From Umwelts to the Idea of a Universe

WE LEARNED earlier of species-specific Umwelts and found that a world-as-perceived is an epistemological as well as ontological statement. For example, a case could be made for an ant universe comprising earth, wood, water, food, enemy, other ants, and an absolute elsewhere. For a mature member of our species the limits of the world may be the confines of a straitjacket, the immense reaches of the astronomical universe, or the kingdom, the power, and the glory of God, including heaven, earth, and hell. Talking about ants in this context is not altogether fair, however, because although, as we see it, they live in the same world as we do, they do not have a self to be contrasted with an aggregate of nonselves which would, then, define their universe. But man, as an individual and as a race, does display a developing identity, hence must regard the external world as a partner, albeit a gigantic one, equal to himself. While trying to establish what we mean by such a partner, we may consider some of

141

the following views: (1) The universe is the subject of my contemplation and, except for some negligible interference, it is independent of me. This view is open to attack because of infinite regress;[1] also, although the interference may be negligible in its effects on the universe, it might be nonnegligible for my view thereof. (2) There exists an intelligible structure that binds together my self with all the nonself exemplified by my contemplation of the universe. But to whom, then, is the structure intelligible? (3) The universe is a hierarchy of my Umwelts, appropriate to my psychobiological organization. This acknowledges that my existential tension is an essential part of my cosmology and whatever that cosmology may be it is only one of the many possible ones because its features themselves are of an evolutionary nature. I shall follow this third view since it appears to be the least undesirable.

Preliminary guidance about what we seek when we search for a universe in a cosmic chaos may be found in the origins of such words as *universe, cosmos,* and *chaos.* The Greeks held that man and the world are analogous, and they took this analogy very seriously. For Plato, "this world came to be, by the gods' providence, in very truth a living creature with soul and reason." [2] The god, when he set about to make the universe, made its body from all the four primary bodies (fire, water, air, and earth) into a sphere without organs, or limbs, rotating on its axis but nevertheless, into the body of a Living Creature. It had no eyes, for there is nothing external to it to behold; no mouth for there is no surrounding air; perfectly uniform (spherical) for demiurge "judged uniformity to be immeasurably better than its opposite." [3] The organismic view of the world as a fellow being held sway for almost two millennia, until its demise through the intellects of Goethe, Hegel, Fechner, and others. Following the Pythagoreans, Plato held that the preeminent characteristic of the world is its ordered beauty. The soul of man, being rationally ordered, was then capable of becoming the world in miniature. The concept of *kosmos* was first applied to a living thing, namely, to a well-ordered society, and only later to the orderliness in the physical world.[4] Besides meaning "order," kosmos also means "good behavior, decency," also "ornament, decoration, embellishment," and in women, the art of being skilled in adornment, *kosmetikos.* The Latin translation of kosmos is *mundus,* "order, ornament, decoration" and as an adjective, "clean." In Italian and in Spanish *mondo* also means "neat" and "pure." The Latin *uni-versum* has the literal meaning of "turning into unity" suggesting the mental labor of changing the multitudes of our experiences into one consistent whole. This is an activity of creation, a making of harmony, the creation of the world as remembered by the Voice in the Whirlwind: "When the morning stars sang together and all the sons of God shouted for joy." (Job 38:7)[5]

The Greek noun *chaos* connects with the verb *chaino,* "to yawn, to gape," derived from *chasma,* which survives in the English *chasm.* In the language of Hesiod and that of the poets Ibycus, Aristophanes, Euripides, and others, chaos meant an empty yawning gap.[6] In the Book of Genesis, "the earth was without form, and void; and darkness was upon the face of the deep. And the spirit of God moved upon the face of the waters" (Gen. 1:2).[7] These metaphors remained emotionally loaded: Chaos as the primordial element of the cosmos stands for the waters of creation which are deep and dark. Perhaps the imagery engenders a prehuman memory of life on earth as having climbed out of the ocean, or perhaps it is a reminder that man enters the world from the dark but life-giving waters of the womb; or perhaps both. The separation of the self from the nonself, and with it the potentiality of a cosmos, begins when the infant leaves the darkness and sees the light. If the "light which shineth in the darkness" is not comprehended then the prerequisite is missing for a universe as known to man. In great detail and with brilliant insight Ernst Cassirer showed how, in the creation legends of nearly all peoples and religions, the process of creation merges with the dawning of the light.[8] Serious inquiry into linguistic evidence connecting creation and light goes back at least to the middle of the nineteenth century, but it is open to anyone who speaks several languages. Equating the world with light is explicit in Slavic languages where *svetu* means both "light" and "world," as does the Hungarian *világ*. In some other languages the world is more inclusive, including light in its center and darkness elsewhere. In German and Old English the visible world is the "middle yard" (Gothic *middjungards* and Old English *middan-eard*)* between the "mist-home" (*Nifl-heimr*) of darkness below (such as that of the Nibelungen) and, presumably, the shiny home of the gods above.[9]

In an inquiry into the connection between language symbol and experience, Eric Voegelin distinguishes two types of relations of man to the cosmos, corresponding to two historical strata.[10] The first one concerns entities directly involved in fate, life, and death and is expressed in the language of the mythical tale and its personnel. The second derives from the differentiated experience of existential tension, in which cosmic temporality is polarized into time and the timeless, and is expressed in the language of noetic and spiritual life. He warns against the dropping of traditional myth in favor of noetic myth, for "If we let any part of reality drop out of sight by refusing it public status, it will lead to a sort of underground life and make its reality felt in intense moods of alienation, or even in outright mental disturbance." In an extensive scholarly survey of the traditional myths, G. de Santillana and H. von Dechend concluded that the places referred to in the world's great cosmological myths are in the heavens and not on earth.[11] Since evidence about the contents of the heavens was not available to the ancients and if we take the de Santillana-von Dechend conclusions seriously we must ourselves conclude

---

* Popular in J. R. R. Tolkien's epic of the Hobbits who inhabit Middle Earth in *The Lord of the Rings*.

that the world's cosmological myths are projections of man, appropriate to particular stages of development.

The ordinary usage of the noun *cosmology*, especially in the physical sciences, tends to convey the view that the subject of its inquiry is something already "there," to be discovered by "research." That this is incorrect, and that cosmology in its deepest sense is a search for universal order of which the cosmologist is a part, may be seen when cosmologies are understood as admixtures of three distinguishable activities of very different intellectual content and emotional coloring. Cosmological teachings that concern beginnings are mostly explicative attempts to recover in a universe of conflict and complexity a world of peace and simplicity. Teachings concerning the "long present" of the cosmos originally dealt with social order and offered guidance in behavior as well as reasons for the experiences of joy and sorrow. Later they split into two subcosmologies, the religious and the scientific—or, poetic and noetic. Cosmological teachings concerning an ending tend either to justify earthly happenings or to offer an evaluation of the role of death in the life of man. We shall examine in their turn some aspects of these three constituents of cosmologies: teachings about beginnings, about the long present, and about endings. We shall keep in mind that each interpretation implies one or more Umwelts. Also, that it is his existential tension which keeps man searching for something he cannot even name, so powerfully expressed in the simplicity of a Maori creation myth: "The light, the light, the seeking, the searching, in chaos, in chaos."

2. Beginnings: From Chaos to Conflict

THE CONVICTION that the world had a beginning, that is, an epoch which was fundamentally different from the present, could not have been derived from examination of nature at large. In the lifetime of an individual, and even in the lifetimes of many generations, nature appears to be, on the average, unchanging. Cyclicities, whether those of the stars or of the seasons, only emphasize timelessness by demonstrating predictability. Nor could progress from an original chaos to a present world have been suggested by vast events of nature because these are likely to be viewed as calamities and tend to inspire thoughts of destruction rather than creation.

The Inner and the Outer Landscapes

Beginnings of individual lives of man, beast, and plant must have been known since life began. But, because of the apparent long term balance of nature, the awareness of biological beginnings is not in itself sufficient to account for search for universal beginnings. While Paleolithic art displays

much concern with the coming into being of new life, simultaneous repre-
senation of the heavenly bodies: the sun, the moon, and the stars are almost
completely absent.[12] The idea of a cosmic beginning might have arisen in
the mind of early man in an inarticulate, unnamed form as his self-
awareness emerged and he, as an individual, was born. To give this habitation a name he had to grasp for metaphors in his store of symbols. So as to
gain some insight into the significance of these symbols as they pertain to
beginnings I will take as true Thass-Thienemann's principle that "the naming of an idea never before met is similar to the naming of an inkblot never
before seen" [13] and follow his general line of reasoning. That is, I will take
that the name itself has the potentiality of disclosing some features of the
thought processes of the name-giver even if those processes are not recognized by him.

According to the Book of Genesis, "In the beginning God created the
heaven and the earth." A comparison of various Bible translations reveals
that the concept of "beginning" is expressed in different languages with
different metaphors which then can be at least partly elucidated through
their associated pictograms. Thus the Hebrew *be-reshith* and the Greek
en arche both translate "beginning." The Hebrew word also denotes "head,"
"first fruit," and "first place" in social rank; the Greek implies "to lead, rule,
govern," and in general "the first" among persons and things. There are
some reasons to believe that the now lost objects of reference in both are the
emergence of the head at birth. The Latin *in principio* reflects in its origins
the meaning "to take in hand, take hold, seize, take by force" as does the
German *An-fang*, "catching with the hands." The Hungarian translation of
Genesis begins with *Kezdetben*, "in the beginning," from the noun *kezdet*,
"beginning," which comes from *Kéz*, "hand." The pictograms that suggest
themselves and which have at least some support from etymology are those
of grasping a woman for the purpose of making her conceive new life; also
of creating something by hand, as the potter creates pottery from formless clay. An imagery of "to go in, to come" is found associated with the
French *commencement*, the Italian *incomincare*, "to begin," and with the
old English *be-cuman* (surviving in "becoming"), having a clearly defined
use in Biblical language relating to beginning. "And he went in unto Hagar,
and she conceived." Finally, words describing origins often refer to the first
leap and spring of the infant in the womb; as in the English "offspring," the
German *Ur-sprung*, "origin," literally, "primordial jump." In English the
verb *to start* has a primary meaning of "to move suddenly with a leap."

Our words for a "beginning" may thus be seen to hail from imagery
of birth, work society, and carnal knowledge. They imply the travail of our
not-so-distant ancestors who reached out for known images so that they
might describe through them some newly discovered ideas: those of the
coming into being in time. Although the concept in its modern use is a
general one, the origins of the words are specific: they pertain to the
intimately narrow regions of man's everyday life and imply that cosmogonies

(and probably cosmologies as well), whatever else they do, will also speak about a model of the world in the mind, which may best be described as an internal landscape.[14] We can duly expect, then, that some elements of this inner landscape will appear in our cosmologies. But cosmologies are also conditioned by the outer landscape surrounding the cosmologist himself, that is, by the physical features of his geographical location.

In Egyptian cosmogony the primordial state was a motionless, watery chaos. The primordial situation was imagined as something regular: the annual inundation of these featureless waters, gradually uncovering dry land upon which the divine creator of all other gods could appear and begin to work. The actual process of creation was the work of a second line of gods. The details of their labor accounted for cosmic phenomena, for the state, and for man. Egyptian cosmogony was described by S.G.F. Brandon as unique in the history of human thought as "one of the two earliest attempts by man to abstract himself from immersion in present experience and to conceive of the world as having had a beginning, and to make a sustained intellectual effort to account for it."[15] The other attempt was that of Egypt's sister civilization, ancient Mesopotamia. Their creation epic, Enuma Elish ("When Above"), probably dates from the First Babylonian Dynasty (1894–1595 B.C.). It identifies the beginning of the universe with the birth of the gods.

> When above the heaven had not yet been named
> And below the earth had not yet been called a name,
> When Apsu primeval, their begetter,
> Mummu, and Tiamat, she who gave birth to them all,
> Still mingled their waters together,
> And no pasture land had been formed and not even a reed
>   marsh was to be seen;
> When none of the other gods had been brought into being,
> When they had not yet been called by their names and their
>   destinies had not yet been fixed,
> At that time were the gods created within them.[16]

The conviction that the world had a beginning, that is, an epoch which was fundamentally different from the present, could not have come from examination of nature at large. Cosmogonies seem to have derived, instead, from the fusion of the inner landscape of the mind with the outer landscape of geographical regions.

In Chinese cosmogony P'an Ku is the legendary great architect of the universe. He is said to have come into life endowed with perfect knowledge and created the universe by chiseling the stars and planets from the cliffside of Chaos. The beginning of time is counted from the beginning of this undertaking which took 18,000 years. After P'an Ku had finished his work he vanished and with his disappearance suffering appeared on earth. Illustration from C. A. S. Williams, Outlines of Chinese Symbolism *(Peiping, 1931). Courtesy, Silvio A. Bedini, Smithsonian Institution, Washington, D.C.*

"Them" refers to Apsu, the personification of fresh water and to Mummu-Tiamat, the personification of sea-water. The epic describes the conflict between an inactive chaos and the active gods. As the struggle progresses the emphasis of description shifts from specifics to abstraction, from the astronomical universe to the powers which are felt as operating them. Final victory is achieved when the cosmic order of Mesopotamia is established, then the affairs of the universe become unimportant because the state, which is a living thing, has come about.

Unlike the cosmogonies of Egypt and Mesopotamia, the cosmogony of Israel became an important factor in an ethnic religion. The details of the Judeo-Christian Genesis reflect the peculiar Hebrew interest in Heilsgeschichte and in the totality of time. Man enters with responsibility, which derives from his free choice between obedience and disobedience; the future of a linear time is open for individuals, though for the long term it is under the controlled plan of Yahweh.

The Enuma Elish is only one example of the world parent myth (with Apsu the father, Mummu-Tiamat the mother); the cosmogonies of some Polynesian tribes and of the Zuñi people in New Mexico belong in the same category. In contrast to these, the *Theogony* of Hesiod and the Book of Genesis picture creation ex nihilo as do the Rig Veda [17] and some Mayan myths. Certain Finnish, Japanese, and Tahitian myths see the world as having emerged from chaos, with chaos conceived as confusion and darkness. Other myths found along a wide belt from Northern South America to Sweden see the world emerging from a cosmic egg. Certain Hindu, some Altaic and other Asian myths are those of earth divers; in these a divine being, usually an animal, dives into the water and brings up particles of earth from which the entire universe grows. Perhaps the most widely spread cosmogony invokes the eternal woman or Great Mother who then gives birth to the known universe.[18] Whatever the imagery of these wide-ranging reports, they seem to fuse the inner landscape of man with the outer landscape of geographical regions.

Shift to the Outer Landscape

The beginning of the universe as depicted in Ionian cosmogonies may be reasonably thought of in terms of stages.[19] First, there was a primal unity or indistinction; out of this emerged, by separation, parts of opposite things leading to the formation of the world order and finally the opposites interacted producing thereby the world of individual, living things. Within this general scheme, patterns of certain imageries may be identified.[20] The world would begin with a yawning chaos between the sundered opposites of heaven and earth, thereby permitting the appearance of light between them; this was the birth of light out of darkness. Then came Eros, personifying the mutual attraction between heaven and earth and bringing about their reunification. But no sooner than the gods completed the cosmos, their supernatural personalities would change into human personalities with all the violence, pain, and joy that characterizes the flesh, foreshadowing the change of cosmogonic attitudes from myth to the rationalism of Greek philosophy. Judging from the writings of Herodotus (c. 430 B.C.) educated Greeks of his epoch, if interested in cosmogony (theogony), learned from Homer and Hesiod, for these two "were the first to compose theogonies, and give the gods their epithets." [21] In Plato's cosmogony in the *Timaeus*, God, the gods, and the Demiurge are often invoked, but they stand only as agents of eternal ideas of which the sensible aspects of the world are projections. The earlier religious factor diminished as a secular rationalism emerged, in which the designs of the gods were replaced by ideas in the form of laws. But laws can speak only of predictable events and not about happenings which must remain unique and unpredictable, such as that of a cosmic creation. This epistemological difficulty survived to our own days.

Through the history of Christendom the Genesis account of creation remained the commanding opinion, even though modulated by surviving pre-Christian myths varying among ethnic groups and among geographical areas. It seems that the Platonic Living Creature did not survive explicitly in cosmogony but did so in the organismic view of the world which came to be challenged only with the emergence of the new mechanical philosophy and natural theology of the seventeenth century.[22] When it was recognized that the understanding of an inanimate world is hampered by regarding it as having a soul and a body, the Greek talent for geometrization was there for the asking. The new metaphor was the clockwork universe. If there is any single feature that characterizes the Cartesian faith in mechanistic explanation, in Laplacian determinism, and in the power of the clock metaphor, that feature is accuracy. This spirit is reflected in the work of James Ussher, Irish archbishop, contemporary of Descartes, whose biblical chronology placed the creation of the world on Sunday, October 23rd, 4004 B.C. [23] Smiles are inappropriate reactions to this claim. Bishop Ussher did not put forth this date as a proof of his faith in a creator—this was not necessary—but as the opinion of a learned citizen of the Age of Reason, as did Newton, who sided with Bishop Ussher in the validity of the latter's chronology. The underlying view of a finity of time was defended by Newton's famous spokesman and exponent, Dr. Samuel Clark, who held that although it was no impossibility for God to create the world sooner or later than He did, yet "the Wisdom of God may have very good reasons for creating This World at That Particular Time He did." [24]

A delicate retrospect must be invoked if we are to understand that mixture of superstition, religion, myth, and science which characterized the age of Newton. Of all epochs, the age of mechanical philosophy and natural theology could boast of many men who sought to interpret the coming about of the natural world through whatever expressions they might command: mathematics, the letters, or the arts. The point was that they wanted to satisfy the demands of the spirit as well as the demands of science. Knowingly or unknowingly, they demonstrated the opinion of St. Thomas, formulated some four centuries earlier: that the world began is an article of faith.[25] But, the world whose beginnings they sought was not that of the local rivers, or even that of a holy city, but that of the astronomical universe, governed by nonorganic laws. The new process of decision itself whereby rational conclusions were sought regarding the possibility of a beginning or no beginning of the universe, came under criticism in Kant's *Critique of Pure Reason* (1781). He believed he had developed indisputable arguments for rejecting both that the world had, and that it had not a beginning in time. Since he regarded these two statements as a true disjunction, he could not permit their mutual truth to coexist. He concluded, instead, that the idea of time does not (for it cannot) apply to the universe, but it is prescribed to it by the mind. Kant's reasoning had a profound and lasting effect on philosophical ideas of a temporal beginning of the universe both because of its conclusion and, perhaps more importantly, because it implied that the problem is one of logic. The subsequent preoccupation of thinkers with the logical (and associated mathematical) solutions to the Kantian antinomies diverted attention from the empirical issues involved; to wit, the hierarchy of temporalities hidden in the argument. To these we shall shortly return. Although Kant was emphatic about the importance of the empirical world, his specific reasoning concerning a beginning of time was almost entirely an appeal to the inner landscape.

A diametrically opposite stance to cosmology as a derivative of the inner landscape is that of scientific cosmology. The cosmology of relativity theory is a distant offspring of the Egyptian Rhind Papyrus (circa 1700 B.C.) which states its own subject as the "rules of enquiring into nature, and for knowing all that exists, [every] mystery . . . every secret."[26] Although mathematical physics is more modest than to claim knowledge of "all that exists," the unstated assumption of physical cosmology is that in some yet unknown ways it nevertheless does so. Yet, to the question of a beginning of time the relativistic claim that the universe was created 5×10^{17} seconds ago is no more help than Bishop Ussher's 5,978 years. We have already discussed some of the schemes designed by physicists to eliminate the necessity of accounting for a beginning of time or at least accommodating it within the power of scientific cosmologies. The great confidence in scientific cosmologies derives from the awesome power of mathematics that makes complex things appear simple and suggests, thereby, a correspondence between mathematics and the very fundaments of reality. But this connection is through the cosmologist, and about him relativity theory says nothing. Thus, the problem is opposite to that of the Kantian antinomies: the idea of creation appears as a feature of the external landscape, with little or no reference to man.

Cronus versus Faust

Gerardus van der Leeuw, writing on "Primordial Time and Final Time," remarked that the riddle of time is the riddle of the beginning.[27] We have briefly considered this riddle from the points of view of language, Kantian thought, religion, physics, and the manifest content of myths. I wish to consider it now from the viewpoint of philosophical anthropology. Specifically, I want to admit as valid the thesis of Mircea Eliade that "if one goes to the trouble of penetrating the authentic meaning of archaic myth or symbol, one cannot but observe that this meaning shows a recognition of certain situations in the cosmos and that, consequently, it implies a metaphysical position." [28] I think that archaic metaphysics helps to shed some light on the riddle of beginning.

In his study of the myth of the eternal return, Mircea Eliade stresses that among the almost endless variety of practices in primitive societies "there is everywhere a conception of the end and the beginning of a temporal period, based on the observation of biocosmic rhythms; and forming part of a larger system—the system of periodic purifications (cf. purges, fasting, confession of sins, etc.) and of the periodic regeneration of life." [29] The practices are motivated by the desire to abolish the reality of temporal passage as expressed by the emergence of novelty. It is as though time itself were continuously regenerated by recurring rituals which celebrate archetypal stories of creation and death. Such celebrations as that of the New Year, or birthday, or the cyclic events of the civil or ecclesiastical calendar are familiar to modern man. There are also events which first appear as though they were unique, but they are changed into repetition by rituals. Baptism, for example, is "the ritual of death of the old man followed by a new birth. On the cosmic level, it is equivalent to the deluge: abolition of contours, fusion of all forms, return to the formless." [30] We should recall here the role of water in narrative cosmogonies as well as the rebirth of a new world out of the ashes of the old, in modern cosmogonies: "It is essential to suppose that the ashes of the previous cycle would have been reprocessed back to the hydrogen required for the stars in the next cycle." [31]

In Eliade's conclusion, the regeneration rites, both individually and communally, by their continuous repetition of what is archetypal, signify a refusal of man

> to accept himself as a historical being, his refusal to grant value to memory and hence to unusual events (i.e., events without an archetypal model) that in fact constitute concrete duration. In the last analysis, what we discover in all these rites and all these attitudes is the will to devaluate time.[32] This eternal return reveals an ontology uncontaminated by time and becoming.[33]

I would like to give an alternate interpretation to the facts superbly demonstrated by Eliade. I would not expect archaic man to have possessed historical consciousness which needed denying, for time ignorance as a mode of comprehension is older than the knowledge of time, both ontogenetically and phylogenetically; periodicity is instinctually known also to animals. The discovery that there exist unique events not representable through archetypes was probably the product of the dawning idea of creative processes in nature. Eliade's examples illustrate the birth of individuality in archaic man as demonstrated in his suspicion and fear that periodicities cannot be made to account for all of experience. The conflict between the beinglike cyclicity favored by archaic man (and by the archaic man surviving in each of us) on the one hand, and, on the other hand, the becominglike, unpredictable aspects of reality is illustrated by the contents and relative ages of the sagas of Cronus and Faust.

The personal history of Cronus, god of harvest, is a confused one, and only cumulative misunderstandings made him the god of time during the early Renaissance. But the association did come about and did remain, and his fear of novelty made him, as it were, the patron saint of all conservatives.

> Cronus married his sister Rhea, to whom the oak is sacred. But it was prophesied by Mother Earth, and his dying father Uranus, that one of his own sons would dethrone him. Every year, therefore, he swallowed the children which Rhea bore him: first Hestia, then Demeter and Hera, then Hades, then Poseidon.[34]

We may contrast this with the Faustian-Spenglerian praise of novelty, a prerequisite for progress, in Herman Hesse's *Magister Ludi*,

> At life's each call the heart must be prepared
> to take its leave and to commence afresh
> courageously and with no hurt or grief
> submit itself to other, newer ties.
> A magic dwells in each beginning and
> protecting us it tells us how to live.[35]

The essence of any ritual repetition is predictability without surprise, a state of being, appropriately symbolized by a straight line, a circle, or a sine wave. There is no suitable geometrical image that could symbolize unique events, because becoming precludes lawlike representation; coming into being is the "magic [that] dwells in each beginning." Cosmogonic myths of the eternal return act out the repetitive character of the conservative Cronus; this status quo is challenged by the emergence of unpredictable, unique events celebrated as well as bemoaned by the progressive Faust, and sought in the more sophisticated cosmogonies. I have postulated earlier that the tension that man feels because of the unresolvable conflicts between being and becoming is at the source of his sense of time. Symbolically, then, that sense of time is the struggle between Cronus and Faust. But as such, these symbols can stand for a much larger family of unresolvable conflicts than those of man alone—but more about that in due course. Right now we must attend to this question: Once the conflict between Cronus and Faust does exist in man, how does he or can he deal with it?

3. The Long Present:
How to Deal with Conflicts

GRANTING that the universe exists and it is knowable, what aspects thereof should we regard as most significant? Writing during the first century B.C. Lucretius perceived that under the wheeling constellations of the sky all nature teems with life, both the sea that buoys our ships and the earth that yields our food. . . . So throughout seas and uplands, rushing torrents, verdurous meadows and leafy shelters of birds, into the breasts of one and all you [life-giving Venus] instill alluring love, so that with passionate longing they reproduce their several breeds.[36]

This is quite a lot to explain: the astronomical universe, matter that makes up the seas and the earth, the behavior of nature at large, life and its preservation, and the mind of man and its cares. It is this world of vastness and variety which is confronted by the observer and knower who, for Protagoras, "is a measure of all things, of things that are that they are, and of things that are not that they are not." We have concentrated so far on the cosmogonic content of cosmologies. But the myths and, with the Greeks, the speculations, did not stop there. They were carried sufficiently far to account for the state of the world and man, such as understood and implied, for instance, by Lucretius and Protagoras and expressed in terms meaningful for the people to whom the cosmologies were addressed. In the course of centuries these universal cosmologies became differentiated into the disciplines of physical cosmology, geology, and history. They became the carriers of man's concern with the "long present" in which operate the conflicts that arose out of chaos.

Universal Cosmologies

Plato's *Timaeus* tells us that, since the world and its maker are good, the Demiurge had to fashion the world after what is lasting, what is "always real and has no becoming [and not after the opposite] (which cannot be spoken of without blasphemy), [that] which is always becoming and is never real. . . . Having come to be, then, in this way, the world has been fashioned on the model of that which is comprehensible by rational discourse and understanding and is always in the same state. Again, these things being so, our world must necessarily be a likeness of something." [37] There are features to this cosmology which unabashedly pertain to the inner and to the external landscape of the cosmologist: it contains ethical and aesthetic judgements, it insists that the world must be comprehensible and that it be the likeness of something (else). These features, in various combinations, may be found in all great cosmologies; albeit, what is admitted as "comprehensible" and of what kind of thing the universe is the likeness, vary rather greatly as do the ethical and aesthetic judgements. We shall describe a number of cosmologies and note some of the ways they see man's conflicts emerging from the precreation conditions.

Not all universal cosmologies are as elegant as Plato's. In *Enuma Elish*, Marduk, the god of Babylon, made the sky and the earth from the two halves of the body of Tiamat, Dragon of the Chaos:

> The lord trod upon the hinder part of Tiamat
> And with his unsparing club he split her skull. . . .
> Half of her he set in place and formed the sky therewith as a roof.[38]

Later he created Homo sapiens:

> Blood will I form and cause bone to be;
> Then will I set up Lullu, "Man" shall be his name.[39]

Thoughtful of his responsibilities

> He created stations for the great gods;
> The stars their likenesses, the signs of the zodiac he set up.
> He determined the year, designed the divisions;
> For each of the twelve months he set up constellations.[40]

The last two tablets of the poem read somewhat like a liturgy in which fifty names and functions of Marduk are recited, corresponding to that many aspects of mortal concerns. But this mixing of images of violence and those of the natural landscape are only the poetic elements of the story. J. G. Gunnell studied this long poem and noted that

> the "political" character of the creation epic is obvious. The nature of the cosmos elaborated in *Enuma Elish* corresponds to the developed form of kingship existent during the periods of empire, [whereas] previous cosmologies conformed to earlier political configurations.[41]

By "political" is meant the role of society as a mediator between the creative heavens and man who searches for order on earth.

In Hesiod's *Theogony* out of the Void came Earth and Eros; also Darkness and black night, and out of Night came Light and Day, her children conceived after union in love with Darkness.[42] Earth first produced the sky, then the mountains; in due course came Law and Memory and a whole family of gods, and with them began creation and destruction produced by man, and the explanation of things of concern to the Ionians. But the *Theogony* also incorporates a great deal of rational speculation about the world under the supreme and irresistible power of Zeus, who bridges the gap between cosmic order and social disorder, of which Hesiod's *Works and Days* is a personal account. Man may choose between the Horae, the goddesses of orderliness in nature and society, * or select hybris, that is, recklessness. But, as he warns, be careful to avoid the anger of the deathless gods, by laboring against injustice and poverty.[43] The source of such moralism and admonitions was his conviction that society is the mediator between the heavenly landscape of eternal order and the conflicting drives of mortal man.[44]

I remarked earlier that the clock dial is a Platonic image of time, while digital clocks are Aristotelian images. These two displays also symbolize two cosmologies. Plato's *Timaeus* is an exposition of astronomy, physics, and biology; it describes man, his life, and his world in a combination of Pythagorean mathematics with Empedoclean biology. Though it does remain a cosmology, it is also a necessary introduction to an understanding of the social cosmos of the *Laws*. From opinions about the universe we are

---

* Dike, goddess of justice; Eirene, of peace; and Eunomia, of wise legislation.

led to instructions on ethics, education, and jurisprudence, to practical rules of the state. But time always remains the time of the cosmos and does not turn into the "time of the tale" (to use Voegelin's expression) [45] which is time on earth. Perhaps it is not unfair to say that for Plato time remained an item of beauty uncontaminated by the rise and fall of political order. Aristotle contracted the cosmic myth of Plato into a rational universe which could accommodate, among other details, his profound interests in biology and in the growth of things in general. The eternal circular motion of the heavens became eternity detached from the sublunar world, in which the rule of time was before and after. Time was not only a creator but also a destroyer, for in time all things pass away. The science of politics was not primarily an attempt to copy eternity, but rather a technique of security against the tyranny of passing.

The Mosaic Genesis reveals all things which were understood as essential to nature, and positions all these things along the steps of a ladder of values, reflecting the judgement of God. We have light and heaven; earth and grass; days, nights, seasons, and years; the sun and the moon; sufficient variety of living things to include seemingly all creatures; and finally, the male and female of our species equipped with free will to disobey God's unchanging law and, consequently, be ready for the Fall. From the story follow the essential parts of the Christian world-view concerning man's position in the world vis-a-vis the skies and the divinity, including the rationale of Christian theology, and all the instructions for behavior which generations of men could derive from it. The potential powers of this universal cosmology unfolded as the West built its civilization by combining the teachings implicit in the Genesis myth of creation with the method of philosophizing invented by the Greeks.

So much for some cosmologies which remained truly universal. We should backtrack now to the early developments of astronomy because, in retrospect, they may be seen as the intellectual ancestors of scientific cosmologies.

The origins of astronomy, as do the origins of all our ancient departments of knowledge, vanish in a nebulous past. The Mesopotamians are believed to have recognized a number of prominent constellations by 3000 B.C. and transmitted their knowledge to the Greeks. Aristotle speculated in detail about the construction of the heavens. According to Archimedes' *Sand Reckoner*, Aristarchos of the Ionian island of Samos (third century B.C.) hypothesized that

> the fixed stars and the Sun remain unmoved, that the Earth revolves about the Sun in the circumference of a circle, the Sun lying in the centre of the orbit, and the sphere of the fixed stars situated about the same centre as the Sun is so great that the circle in which he [Aristarchos] supposes the Earth to revolve bears such a ratio to the distance of the fixed stars as the centre of the sphere bears to its surface.[47]

In the second century B.C. the great Alexandrian astronomer and mathematician, Claudius Ptolemaeus, summed up the astronomical knowledge of antiquity, adding to it the insight of his own observations and genius. His *Almagest* is an encyclopaedic work which deals with trigonometry, motions of the sun and the moon, problems of the calendar, motions of the planets, mapping of the fixed stars, and with the precession of the equinoxes allegedly already known to the Babylonians. His system of circular planetary orbits eccentric to the earth, and circles rolling upon these circles to generate epicycles, permitted consecutive corrections to his model of planetary motion to account for observed irregularities. Although the Ptolemaic system became prohibitively complex as observational data increased, it did explain a sufficiently large portion of astronomy to give it not only acceptability but an absolute reign over competing ideas. Furthermore, its earth-centeredness and therefore its man-centeredness was sympathetic to Christian theology. As a result, Ptolemaic views were dogmatically asserted as correct in Western Christendom until the Copernican revolution at the turn of the fifteenth century. We shall follow up that story later.

That we can detect some important differences between the narrative and prescientific elements in classical cosmologies, such as, for example, those separating the astronomy of Plato from that of Aristarchos, is a gift of hindsight. In an uncharitable evaluation of Plato's universal cosmology, Sarton berates *Timaeus*, "which many commentators have considered for thousands of years as the climax of Platonic wisdom, but which modern men of science can only regard as a monument of unwisdom and recklessness." [47] This contradicts Sarton's own view stated in a preceding paragraph, namely, that "Science, as we understand it, concerns itself with limited objects, and it owes its success and immense fertility to its deliberate and severe restraints."

Are we then, to regard man as a part of, or apart from the universe, whose study is cosmology? The splitting off from universal cosmologies of such disciplines as history, geology, anthropology, psychology, biology or physical cosmology gives witness to methodological necessities. Neither severally nor as a group do they fill the demand for a unified vision of existence of the type offered by their common ancestors. As the subject of cosmology narrowed to the physics of the astronomical universe, beginning perhaps with Kant, responsibility for explaining the "long present" was transferred from concern with existential tension to the tensionless, timeless world of number and law. This process enhanced our understanding of an increasingly limited segment of existence while it removed first man, then life itself from among its concerns. But, as cats, which, when thrown out of the door, come back through the window, some of the ancient problems of cosmology reappeared in unexpected ways.

Cosmic Time as Geometry

In scientific cosmology the idea of time spanning the long present between a beginning and an end, or else between no beginning and no end of the universe, is tied to concepts of finity and infinity and to that of space-time geometry.

The same Lucretius whose invocation to "lifegiving Venus" was quoted earlier, also held that the universe is spatially unbounded and temporally infinite; also, that it is a random interplay of atoms and the void and is not due to deliberate design. Opposing the Aristotelian reasoning that the universe must be finite in size, he held that if it were finite, it would have a limit; but a thing cannot have a boundary unless there is something outside the boundary. He wrote that

> nature is free and uncontrolled by proud masters and runs the universe by herself without the aid of gods. For who . . . can rule the sum total of the measureless? Who can hold in coercive hand the strong reins of the unfathomable? Who can spin all the firmaments alike and foment with the fires of ether all the fruitful earths? [48]

Unbeknown to Lucretius, similar questions were addressed to Job by the Voice out of the Whirlwind:

> Where was thou when I laid the foundations of the earth? declare, if thou has understanding. Who hath laid the measure thereof? Or, who shut up the sea with doors, when it brake forth, as if it is issued out of the womb? Canst thou bind the sweet influence of Pleiades, or loose the bands of Orion?

Job gave the answer of the Christian world: "I know that thou canst do everything." But, while for Lucretius and for Job the immensity of the world were sources of inspiration, to the inquisition judging Giordano Bruno in 1600 A.D. these were evil thoughts. Yet, Nicolas of Cusa did take the infinity of the world seriously and helped loosen the hold of geocentric cosmology on the mind of man in his philosophy of the coincidence of the contraries. He held that the center and the circumference of a sphere are coinciding contraries for the world at large, a tenet "unintelligible without God as its center and circumference. It is not infinite, yet it cannot be conceived as finite, since there are no limits within which it is enclosed." [49] In this explicitly non-Aristotelian fashion his thoughts, judged in retrospect, described cosmologies to come.

On the way to modern ideas the Copernican system constitutes an intermediate revolution. The Canon of Frauenburg was not known to have been a wild social innovator: he was a mathematician, a linguist, and a physician and conducted a generally retired life; his arguments in favor of a moving earth retained the dogmatic use of circular orbits and of the Ptolemaic device of epicycles. His dissatisfaction was intellectual, and it was in the self-evaluation of man that he produced profound changes. By his time, the Ptolemaic account of the motion of the planets demanded a plethora of epicycles. Compared to these, his explanation involving the threefold motion of the earth,* once understood, was surprisingly simple, hence dangerous

---

* The earth's spin about its axis, its revolution around the sun, and the virtual precession of its axis because of its fixed inclination to the ecliptic.

to the established views. Apparently, Copernicus expressed concern with adverse criticism of his theory of the earth's motion. In a letter addressed to him by Andreas Osiander, a man we would describe today as his literary agent, subsequently published by Kepler, Osiander writes that for his part, he always felt about hypotheses that they are not articles of faith but bases of calculation, so that, even if they be false, it matters not, so long as they exactly represent the phenomena of the celestial motion.[50] When Osiander took upon himself the responsibility of inserting a preface stating these views in *De Revolutionibus Orbium Coelestium* in 1543, he gave a mid-sixteenth century expression of the principle of parsimony, though this was definitely not his purpose.

The Copernican heliocentric theory began gaining currency by its publication, in 1576, by Thomas Digges under the title *A Perfit description of the Cealestial Orbes*, depicting the "orbe of starrs fixed infinitely up . . . garnished with perpetuall shininge glorious lightes innumerable . . ." suggesting a purity of faith, and great enthusiasm for a world having the sun at its center, and extending into infinity in all directions. With the Copernican system of a sun-centered and possibly infinite universe, the relative size of the earth with respect to the astronomical world and, with it, the relative importance of man in that universe began to decrease radically. This meant the end to the usefulness of classical universal cosmologies. The crisis of this revolution was acted out in the drama of Galileo, who was both an instrument and a victim of the change. In 1610 in a pamphlet called *Sidereus Nuncius*, he announced that he enlarged the universe a hundred and a thousand times from what wise men of all past ages had thought.[51] This was disturbing news for those who were dogma-bound to regard man as the center of the universe, especially if it was combined with the threat of a moving earth. This effrontery to the Roman evaluation of man (together with a kaleidoscope of other matters) was the central issue in the epic of Galileo. No one seems to have been disturbed by Galileo's mathematization of physics. But this is ironic. For the substantial later decline in the power of traditional religions is not at all the consequence of the discovery that man is of insignificant physical dimensions; on the contrary, this fact is likely to awe and even inspire all but the very dull. The declining power of the churches derives, instead, from materialistic philosophies with a common belief that the predictability which mathematized natural science implies is intrinsic in nature and makes the idea of transcendental agencies an unnecessary, hence erroneous assumption.

The path from Galileo to materialistic philosophies is a continuous but not short one. Kepler sought mathematical order and harmony in the world as a manifestation of God's creative action; Newton saw in mathematics the only sensible way to express mechanical relationships and was convinced that there was a mechanical pattern in any conceivable description of motion. He put the "whole burden of philosophy" to the task of finding mathematical formulations of mechanisms that underlie, he said, all of nature.[52]

Kant, in his *Universal Natural History and Theory of the Heavens* (1755), recorded his reaction to the contemplation of the immeasurable greatness of the universe:

> If the presentation of all this perfection moves the imagination, the understanding is seized by another kind of a rapture when, from another point of view, it considers how such magnificence and such greatness can flow from a single law, with an eternal and perfect order.[53]

Then, arguing for a "Systematic Connection" or "Universal System" as underlying the totality of the world, he proceeded "to trace out the construction of this Universal System of Nature from the mechanical laws of matter striving to form it." The result is the cosmogenic system known today as the Kant-Laplace theory of the creation of the solar system.

In the same work Kant hypothesized that the "nebulous stars" described by de Maupertuis from personal observations may be analogous to the stellar system in which we find ourselves, and "are just universes and, so to speak Milky Ways." This early suggestion of extragalactic nebulae helped direct attention away from the earth and the planets as the commanding inhabitants of the universe, to a world of stars as primary units and eventually to a universe of galaxies as its particle-members. Not unlike the Renaissance, which witnessed the end of the ability of classical universal cosmologies to accommodate a sun-centered and possibly infinite universe, so the turn of the nineteenth century witnessed the insufficiency of Newtonian principles to accommodate the laws of an immense universe.

The new metaphor that proved to be useful is a Pythagorean one, expressed with conviction in relativistic cosmologies. We have already learned about its basic features: the laws of geometry are subject to empirical tests; in a space-time continuum the Egyptian harpedonaptai or rope-stretchers are replaced by "light-beamers," and the deviation of the geometry of a space from that of Euclidean space as determined by the light-beamers may be measured by a quantity known as curvature. We may, therefore, attend directly to the cosmological significance of that quantity. H. P. Robertson in a paper on "Geometry as a Branch of Physics" put forth certain warnings against misinterpretation of the curvature of four-space. In relativistic context this curvature deals exclusively

> with properties intrinsic to the space under consideration—properties which in later physical applications can be measured within the space itself—and are not dependent upon some extrinsic construction, such as its relation to an hypothesized higher dimensional embedding space. We must accordingly seek some determination of K—which we nevertheless continue to call curvature—in terms of such inner properties.[54]

But if geometry is a branch of physics because its laws are subject to empirical tests by astronomers and physicists, then geometry must always have been empirical even though it was judged to have been a priori. If laws of geometry may be refined and corrected by findings obtained through our extended senses (such as telescopes and radars) then—and this will sound obvious to some and unacceptable to others—the original laws must have reached us through the services of our unextended senses. True, our knowledge of how postulates of geometry and mathematics derive from sense experience is very poor, but the link is nevertheless there. But if so, then geometry and mathematics are legitimate domains of inquiry for psychology, biology, sociology, and whatever other fields may contribute to their understanding. Therefore, because of the empirical nature of geometry discovered by Gauss and by the arguments we quoted from Robertson, we can have a Gauss-Robertson program. It authorizes a search for those inner properties of space-time which determine the curvature of four-space but which properties themselves are not within the purview of physical science.

Eddington has remarked that

> When we perceive that a certain region of the world is empty, that is merely the mode in which our senses recognize that it is curved no higher than the first degree. When we perceive that a region contains matter we are recognizing the intrinsic curvature of the world.[55]

This statement bridges mathematical physics and perception, hence it may be frowned upon in terms of the academic compartmentalization of knowledge. Yet it asserts only that what we see in things is what we know of them; the sun used to be a chariot but now it is a store of hydrogen bombs. Eddington's brief remark is a hint of the Gauss-Robertson program. Earlier, following Einstein's idea of the "spherelanders" we speculated about multidimensional creatures who could determine from measurements performed entirely within their spaces the degree to which their geometry differs from Euclidean geometry. That there is a genuine mathematical distinction among spaces of different dimensions has been known since L. E. J. Brouwer first showed that it is impossible to establish one-to-one correspondences even between spaces of different dimensions; [56] each new dimension introduces some new features not available in the lower dimensions. It is the discovery of relativity theory that the unique feature of cosmic four-space, not available in the lower spaces, is temporality. This discovery finds expression in special relativity theory through the temporal character of proper time, the absolute invariant of the theory. In the cosmologies of general relativity theory it is expressed through the privileged position of time in determining the geometry of space.

The determination of spatial curvature is a complicated process involving a continuous dialectic between theories and measurements. First a line element of the schematic form of equation (3) (in chapter 4, section 2) must be found to satisfy the cosmological equations of general relativity theory. Then the physical consequences of the use of that line element must be explored in terms of measurable parameters. Subsequently, one looks for correspondences between the model built on the particular solution and the universe as seen through our extended senses. The curvature of space may then be determined from a functional relationship among the macroscopic density of energy (matter), the quantity known as the radius of the cosmological model, and the curvature K.[57] The average density of matter is an unambiguous quantity unless continuous creation ex nihilo is assumed, but the radius of the model is not; it is an easily misleading and sophisticated concept.[58] But in all formulations it is a distance obtained by multiplying the velocity of light by a measure of time elapsed since the expansion of the world began. This quantity has been calculated from such measurements as counting nebulae per unit coordinate volume and determining magnitudes of red shifts of distant nebulae.[59] Its value is estimated as about eleven billion years, in remarkably close agreement to the best estimates of the age of the galaxies (about 15 billion years). This agreement, assuming that it is not a coincidence, is a crucial point for it implies that the curvature of space-time, a quantity essential to general relativistic cosmology, may then be determined right here, without going elsewhere, by any sufficiently old man with a clock. That is, a man who knows before/after, future/past/present, can distinguish between stationary and creative processes—and was there "when the morning stars sang together." He need not measure any distances, but only times.

But if temporality is truly central in general relativistic cosmology, then the relativistic universes could be expected to exhibit some problems which have bedeviled earlier students of time. This, I believe, is the case, as may be seen by examining the curious quality of a universe which is finite but unbounded. First we note that it could have been called "bounded but infinite" or even "finite but infinite." [60] This is not a trivial point because the words color mental imagery and guide our attention. Let us compare this with the cosmology of Nicolas of Cusa, for whom the world "is not infinite yet it cannot be conceived as finite, since there are no limits within which it is enclosed." Cusanus and Einstein seem to be saying very similar things, yet Cusanus need not be credited with mystical foresight of relativistic cosmology nor can we exclaim that there is nothing new under the sun. All we need to say is that both theories are plagued with the coincidence of the contraries. For the Neoplatonic cardinal this would have been unacceptable without the idea of God; for relativists it would remain unintelligible without the ideas of non-Euclidean geometries. But whereas Cusanus incorporated in his cosmology a mystical oneness of man, life, the earth, and the sky, the cosmology of relativistic four-space includes man only implicitly. By shifting cosmological formalism away from concern with existential stress, scientific cosmology offers a symbolic structure without overt conflicts and thereby passes all responsibilities to the cosmologist, who must know time. It would seem, then, that the first practical task of the cosmologist is to learn the levels of temporality right here. If he is successful in identifying beneath his intellectual, noetic sense of time the eotemporal, prototemporal, and atemporal Umwelts, he then possesses the knowledge necessary for the interpretation of what he observes as he examines the gradually regressive levels of temporalities at cosmological distances.

In Arthur Koestler's *Darkness at Noon* the protagonist spent his life in trying to build a polis, imitating a heavenly image. He is now in prison facing an inevitable death and the collapse of his city.

> Rubashov stood by the window and tapped on the empty wall with his pince-nez. As a boy he had really meant to study astronomy, and now for forty years he had been doing something else. Why had not the Public Prosecutor asked him: 'Defendent Rubashov, what about the infinite?' He would not have been able to answer—and there, there lay the real source of his guilt. . . . Could there be a greater? [61]

Time as History

The loss of authority of universal cosmologies gave rise to lines of inquiry other than physical cosmology; namely, to the study of man's own history and that of the earth. Here we are concerned with the history of man.

(1) Origins

Archeological record suggests that man has not accepted his existence as do other forms of life but has shown profound concern and paid special attention to the beginning and end of human life. He has shown curiosity toward becoming (coming into or going out of being) while also asserting in primitive eschatologies a type of changelessness that protected his imagination from the threat of final passing. But, as in the life of the infant who awaits the breast but is fed according to the dictates of conditions not under his control, so the cyclic expectation of man encountered contingencies. The resulting existential stress might best be described as an Ur-Sorge or primordial anxiety. Although history in the modern sense did not exist until the gradual formulation of historicity in the Christian West, the sources of man as a historical being must be sought in his primordial anxiety.

Perhaps what early man realized is what Heidegger's Black Forest philosophy teaches in so many words. Man does not ask to be born, but is "thrown" into the world; he strives for identity (called authenticity by Heidegger) because he is driven by a deep-seated dread. This dread is not of anything specific, but of everything, of a final meaninglessness or, as I would prefer to put it, the fear of a discovery that his struggle for identity was not worth the sacrifices which he had to make to rise from animalhood. But, if he can reach authenticity, then he comes to possess destiny and becomes a historical creature.[62] Sidney Hook once described Heidegger's philosophy as a map of man which makes agony a part of the geography. Of course, agony is coeval with identity, and thus with humanity, and was not recently invented. Shakespeare put it more clearly—and more beautifully—than did Heidegger.

> Ruin hath taught me thus to ruminate—
> That Time will come and take my love away.
> This thought is as a death, which cannot choose
> But weep to have that which it fears to lose.
>
> Sonnet 64

There is ample evidence that both as individuals and as communities the ancients had memories. Semasiography or pictures conveying general meaning seem to have appeared with Homo sapiens. Word and syllabic systems are first attested to around 3100 B.C. in southern Mesopotamia, and records of recurring events may be found in the Palermo Annals of the 4th Dynasty of the Old Kingdom (circa 2600 B.C.). But cyclic events can make up only an eotemporal world: the Egyptians had their "befores" and "afters" but no general trend in time. They believed that the conditions of their existence as a people had always been and always would be governed by the gods, whose will and purpose was completely inscrutable.[63] The corpus of Greek mythology, contemporary in its origins with that of the New Kingdom of Egypt (circa 1500–1000 B.C.), refined and enlarged the original myths so as to satisfy an increasing awareness of the complexity of the world. In terms of concepts already considered, prehistorical social time may be described as one in which connectivity and intentionality are as yet undifferentiated: regularities which were thought to have connected events had no legitimate existence apart from the intentions of the divinities.

The origins of historiography are associated with the work of Herodotus in the fifth century B.C. He was the father of a new intellectual discipline in which man is given a story, not unlike the way in which universal cosmologies recounted the stories of the Big Everything.

> Herodotus of Halicarnassus, his Inquiries are here set down to preserve the memory of the past by putting on record the astonishing achievements both of our own and of other peoples; and more particularly to show how they came into conflict.[64]

The emergence of the idea of Heilsgeschichte around the eastern Mediterranean made historicity an essential feature of man's time on earth by discerning an irreversible trend among events within communal memory and anticipating equally irreversible events in the future. It took then sixteen centuries to produce the first secularized salvation history (in the work of Vico) and generally provide a mental scaffolding onto which individual scholars could hang their findings about the past of man. The essentially Christian scheme of history was often challenged regarding some details of its content but not importantly threatened as regards its structure until after it ceased to give a satisfactory framework into which the finding of man's expanding Umwelts could be placed.

(2) Ideas of lawfulness

Salvation history was ill-prepared to accommodate the lawfulness which the early pioneers of science have identified in nature; yet the idea of divine law served as the paradigm for those very laws of nature, and later also as the paradigm for connectedness in history, as separate from intention. In spite of their common ancestry, however, laws of history and laws of nature have retained their separate status up to and through the positivistic challenges of our own epoch. The difference is an epistemological one, epitomized in Ortega y Gasset's famous remark that man has no nature, only a history. If taken literally, this might imply a mode of communal existence which is completely unpredictable. But hardly anyone argues against the existence of some general laws; most Western thinkers proceeded to employ the principle of salvation history to their particular field of training. One example will elucidate this point. When Marx held that since the course of history cannot be altered, the overthrow of capitalism once and for all is inevitable, he sounded like St. Augustine, who held that Christ saved the world by dying but once. But Marx and St. Augustine sounded very unlike Aristotle, for whom the Trojan War was both before and after his time.[65]

But whatever the specific suggestions may be concerning the motive powers of history (and there are many), they all assume that it makes sense to talk about general laws. Whether or not such general laws may exist, and what their nature and limitations may be, has been debated by philosophers of history—who follow two main intellectual allegiances.[66] Those with a critical bent analyze the concepts and the structure of historical knowledge; those with a speculative bent attempt to construct views of the whole of reality. Both these and others, not easily classifiable, have been laboring to separate stationary from creative processes in the nature of historical time or, if we wish, in the existential tension of man's history.

(3) Speculative perspectives

There is no definite earliest starting point to speculative views of history; they more or less grew out of the tradition of universal cosmologies. Here is a sampling.

Giambattista Vico sought empirical evidence for steps of progress from the use of force to use of principles of justice and from privilege to rule of law. Kant formulated nine propositions in his "Idea of a Universal History" which he believed to contain the essential features of time and history. The fundamental statement is that of the eighth proposition.

> The history of the human race, viewed as a whole, may be regarded as the realization of a hidden plan of nature to bring about a political constitution, internally, and, for this purpose, also externally perfect, as the only state in which the capacities implanted by her in mankind can be fully developed.[67]

The most powerful modern salvation histories are those of Hegel, Schopenhauer (1788–1860), Karl Marx (1818–1883), Nietzsche (1844–1900), and Arnold Toynbee. We have already considered Hegel and recall that, to him, history is the development of the spirit in time, nature the development of the idea in space. For Schopenhauer history is the ethical passage of man, whereas Marx understood history as the successive emergence of levels of different economic conditions, causally connected. For Nietzsche time and history were those of the eternal return, yet at the "midday" of this history man was to go down and superior man was to arise. The reconciliation of the periodicity of the eternal return with unique appearance of superior man is resolved by an association of the final state of history with the desirable status quo.[68] The imposing universality of natural selection suggested to some that the true connecting principle between historical events is the struggle for survival; this is expressed in the tenets of social Darwinism as represented, for instance, in contemporary America by apologists for white supremacy.

Arnold Toynbee sees the universal laws which control the dynamics of history as qualitative rather than quantitative; the units to which these laws apply are not nations, states, or periods, but societies. He sees the rise and fall of civilizations as a challenge and response interaction between societies and their environments. His theme is "the possibility that man achieves civilization, not as a result of superior biological endowment or geographical environment, but as a response to challenge in a situation of special difficulty which rouses him to a hitherto unprecedented effort." [69] Growing civilizations are brought about when the nature and intensity of challenge are appropriate to some initial conditions of the society—a qualitative principle resembling distantly that of natural selection. Toynbee regards a society as growing when there is a gradual accumulation of power which can be and is directed toward the meeting of internal and spiritual challenges, rather than needed to meet external and material challenges. Such a society is likely to rise to a universal state recognized by a unity of laws, beliefs, and purpose. Toynbee marshals what appears to be a bottomless store of examples in support of his idea of the rise of societies, as well as evidence for patterns in the breakdown of societies. These lead from dynamic growth to stagnation, and then to conditions when societies become victims of external and internal disruptive forces. To Toynbee the time of history is a divine comedy, in the original sense of Dante's *Commedia:* something that begins fearfully and ends happily.

> While civilizations rise and fall and, in falling, give rise to others, some purposeful enterprise, higher than theirs, may all the time be making headway, and, in a divine plan, the learning that comes through the suffering caused by the failures of civilizations may be the sovereign means of progress. [70]

Criticisms of Toynbee's monumental work are as extensive and complex as the work itself. [71] Their main thrust seems to be that the regularities he perceived in the rise and decline of civilizations are only partly valid; more accurately, the hosts of exceptions and variations make these laws not truly universal.

(4) Critical perspectives

Critical philosophy of history is an offspring of the enlightenment. Broadly speaking, critical philosophers of history are either of positivistic or of idealistic persuasion. The positivists generally hold with August Comte (1798–1857) that when the study of history finally matures, it will become an exact science and permit explanation in terms of precise laws of nature. A concise statement of this view is that of the British economist

J. S. Mill (1806–1873). He agreed with Comte on the three historical stages of human knowledge:

> in the first of which it tends to explain the phenomena by supernatural agencies, in the second by metaphysical abstractions and in the third or final state confines itself to ascertaining their laws of succession and similitude.[72]

Additionally, he maintained that the collective series of social phenomena, in other words the course of history, is subject to general laws, which philosophy may possibly detect. With Comte positivistic science itself became a religion. He developed a plan of social reorganization based on a utopia where the Catholic God was replaced by a Great Being, namely, humanity past, present, and future, leading to a new religion of humanity. The priesthood (with headquarters in Paris) was to be made up of secular sociologists responsible for social order and mental hygiene. Temporal power would be handed over to the captains of industry, especially businessmen and bankers. In his influential writings Comte both foresaw and prepared the way for late twentieth century society and anticipated, in the character of his thought, the scientific mysticism of our own epoch. His utopia is no less fanciful than the New Science of Vico; but because of its imagery and language Comte was acceptable to an age of science as religion, Vico to an age of religion as science.

William Dilthey (1833–1911), a philosopher of history with Kantian inspiration but influenced by British empiricism and French positivism, held that the central concern of philosophy should be the theory of knowledge informed by an understanding of cognition and thought processes. He distinguished two kinds of sciences: the *Naturwissenschaften* or natural sciences and *Geisteswissenschaften* or humanities. (The German terms are more expressive than the English ones for they imply a Hegelian distinction between Nature and Spirit). While the natural scientist studies regularities in nature, the historian must be a humanist who searches for the unique, superimposed on the regular.[73] R. G. Collingwood, elaborating further the philosophy of Dilthey and adding a great deal of his own, insisted that history is an autonomous discipline which must evolve its own methods and not depend on scientific explanations which see events only as specific manifestations of universal laws.[74] The problem of historical explanation during the early decades of this century would generally favor the idealist approaches of Dilthey and Collingwood.

Opposing the idealist views, Karl Popper, Carl Hempel, and others have held positions inspired by natural science. Hempel believes that there cannot be such things as laws of history but only laws of different disciplines useful for history, such as psychology, economics, sociology, etc.[75] This has been called the "covering law model" of historical explanation which, because of the nature of the epistemology of its constituents, must remain deductive. Karl Popper's model of historical law is that of an (eventually) mathematical expression which describes specific events only upon the application of appropriate boundary conditions.[76] He stresses, however, that the peculiarity of history is not so much in any such law, but in the initial conditions which cannot be subsumed under the law.

(5) Epistemic hurdles

The many and colorful approaches to time as history all appear to be attempts to find Aristotle's "consistent plot," the universal that would have us understand and thereby master the fate of man on earth. Historians of all epochs, and here we must include epic poets and myth-reciting magicians, if they were gifted observers, were able to perceive connecting principles in the social behavior of man unknown to those whose behavior they described. Modern working historians can successfully demonstrate past facts which contemporaries could not have known, for in their times the paradigm of the connecting principle had not yet been formulated. Thus, an Egyptian could hardly have imagined that he was acting according to the Marxist laws of dialectical materialism; the soldiers of Rome would not have dreamt that they were following the Hegelian dialectics of the spirit; and I doubt that Thucydides thought of the Peloponnesian War as an illustration of social Darwinism. We are faced here, I believe, with a logical and an epistemological problem. The logical one is well known: the historian must apply his principles ex post facto. The epistemological one is, I believe, the operation in the history of man of a principle of emergence of first principles, or, as I have called it in the context of atomicity and continuity, nomogenesis. That is, no meaning may be given to any regularity before the variables of the regularity come into being. Yet, the principles mentioned have very often been applied retroactively, though maintained totally only by scholars of very pure confessions. It is clear that we are applying connecting principles which have emerged in the course of time to epochs when they might not have existed. In other words, I suspect the operation in the domain of human life of an equivalent of the principle of emergence of first principles, already proposed in the context of physical probability.

In his *Reconsiderations*, Toynbee assesses the evolution of his own labor. It "began as an analytical-classificatory comparative study of human affairs and turned into a metahistorical inquiry." [77] He explains his idea of metahistory by recalling that Aristotle, upon completing his work in physics, was left with a number of unanswered questions prompted by physics but not about physics itself. They pertained to "some hitherto nameless subject which was apparently the setting within which the inquiry into physics had been conducted. Thus the intellectual conquest of the field of physics opened up a further field of inquiry beyond it, and Aristotle labeled this 'what comes "after" or "beyond" physics' (tà metà tà physiká)." [78] Metahistory, then, bears the same relation to history as metaphysics (did originally) to physics. All this is sound epistemology. The rules of a field of knowledge, when that field is understood as a portion of a larger scheme, cannot be found within the field itself. This is isomorphic with the reasoning which led semantics to the formulation of metalanguages and, as one specific exemplification thereof, to the Gödelian structure of the hierarchy of undecidable propositions.

Although I cannot share the conviction of those, such as Hempel, who believe that history can be pieced together from interpretations of sciences, I do maintain that the sciences as well as the total tradition of humanities are valid sources of research for historical understanding. This alone would make matters difficult enough because of the overwhelming quantity of material, yet beyond this there are deeper problems. Critical history on the positivist side must assume a metaphysical factualness of past events, otherwise its search for connecting principles remains without foundation; but the status of "facts" is a very shaky one because it is a function of a posteriori speculations. Critical historians of idealistic persuasion must accommodate in their schemes even the narrowest positivistic interpretation of history, because disclaiming the existence of physical reality would leave them alone on strange shores. After all, even the beautiful body of Helen of Troy was made from the atoms of the periodic table. Is then, the problem of formulating covering laws one of the unmanageable amount of input? I think not, for the hallmark of lawfulness is the ease with which it isolates the relevant from the irrelevant material.

The deeper problem is, I believe, in the heterogeneity of the many levels of lawfulness which manifest themselves in the behavior of historical man, coupled with the nomogenesis of the various evolutionary processes. Thus, when critical historians (or sociologists) of positivistic persuasion seek the stationary processes of history in the form of laws resembling the laws of physical science, they are applying methods appropriate only to more primitive integrative levels than that of society. Statistical laws, for example, imply prototemporal or eotemporal Umwelts, certainly present in man and society but not representing what is more interesting in Homo sapiens. Whether Toynbee's handling of the army of details at his disposal is to the liking of professional historians is not for a philosopher to decide; whether the spiritual guidance he claims to detect in history is admissable or not is irrelevant. But that the lawfulness of history must come from a metahistory rather than historical research per se, appears to be quite self-evident.

I have placed this discussion of time as history directly after the idea of time as cosmic geometry not only to stress their common ancestry in universal cosmologies, but also to contrast the widely separated levels of reality with which they deal. Scientific cosmology is an abstract Platonic image of the beauty of the astronomical universe; history is a search for order in the drama of man with all its dynamic mess. Although the latter is played on the stage of the former, those features of the latter which are most interesting are the ones that derive from the functions of levels higher than the atemporal, prototemporal and eotemporal Umwelts. Thus, although the possible actions of history are limited by matter and life, the rules of the game (such as they are) are those of the minds of clockwatchers, individually and communally.

4. Endings:

Estimates of Death

IN VOLTAIRE'S candid opinion, history is a trick played by the living upon the dead. This is a concise criticism of the practice of assigning inappropriate connecting principles to past events. This section deals with the problems of assigning laws to future events, a practice prompted by tricks that death plays upon the living. The origins of such tricks in man's sense of time were stated with uncanny beauty and clarity by Hamlet:

> For who would bear the whips and scorns of time . . .
> When he himself might his quietus make
> With a bare bodkin? Who would fardels bear,
> To grunt and sweat under a weary life,
> But that the dread of something after death
> The undiscover'd country from whose bourn
> No traveller returns, puzzles the will.

> (III, i, 70–80)

The fear and fascination of matters after death, whether the death of the astronomical universe, of the earth, of nations, beasts, or men, is the subject of eschatology (from *eschatos*, the last or farthest). The imagined last events themselves may be called the *eschaton*, or end-time.

We have already noted that inspection of the world at large gives no immediate indication that there has been some sort of evolution, or devolution from an original condition different from the present. Neither the cyclic rebirth of man and nature, nor natural calamities suggest that in the long run the world is anything but unchanging. But for reasons that may best be understood through the psychology of time sense, memories of the destructive aspect of time tend to be more overwhelming than those of the creative aspects. This asymmetry manifests itself in the distinct difference in atmosphere and literary flavor between legends of world creation and world destruction. While myths of creation tend to appeal to the intellect by way of explaining things and actions seen in the present, visions of an end are usually moralistic, restrictive, and suggestive of preferred behavior. To the extent that views of the universe influence behavior, they do so through eschatologies rather than cosmogonies or cosmologies. This fact adds to the store of evidence that the future (as compared with the past and the present) is the most easily reachable category of time, possibly because it is the most primitive one. Also, it suggests certain intimacy between the end of the self and that of the world, not paralleled in thoughts about creation.

The idea that the end of the self and that of the nonself are somehow the same is expressed in the melancholy poem of A. E. Housman:

> Good creatures, do you love your lives
> And have your ears for a sense?
> Here is a knife like other knives,
> That cost me eighteen pence.

I need but stick it in my heart
And down will come the sky,
And earth's foundations will depart
And all you folks will die.
                    "I Counsel You Beware"

It is impossible to demonstrate to an individual that the world will not "really" come to an end upon his death. If he is asked to fancy that it will not, then he must assume the survival of an observer which is somehow part of his identity now, which is against the original assumption of his final demise. The riddle of time is as much a riddle of its ending as it is one of its beginning. In some ways it is more difficult to explore the idea of an end of time than concepts of its beginning, because there is no established body of inquiry into questions of the eschaton, comparable in scope and magnitude to history, cosmology or cosmogony. Information must be sought from universal cosmologies, religion, physical cosmologies, and from estimates of afterlife.

Eternalistic Peace

The doctrine of the destruction of the world (pralaya) is implied in the sacred literature of the Hindus, usually in connection with beliefs and rituals connected with dying and the dead. In the Atharvaveda, written sometime between the fifteenth and fifth century B.C., Agni, the fire-god, consumes the body of man, preparing it for the heavenly state which it is to enter after consummation. We read that "As between heaven and earth Agni went burning on, all consuming. . . . Into the floods had Matarisvan entered, the deities passed into the water." [79] The fate of the deities is the reverse of those of ancient Egypt where they emerged out of the waters. Writing about the Hindu evaluation of time, Brandon remarked that

> from the Vedic period the universe has been seen as the product of an ambivalent character, according to human estimate, but as such by an intrinsic logic that links creation with destruction, life and death.[80]

Somehow this balance prohibited a Final Judgement from taking over the world and, in spite of the horrors of destructive time, new epochs emerged from the ashes of the old. In chapter 2 we noted the bewildering calendrical system of ancient India, made up of cycles upon cycles, from 10^{15} years down to milliseconds. The cessation of the Hindu universe is also stratified in a way which resembles the hierarchy of cycles. According to Vishnu Purana [81] the dissolution will take place in three major steps of increasing importance: incidental, elemental, and absolute. Each of these contains a multitude of steps, with eons which are progressively shorter and less blissful than those preceding it. Then follows a stunning and occasionally horrifying provisional end: the appearance of seven suns, fire storms, deluges, vast elephants, and darkness. But, *in saecula saeculorum* (forever and ever), the world is created anew by the reposing goddess.[82] The conflicts of the present may thus be made more bearable by knowledge of the cleansing horror of its ending, followed by the peace of eternity.

Universal regeneration is embodied in cyclic theories of practically all eschatological myths, many of them more recent in origin than those of ancient India. For instance, the struggle of the Aztec gods gives rise to the death and rebirth of the universe. The rebirth is accompanied by the consecutive appearance of five suns, in manifest resemblance to the suns of Vishnu Purana. The end of each world, ruled by a sun, is also the beginning of another. Finally "the fifth sun is born and begins to dance at the center of the universe, the four corpses [of prior suns] rise again, and dance too." [83] The first four are thought to represent fire, earth, air, and water; [84] the fifth one is a spiritual sun whose image survives through the medium of the dance. Somehow incorporated in the cycles of suns, deaths, and rebirths the miracle of life goes on in eternal peace, as attested to in the poetry of ancient Mexico:

> In the sky, a moon:
> in your face, a mouth.
> In the sky, many stars:
> in your face, only a pair of eyes.
>
> In the dewdrop shines the sun:
> the dewdrop dies.
> In my eyes, which are my very own, your eyes are shining:
> I—I live.
>
> The endless river
> goes on its way.
> On and on
> goes the wind, never-ending.
> Life passes,
> never returning. [85]

In Scandinavian mythology Ragnarök is the name for the end of the world of gods and men. It is described in the thirteenth century Prose Edda (written by an Icelandic chieftain). In it the Fenris Wolf, the World Serpent, and the hellhound Garmr slay Odin (father of all gods), Thor (god of thunder) and Loki, the mischief maker. The external landscape is Nordic: Monstrous Winter, wind and frost with no summers, with stars that vanish and the earth that sinks into the sea. But the earth rises again, Balder the innocent returns, and the host of the just lives forever in beauty and love.

> There will arise out of the sea . . . another earth most lovely and verdant, with pleasant fields where the grain shall grow unsawn. . . . From a woman named Lif and a man named Lifthrasir the sun shall have brought forth a daughter more lovely than herself [Lif] who shall go in the same track formerly trodden by her mother. [86]

Liberation from the shackles of time on earth need not take, however, the grand violence of the Vedas or the romantic beauty of Nordic saga. After the decline of the colonial order, cults of cargo evolved on New Guinea and the islands of Melanesia. Their beatific vision is the comfortable life of Europeans; at the end of time equal comforts will be provided for the natives to last henceforth forever and ever. Their eschatology, summed up in Pidgin English is *rot belong kako* which is "God prepares the cargoes." [87] God, or Jesus, or the angels prepare the food and goods: beads, guns, bolts, aspirin, china, and rice and make it ready for shipment by sea or even by air. The arrival of the cargoes is sometimes associated with the Second Coming of Christ, accompanied by the sound of the "last trump" that initiates a new and eternal era of peace and plenty. It is uncanny to observe here the eschaton forming under our own eyes; the genesis of myths of this type, in the case of other faiths, has taken centuries and even millennia. Yet, the yearning of man for a final solution of his temporal problems finds an even more up-to-date expression in the cult of the flying saucers.[88]

The examples could go on, for instance with reference to folk literature,[89] but a few important points may be made based on these representative examples of eternalistic eschatologies. As a genre they belong with the creation myths of universal cosmologies; the destinies of man and the universe are one and the same. Some myths provide transient endings which lead to the rerun of prior evolutions and devolutions, hence they are only pseudo-eschatologies. Some depict conditions better than ours, existences without pain and injustice, yet the elements of self-preservation and those of the preservation of the race are retained. They are "in illo tempore" in the future, and with residues of the conflicts of earthly life.

Apocalyptic Tension

The absence of an ending of world-time is not equivalent to a prohibition against the desire of the individual for atemporal states. Thus, for example, the desirable end-state of the individual in Indian thought is nirvana, a condition "where there is neither this world nor a world beyond, nor both together, nor moon, nor sun," [90] the summum bonum counseled by the Upanishads, a beatific vision of timelessness. This state may be reached through training in ecstasy here and now, and permanently, after death; the individual may also extricate his soul from the sorrowful wheel of endless misery by self-elevation through metempsychosis (the transmigration of souls). In Zoroastrianism an optimistic coincidence of time on earth ("Time of Long Dominion") and the time of the universe ("Infinite Time") probably also represented a blessed state of atemporality.[91] In the Egyptian Book of the Dead the deceased who has been assimilated to Osiris exclaims: "I am Yesterday, Today and Tomorrow" [92] expressing thereby a universal desire of man: that of finding the timeless. Eschatologies have given powerful expression to this desire and found themselves up against the difficulty of reconciling the temporal and the atemporal within a single framework of thought.

A practical integration, or intellectual reconciliation—more or less—of time with the timeless is the merit of the apocalyptic vision of Christianity.

In this vision the end-time of the individual and that of the world are identical and final, save for the fact that the postmortem portion of man's self must await (in time) the end of the universe. The scenario comprises two steps. First comes the Day of the Lord, which is the day of Last Judgement for the living and the dead; this is also known as the Second Coming of Christ (and the Last Trump of the Cargo cults). The Sign of the Son of Man will appear in heaven and Christ will come in his majesty, power, and glory. Apocalypse, from the Greek "uncovering, unveiling" is sometimes the very final ending alone, sometimes the complete process of Judgement and ending.[93] Although the staging resembles other world-endings, the Christian apocalypse has guided the faithful to practices very different from those originating in non-Christian eschatons.

The dogmatic possibility of an imminent end to the world (and the attendant salvation of Jew and Gentile), coupled with complex historical and ethnic determinants, resulted in an undisguised urgency which came to characterize Christianity beginning with the very ministry of Christ. The Old Testament view of life was closer to pantheism than to apocalypticism: "In the sweat of thy face shalt thou eat bread, till thou return unto the ground; for out of it wast thou taken; for dust thou art, and unto dust shalt thou return" (Gen. 3:19). Postmortem existence seems to have been of no particular concern except in the original apocalypse, the vision of Daniel, "And many of them that sleep in the dust of the earth shall awake, some to everlasting life and some to shame and everlasting contempt" (Dan. 12:2). It is generally held however, that in spite of its origins in Judaism, the apocalypse was not in the mainstream of practicing Judaic thought. This might have to do with what has been called the "this-worldness" of Judaism,[94] although it is difficult to say whether the "this-worldness" is a cause or an effect. Be that as it may, the Jewish apocalyptic writings became popular with the early Christians, who found in them much that was significant to their own fate as Zealots as well as pacifists.[95] Not content with the Jewish apocalypse and feeling that stress of immediacy which informs the Pauline gospels, later Christian writers interpolated the early apocalypse with specifically Christian teachings; this movement reached its most significant expression in the canonical Book of Revelation. Its vast cosmic drama inspired, and was imitated by, many of the apocryphal apocalypses but seldom, if ever, was matched for grandeur. The apocalyptic feeling expressed therein runs deeply and continuously in Christian religious life, and although it tends sometimes to fall out of (and then back into) favor, it survives to our own days both in religious and in secular forms. The hymn of Thomas de Celano, from the turn of the twelfth century, is a majestic statement of the Christian eschaton:

> Dies irae, dies illa!
> Salvet saeclum in favilla,
> Teste David cum Sybilla . . .

Lacrimosa dies illa
Qua resurget ex favilla
Judicandus homo reus;
Huix ergo parce, deus! *

This hymn may stand as a concise expression of some of the features of Christian Apocalypse which formed the modern Western mind. It suggests that the free will of the individual (for only an agent free to choose may be justly sentenced) coexists with an all-knowing God. Thus, time presumes being and becoming, the unpredictable and the lawful. The world and time may come to an end at any next moment, hence one should do what one can now because tomorrow may be too late; the moment missed will be mourned forever. Someone, somewhere, keeps a record of one's daily and nightly activities which will come to light when all the living and the dead are called for. The apocalyptic end of the world itself is, thus, an expected/unexpected event whose permanent threat implanted in the faithful a feeling of tension, which through the centuries, became a characteristic feature of Christianity. It assisted the emergence of the Protestant era with its keen consciousness of the importance of temporal passage, and has informed Western literature, and through it Western intellectual life, with a continuous "sense of an ending." [96]

The Eschaton in Natural Philosophy

Universal cosmologies, as we have seen, have two full-grown offspring: scientific cosmology and history. The Apocalypse, while surviving in its own right, also gave rise to two intellectual derivatives: the eschaton (end-time) as seen in, or extrapolated from the natural philosophy of time, and the issue of death in existential thought. Here we are concerned with the former. For many reasons, among which apprehension about issues which have to do with death might well be the most important, science and natural philosophy paid much less attention to questions concerning an end of time than to problems related to a beginning. Let us fall victim to the same temptation and begin with one well known view about the past. Kant's famous First Antinomy comprises simultaneous arguments designed to prove that the world did (thesis) and that it did not have a beginning (antithesis).[97] His conclusion was that, since they cannot both be true, temporality is not a feature of the universe but is assigned to it by the mind.

Kant's antithesis is isomorphic with Lucretius' argument about space: if there is a boundary in space, there must be space external to that bound-

---

* Day of wrath, that day of mourning,
All the world to ashes burning,
David and Sybil spoke concerning . . .

Day of tears and late repentance
From the dust of earth returning
Man shall rise to hear his sentence
Spare, O God, in mercy, spare him.

ary; if there is a beginning, that must be preceded by "empty time," but, upon a second look, empty time cannot be different from time. Hence, time has no beginning. Kant's thesis may be summed up in three steps. (1) He identifies infinity with the idea of "an infinite series of successive states." (2) He stresses that by an infinite series of states, or past acts, we can only mean the impossibility of completing it in finite time. Yet (3) we do experience a present which, according to (2) cannot be the case if it were preceded by an infinite series of acts. Hence, time does have a beginning.

Current debate about this antinomy focuses on disentangling the idea of an elapsed infinity of acts from enumerating those acts. Whitrow writes that "It is remarkable how this argument [Kant's First Antinomy] has been misunderstood by many acute minds. The misunderstanding is due to the belief that Kant's antinomies can be automatically disposed of by appealing to the modern theory of infinite series." [98] But, Whitrow says that all references to time as such have been purged from the mathematical theory of sets and series, whereas Kant's arguments essentially concern successive acts occurring in time.[99] Thus, an elapsed infinity of acts may well remain a self-contradictory concept, judged on its own merits. Opposing this view, J. D. North holds that an actual infinity of past acts cannot be denied, and that a completed infinity of events can be accommodated if the concept of potential infinity is admitted.[100] I submit that the difficulties of the argument may be resolved by taking into account the hierarchical structure of temporality.

I believe that the idea of a beginning should reflect the hierarchical ordering of time, and that many different meanings must be assigned to a "beginning" according to the Umwelt in which we seek to identify it. Numbers and actions will then be seen to belong to different worlds. Certainly, in an atemporal world a beginning or an end cannot be thought of. This is Kant's "no beginning" or "infinity of time." In a prototemporal world a beginning or an end must be of a probabilistic character. By this I do not simply mean that a beginning may or may not occur, but rather that it is something the clockwatcher might only infer but could not possibly locate in time—even in principle.

Let us recall what we learned about the eotemporal world. In those equations of physics which describe that Umwelt the "physicist's t" always appears squared or otherwise hides a preferred direction of time. Hence, while identification of change is possible including, presumably, the emergence or disappearance of the possibility of ordering, we still could not tell a beginning from an end. This is difficult to imagine but we must remember the phenomenon of nomogenesis and with it the fact that some Umwelts simply cannot support certain types of laws—or temporalities. Consider, for instance, the models of oscillating universes. If such a model is taken seriously, it must involve the collapse toward an original quantum state out of which it will rise again. This is a regression from eotemporality to prototemporality and atemporality. Thus, oscillating universes cannot correspond to the Aristotelian vista where the present of a sublunar world may be both before and after the war of Troy. Rather, it is temporality itself which reduces to an Umwelt where beginning and end ought to be described by the same word.

Finally, successive states as recognized by man, and this is what Kant must have meant by a series of successive states, belong to a world of noetic temporality or, as we shall call it, nootemporality. We are yet to identify the salient features of this, the presently highest level of temporality; for now it will be sufficient to take it as equivalent to an unexamined image of "man and time." On the level of nootemporality we recognize a beginning (and an end) after its paradigm, the coming about and the cessation of our identity. Thus, the Kantian conclusion that the world had no beginning is true only for the atemporal world of pure radiation. The Kantian conclusions that the universe did and did not have a beginning are seen as statements in many parts, all true, with proper qualifications. The Kantian thesis and antithesis, however, do not form a true disjunction; hence they are not mutually exclusive. Therefore all conclusions which might have been drawn from their mutual exclusiveness collapse.

We shall return to our particular Fragestellung regarding the beginning and end of time, in terms of the theory of time as conflict (in chapter 8, section 5).

There exists a problem relating to the end of time which is the mirror image, as it were, of the thesis of Kant's First Antinomy. This is the Tristram Shandy paradox of Bertrand Russell.[101] In Laurence Sterne's novel, *Tristam Shandy*, the title character complains that it took him two years to record the events of the first two days of his life, and it seems to him that at this rate he will never finish. Russell argues otherwise. Since for every day lived, says Russell, there corresponds a day written, all we must give him is an infinite number of days and he will complete his task.[102] Whitrow, opposing Russell, argued that there is nothing debatable about saying that *any* day you may pick will be covered by the biography; the problem arises only if we assume that *all* days will be recorded by Shandy.[103] (If I were to argue this point, I would add that Russell's reasoning demands a day when the diary is both complete and incomplete, rather a contradiction). Yet the curious quality of infinity, namely, that it is not to be regarded as a number, could still be used to maintain Russell's argument. Using Cantor's concept of transfinite numbers, the infinity of Shandy's days on earth may be identified with $\aleph_0$, the cardinal number of the set of all natural numbers.[104] It is the property of $\aleph_0$ that any of its proper subsets equals the set in cardinality. All that is necessary is to assure us that the number of days recorded is in one to one correspondence with such a proper subset, and Russell's claim has been mathematically proven.

Yet the total argument is invalid, for it includes a category mistake. (And I wish to neglect any other arguments such as that Shandy will not live long enough, or, for that matter, that the universe might not exist forever). As we shall see, the "unreasonable effectiveness" of the use of mathematics in natural science derives from the fact that numbers are rooted in the atemporal, prototemporal, and eotemporal Umwelts. But these "lowly" origins also place a limit on the use of mathematics, one that cannot be transgressed, no matter how sophisticated the mathematical operations may be. Simply, there is no meaningful way of establishing one-to-one relationships between series of numbers and events in the nootemporal, or even biotemporal Umwelt. These Umwelts, those of numbers and those of living and thinking things, are qualitatively different. All that Russell's argument demonstrates is that the proto- and eotemporal worlds display some strange qualities—and this should not come as a surprise to the reader.

What, then, about an end of time? Perhaps the writer should be excused from holding forth about apocalyptic matters, but he might be permitted to speculate along lines which he feels are consistent with the theory of time as conflict. Let us first take seriously our reasoning that levels above the eotemporal involve the coexistence of entropy-increasing and entropy-decreasing processes. This leaves us two alternatives. One is that long before the universe at large reaches thermal equilibrium (which is an "if-then" statement expressed in terms of the future-past-present mode of nootemporality), all systems that gave rise to higher-level temporalities will have vanished. This would amount to a regression through different types of endings; not a bona fide apocalypse, but a slipping into oblivion. The "final" state would be atemporal and light would again cover the face of the deep.

The other alternative is equally speculative. Namely, we cannot rule out the possibility of the continued emergence of higher-order integrative levels and the spreading of these advanced states of matter to all reaches of the universe. Although biomass, the only truly open system we now know, is insignificant in its spatial extension, its power of emergence by complexification is rather great as demonstrated by the history of life, and mind, on earth. Thus, entropy-decreasing systems may, in principle, become sufficiently powerful to alter the fate of the inanimate universe. Admittedly, this is a vision reminiscent of those of Blake, or of the prophets who awaited the final victory of disembodied spirits. It is also akin to the idea of a spreading noosphere described by Teilhard de Chardin, as well as to the vision of Arthur C. Clarke about the end of the earth in *Childhood's End.* In that modern day epic, creatures more advanced than man come to rule over increasingly larger portions of the world in a nightmarish victory of mind over body. There and then, the whole idea of an ending of time might appear in forms which are presently inconceivable.

VI

TIME EXTENDED:
LIFE

A VERY SMALL PORTION of the universe, whose large-scale features we have just considered, is inhabited by living matter, distinguished from inanimate matter by a constellation of peculiar functions. The historical and geographical continuity of these functions is called life. In the preceding chapters we have taken life for granted. Here we shall consider life as a continuation of certain trends already evident in the physical world, an extension of lower-order temporalities, and the determinant of a new Umwelt.

1. The Cyclic Order

THROUGHOUT the career of organic evolution the lives of men, animals, and plants have been subject to an unceasing variation of light and darkness, heat and cold, and other regularities of the physical environment. These cycles are complemented by periodic phenomena of other lives external to the organism, and further complicated by the necessity of keeping the multiplicity of internal rhythms coherent so that the organism might function as a unit. The harmony between internal and external rhythms has been a source of inspiration and a lasting example of beauty.

> For winter is now past,
>   the rain is over and gone
> The flowers have appeared in our land,
>   the time of pruning is come:
>   the voice of the turtle is heard in our land:
> The fig tree hath put forth her green figs:
>   the vines in flower yield their sweet smell.
> Arise my love, my beautiful one, and come.

In the Song of Solomon the question was not asked whether the clock was in the fig tree, in the skies, or in the lover: tree, man, and the heavens were one.

Certain rules of preferred behavior were intimately connected with natural cycles long before the religious instructions of the Bible were recorded. By custom immemorial, the women of the Chinese emperor approached him according to the cycles of the moon because of the belief that the virtue of the offspring-conceived related to the moon's phases,[1] and there is strong suggestion in Renaissance poetry that it is foolish to make love after dawn.[2] There is evidence that diurnal movements of tree leaves were known and recorded during the campaigns of Alexander the Great in the fourth century B.C.;[3] and, from the gardens of Shakespeare we receive

> Hot lavender, mints, savory, marjoram;
> The marigold that goes to bed with the sun,
> And with him rises weeping.
> *The Winter's Tale* (IV, iii, 103)

Dictators have not been ignorant of the fact that, for reasons unknown to them, man is most subject to traumatic terror very early in the morning.[4] Executions traditionally take place early in the morning, which is also the time for the knock of the policeman to be heard on the door and for the launching of surprise attacks.

The connection between cyclicity in nature and in the living were accepted by civilizations which regarded man as part of that overall world, but rejected with the Judaeo-Christian view that man is something apart from nature. After the Renaissance, however, the issue of biological cyclicity was slowly pulled out of its metaphysical oblivion. Science rediscovered the obvious and began to supply details, unknown and quite unsuspected previously.

The Physiological Clock

The idea that there may exist nonphysical clocks was itself a discovery. The idea was implicit, but very far from being explicit, for example, in Galileo's pulsilogium. This was a pendular device with adjustable period, employed for the measurement of pulse rates; subsequently it was refined by others so as to be suitable for the measurement of breathing rates of patients and of the systole and diastole of the heart.[5] The device was produced shortly after his alleged observation of the isochronism of the pendulum in 1582. Galileo was the son of a musician and, at that time, a medical student; the pulsilogium resulted from a fortunate combination of his knowledge of pendular rhythm, his awareness of biological rhythm, and his desire to quantify these rhythms for the benefit of medicine. In retrospect, the pulsilogium was the first mathematical model of the physiological clock. Some seventy years later, Descartes, in a letter to the Marquis of Newcastle opined that "doubtless, when swallows come in the spring, they act [by force of nature] like clocks. All that honeybees do is of the same nature."[6] This was not a metaphorical use of "clock," because, unlike Kepler for whom the world was designed as accurately as a clockwork, Descartes suspected the swallows to *be* living clockworks. That organisms on all levels of complexity have innate capacities for rhythmic behavior was not demonstrated, however, until the early decades of this century. Current understanding regards periodic behavior which is in response to external periodicities as exogenous in character hence not truly clock-like; however, periodic behavior which is demonstrably independent of the environment is believed to be the manifestation of truly internal, endogenous, physiological clocks.

To understand the concept of a physiological clock (also called biological, biochemical, biophysical, and living clock) it is recalled that a clock has a display (hands, numerals, scents, tunes); a rule which is regarded as reliable by the clockwatcher (rates of water flow, burning rates, controlled vibrations of small bodies), and the sense of time of the clockwatcher which assists in distinguishing before from after. The display of a rat clock consists of such activities as running, feeding, and defecating. The rule is the prior conviction that such activities repeat themselves with great accuracy within each twenty-four hour period. The decision about what came earlier and what came later as far as events within these periodicities are concerned, must derive from the clockwatcher's sense of time. We need not limit ourselves to rat clocks, however. Identical arguments hold for the closing and opening of marigolds or for the onset of menses.

Our knowledge of physiological clocks is best described as phenomenological. We call them clocks because they display "selected regularities." Although the number of carefully observed physiological clocks is very large, our understanding of their workings is rather poor, partly because of experimental difficulties and partly because of confusion as to what we mean by a physiological (or by any other) clock. We shall review the spectrum of the known physiological clocks as a way of preparing our arguments in support of the thesis that biological rhythm is coterminous with, and necessary for, the functioning of all life forms.

Perhaps the most obvious physiological clocks are those whose periods approximate the daily rotation of the earth. In 1960, Pittendrigh formulated a number of generalizations about these clocks. Their periods are, indeed, circadian (about twenty-four hours); they are ubiquitous and endogenous, their exact range tends to be species-specific; they are resistant to temperature and chemical influence, yet entrainable.[7] The diurnal rhythm and its importance in animal ecology has been of increasingly explicit interest to biologists since the turn of the century. Early studies dealt with primitive, mainly nocturnal creatures, such as cockroaches, silverfish, bristle-tails, stick-insects, crickets, spiders and scorpions [8] which were all found to follow regular and reliable routines. Equally reliable routines were subsequently observed for nocturnal birds and mammals of nocturnal habits. Circadian regularity reaches down to the locomotion rhythm of protozoa; in the ciliate paramecium, division rates follow a circadian rhythm even in total darkness. Microfilariae larvae of Wuchereria bancrofti, the organism responsible for elephantiasis, display diurnal periodicitis which they maintain even when isolated from external periodicities for ten years or more.[9] The circadian migrating habits of free-swimming plankton are very well documented [10] as are the diurnal rhythms of many aquatic and terrestrial animals. Already in lower animals, however, the daily rhythm becomes complicated by the presence of other cycles, such as those of 48, 36, or 12 hours.

Among the primates, squirrel monkeys possess a highly accurate and reliable inherent clock with 24 as well as 12 hour periods. C. P. Richter demonstrated that several periodicities were retained in blinded squirrel monkeys and rats kept in a controlled environment for over a four year period.[11] The blinding of these animals, as well as their isolation in environments of uniform temperature, or light, or darkness are attempts to isolate the physiological clocks from external influences and thus demonstrate the endogenous nature of their rhythms. In one case, some 350 rats, blinded by enucleation before they ever opened their eyes, then kept under constant illumination, tended to retain the pattern of their daily, spontaneous running activities throughout their lives. When so isolated from external cues, the rats natural period would tend to deviate, however, from 24 hours but otherwise remain constant. Also, in Richter's experiments, the natural acitivity periods of two blinded, female squirrel monkeys were found to be 24 hours and 50 minutes, superposed on an activity pattern cycle of about 30 days. The coexistence of these two periods suggested to the experimenters that the evolutionary origin of the clocks is related to the lunar cycles; the synodical month in our epoch is 29.5 days, while tides appear each day 50.5 minutes later than the prior day. A graph depicting almost four years of data shows a rather stunning coincidence, during that period, between the onset of active periods in the monkeys and the times of appearance of the new moon.[12] Lunar and tidal rhythms are widely spread. Various simple worms that live in tidal regions retain their patterns of periodic motion or color change, corresponding to the tides, even under laboratory conditions. Some anticipate high tide with their diurnal periodicities even in aquarium water.[13] Fiddler crabs show a daily rhythm of gross color change corresponding to a base period of the tides. Certain mussels show tidal rhythm in the rate of their water propulsion; some, collected on Cape Cod and transported to California remained in phase with their original lunar periods for over four weeks. The spectrum of menses in the human female centers at about a lunar month, as do mood changes in the human male, though both are distributed in phase.

Circadian rhythms in man have been studied by many investigators.[14] Some of the easiest parameters to monitor are sleep-wake cycles, body temperature, composition of secreted urine, and mental readiness. These, and a large number of other variables in man display an approximately 24 hour frequency even under conditions of careful isolation from external cues. Aschoff and others suspect the operation of a single inherent 24 hour oscillation, although this assumption is not a necessary one.[15]

The systematic study of circadian clocks did not begin, however, with those in men or animals, but with those in plants. By mid-eighteenth century circadian leaf movements of beans were shown to be independent of variations of dark and light. Carolus Linnaeus, inventor of the binomial classificatory scheme for naming plants and animals, studied the times of the day at which various flowers opened and shut. He intended to plant a floral clock "by which one could tell time, even in cloudy weather, as accurately as by a watch." In 1727, while in Skane, near Lund, he observed two plants, the *Crepis* (hawk's-beard) and the *Leontodon* (hawkbit):

> The *Crepis* began to open its flowers at 6 A.M. and they were fully open by 6:30. The *Leontodon* opened all its flowers between 6 and 7 A.M. In the evening the *Crepis* began to close its flowers at 6:30, and by 7 all were closed; but the *Leontodon* shut all its flowers between 5 and 6 P.M. Experiment was then made with the same flowers indoors in a vase of water. The *Crepis* now opened at 6:30 A.M. but the *Leontodon* not before 7 A.M. The *Crepis* closed at 7 P.M. and the *Leontodon* at 6 P.M.[16]

But the first summary of all prior work, with a great deal of original contribution, appeared only in 1958; it is Erwin Bünning's, *The Physiological Clock*.[17]

The spectrum of observable physiological rhythm in normal individuals is continuous from the muscular fibrillation of insects (20–2000 cycles per second), through cellular periodicities (seconds, minutes and hours), lunar periodicities of fish, bird, and man, to circannual rhythms and the flowering programs of bamboos.[18] Representative circannual rhythms are the hibernation of animals with attendant periodicities in body weight,[19] and the *Zugunruhe* or migratory restlessness of birds.[20] Circannual testicular growth culminating in May has been observed in the European starling,[21] in what one might suppose to be the bird's equivalent of Tennyson's

> In the spring a livlier iris changes on the burnished dove;
> In the spring a young man's fancy lightly turns to thoughts of love.
> "Locksley Hall"

With the scientific acceptability of physiological rhythm research inquiries proliferated; we find stated in cold scientific prose what the Bible and the Renaissance had presented in verse. We learn of the regularities in the distribution of coitus in the menstrual cycle,[22] the effects of sexual activity on beard growth in the man,[23] the bimensual variation of absolute pitch in the voice of a professional female singer.[24] The interest spilled over into psychology and sociology. The interaction among patients in a noncompetitive workshop for the physically and mentally handicapped was shown, in one experimental situation, to follow a rhythm apparently independent of external cues.[25] Finally, the groundwork was laid for systematic study of social behavior in terms of temporal pattern.[26]

In the opinion of the British biologist Brian Goodwin, the living cell is a resonating unit which cycles continuously through a set of states.[27] The oscillating activity represents a type of biological energy which he called "talandic" (from the Greek ταλαντωσις "oscillation") and quantifiable as a type of temperature. Talandic temperature is analogous to absolute temperature in thermodynamics in that it has a zero point; but its magnitude is a measure not of the kinetic energies of particles but rather of the oscillatory energies of the cell. As Goodwin sees it, the existence of a large number of resonant modes could explain the remarkable intrinsic coherence and stability of the biochemical processes in cells. The stability of healthy cells might then be understood as the constancy of temporal relations among biochemical processes. Combinations of shorter periods lead to beat frequencies which correspond to long periods, such as circadian rhythms.

All rhythms mentioned so far are those of healthy individuals of various species, from insects to palm trees. Under stress or in disease, however, beneath these already numerous biological cycles, there appear other cycles detectable only when the overriding periodicities are repressed or disturbed. Thus, after various forms of experimental interference, or in disease, we encounter in rats periods of 40–60 days, 160–180 days, and nearly 365 days; in the normal rat these are unobservable. Somatic illnesses in man show periodicities of 12, 24, 48 hours, 7, 14, 20, 24, 26, 60 days, 4–5 weeks and even 10 years,[28] and there are numerous illnesses whose periods are identifiable with those of the causative agents. Mental and emotional periodic illnesses show a spectrum from 24 hours to 10 years. Patients who harbor different personalities often regularly alternate these in 24 hours, and, in recorded cases of stigmata, trancelike states of ecstasy have been observed at regular weekly intervals.[29]

It should be evident, or at least strongly suggested by the foregoing material that in the physiological clock we are not faced with tangential side issues but with something ubiquitous in biological processes. We also note that a strictly cyclic process—and the reader may envisage a sine wave as a suitable representation—determines an eotemporal Umwelt, for it can display conditions of earlier-later (before-after) but cannot display a preferred direction; a purely cyclic process is one of pure succession. In the following two subsections I shall point to the importance of the cyclic, eotemporal order in the nature of life.

Physiological Clocks and Their Zeitgebers

We learned in chapter 2 that time measurement comprises the comparison of readings between two or more clocks. We gave the name "time scales" to the transformation rules and stressed that a complete network of time scales weaves into unity the aggregates of all conceivable timekeepers. An analogous complex network of time scales also weaves into unity all biological clocks as well as all conceivable physical clocks and forms, thereby, a temporal substratum of living and inanimate matter. In biology the substance of this claim is discussed and known as the phenomenon of entrainment of physiological clocks. By entrainment is meant the manipulation of the transformation between the readings of one or more physiological clocks and one or more external clocks; for instance, the adjusting of the periods of the digestive system to a new circadian referent following rapid East-West flight. The external referent is usually called by the German name *Zeitgeber*, which means a reference period, synchronizer, or cue. To effect the entrainment a link between the bioclock and the Zeitgeber is, of course, necessary.

A summary of our knowledge concerning environmental entrainment was given by V. G. Bruce in 1960, who tabulated the major evidence then available on twenty-six living organisms as different as hamsters, cockroaches, humans, and lemon cuttings.[30] According to this survey, the most universal link between Zeitgebers and physiological clocks is light appropriately modulated, such as by periodic repetitions of darkness and light. We have already seen the importance played by light in the physical world, its role (through vision) in establishing identity, and its position among our ideas of creation and annihilation. To these functions of light we may now add its role as the foremost link between the environment and physiological clocks.[31] Light, however, is not the only link. Temperature cycles have been used to produce entrainment in certain plants and lizards, and successful entrainment of the circadian locomotor rhythm of the house sparrow by means of tape-recorded bird songs has been reported.[32] Circadian activities of mice were entrained by varying the gradient of electrostatic fields; and some parameters of the circadian rhythms in man were entrained by means of low frequency electric fields.[33] Other members of the species under test may also be Zeitgebers, with social interaction as the link. Thus, young women living in a college dormitory were shown to have involuntarily synchronized their menstrual periods.[34] Mutual entrainment among physiological clocks of very short periods is known from studies in speech communication: it was demonstrated, for instance, that the informative portion of speech resides in accurate time-patterning of sounds between a Zeitgeber (the speaker) and the physiological clocks of the listener.[35]

The sophistication and unobviousness of links between biological clocks and Zeitgebers were revealed by recent findings that biological rhythms in birds may be entrained by nonvisual light perception.[36] Seasonal testis growth, as well as daily activities of blinded house sparrows were shown to be entrainable by a periodic light and dark regimen. Since the degree of entrainment could be inhibited by light-absorbing paint on the birds heads, it is suspected that the place of entry of the light signal into the organism is through some part of the brain.

It is difficult to avoid the conjecture that all physiological clocks are entrainable within certain limits, given the appropriate channel of communication to the Zeitgeber. This points to a plasticity of the physiological clock and suggests entrainment as an important means of selection in organic evolution. The adaptive functions of biological rhythm reside in the advantages they confer upon the individual, such as by providing synchronization with the environment or in the maintenance of homeostasis against the rhythmically varying changes in the environment. Cloudsley-Thompson gave many illustrations of the adaptive advantages of biological clocks: for instance, endogenous circadian rhythms in certain aquatic animals and parasites make possible their survival and propagation under the harsh conditions of desert life;[37] seasonal changes in animals, such as those in colors, assist survival within the changing ecology of the tropics.[38] Physiological clocks have also been shown to be essential for animal navigation: bees can follow certain routes reported by returning scouts by employing a capacity to associate the time of day with the position of the sun;[39] a sun compass combined with a clock may also be postulated as operating in homing pigeons.[40] Bird navigation by stars also seems to depend on endogenous clocks.[41] Green sunfish trained to magnetic compass direction under the sun at Madison, Wisconsin, when transported to the southern hemisphere moved according to the sun's position corresponding to their training environment.[42] But the sun is not the only referent which may be combined with the functions of an internal clock. The famous dance of the bees employs the direction of gravitational gradient as a referent in symbolizing the distance and direction information which the bees want to convey through rhythmic motion.[43]

A very sophisticated employment of physiological rhythm in the growth of the embryo was suggested by B. C. Goodwin. He considered embryological development in terms of fields, that is, in terms of a system of locations of certain organs, for example, the eyes, or the limbs at points in space, and at instants of evolutionary time. He postulated that the coordinate location necessary for the evolving spatial organization is effected by means of periodic intracellular events which propagate at different rates from cell to cell in the developing tissues.[44] If Goodwin's model corresponds to fact, it might also explain why that portion of our anatomy which, even in the adult, retains the developmental capacity of the embryological process, namely, the central nervous system, functions as the basic temporal control in the mature adult.

Enough has been said and implied in support of the view that physiological clocks are ubiquitous among the living; also, that at the level of evolutionary development of which our own life is a part, they do show genuine, endogenous features. The generally accepted opinion is that the ultimate controlling system of rhythmic organization will be found within the physiological clock, and that such an ultimate control will be discovered to be some autonomous process in need, nevertheless, of occasional phasing from the external world.[45] But, although the physiological clock may be endogenous in an experimentally acceptable sense, it cannot possibly be divorced in any conceivable situation from genetic memory and, in some cases, not even from ontogenetic imprinting. Hence, all endogenous physiological clocks are also, in a fundamental way, exogenous.

Physiological clocks cannot be separated from the environments any more than can atomic, gravitational, or spring-driven clocks. It is not surprising, therefore, that in the spectral content of physiological clocks we can detect the inherited memory of cycles of the evolutionary eotemporal Umwelts: those of the sun and the moon, of the tides, of the changing seasons, periodicities, of predators, and of food supply. We will see later that, as do all complex oscillators, organic systems also develop beat-frequencies among their basic, entrained frequencies; that is, they generate harmonic combinations which do not necessarily correspond to any external cyclicity. B. C. Goodwin remarked that "biological time is inextricably woven into the fabric of biological organization throughout the levels of complexity that have arisen in the evolution of the metazoa." [46] Indeed, the capacity to create physiological clocks in response to environmental rhythms appears to be a necessary feature of all life forms.

> The dance along the artery
> The circulation of the lymph
> Are figured in the drift of stars.
> T. S. Eliot, *Burnt Norton*

Periodicity, Primitive Life, and Existential Tension

A recent introduction to the biochronology in plants is dedicated "To the students of the future who will reveal the balance wheel of the biological clock." [47] This amounts to an assumption widely held by working biologists that such a timekeeper can eventually be found. Certainly, processes characteristic of the physiological rhythm of specific organisms will be identified; but our earlier inquiry into the clockness of clocks suggests that such discoveries can reveal only provisional, rather than ultimate, timekeepers. By dissecting an organism one can hope only to find things; only by "timesecting" an organism, that is, by understanding the hierarchical structure of its temporalities, can we hope to come to understand the role of the physiological clocks in life. But it is evident that the more universal a basis is found for physiology the more likely it will be something quite removed from time and life until, in successive stages, we eventually arrive at the prototemporal world of indistinguishable particles.

We learned in chapter 2 that man-made clocks are selected processes whose regularity, as judged by the clockmaker, compares satisfactorily with processes better established or thought more reliable; also, that the history of clockmaking is one of successively improved approximations to something that seems to exist only in the mind of man. Clockmaking through successively improved approximations of environmental rhythms has also been the strategy of organic evolution. Our first evaluation of physiological clocks suggested that they are necessary for the survival and propagation of the organism; according to that view physiological clocks link the organism to its environment. A closer look reveals, however, that the first view is inadequate. Biological clocks do not assist but, to a large degree, make up the individual. I perceive at the basis of life processes a host of programs (temporal analogues of spatial models) composed of selected periodicities of the environment. "Selected" does not assume a telic selector but only the operation of natural selection which favors behavior well-adapted to external Zeitgebers. Thus, unlike a clockmaker who fabricates clocks so that they may serve man, the physiological clocks created by evolution comprise man—and life in general.

We may imagine that the simplest primordial organism, comprising perhaps hardly more than models of daily temperature and salinity variations of the sea, already experienced existential tensions. First of all, the cyclicity of the organism could never have reproduced with sufficient accuracy such astronomical cycles as the rising and setting of the sun. Hence the clocks had to remain imperfect. Furthermore, the drifting clouds in front of the sun, a current in the ocean or eruptions of volcanoes were unpredictable elements even in principle. Hence the cyclic changes they anticipated had to differ continuously from conditions actually encountered. It is probably here that the becoming-type and being-type elements in behavior became first differentiated as the distant ancestors of the existential tension of later life.

(1) A detour to pathology

To deal further with the relation of the physiological clock to life we will find it useful to consider primitive disease and death; and to do so, we must detour to pathology. Opinions as to what causes and what constitutes disease have changed greatly. For the Pythagoreans, diseases were caused mainly by opulent living. In the Greek prognostic doctrine, health was synonymous with balance (material, fluid, or activity), disease with the disturbance of that balance.[48] The revolution in the general estimate of health and disease came with the Hippocratic corpus. This reflects the conviction of the father of medicine that diseases are phenomena understandable in terms of natural laws on the basis of objectively recorded information and hence, in modern terms, may be dealt with by therapeutic procedures developed on a rational basis. His principle, often stated in Latin as "Natura vix Medictrix" (nature is the physician of disease) is an article of faith attesting to the harmony of all processes. His *Airs, Waters, Places* would be classed today as an essay in ecology. The work stresses the central importance, for the physician, of the items listed in its title, as well as his need to know the personal habits of his patients. The physician should also be knowledgeable about changes in temperature, humidity, and the winds, position of the sun and the like. And he wrote that,

> If it be thought that all this belongs to meteorology, [the physician] will find out, on second thoughts, that the contribution of astronomy to medicine is not a very small one, but a very great one indeed. For with the seasons men's diseases, like their digestive organs, suffer change.[49]

Reading the proceedings of the *International Biometeorological Congresses,* or those of the *International Institute for Interdisciplinary Cycle Research* would have delighted the Hippocratic heart.

Bypassing the history of medicine between Hippocrates and the turn of the sixteenth century, a world unto itself, we must mention René Descartes (1596–1650). His direct contribution to medicine is very little, if any, but his general view of man as a machine remained influential to our own days. Even more profound has been the indirect influence of his analytical method on our views concerning life. The explosive growth of biological and medical knowledge, as the growth of other sciences, may be attributed greatly to the power of that method. When in twentieth century medicine and biology the investigator proceeds from organism to organ, then to cell, to cell fragments, to molecules, atoms, and, perhaps to Maxwell's equations, he is reaping the reward of Cartesian analytics. The price paid is that in this process of elimination the central importance of life is irrevocably lost as biological research flees from existential conflict into physics and chemistry. As Dubos put it,

> Many scientists who dedicate themselves to medical research tend to shy away from the problem peculiar to man's nature, and even from those posed by other *living* organisms . . . and they deal by preference with questions pertaining to lifeless fragments of the body machine.[50] [italics his]

About eighty years after Descartes' death the great French physician Pierre Bretonneau formulated the dogma of specific etiology. It states that each disease is caused by a specific organism; this makes it transmissible and, if the organism is controlled, controllable. Philosophically, this is a belief in single-valued causation; medically, it foreshadowed the germ theory of disease and remained a guiding principle of the working physician. With the increasing body of experimental knowledge it became clear, however, that the situation is not that simple.[51] A noxious agent may cause a variety of pathological states, while different agents may elicit similar reactions; animals or man completely isolated from external infections can develop abnormalities which, if the experimental conditions are maintained, can lead to death. The organismic approach which appeared so clear-cut and useful for two centuries seems to have been based on an oversimplified view of the organism as apart from, rather than as a part of the environment. In life phenomena one encounters general systems behavior: each organ or cell, and each detail of the internal and the external landscape, loses its individual significance and remains important only insofar as it is part of the on-going life *process.* Therefore, it is in the temporal organization of the simplest of life processes that we ought to search for the origins of what, in complex organisms, we know as aging, disease, and death.

(2) Unicellular organisms

Consider, for instance, bacteria, which are among the smallest living creatures known. Since 1964 oscillations in cultures and in individual bacteria have been observed by several investigators. Earlier it was believed that, for reasons unknown, bacteria were exceptions to the universality of physiological clocks in the living. Currently, evidence has been gathering that the clocks are endogenous and entrainable.[52] A suspension of bacteria, as far as we know, is unaging, hence in a way immortal. Cell division in bacteria will continue indefinitely, limited only by the size of the vessel, though there is some evidence that some must occasionally undergo a form of sexual reproduction.[53] Under unfavorable conditions, of course, individuals or the whole colony may die. B. L. Strehler sees sufficient evidence to declare, in partial endorsement of the work of the nineteenth century German biologist August Weisman, that "the essential absence of deteriorative processes in the germ line may be accepted without serious reservations." [54] In Alexander Comfort's study of aging, he points to a certain indeterminacy of what is meant by aging in unicellular organisms. But he concurs with the prevailing view, which is still generally held, that the outcome of a protozoan cell division is a pair of rejuvenated and infant cells rather than a mother and daughter of different seniority.[55] In some other creatures, such as the hydra,[56] and in certain strains of paramecium,[57] there is a potential immortality in that the individuals may be kept in continuous, vegetative reproduction. For numerous microorganisms above the level of bacteria, such as the euglena * persistent circadian rhythms have been observed.[58] Above the level of bacteria the issue of death by aging becomes increasingly complicated, because some unicellular organisms age but others do not.

It would seem, then, that the simplest life processes comprise hardly more than physiological clocks which are purely cyclic and determine an eotemporal Umwelt. We have learned that eotemporality is the world of pure succession, but with no preferred direction. It is not surprising, therefore, that entirely periodical creatures do not age, or, in any case, that they have an indeterminate life span. The distinguished dean of German neurology, G. Schaltenbrand has noted that living systems made up of periodic processes alone can change only through creative acts, such as growth, differentiation, or the loss of some process.[59] Indeed, aging, and with it an adumbration of a preferred direction of time, appears among the living only upon the assumption by the individual cells of some specific tasks within the integrated functioning of higher-order organisms, of which some more anon.

In an ideal abstract case, an organism comprising periodic processes only would find itself in complete sychronicity with its Zeitgebers, and would command precise coherence among its internal clocks. In actual cases, however, there will always be a difference between the periodical conditions expected and those encountered, quantitatively expressible as a phase-angle difference. The same type of expected/encountered conditions must hold for the inner coordination of physiological clocks, even in the

---

* An interesting creature, claimed as an animal by zoologists and as a plant by botanists.

simplest organisms. Such phase-angle errors amount to simple existential tensions. In this philosophical-schematic view one can say that as long as these tensions do not interfere with the integrity of the organism, the organism is alive and healthy. For many unicellular creatures happiness is a regular time rate of change signifying that the organism is adapted to its environment. When the tension between the expected and the encountered becomes too large, the organism may be disrupted and become diseased, or die. The expected is clearly a beinglike component of this stress; the unexpected, which is the difference between the expected and the encountered, is becominglike.

The concept of existential stress in unicellular organisms, expressed in terms of being and becoming, is certainly not as familiar as the stock "inability to adapt to the environment." Yet the former is in some ways more specific and quantifiable. We are speaking of the organism being out of phase with its environment according to a program of rhythms. It is well known, though not extensively studied, that pathological phenomena accompany the loss of coherence between internal and external clocks, or the loss of integrated coherence among internal clocks. Attempts to impose nondiurnal rhythms on the circadian clocks lead to disturbance in plants, animals and man, as do attempts to confuse by dissociation the coherence of physiological clocks within an organism.[60] I have already mentioned periodic diseases which show either the periodicity of the pathogenic agent or else the periodic susceptibility of the host. Removal of normal synchronizing stimuli also often leads to disease. Perhaps pathogenesis itself, certainly at the primitive levels of life, may best be understood as a temporal mismanagement. The intensity of the existential tension between physiological clocks and their Zeitgebers seems to be a crucial parameter in the fate of the living. It may be sufficiently large to disrupt the organism, and we speak of death. Under different conditions, and/or at different intensities, the tension may lead to aging or, it may cause the organism to evolve by natural selection through successive stages of adaptation.

(3) Improved bioclocks

Subsequent refinement of the complement of physiological clocks may be imagined as an increasingly sophisticated modeling of environmental rhythms in complete analogy to the increasingly refined theoretical models advanced by man to account for his Umwelt.[61] Inevitably, the coexistence of even a few clocks of different frequencies within an integrated organism must produce new periodicities by linear superposition and by nonlinear generation. Such new periods do not need to correspond to any preexisting external periods. Thus, an inner landscape of species-specific and private rhythms must evolve with their own strong inner criteria for selection, namely, viability through coherence. Inner temporal conditions are thereby added to external temporal conditions as selection pressures. Refinement in the inner imaging of external periodicities amounts to improved adaptation, because it produces a better fit to external regularities. Each surviving new rhythm adds a new component to the harmonic analysis of the environment. But improved adaptation provides a necessarily increased region of interaction between the organism and its environment; there is now an increased domain of experience wherein the expected and the encountered may differ. The inevitable result, on the average, must be an increase rather than a decrease in existential tension. Furthermore, if the increasing complement of physiological cycles involves an extended population, then the environment to which the organism has to adapt must itself be regarded as having complexified.

That improved adaptation leads to an increase rather than to a decrease in existential tension is an important part of the theory of time as conflict, as it applies to organic evolution. To elucidate this curious condition we must first appeal to that feature of the Umwelt which assures that whatever an organism does not perceive is something which, in the Umwelt of that organism, does not exist. It is this fact which is exemplified, for instance, by the model of child development described by Erik Erikson.[62] Erikson and others see in the developing personality of the child a series of well-defined stages. Each is entered through a crisis which arises when the child first encounters something which, up to that age, he was unable to encounter. Child psychologists describe these encounters as sudden and radical changes in perspective. While the growing child becomes better and better adapted to his environment, each step brings with it a new and previously unknown and unknowable store of concerns.

Another homocentric example may be the learning of a language. When one first hears a foreign tongue, say Turkish, it all sounds completely unstructured. One could not react to a Turkish announcement be it one of gloom or plenty. Subsequently, though much better adapted to the Turkish environment, one will begin worrying about many things which earlier were of no concern. Metaphorically speaking, a magnifying glass makes it possible to construct a better square plug for a square hole, hence it helps adaptation; but it reveals a need for such further refinements as were quite unknown earlier. I believe that on all levels of natural selection, improved adaptation also means increased existential tension.

Existential tension derived from the differences between cyclically anticipated and actually encountered conditions saturate all of life, including that of man. They must not be mistaken, however, for the source of our sense of time, that is, of nootemporality.[63] At this level one can accommodate only the rhythmic origins of life with their unsophisticated tensions, such as may disrupt a culture of bacteria. In the philosophy of time as conflict we may perceive here the emergence of biological temporality just above the eotemporality of macroscopic matter. If we are to go to higher levels of temporality, we must advance from eotemporal physiological clocks to the irreversibilities of aging and death.

2. Aging and Death

Aging of Others

The cyclic order extends beyond the private life of organisms; genetic systems, for instance, can also have oscillating properties. Successive generations of organisms sometimes follow cycles with periods longer than the individual life spans, such as the cyclic appearance of spring and summer forms of butterflies or the life cycle of metamorphic insects.[64] Social rhythms of animal colonies, such as those of migrating birds, bees, fishes, and certain mammals, are examples of aggregate periodic behavior. Thus, rhythm may embrace generations of individuals in time and multitudes of individuals in society.

The changing environment of primordial life with its inanimate as well as living Zeitgebers must have favored the formation of multitudes of cells associated in integrated groups, taking advantage of the possibility of a division of labor. The Portuguese man-of-war is an extant example of a colony of specialized animals that form a unit which assures an increased chance of survival for its members. Specialization makes possible the seeing and sensing of events and things at greater distances, and this is equivalent to foreseeing more distant futures. To achieve this, however, in the new autonomous society, the members have to yield their former autonomy and assume specialized roles useful for the aggregate as a unit. For instance, in sexually reproducing organisms the power of reproduction, as one function, becomes concentrated in selected cells which remain the only ones potentially ageless in the sense explained earlier. Other cells of the body, the somatic cells, exhibit, in addition to periodical behavior, the phenomena of aging, or senescence and death.[65] That some cells which make up a complex individual do age while others do not is in itself interesting; even more interesting is the fact that the organism as a unit ages with its somatic cells and does not remain of an indefinite age with its germ cells. While the germ plasm lives in its eotemporal continuity, the soma passes on; while Jesse dies, "a rod of the stem of Jesse" may come forth "and a branch grow out of his roots." Less poetically, senescence is not so much a characteristic of a species or line of animals as it is of individuals representative of the species. Accordingly, theories of aging generally concern the biology of the soma. But there is no generally accepted, universal biology of aging and death. Insight comes from many approaches, presently uncoordinated.

In sufficiently complex organisms aging is sometimes attributed to various cellular activities. The somatic mutation theory holds that spontaneous, irreversible mutations in the soma curtail cellular functions and lead to the eventual death of the cell.[66] Another view holds that cells become autoimmune, that is, they step up their immunological activities against other components of the body's own tissues.[67] Either of these reasons would render the individual more likely to die from disease or accidental causes, and this feature itself is one possible definition of senescence. In the early decades of this century G. P. Bidder suggested that senescence derives from the need of the organism for optimum size, as determined by physical requirements. He postulated the existence of a regulator that controls the rate of growth to make certain that individuals do not overgrow. He understood senescence as the result of the continued action of the regulator after growth was stopped.[68] It has also been suggested that confused and unnecessary cross connections which interfere with the operation of the central nervous system (in organisms which possess one) are responsible for aging.[69] Recent work on the origins of cell tissue aging suggests that senescence originates when the cyclic activity of cell tissues is inhibited.[70] In this scheme of cell tissue aging, two potential gaps are postulated in the recurring cycle of nuclear DNA synthesis—mitosis—nuclear DNA synthesis—mitosis . . . namely, between synthesis and mitosis and between mitosis and synthesis. If blockage at either of these points occurs, the cell begins to age and, unless the blockage is removed, the cell senesces and dies.

There are, of course, as many potential sources of aging as there are different types of biological functions. Hence, appropriate domains of inquiry into aging include not only biology (molecular, genetic, etc.) but psychology, sociology, and evolution. I believe that the most interesting starting point for an understanding of aging and death is the examination of their possible adaptive advantages. In this respect Comfort suggested that senescence and death are manifestations of (1) the accumulation of lethal genetic effects delayed earlier by selective pressure so as to interfere less with reproductive activity and/or (2) the decrease of evolutionary pressure toward homeostasis with increasing age, a running out of program, as it were.[71] It is certainly true that the average life of wild animals is short compared to their potential age, hence biological control of the post-reproductive period is useless; this is consistent with Comfort's two points. There is a danger of circularity, however, if short lifetimes are taken as consequent on senescence and senescence consequent on the necessity of short lifetimes. For human societies the issues are further complicated because adaptive pressures in favor of survival beyond the time of reproduction can remain high. Yet, aging remains an incontrovertible fact.

Although modern science has succeeded in increasing the average life span of Western man, man's maximum lifespan probably has not changed in recorded history.[72] Old age itself has remained, as it were, a disease:

> Last scene of all,
> That ends this strange, eventful history
> Is second childishness, and mere oblivion,
> Sans teeth, sans eyes, sans taste, sans everything.
>
> *As You Like It* (II, vii, 163)

These changes which are either correlates, or derive from elemental changes in the subcellular structure, have been described by B. L. Strehler as of two contrasting types. The first type includes stochastic processes, that is, those resulting from improbable events that have no structural counterparts (for they could not) in the genetic makeup of the organism. The second type includes the genetic process. He distinguishes between the two by observing first that evolution favors the development of forms which minimize the probabilities of demise by internal or external accidents. In aging complex organisms

> the genome tends to compensate for and adapt itself to the regularly recur-
> ring events in the environment. Extremely rare or unique events would only
> possess an antagonizing counterpart in the genome by chance, since natural
> selection would have had no prior opportunity to select for the compensated
> variant.[73]

Thus, in the chromosome complement of the gamete or zygote on the one hand and, on the other hand, in the soma, we come upon the conflict between being and becoming. Strehler's "genetic process" represents cyclic predictability in contradistinction to the unexpected, stochastic world of the soma.

I have reasoned that cyclic systems evolved under the selective pressure of the internal and external environments through the addition of new functions with new frequencies. The environments of cells, animals, and men must be thought of as made up of random noise until such time that components of that noise are identified by the living unit as cyclic regulari-ties; furthermore, this dynamic condition is a permanent one because the environment will always contain regularities yet to be identified. Environ-mental regularities are not only practically inexhaustible in the first place but they constitute a store whose cyclic content increases through the very action of evolutionary adaptation. After the crudest cycles of solar and lunar rhythms, as well as rhythmic variations in food supply and predator activity, become mapped into the physiological cycles of the organism, there will remain (among other cycles) some with very long periods which will appear linear and some changes which are truly linear. Examples may be geological periodicities (such as weather cycles); geological, linear changes (such as the cooling of the earth); long term cyclic and linear changes in the living environment (such as the increase of biomass) and, importantly, the enduring nature of the organism itself. The adaptive process of the organism to the cyclic, quasi-linear, and linear changes in its environment, through the reconstruction within itself of the periodicities of these changes, is a filtering, abstracting and differentiating process on its way to becoming self-referential.

Just as an increasing existential tension had to accompany the increase of the spectral complement of the physiological clocks, so the modeling of quasi-linear and linear processes by the organism must also result in increased existential tension. Namely, while the slower components do make for better adaptation because of the increased capacity of the organism to foretell future conditions, they also increase the domain wherein conflict between the cyclically *and* linearly predicted conditions may differ from the cyclically *and* linearly encountered ones. Furthermore, organisms that age must possess features which secure the viability of their internal organization, producing some new and further inner cyclicities by linear addition and nonlinear generation not originally present in their environment. In the philosophy of time as conflict the increased tensions are seen to correspond to a new level of temporality embracing all organisms that senesce: Portuguese men-of-war, frogs, horses, and men. We shall describe this new Umwelt as biotemporal.

Aging of the Self

The concept of the organism as a temporal unit, that is, as a single event rather than a static spatial entity (the anatomy of the adult phase) has been gaining a foothold in developmental biology. It is the entire life cycle which undergoes evolutionary change, and phylogeny is a succession of changing life cycles, each with its "immortal" germ cells and mortal soma. In this view "I" am a multicellular organism whose soma scenesces while its germ cells, or in any case their programs, keep on functioning in "my" children. But, whereas I do see my children and I know that I am aging, I do not feel any of these very complex goings-on. There is evidently a gap between understanding aging as a biology of senescence and experiencing aging. What to others appears to be a gradual impairment of sensory functions constitutes, for the individual, a change of his Umwelts. Correspondingly, there are two basic methods of approaching the question of aging: those of others starting from biological principles, and those of "me," starting with the phenomenology of human aging in its biological, psychological, and sociological settings. These two approaches might conceivably meet through the clarification of the concept of identity as a common theme.

Unlike the offspring of most animals, the human infant is unfit to fend for himself and comes to complete maturity only after many years in the rough womb of society. As anthropologists and biologists have correctly stressed, the young child retains some of the plasticity which characterized his prepartum existence, hence man, more than other creatures, is a product of society. He is what in birds would be called nidiculous, that is, he must be reared in a nest. Therefore it is not possible to discuss the meaning of aging without reference to its social setting. Attitudes of society to aging and to postmortem existence are part and parcel of the individual's view of his own aging.

As the child grows he discovers futurity and comes to understand that he is getting older. Then he discovers mortality as a part of futurity, a piece of information which through his life undergoes a number of successive evaluations. His childish "growing up" changes to "getting older" and, eventually, to thoughts about time.

> Though lads are making pikes again
> For some conspiracy
> And crazy rascals rage their fill
> At human tyranny,
> My contemplations are of Time
> That has transfigured me.
>
> W. B. Yeats,
> "The Lamentation of the Old Pensioner"

Still later, as the inevitability of death makes itself felt through memory and expectation, the motive of "getting older" returns as the aged does fewer and fewer things for the first time and more and more things for the last time. As the uncertainties of life (which in the past were feared but now are coveted) change into the certainty of death, man retreats along his evolutionary Umwelts.[74] C. G. Jung often emphasized that the psychology of the latter half of man's life is a preparation for the inevitability of his death and that all of man's actions are increasingly so informed. This period is somewhat the reverse of the neonate's coming of age outside his mother's womb; the elderly are nidiculous like young birds. It is during this period that the unity of the experience of aging becomes redifferentiated in the increasingly sharp separation of life from death. Although this process of polarization affects society because it guides the behavior of the individual, the tension between the trend to rise out of and also return to the inorganic, of which the individual is increasingly aware, remains a privileged condition; it is known only to him.

In the sociopsychological vistas of aging there appear to be only a few theories and none very new; they are rather old ideas in new forms. The "disengagement theory" emphasizes the communal aspects of aging. It holds that aging is a process of mutual withdrawal between the self and the nonself.[75] Whether originated by the self or forced upon it, it assists in the withdrawal of cathexis (the energy inspired by a goal) from external goals to those related to the self. Hopefully, this leads to a new equilibrium with a limited external scope, corresponding to the reality of less available time. That this view, which is popular among gerontologists, has many grains of truth should be evident from personal experience.[76] Yet it does have two shortcomings. One is that it is too Western and, even more specifically, American. In another country, where age is venerated, disengagement may consist of more rather than less involvement. Also, differences in time span make for qualitative difference: disengagement through a period of a decade is quite different from disengagement two minutes before certain death. This time-scale problem bedevils the whole growing field of gerontology which deals with aging of the aged rather than with aging in general.

The "developmental theory," a conceptual cousin to developmental biology, was described by Kastenbaum as one in which the person is seen as comprising a number of subsystems.[77] This permits an interpretation of aging in terms of several structural functions, held together by the time perspective of the individual. Research into problems of aging will doubtlessly continue under the political and social pressures of an increasingly older population, and the potentially ruinous prejudice against the aged in the industrially advanced countries. There are, for instance, studies directed to the possible elongation of the average human life by means of recoding the DNA molecule, or by nongenetic means, such as the prevention of the formation of molecular aggregates characteristic of aging. Such efforts tend to be examples of scientific naiveté, apparently unconcerned with the psychological and social consequences and ramifications in the status of aging of the self, consequent to biological meddling. How would this self be affected by the social manipulation of its age? Would extended aging amount to longer vigor by making adolescence last for thirty years? Are we to strive for a population of nonsenescent vegetables with senile, television smiles? Or would we come up with long-lived Frankensteins? Or, produce "excellent Strudlebugs, who, being born exempt from the universal calamity of human nature, have their minds free and disengaged, without the weight and depression of spirits caused by the continual apprehension of death" as those in the Land of Luggnaggians found by Captain Gulliver? [78] What qualitative differences in the identity of the individual might result from the continuously increasing spectrum of age differences? Perhaps all of man's most noble qualities come from resignation to or revolt against death. What character other than that of green beans (a vegetable) could we expect to find in a deathless man?

While examining the aging of others: cats, horses, or neighbors, it is always possible to escape from the conflicts of life to the timelessness represented by science or religion. But "my" experience of aging cannot ignore certain unresolvable conflicts implicit in the noetic form of time except perhaps by regressing to the experience of timelessness characteristic of saints, savages, and children. I do not believe that a suitable theory of aging can be formulated until we permit the privileged knowledge of the individual concerning his aging, and conventionally expressed through the humanities (i.e., as knowledge felt) to form a bond with the views of aging as a biochemical process (i.e., as knowledge understood).

Death of Others

Various evidence cited earlier suggests that men of ancient civilizations and, before them, paleolithic man already wondered about death long before the Renaissance taught us to wonder about life. In a primitive, panvitalistic world-view, life is the unquestioned mode of existence, the universal way things are; the significance of death, profound as it may be, remains within

that scheme of things that are in the design of life. Even in Plato's highly
sophisticated mind the world was a Living Creature. Religious evaluations
of life, spanning as they do the history of man, cover a rather broad spectrum of views from attitudes to human life as a gift of God to judgements
of life as only of transient value. But the articulate opinion that life is rather
the exception in an overwhelmingly inanimate world, hence deserving of
special attention, is only a few centuries old. With the rise of scientific
understanding, stunning discoveries were made about life and the living;
compared to these, death remained unchanged, gray, and negative; the flight
to (and increasing rush toward) exactness and timelessness, which came to
be expressed in the high esteem for scientific thought, reinforced and
welcomed this view. Death itself came thus to be degraded two ways: it was
less significant than life, hence not worthy of attention; and, with the
demise of the organic view of the world, it was also removed from the
scheme of life. In the unwritten but binding laws of industrialized societies
this evaluation came to be expressed in the form of a social taboo about
death, a stance generally unchallenged until the appearance of existential
philosophy. And, not until the last few decades, cajoled by the humanistic
atmosphere of romantic and youth revolutions, did science turn to an
examination of death. Accordingly, theories of mortality had to begin the
way new sciences usually do: as mixtures of anecdotal knowledge, common
sense approval by peers, and some rudimentary quantitative relations.

Since the early work of British actuaries it has been known that there
exists a lawful relationship between the years of life as an independent
variable and the relative mortality rate. This logarithmic relationship is
known as the Gompertz curve, after its discoverer in 1825, Benjamin
Gompertz. With some modifications it has been successfully extended to
many species and to various types of populations.[79] Many attempts have
been made to interpret its meaning for a human population. The "loss of
function" view of mortality holds that each age has its own specific ailments
which eventually result in death, leading up to it according to the lawfulness
of the Gompertz plot.[80] The Brody-Failla theory replaces mortality rate by
its reciprocal, the vitality rate, in hope of finding better understanding. The
Simms-Jones theory attributes the logarithmic lawfulness to the accumulation of damage and disease; the Simms-Sacher theory invokes stochastic
processes whose random fluctuations carry individuals to the lethal limits of
their physiology. The Strehler-Midwan theory sees environmental challenges as random attacks on the linearly decreasing functional capacity of
man.

It seems to me that those who propose to determine the mechanism
underlying aging and death by beginning with statistical findings, such as
those embedded in the Gompertz plot, set for themselves a nearly impossible
task. By analogy, it may be compared to the proposal that we derive the
field equations of Einstein by trying to find analytical fits to ballistic curves.
Instead, let us take a hint from the fact that the lawfulness of the Gompertz

curve is an example of the ergodic theorem. The subjects, for the purpose of the plot, are indistinguishable; it is this condition that makes possible the going from the mortality rate of a species (an ensemble average) to the physiological age of an average individual (a time average). The demand for indistinguishability suggests that we first improve our perspectives by retreating from man and recalling that mammalian germ cells are eotemporal. They are "immortal" in that upon division no dead bodies of the parent remain. The same holds for fungi and for some plant cells. Annuals grow according to genetic programming which includes death as the natural end of a continuous existence; the senescence and death of trees, however, is usually due not to genetic programming but to mechanical difficulties. Thus, more advanced plants return to dust and get, perchance, recycled. In mammals the soma senesces while the zygotes leave the aging body to carry on their eotemporal existence. Thus, death through aging and the residue of a dead body are not complements of all life, but only of more advanced types of life.

Let us hold now uncritically that Darwinian natural selection is the final arbiter of the shapes and functions of the living, and that it operates mainly on purely random genetic variations. What do we find in organic evolution that could give some hint of the advantage of death? Longest living organisms, approximating the "immortality" of bacteria, are trees. The bristlecone pine may live for 4,600 years, the giant sequoia for 3,000 years.[81] Fossil records suggest, however, that trees in general are on the descendancy. Woody plants have existed for perhaps 350 million years, while grasses are only about 70 million years old. The evolutionary trend is toward shorter-lived plants, for reasons which Darwinian selection can explain with remarkable clarity. Namely,

> the most aggressive species of plants, those which adapt most readily to the environmental niches—the weeds—are predominantly annuals, in which the over-all senescence imposes a rapid turnover of individuals in the population.[82]

Rapid turnover is favored by natural selection because it makes possible in a shorter time the incorporation of beneficial genetic changes in a large segment of the populace. (Because of the difficulty of regressive sharing we do not mourn the simultaneous death of innumerable individuals as we contemplate the wheat fields ready to be fallowed. Though we may share Isaiah's sorrow that "the grass withereth, the flower fadeth," we would still praise the victory of next year's crop in our bid for immortality, "but the word of our God shall live forever"). The advantages of high numbers of generations per unit time are also known in the evolution of animals.[83] Although, as Comfort observed, adaptive modifications of life span in phylogeny have many examples pointing in the opposite direction,[84] the shortening of individual lives for the advantage of increased viability of the species seems to be the policy of evolution.

The thrust toward shorter-lived organisms below the level of man suggests a "time economy" to organic evolution. Let us assume that we are ignorant of temporality and know only organic evolution by natural selection. The trend for shortening individual lifetimes, favoring the survival of the species, would then appear to be something of a minimal principle. (*Not* a principle of minimal time, for that would be circular). We would connect this quality of nature with the phenomenon of life and see in it a principle which, in itself, does not guarantee change but only specifies its direction.

Let us return now to our understanding of the origins of organic evolution as an Aufbau Prinzip of physiological clocks, mapping external periodicities and generating some of their own. It would appear that such an enterprise demands complexification which might lead to some unmanageable conditions. As generations of organisms are selected for increasingly sophisticated cyclic structures, new cyclicities must appear in the internal and external landscapes of the organism leading, one would think, to limits intrinsic in the biological functions of matter. There will be a necessity, for instance, to extend the spectrum of periodicities to very long and very short periods. The short periods will have to reach, eventually, atomic frequencies. This, of course, is the case with light; in the functions of the eye the physiological clocks reached the boundaries imposed by the atomic character of light. The rods and cones do not vary with the size of the animal, as D'Arcy Thompson noted in his famous study on the proportions of living things [85] but have their dimensions limited by the optical interference patterns of light waves which, thereby, set the boundaries to the production of clear retinal images. To know a hawk from a handsaw the eyes must distinguish among atomic frequencies; thus retinal partitions coincide with the limits of interference patterns.[86] But the physiology of biochemical clocks responding to periods around 10^{-15} seconds can bear very little resemblance, if at all, to the physiology of, say, circadian clocks; specialization therefore is forced upon living matter by the differing properties of the physics of different frequency domains. As to the lengthening of periods, we learned that rhythmic functions may be shifted from the physiological clocks of the individuals to cyclic behavior of the species. The clock may thus be passed to the genetic system, doubtless, with its own phylogenetic limitations. Alternately (and opposing the trend discussed a few paragraphs earlier), the evolution of higher animals may favor longer lives for the soma so as to enhance the viability of the species by communal foresight through further specialization.

A widening spectrum of physiological clocks also necessitates vastly increased correlating and integrative activities within the organism. Such demands must eventually lead to conditions which might be called an infinity catastrophe; that is, the impossibility of becoming infinitely complex. It seems, therefore, that if the existential tension of the cyclic order is resolved solely by continued extension of the spectrum of biological clocks, we must arrive at an equilibrium of attrition by specialization and thereby to a cul de sac of organic evolution. This reasoning suggests that death by senescence is not a necessary correlate of life, but a demonstration of the limits beyond which conflict-solving through biological functions is not possible. Death then, signifies the boundaries of a particular method of conflict-solving in the strategy of existence. The wise Luggnaggians of Gulliver, instead of creating their immortal Strudlebugs, would have been better advised to cultivate their minds.

Death and the Self

Psycholinguistic analysis of verbal expressions of death suggest that "natural" death, that is, death by senescence, is a fairly recent idea. Conditions of life for our not too distant ancestors, and for many of our contemporaries, were and are in many ways no different from those of wild animals. On the average, people were killed violently or perished from hunger or disease long before they would have perished by senescence. The necessity of death emerged as a rather late and painful insight, Thass-Thienemann observed, and because this insight is painful, man avoided calling it by name. He resorted to cover words and substitute expressions.[87] Two such categories may be distinguished: one that perceives death as "killing" by an outside agent, and one that suggests an internal, almost voluntary process, death by "departing" or "passing away."

Infantile fantasies interpret the beginning of life as the arrival of a newborn from wherever it lived before; similarly, the end of life is perceived as the dead leaving this abode for another one. The fantasies of death as a journey to a simpler or better existence are all too well known; their symbolism permeates the religious and secular literature of passing and suggest going home, often to mother. The utterances of the dying tend to return to childish exclamations of fear or love, to phrases of long unpracticed mother tongues, or cries for eternal peace without conflict. In the beautiful Passion legend, *The Last Temptation of Christ*, Kazantzakis writes of the hallucinations of Christ dying on the cross.

> The moment he cried Eli, Eli and fainted, Temptation had captured him for a split second and led him astray. The joys, marriages and children [which he imagined] were lies. All—all were illusions sent by the Devil. His disciples were alive and thriving. They had gone over sea and land and were proclaiming the Good News. Everything had turned out as it should, Glory be to God!
> He uttered a triumphant cry: IT IS ACCOMPLISHED!
> And it was as though he had said: Everything has begun.[88]

By the making of the end into the beginning, old age and death can be accepted in the mind as the timelessness of childhood regained. One of the great difficulties in conceptualizing "my death" is the requirement of regressive sharing implicit in this return.

Although mortal terror in the face of imminent death is shared by man and beast, the importance of continued anticipation of death, expressed in a diffused anxiety, is unique to man. It is as though through the knowledge of the inevitability of death the idea of "inevitability" itself came to underlie the behavior of man and send our species scurrying for eternal biological life. This anxiety enters through the individual in the form of distinctions between the death of others and the death of the self. Ingenious reflections on the possible sources of this separation may be found in Freud's *Thoughts for the Times of War and Death*. Primeval man, said Freud, held radically different views about his own death and about the death of others. This was a convenient stance which permitted him to defend himself and kill others without self-criticism. But this stance became disturbed at the death of someone whom he loved or cared for. Such a death was horrifying as the death of the beloved, but also irrelevant, for even the beloved must have remained, as nonself, somewhat of a stranger.[89] Because he had accepted uncritically his own desire to kill (others), an inarticulate sense of guilt appeared: "Have I killed him?" and with it, a revolt against death. Thus, as Freud sees it, man's concern with death did not derive from simple observation of the death of others, but from a complex mental process which presumed the emergence of his identity. In terms of this Freudian understanding the polarization of the world into the deathless unconscious and the passing (the conscious) self is coemergent with ancient guilt or original sin. The identity of the self is thereby tied to the unresolvable stress between the knowledge of becoming (going out of being in death) and being (limitless continuation in life). But then, primeval man cannot be said to have held, a priori, different views about his death and those of others. Rather, he should be thought of as ignorant of his own death until he learned to confront the knowledge of his inevitable demise through the experience of the death of another.

We learned earlier that death evolved in the form of a throwaway soma versus surviving germ cell (or, biotemporality versus eotemporality), when procreation was relegated to a specialized part of the body. Now, the doctrine of the First Intercourse, otherwise known as the original sin and expressed by Saint Paul in "the wages of sin is death" became so articulated only some 1,000 million years ex post facto, and remained one of the more successful dogmas of Christianity for two thousand years. I cannot but wonder what racial memories of the evolutionary coemergence of sexual procreation and individual death underlie the power of this dogma transmuted, as it has been, into innumerable secular forms in our secular cities. Perhaps even the ancient guilt, postulated by Freud for our early human ancestors, was a phylogenetically late echo of the simultaneous birth of carnal knowledge and carnal death.

Even a brief examination of the phenomenology of "my death" will reveal that in it a curious duality still survives. The living self is always imagined as the observer of the dead self, even though the dead body is presumably without powers of observation. (Hence the logical need for postmortem existence). This intellectual separation of the self into an observer and an agent can only be eliminated if meditative introspection removes from awareness the structural and analytical concepts of the world. When this is attempted, however, the result is likely to be a state of mind best described as a falling into the infinite abyss, an encounter with the final, absolute nothingness of Kierkegaard, Heidegger, Sartre, and Kitaro Nishida. Socrates had already held that philosophers are in love with death, and the Psalmist knew of the need of courage ("though I walk through the valley of the shadow of death, I will fear no evil") but it was modern existentialist thought, exemplified by the spirit of Kierkegaard, which came to stress its central role in the affairs of man. For Heidegger and his many interpreters, man's identity is defined by the anxiety which arises from his awareness of death and the related experience of the contingent.[90] They see the source of man's freedom in his potential to act and choose and thereby change his dread into courage. This is, as I see it, a way of putting our distributed mortal terror into the service of life. An inspiring summary of such an existential view is set forth by Paul Tillich in *The Courage to Be*.

> The popular belief in immortality which in the Western world has largely replaced the Christian symbol of resurrection is a mixture of courage and escape. It tries to maintain one's self-affirmation even in the face of one's having to die. But it does this by continuing one's finitude, that is one's having to die, infinitely, so that the actual death will never occur. This, however, is an illusion and logically speaking, a contradiction in terms. It makes endless what, by definition, must come to an end. The "immortality of the soul" is a poor symbol for the courage to be in the face of one's having to die. The courage of Socrates (in Plato's picture) was based not on the immortality of the soul but on the affirmation of himself in his essential, indestructible being.[91]

A Heideggerian bid for authenticity in the face of death came from Sigmund Freud. "Life is impoverished," he wrote, "when the highest stake in the game of living, life itself, may not be risked." [92]

Descent and Suffering

When death occurs because some common pathogen has become uncommonly effective against an aged body, a regression unto death can often be observed. As senility sometimes takes over, the person loses not his life but his personhood. A large variety of cases of dementia are known, exhibiting something of a reversed journey along the ontogenetic Umwelts of nootemporality, biotemporality, eotemporality, prototemporality, and, finally, atemporality. In a way this is also a reversed journey along the temporal levels of phylogeny. No neat arrangement is implied thereby; no pointers are claimed to move along a table of physical quantities. Instead, I

am talking about a bloody, mixed-up mess whose declining attributes relate
to the abstract idea of a reversed journey along the ontogenetic Umwelts, as
the attributes of two twigs of equal lengths relate to the abstract idea of
equality in the Socratic dialogue. Phlebas the Phoenician had such a
journey.

> A current under sea
> Picked his bone in a whisper. As he rose and fell
> He passed the stages of his age and youth
> Entering the whirlpool.
>   T. S. Eliot, "Death by Water"

The mortal descent does not involve an unchanging observer and a declining observed, but constitutes rather a common journey of the observer and
the observed along the devolving Umwelts. "My death" does not include
burial.[93]

Dying for man begins with Dylan Thomas's father,

> . . . there on the sad heights,
> Curse, bless, me now with your fierce tears, I pray
> Do not go gentle into that good night
> Rage, rage against the dying of the light.
>   Dylan Thomas,
>   "Do not go gentle into that good night"

and ends with

> HOTSPUR: And time, that takes survey of all the world,
>   Must have a stop. O! I could prophesy,
>   But that the earthy and cold hand of death
>   Lies on my tongue. No, Percy, thou art dust,
>   And food for —
> PRINCE: For worms . . .
>   Henry IV Part 1 (V, iv, 82)

Identity is corollary to the "sad heights," loss of identity to the "food for
worms"; the wending path between the two is along devolving Umwelts.
Identity must often be yielded up by people in love, in face of common
danger, among cloistered nuns, or in traditional armies; the desired result
is the replacement of the selfhood with a simpler, undifferentiated character.
Such a condition can sometimes be reached by an orderly process, by which
I mean that the somatic and mental functions remain compatible with the
Umwelt. (The feeding, sleeping, and sexual schedule of the body, for
instance, may be adjusted to benefit the group, as are the person's plans or
hopes). When, however, there is a serious discrepancy in the consistency of
the devolving Umwelts (when the soma devolves from biotemporal to
eotemporal forms of matter while the ego insists on nootemporality)
suffering obtains.

Admittedly, suffering is an elusive concept. While it has been important in traditional philosophies and central to religions, it has been of

little use to logic, methodology, or to the philosophy of science. I wish to consider now temporality in the phenomenon of suffering. In somatic disease and in accident the overpowering characteristic of pain is its present-ness; neither past nor future appear real or relevant.[94] But this presentness is not that of the timelessness of the healthy child; it is not an ignorance of "yestermorrow" but a demand for the reintegration of the mental present in terms of the most comprehensive level of temporality in man, one which we have been calling nootemporality. In case of terminal illness or deadly wound, there seems to be a radical increase in the polarization of life and death until in the finality of death the expected and the encountered coincide, their conflict removed, and the body becomes totally adapted to a mode of eotemporal or prototemporal existence.

Let us revert to the notion that in "my death" the observer and the observed regress together; this would imply that, before the food-for-worms stage is reached, the Umwelts devolve back to the eotemporal and proto-temporal. Leonardo must have sensed something of this sort when entering these reflections in his diary, talking, as it were, to himself.

> Now you see that the hope and desire of returning to the first state of chaos is like the moth to the light, and that the man who with constant longing awaits with joy each new springtime . . . does not perceive that he is longing for his own destruction. But this desire is the very quintessence, the spirit of the elements, which finding itself imprisoned with the soul is ever longing to return from the human body to its giver. And you must know that this same longing is that quintessence, inseparable from nature, and that man is the image of the world.[95]

Along the evolutionary ladder organisms seem to bear certain affinities according to their complexities: plants with plants, frogs with fish, dogs with cats, man with man. Some people believe that the agony of the shrimp when dropped in scalding water is picked up by nearby flora.[96] One may easily doubt the appropriateness of instrumental measurements on which such claims are based, but there can be little doubt that communication among social animals from insects up abounds in examples of sympathy in the presence of danger. And I have no doubt about the human univer-sality of Whittaker Chambers's confessions in his "Letter to my Children." Why would a sound and dedicated Communist leave the party? Because

> one night he heard screams. . . . He says to himself, "Those are not the screams of man in agony. Those are the screams of a soul in agony." He hears them for the first time because a soul in extremity has communicated with that which alone can hear it—another human soul.[97]

Empathy is a feature of horizontal classes of similar complexities; along the vertical rise of complexity, empathy is difficult or impossible. The feelings that characterize a downward view towards inhabitants of eotemporal and prototemporal Umwelts may include sympathy, such as for a louse about to be crunched between two fingernails, but it is difficult to feel empathy. Self-sacrifice is characteristic toward peers or toward higher creatures: from dog to man, from man to God, but seldom downward. One might look for a lost cat but self-sacrifice for the cat is rare, except perhaps by a child for whom the cat seems a brother.

One might relate our absence of empathy for amoeba, lice, or fish to the vastly different means of communication that function among amoeba, among lice, fish, and among men. We have certainly not gotten very far with talking with animals, not even with the intelligent dolphins. But this is only part of the story. There exists an asymmetry to the quality of communication among creatures in different Umwelts. Indifference, curiosity, sympathy, appear to describe qualitatively our views downward; empathy does the same horizontally; fear, awe, and subservience describe upward attitudes. This asymmetry is, as I see it, an evolutionary necessity manifesting itself in what I have called the difficulty of regressive sharing. This asymmetry can also be identified in the substantially different attitudes we have toward eo-, proto- and atemporality before birth, versus the same levels of temporalities in the dying or, presumably, after death. The non-being before birth is regarded with indifference or curiosity. I have never come upon a poem inspired by the nonbeing before conception; whereas the nonbeing after death we regard with awe, subservience, and fear.

> Oh build your ship of death
> Oh build it.
> Oh, nothing matters but the longest journey.
> > D. H. Lawrence,
> > "The Ship of Death"

We found a similar asymmetry in cosmic views: cosmogony is an object of intellectual quest and curiosity; the apocalypse is the subject of awe and fear. The asymmetry of our attitudes for prepartum and postmortem levels of lower temporality derives, I believe, from our instinctual awareness of the different types of conflict-resolutions represented by ontogeny and phylogeny on the one hand, and death on the other hand. Whereas the first ones comprise conflict-resolutions through the hierarchical emergence of higher-order conflicts, the latter amounts to conflict-resolution by regression, namely, through the mutual annihilation of the coexistent opposites of life: the trends to rise out of and to return to the inorganic.

Righteous Life and Sinister Death

The words which designate the spatial directions of "right" and "left" are much richer in meaning than is necessary for informing people which way to turn at the end of the street.[98] In English "right" is also "correct" and is the opposite of "wrong." In politics "right" is the familiar and the conservative; in the law it carries "power." The "left" is likely to be "sinister" (from the Latin *sinister* meaning "on the left side, unlucky, inauspicious"); in politics it is disturbing, and a left-handed compliment is an insult.

Christ, having ascended into heaven, "sitteth at the right hand of the Father," and on Judgement Day

> he shall set the sheep on his right hand, but the goats on the left. Then shall the King say unto them on his right hand, Come, ye, blessed of my Father, inherit the kingdom prepared for you from the foundations of the world. . . . Then shall he say also unto them on the left hand, Depart from me you cursed, into everlasting fire. (Matt. 25:33–41)

Life well-lived is righteous until cut short by sinister death.[99] Right and left in space are the opposites of the Heraclitean road which goes both ways and yet is the same; it is curious to meet it here, implying a unity of life and death as rights and lefts of the same road. We have already seen earlier why it is erroneous to associate the direction of time with the thermodynamic arrow of closed or of open systems alone, and suggested that biotemporality should, instead, be associated with the existential stress between self-organizing matter on the one hand and inanimate matter on the other hand. We intend, now, to carry this view somewhat further in an attempt to integrate "righteous life" and "sinister death" into one, single process.

In a classic contribution to the problem of evolution and thermodynamics Joseph Needham distinguishes two trends.[100] Physical systems tend from random separatedness toward random mixed-upness as illustrated, for example, by the increasingly uniform distribution of milk molecules after a drop of milk is added to a cup of coffee. In contradistinction, the processes of life, growing social organization, order by artistic creation or speech, or the gaining of knowledge, constitute change from an initial level of random mixed-upness to a patterned mixed-upness, illustrated by the combination of grass, air, and water into a horse. The law of evolution, then, is a kind of converse of the second law of thermodynamics, equally irreversible but contrary in tendency.[101] Description of physical systems has been the task of classical, or closed-system thermodynamics; description of self-organizing systems is that of open-system thermodynamics.[102] It is true that the application of the concept of entropy to living systems and cognitive functions is difficult because of the great complexity of life [103] and the elusiveness of the mind; yet, we have reasons to assume that entropies of living systems and cognitive learning are, in principle, measurable quantities—and that is all there is of interest to us here.

To give the random-to-patterned-mixed-upness a mathematical name Watanabe formulated an "inverse H-Theorem." He applied his theorem to examples of learning through the formulation of an entropy function called inductive entropy, quantified by measures of a posteriori probabilities, called credibilities. His inverse H-Theorem claims that, on the average, entropy in cognitive learning tasks decreases with increasing numbers of tests.[104] His attempt to quantify the entropy of such complex processes is only the beginning of an enterprise and not its conclusion. The growth of knowledge and argumentation is alive in the pages of journals, and this is not the place to display its growing pains.

C. A. Muses has argued from the action laws of physics that all closed systems change so that entropy increase be minimal.[105] Ilya Prigogine reasoned convincingly that living organisms in their mature states of equilibrium are characterized by minimum entropy production.[106] This theory led him and Wiama to suspect that minimizing entropy production, to give a single example, might correspond to the greatest economy of metabolism per unit weight. This theory was successfully tested for bacterial metabolism.[107] These findings, combined with what we have learned (in chapter 4, section 1, about the thermodynamic arrows of time), lead to interesting conclusions.

There seems to be a universal trend in nature to minimize the effects of the second law of thermodynamics by keeping entropy production rates of physical processes minimal. We must hold, therefore, that as new phenomena evolve they will also abide by this trend and that laws of the new phenomena will also restrict entropy increase. But what type of new phenomena would be more efficient in doing this but such as would assist in actually decreasing entropy, rather than increasing it, when incorporated in already established processes? A closed system, already following the rule of minimal entropy production, could further decrease its entropy production if it were to contain new self-organizing systems, such as life, social organization, artistic creation, or the gathering of knowledge. Thus, life itself may be seen as a continuation of a trend already manifest in inanimate matter. But now we have a new conflict, that between growth and decay, between the open system and the closed system of which the former is a part. I shall identify this conflict as the basic feature of life and a determinant of the biotemporal Umwelt. What I have just described is the emergence of biotemporality out of eotemporality. Aging and death may be seen as manifestations of a single underlying tension which, metaphorically, we sometimes describe as the strife between righteous life and sinister death.

3. Organic Evolution

In this section we deal with the ways of nature which made the cat so much like the moon.

> Minnaloushe creeps through the grass
> From moonlit place to place,
> The sacred moon overhead
> Has taken a new phase.
> Does Minnaloushe know that his pupils
> Will pass from change to change,
> And that from round to crescent,
> From crescent to round they change?
>
> W. B. Yeats,
> "The Cat and the Moon"

Life in the World at Large

The position of life in the structure of the world at large has called for illumination throughout intellectual history, even if life itself was the unquestioned mode of existence. Democritus in the fifth century B.C. perceived a world in which reality comprised only atoms and the void. The atoms of life were nevertheless something special: they were fine, round, and smooth (identical with the atoms of fire), penetrating all the organs and causing them to move. Plato was antimechanistic in biology and, through Socrates, berated philosophers who believed only what they could grip with their hands. Aristotle took living itself for granted, identified soul with life, and concerned himself in great detail with the behavior of animals and with the organization of things alive. In *De Anima* he defined soul as "the first grade of actuality of a natural body having life potentially in it." [108] He theorized that all living things have nutritive souls which enable them to feed; in addition, animals have sentient souls which enable them to feel and man has a rational soul which enables him to think.

> Nature proceeds little by little from things lifeless to animal life in such a way that it is impossible to determine the exact line of demarcation, nor on which side thereof an intermediate form should lie. [109]

Lucretius, during the first century B.C. argued in favor of the materialistic view of life of Democritus but completed it with the idea that mind was a directing agent of the body.

Fifteen centuries of Christianity added mostly only philosophical debate to the pre-Christian views until Paracelsus of Hohenheim, (1493–1541) a man of many and great talents, came to regard the body as a conglomeration of chemical matters, and the life of the body a coordinated action of "archei" or little demons. This makes Paracelsus, as Needham remarked, the first biochemist as well as the first vitalist. [110] The duality adumbrated by his two-pronged approach is still with us; its polarization, instead of lessening, became sharpened by the advance of the scientific method.

The path from Paracelsus to our own days is continuous, if not smooth. The profound mechanistic influence of Descartes, summed up in the often abused epithet, "man is a machine," though discredited even in Marxist philosophies of biology, such as that of Oparin, is still around in cloaks of subdued color such as molecular biology or biological cybernetics. The vitalist line of Driesch, Bergson, and Lecomte du Noüy survives only in the writings of the followers of Teilhard de Chardin; otherwise it has withered away for the simple reason that a life force as envisaged by the vitalists did not prove to be necessary as an assumption, nor was it to be found anywhere. Now, the duality of vitalist versus mechanistic biologies did not stem, however, from data alone, but only from data filtered through the minds of living men; the division represents differences in personalities and world-views. (I shall argue this issue in detail in the chapter on epistemology). But discoveries do not eliminate personality differences and two attitudes which are, respectively, sympathetic to, and antagonistic toward transcendental values, remain. It is probably for this reason that journals of biochemistry or treatises on the foundations of biology (in the philosophy of science) exude mechanical philosophy. It is also probably for the same reason that the works of Joseph Needham, Theodosius Dobzhansky, Adolf Portmann, and others emphasize a healthy reverence for life which is irreducible to quantitative relations. What seems to be needed is a sufficiently broad evolutionary understanding of life in the general scheme of things, one which can satisfy those yearnings of man which, beyond the power of scientific truth, demand something of the power of qualitative principles.

I wish to think of evolution as a creative process of successive forms of existents and levels of integration, such that each successive form or level may be associated with the idea of being more advanced than the form or level preceding it.[111] This definition takes two issues for granted. "Successive" presumes a sense of time. "More advanced" assumes the existence of standards whereby a determination can be made as to what constitutes more advanced. Such standards must logically comprise a hierarchy of features described in any one of many ways: by means of physical parameters, by means of some pragmatic elements useful to object or subject, or even by means of ethical or aesthetic values. The a priori qualities of "successive" and "more advanced" permit a free selection of what we mean by time or by a hierarchy; hence it permits the incorporation of evolutionary theories which at the moment apply separately to matter, life, mind, and society. If these separate theories are ever to be merged, prior agreement will have to be reached about what we mean by temporality and by a hierarchy common to such theories. Here we confine our attention to hierarchy and temporality in organic evolution.

The hierarchical nature of biological order is one of the immediately obvious features of living matter. Artistic and religious expressions of archaic man suggest awareness of ranking in the animal kingdom. Aristotle devoted a great deal of thought to the problem of biological levels and it is suspected that he intended to, or actually did, prepare a table of classification, though such a table has not come down to us.[112] Long after Aristotle, the stratification of the whole of reality into grades became the fundamental feature of Hegel's Naturphilosophie. According to Hegel, the world divides into the inorganic or lifeless and the organic or living which interpenetrate; the mind also has its lower and higher strata but these do not interpenetrate. The famous Hegelian dialectic sees history as the progressive manifestation of the spirit, yet, curiously, the transition from the lower to the higher levels is only logical and not temporal, hence not real. In spite of its stratification, the Hegelian world of levels and dialectic is a static one.[113] Still, Darwinian evolution would fit well in Hegelian dialectic (as it fits into dialectical materialism), but Hegel himself rejected evolution, possibly because it did not reflect the level of scientific preparedness of his day.[114]

The hierarchical classification of plants and animals is based on a system formalized by Carolus Linnaeus half a century before Hegel; it is built on observable anatomical and physiological characteristics of a living world. Consistent with the belief of his day in special creation and the fixity of the species, Linnaeus was convinced that he was classifying Creation as accomplished. As in the Library of Congress catalog system of book subjects, his classification of organisms involves existing features of the classified items with no demand that it reflect the process which created them. Nineteenth century systematics, though evolutionary, retained many preevolutionary practices by picturing the diversity of life as a series of idealized, static types. Slowly, however, this gave way to systematization based on group-related features; judgements as to where an individual belongs were now made in the context of population. With the evolutionary revision of systematization arose the study of average characteristics under generic terms such as homology, homoplasy, transformation, convergence, divergence, and parallelism. In the new scheme, characteristic patterns were not conceived of as those of real individuals but those of artificial individuals constructed from statistical averages. These new dynamical methods of classification successfully complemented the pliable character of organic evolution by natural selection, even though in themselves they gave no more guidance about the origins of hierarchical organization among the living than was given by the Aristotelian awareness of the hierarchical fact.

In search for hints regarding the necessities which might be responsible for the hierarchical structure of life, I wish to turn away from systematics which simply acknowledge it and turn to a process which displays the dynamics of hierarchies. Such a process is the relationship between the phenotypes and genotypes of organisms. The phenotype of an individual is the sum total of traits attributable to it, be they morphological or physiological; it comprises all of its manifested structures. The genetic constitution of an individual is his genotype. The former are usually observable by the unaided senses in a single individual; the latter are derived qualities obtained from the examination of many members of a group through scientific arguments, and usually involving sophisticated instrumental determination.[115] We may think of a hen as a display of the phenotype produced from the genotype of the egg from which it came, in the specific environment in which the hen was raised. In a different environment the hen might have looked different, for identical genotypes can lead to different phenotypes. Also, identical phenotypes may be produced from different genotypes. Hence, neither the phenotype nor the genotype is the obvious master of the other, and in that sense they are on equal footing. From the temporal point of view, however, there are some clear differences between the two.

The phenotype changes in response to environmental conditions at rates necessarily comparable with the life of an individual; it is here for instance, that the shortening of the individual lives becomes evident (selected to facilitate faster spreading of beneficial genetic change). Compared to the phenotype, the genotype is remarkably stable; changes in it have periods comparable to the life of the species. The living process itself is even older then any specific genotype, and has been stable for perhaps 4,000 million years in the sense that it functions as a distinct integrative level. In lowly organisms, such as viruses and bacteria, the phenotype is generally not very different from the genotype; as we look backward along evolution we see them merge. Beneath the long-lived genotype, we find the physical world with its forms almost eternally stable. We have wandered from biotemporality to eo-, proto-, and atemporality.

I have speculated earlier that organic evolution creates an increasing variety of physiological clocks with a widening spectrum of periods, so as to better image external periodicities—which themselves also change and complexify; we found that this cyclic order extends down to atomic frequencies, and up to periods longer than the individual's life time. We see now not only that organic evolution had gathered physiological know-how about the temporal behavior of the inanimate and the living environment of the organism, but also that the biochemical and biophysical properties of matter set certain limits to what may be done, and how. The complexity of living matter apparently delimits the functioning of the soma (as we have already speculated); this limit is overcome by systematic regressions to the genotype, a simpler and very slowly changing structure, from which the phenotype is again and again rebuilt in the evolutionary processes of phylogeny and ontogeny. What we may further gather from this is the suggestion that one reason for the hierarchical organization of life resides in the continuously widening spectrum of adaptations in the temporal domain which, in its turn, lead to a plethora of techniques and forms. Thus, the position of life in the structure of the world at large seems to be fundamentally tied to temporality.

Time and the Origins of Life

Although we can observe the coming about of lasting changes in the form of living matter, and we are able to produce some such changes experimentally, we have not so far produced a living thing directly from inanimate matter. We have never observed nor have we brought about biogenesis. Brief reflection will reveal that it is not even clear exactly what we mean by biogenesis. That human life, for instance, can be created from food, air, water, sun, and the good works of two people is demonstrated by every newborn child; but birth exemplifies sexual reproduction and growth and not biogenesis. It would be a more unusual feat if we could construct, from chemical elements, the zygote of a garter snake such that the grown specimen would be indistinguishable from one normally hatched. But even this would not be enough, for the biologist would only have been a technician with prior knowledge of the structure and program of hereditary material and the morphology of the garter snake which he faithfully copied. Had he created, instead, a serpent otherwise unknown on earth and also following unknown laws of life, he would then be required, besides carefully guarding his creature, to convince his examiners that, granted the necessary conditions, evolution could have led to the serpent at hand. Thus, by biogenesis we mean the coming about from inanimate matter of an organism which would behave according to the laws of biology as they are understood today. Were our understanding of life to change, so might our ideas of biogenesis.

It is believed that biogenesis occurred only once during the last 4,000 million years; but informed opinion is not particularly disturbed by this uniqueness and many ingenious arguments have been given to make it acceptable. Darwin anticipated the currently held view when he wrote that

> it is often said that all conditions for the first production of a living organism are now present, which could ever have been present. But if (and oh! what a big if!) we could conceive in some warm little pond, with all sorts of ammonia and phosphic salts, light, heat, electricity, etc., present, that a protein compound was chemically formed ready to undergo still more complete changes, at the present day such matter would be instantly devoured or absorbed which would not have been the case before living creatures were formed.[116]

We can only add that conditions for the appearance of Sappho, Josephus, or Darwin were appropriate only once, yet this does not make the reports of their births doubtful. Also, we shall argue later that interfaces such as between inanimate matter and life, or life and mind are metastable states; thus, biogenesis need involve a stage of only marginal stability between life and no-life. The greatest difficulty is not in the uniqueness of the first life, but rather in the absence of any useful hints as to how the principles of inorganic, organic (and later, social and intellectual) evolutions may interconnect. There is a need for some principle which would subsume into one scheme the various categories of evolution, now separately valid.

According to Dobzhansky, the existing accounts of the origins of life leave one uncomfortable not because the available information is incomplete but because they involve a petitio principii.[117] Namely, once the existence of life is assumed, organic evolution can be said to bring about some very unlikely forms by natural selection, using such key functions as adaptation to the environment, differential reproduction, geographic separation among species and the like. But none of these functions exist in the inorganic world, hence the begging of the question. Since the origins of life are coeval with those of natural selection, neither may be used to explain the other; how and why natural selection arose is precisely the problem. The same circularity was recognized by von Bertalanffy who concluded that the question of the origins of life will rest unsolved until some essentially new principles are discovered. Other evolutionists have argued for an extension of Darwinian selection to the inanimate world. Variations could then be said to reside in the formation of new compounds by "chance mutations" and natural selection would favor stable compounds.[118] This is an interesting exercise in epistemology; I do not think it has been followed up. In any case, we would again have to account for the emergence of natural selection with the creation of matter.

A. I. Oparin, distinguished partisan of dialectical materialism, perceives the origins of natural selection in a process that resembles what I have called nomogenesis, or the birth of laws. He envisages that when the colloidal multimolecular systems separated from the primeval soup, their further evolution began to be controlled by natural selection; thus a new law which previously did not exist in nature came into being.[119] Dobzhansky is emphatic about the complete breaks in evolutionary continuity between life and nonlife, though he warns against oversimplifying the split; by way of a general scheme, he advocates the use of integrative levels.

> Stated most simply, the phenomena of inorganic, organic and human levels are subjected to different laws peculiar to those levels. It is unnecessary to assume any intrinsic irreducibility of these laws, but unprofitable to describe the phenomena of an overlying level in terms of the underlying ones. [Each level is superimposed on the lower ones and characterized by an increased rate of change]. The attainment of a new level or dimension is . . . a critical event in evolutionary history. I propose to call it evolutionary "transcendence." [By which is meant the going] beyond the limits of, or to surpass the ordinary, accustomed, previously utilized or well-trodden possibilities of a system.[120]

We shall explore the possibility of placing transcendence in the context of temporalities.

Let us first attend to an idea fundamental to the study of the molecular bases of life, that of a molecular configuration. This is ordinarily defined as specific spatial arrangements under physicochemical control. Michael Polanyi has questioned whether these controls are or are not sufficient to account for life.[121] As he sees it, the extreme stability of genetic material derives from its orderliness; but the orderliness amounts to a determinacy which prohibits the occurrence of the very improbable, yet it is the emergence of the very unlikely which marks evolution. Polanyi is thus impelled to seek and identify certain boundary conditions which are responsible for organic evolution but are not themselves under the control of physical and chemical forces. I believe that such boundary conditions are the potentialities of natural selection working in the temporal domain; that is, the boundary conditions are variations in programs.

Consider for example the DNA molecule, in which the spatial structures of organisms are represented by means of one-dimensional messages.[122] The DNA molecule is an attractive candidate for the possibility of programmatical functioning; it prompts a search for clockworks which may be responsible for the DNA synthesis but are themselves independent of that synthesis. Hartwell, using time-lapse photomicroscopy, was able to isolate mutants of two different loci (positions along a one-dimensional message) according to their temporal positions in the process of DNA synthesis.[123] Earlier, Ehret and Trucco outlined a phenomenological model for biological circadian timekeeping in molecular clockwork.[124] Consider now, that it is a fact of spectroscopy that sharp monochromatic responses (at normal temperatures) do not exist.* The reasons for line widening are numerous but, if traced to their origins, they probably all derive from the type of statistical uncertainties of particle physics which we have already discussed. It follows, therefore, that no two spatially determinate structures, even if judged exactly alike from the structural point of view, carry identical programs in terms of a very narrow temporal span. Thus, for instance, even structurally identical genes may provide a range of variations on which natural selection can work, by whatever means natural selection employs in connecting the phenotype and the genotype.

Reverting to our reflections about the cyclic order, it is evident that even the most primitive aggregate of physiological clocks, if it was not to be self-destructive, had to provide some type of coordination among its component clocks; the functioning of the organism was categorically tied to its capacity of maintaining an internal harmony. This peculiar necessity made living matter more vulnerable to external perturbations than matter in the amorphous state. Whereas inanimate matter, if broken to pieces, would still remain matter, a living organism would not remain alive.† The demand for maintaining an internal coordination may be described as the need for homeostasis, or perhaps homeodynamics among the cyclic components of the inner landscape. Such inner balance consists in securing the simultaneous happenings (or not happenings) of certain events, a requirement quite meaningless in the atemporal, prototemporal, and eotemporal world. The point is not that in these lower temporalities simultaneities cannot be defined; they can be, and we discussed some of them. The point is that in the physical world simultaneities cannot be associated with necessities; they are of no specific importance. In contrast, for a living or-

---

* The extremely sharp monochromaticity of recoil-free gamma-ray absorption, known as the Mössbauer effect, achieved its fame precisely because it is a unique spectroscopic phenomenon.
† Matter can be broken to its atomic components; organisms sometimes may also be broken to very small pieces and still remain alive, but not when atomized.

ganism the instant by instant maintenance of simultaneities among its constituent processes is essential.

This feature of life amounts to the creation of a new Umwelt, one with a temporal present; it is the insertion of a "now" in the pure succession of the eotemporal Umwelt. Also, it is a new use for already existing physical structures. The new triadic form of time is secured through a balance between the opposing trends of self-organization and decay; accordingly, it displays the existential stress that corresponds to the unresolvable conflict between them. In the beginning of this section I said that by biogenesis we mean the coming about of an organism which behaves according to the laws of biology as understood today. With a shift in emphasis we may now add that by biogenesis we can also mean the coming about of a system which permits the definition of a present, or, more specifically, what I have called the creature present, as biotemporality emerges above eotemporality. But let us spin this tale further.

Somewhere in the hazy past I imagine the coming about of organisms comprising the simplest cyclic order of variations, modeled after periodicities which reached them as stationary objects, or objects carried around by water or gas. Perhaps a thousand million years ago multicellular organisms became self-locomotive through the use of flagella. Their motion must be seen as more than just the exploration of a region: it was the discovery of space. The modalities of excitation appropriate to this level of life are touch and taste: they give information about the immediate environment involved in the creature present. We have seen how the widening of the spectrum of cyclicities in the primordial organisms provided better adaptation to the environment, and also led to an increased existential stress, by increasing the region of experience where the expected and the encountered could differ. The cyclic expectations for a moving organism would now include a space limited only by conditions unfavorable for survival. Perhaps 500 million years ago at a stage represented today by worms, our ancestors formed light-sensitive regions in their epiderms and since then light has been the foremost link between physiological clocks and their Zeitgebers. Light sensitive spots would surely be of most use in those segments of animals which encounter the unknown world. Like Zeno's arrow, which must be permitted to extend beyond its stationary limits if it is to explore the unknown, these worms began extending themselves into realms not available to touch or taste.* In the wisdom of our unconscious language, they met the future head on.

Those who could see were now endowed with great advantages. They could, for instance, learn (by whatever ways they had for "learning") about regularities in the external world which were not cyclic, such as certain behavioral features of prey, predator, or neutral background. This was a new

---

* It will be recalled that nonvisual entrainment by light has been demonstrated in blinded animals, with the source of entry of the light in some part of the brain.

type of future unavailable to creatures without sight and the ability to get around. But while it meant improved adaptation, it also meant a further increased domain in which the expected could differ from the encountered and in which there was an increased possibility of making erroneous judgements—for the future does not copy the past exactly. But light sensitive regions alone are useless; they must be parts of an associated evaluational system. Hence, seeing, from its very origins, had to include an organ of excitation, a system of processing, and organs of action.[125]

We have gotten now perhaps 500 million years away from the time when the first coordinated physiological clocks began to function on earth. During these eons biotemporality emerged from eotemporality through the potentiality of the definition of a meaningful present, and organisms acquired the capacity to recognize not only cyclic but also noncyclic regularities.

Levels of Causation, Uncertainty, and Undeterminacy

In chapter 4, section 1, I suggested that causality is an example of emergent lawfulness and that it manifests itself on several levels, all of them coexisting in our experience. In the atemporal world causation cannot be defined; in the prototemporal world it is probabilistic; in the eotemporal world it is deterministic but without a preferred direction. Passage between the proto- and eotemporal levels gives rise to rules of connection such as Heisenbergs's uncertainty principle and the H-Theorem. Presently I wish to extend this hierarchy to include final causation and the biotemporal region, and also add a new principle, that of undeterminacy.

When I say that the rules of an integrative level leave certain regions undetermined, I mean that the potential existence of those regions is unrecognized. For instance, the behavior of aggregates of particles, characteristic of the prototemporal world, is left undetermined by the atemporal world. (Rules which guide the interference of light waves are examples of lawfulness in the atemporal world; such laws leave undetermined the equations of state, which regulate prototemporal physics). Thus, the prototemporal world in toto emerges from regions left undetermined by the atemporal world. Likewise, the behavior of aggregates of matter is regulated by such laws which are characteristic of the eotemporal world; these arise from regions left undetermined by the prototemporal Umwelt. In biogenesis, life arises from regions left undetermined by the eotemporal Umwelt. There is, for instance, no physical law that would prohibit the existence of self-reproducing structures, thus self-reproduction was left undetermined in physics. In contrast, relative positions of atoms follow laws determined in physics.

Within the living world itself there is a similar progression among levels of undeterminacy. For instance, morphogenetic determination shows progressive levels of potentialities (which provide progressively larger regions of undeterminacies) as the presumptive fate of living matter finally settles in its adult form. Each higher level arises from the region left undetermined by the lower level. For instance an amphibian gastrula which would normally develop into an eye, if transplanted elsewhere early enough turns into structures characteristic of the new location. If transplanted too late, it differentiates into an eye. Thus, the fate of the early gastrula is clearly undetermined; the eventual eye (or other structure into which it turned) had arisen out of the undetermined potentialities of the gastrula.[126] If we compare the adult forms of various organisms along the evolutionary ladder of complexity, we again find, on the average, a trend of increasing undeterminacy. This will become immediately evident if we think of undeterminacy as a range of potentialities. A plant may open or close; an animal may maneuver, hide, fight, and reproduce; a man can do all of these and in addition may, perchance, dream.

The structure of causation is similarly intricate. Probabilistic causation is a type of connectedness left undetermined in the atemporal world; deterministic causation was left undetermined in the prototemporal world. The nature of higher-level causations is not predictable from those on the lower levels but not in conflict with them. Thus, we cannot deduce probabilistic causation from atemporal no-causation, or deterministic causation from a probabilistic one. With this background, let us turn to biological causation.

Probably the most general representation of an organism is that proposed in biosemiotics (the study of life as a vehicle of signs and signals) which considers the organism as an autonomous system which, upon certain kinds of excitation and after a lapse of time, will take some action.[127] Using the biosemiotic approach, we see that the signal processing is mediated by a repertory of coordinated clocks; the rate of reaction is determined by such variables as the complexity of the organism and the speed of signal propagation. Acts of mediation between excitation and actions are often left undetermined, even on lower levels of temporality. Thus a tidal wave also takes time to travel between distant points. But this delay cannot be interpreted as the function of an autonomous coordinated system, even if the start of the tidal wave is the cause of a much later effect. Biogenesis provides a new type of delayed connection between cause and effect definable only in terms of a life process. Causal connections in the physical world correspond to the Aristotelian idea of efficient cause; the new type of causation corresponds to Aristotelian final cause. What characterizes final cause is that the time delay between cause and effect may be judged with respect to its value for the autonomy of an organism; thus final cause becomes one of the determinants of the biotemporal Umwelt.

Ideas of final causes in the form of teleological explanations were the hallmarks of much of pre-Darwinian evolutionary thought. The logic of teleological reasoning is interesting because, short of reference to revealed truth, it must be based on arguments ex post facto. A crude example will illustrate this. Let us say that God created man (long ago) to be the king of all creatures (as he is today); if I imagine myself witnessing Creation (long ago but knowing what I now know) I would then perceive a goal-directed-ness in the flow of events (yet to come). A scientific defense of evolutionary teleology so understood is not possible for many and obvious reasons. There is nothing wrong, however, with seeing in the bees' dance a final cause, to wit, the desire to direct other bees to pollen-laden flowers. Thus, it is not final causation and teleology which must be dumped out as the baby with the bathwater; it is the validity of its domain which must be carefully delineated.

Working biologists tell us that deterministic causation is all but a useless concept in the study of organic evolution. In an essay on "Cause and effect in biology," Ernst Mayr lists four major underlying reasons. (1) Beneath the undeniable complexification of living matter with time, biological change manifests randomness with respect to significance; e.g., errors in genetic copying occur (one must assume) without anticipation of their effects for the phenotype. (2) Unlike in physics, in organic evolution all entities are unique. (3) Biological systems are extremely complex. This is not a fact invoked in support of excusable ignorance but designates a quality of the living. (4) New qualities emerge at higher levels of integration. "When two entities are combined at a higher level of integration, not all properties of the new entity are necessarily a logical and predictable consequence of the properties of the components." [128]

In evolution, as G. G. Simpson sees it, one cause leads to multiple solutions and not only to a unique outcome. Hence evolutionary causation is both nonrepetitive and nonpredictive.[129] Niels Bohr believed that a degree of unpredictability is a necessary property of life, in analogy to the principle of uncertainty.

> In every experiment on living organisms, there must remain an uncertainty as regards the physical conditions to which they are subjected, and the idea suggests itself that the minimal freedom we must allow the organism in this respect is just large enough to permit it, so to say, to hide its ultimate secrets from us.[130]

Bohr's insight may be made more precise if we distinguish "uncertainty" from "undeterminacy."

Uncertainty arises when we seek to connect certain features immanent on one level of integration with their origins on a lower level; qualities well defined on a higher level may be ill defined on the lower level. For instance, momentum and position are well defined in the eotemporal world but ill defined, hence uncertain, in the prototemporal world. A flying stone determines the overall fate of its particles while also accommodating their particulate nature. Likewise, a total organism determines the overall fate of its component systems [131] while accommodating their biophysical and biochemical character. However, the most interesting regularities characteristic of life are usually not those of the parts but those of the organisms as autonomous units. Uncertainties arise in connection with qualities well defined for a total organism but ill defined for the components. For instance, what do we mean by the birth of whiskers apart from the birth of the cat, or the reproduction of sperm cells apart from the reproduction of the total organism?

A corollary of uncertainty is the impossibility of deducing regularities immanent on higher levels from those on lower levels. For instance, although the laws of biology must conform to laws of physics and chemistry, there is no way of deriving the principle of evolution by natural selection from biochemistry; in fact, both biophysics and biochemistry can well afford to disregard the functioning of the organism. Uncertainties so understood are ontic qualities; they bear witness to true emergence by demonstrating that certain qualities immanent on an integrative level are perhaps adumbrated but not represented on lower levels. The secrets of life, by which I would mean the yet unknown regularities of life, those which Bohr placed in the region of uncertainties regarding physical conditions, I would place in the regions undetermined by the lower integrative levels.

Reverting now to the question of biological causation, as that level of causation which is appropriate to the biotemporal Umwelt, we note that any such emergent quality should be an essential feature of lawfulness immanent to that world. To describe those hypothetical regularities which, although consistent with the physicochemical laws of nature are, nevertheless, unique to the integrative level of life, Elsasser coined the neologism "biotonic laws." [132] If the hierarchical structure of reality is taken seriously,[133] we may then specify some features which biotonic laws must have. They must, of course, accommodate all the lawfulness immanent in the atemporal, proto-temporal, and eotemporal worlds. They must give systematic form to uncertainties so as to guide us from strictly "biotonic" concepts to concepts known on the lower levels of hierarchies. Also, they must incorporate that level of causation which is characteristic of life, that is, one which permits the insertion between cause and effect of a process of evaluation and is wide enough to permit the variability of life.

The Dynamics of Adaptation

In *The Origin of Species* (1859) Darwin observed that variations useful for organic behavior do occur. Consequently,

> individuals so characterized will have the best chance of being preserved in the struggle for life; and from the strong principle of inheritance, these will tend to produce offspring similarly characterized. This principle of preservation, or the survival of the fittest, I have called Natural Selection. It leads to the improvement of each creature in relation to its organic and inorganic conditions of life; and consequently, in most cases, to what must be regarded as an advance in organization.[134]

The general thrust of Darwin's ideas represented in this quotation remains valid; natural selection so defined is held to be the source of the creative advance of evolution. In Dobzhansky's summary evaluation

> natural selection is in a very real sense creative. It brings into existence real novelties—genotypes, which never existed before. These genotypes, or at least some of them, are harmonious, internally balanced, and fit to live in some environments. Writers, poets, naturalists have often declaimed about the wonderful, prodigal, breathtaking inventiveness of nature. They seldom realized that they were praising natural selection.[135]

Natural selection can operate only on distinguishable individuals (as contrasted with indistinguishable particles). Early opinion perceived the source of this variation entirely as random mutation, a view which subsequently became untenable. The most serious objection is that random mutation cannot be oriented toward adaptation, hence, as G. G. Simpson put it,

> if they alone produced evolutionary change, evolution would be the result of mere chance. But that is absolutely incredible. It is just not possible that such thoroughgoing, almost inexpressibly complex adjustment between organism and environment was produced by chance.[136]

What happens to the hen must somehow affect the eggs; there must be feedback between phenotype and genotype, and there is. It operates through such agencies as relation to environment, geographic separation of population segments, relative success of reproduction, etc., the loop incorporates both the hen and the egg.[137]

The importance of random mutation did not vanish, however, but only receded from view. Ernst Mayr concluded that genotypic variations are due almost entirely to gene flow and recombination even if ultimately, somewhere in their earlier and simpler forms, all new genotypes originated in mutations.[138] Recombination expresses a genetic plasticity which makes possible the production of new kinds of gametes over and above simple parental combinations; gene flow is the result of migration and mixing of the populace. Thus, Mayr's modest statement invokes the complete machinery of evolution acting upon individuals and groups.

Evolutionists like to insist that evolution in no sense implies necessary advance, and quote many examples of devolution by natural selection. Yet organic evolution did produce out of a presumed primeval broth some unicellular creatures, then vertebrates, some of them mammals, including several primates including this writer.* This advance of "real novelties" did not, however, come about through the mere existence of selection rules, whether acting on genotype or phenotype; natural selection specifies only the utility of certain desirable states of equilibrium between the individual

---

* Examples of occasional de-complexification make the scheme more rather than less credible. Perfection is a feature of ideas and not of embodiments of those ideas, as Plato might have insisted.

(or group) and the environment. The process itself whereby such desirable equilibria are reached is called adaptation, defined as the "adjustment to environmental conditions by an organism or a population so that it becomes more fit for existence under the prevailing conditions." [139] It has often been said, and properly so, that in organic evolution an organism must adapt or die. That organisms and groups of organisms prefer to adapt rather than die, although the alternative of death is permanently open to them, is an unarticulated assumption suffused through evolutionary biology. Yet the sources of this essential assumption are not sought, its validity not questioned, and even its mere existence goes unacknowledged. Being, as it is, an unquestioned answer, it must be described as a dogma. As such, it is worth the attention of the philosopher.

Ludwig von Bertalanffy was a vocal and brilliant critic of the unexamined use of adaptation as the driving force of organic evolution. His views may be illustrated with reference to his critique of the idea that adaptation and selection direct evolution through such means as (quoting Mayr) the increase and decrease of the probability of successful reproduction. If this be the case, then

> it is difficult to see why evolution ever progressed beyond the rabbit, the herring, or even the bacterium, which are unsurpassed in production of offspring. This doubt particularly applies to the decisive, transitional phases of evolution. The abundant remains of, say, dinasaurs testify to their well-adaptedness and profuse reproduction. The first contemporary mammals and birds (and later protohuman forms), in contrast, apparently were highly vulnerable, adaptively undecided and weak forms [with not particularly high rates of reproduction.] [140]

What von Bertalanffy argues through these and similar examples is not whether selection does take place or whether adaptation is necessary for survival, but whether they are sufficient to account for evolutionary advance. Perhaps they are not, but they do come close to it if we take a somewhat novel perspective of the dynamics of adaptation.

In the ordinary interpretation of Darwinian selection, an organism or, more accurately, groups of organisms, are seen as evolving under the selection pressure of the environment (both living and nonliving); this

....

Life cannot be satisfactorily understood if it is thought of as a single evolutionary thrust of some sort, a process which tends to lift matter from its inorganic to its organic mode. Instead, life must be thought of as a conflict between two processes: that of growth and that of decay. Life lasts only as long as this conflict lasts, hence the conflict itself, at least on the integrative level of life, is unresolvable.
Kaethe Kollwitz, Death and Woman. Etching, 1910. Courtesy, Galerie St. Etienne, New York.

....

process is summed up by saying that Darwinian natural selection replaced purpose by method. But, curiously, whereas the organisms remain alive and evolve in spite of themselves (for they are in no sense masters of their evolutionary development), they do become active agents in the evolution of others—by eating the right kind of moth or copulating with the right kind of giraffe—for then they are considered as parts of the environment. Adaptation as generally defined and as ordinarily interpreted is, thus, a frightfully lopsided process. To understand the full significance of adaptation we must learn instead, to take the sea horse as a fully equivalent and equal partner to sea and seek the dynamics of mutual adaptation: the organisms to their environment and the environment to the organism. I wish to call this the principle of the mutuality of adaptation. I prefer to perceive in the paramecium seeking to escape a nontoxic dye, in the migrating geese over James Bay wending their way toward better climates, or in the sad monkey pressing its lever for food, the self-same spite "to take arms against a sea of troubles, and by opposing, end them."

One may claim that such a mutuality has been incorporated in evolutionary thought. Except for man and his environment, however, this has not been the case, although mutuality is implicit in the idea of natural selection. Our image remained that of a species adapting.[141] So as to be able to deal with the dynamics of adaptive change in its two-way character, I wish to distinguish in the structure of the usual concept of adaptive pressure two components which both must be simultaneously present. I shall designate them as one, or the opposing, member of pairs of terms such as efferent/afferent, centripetal/centrifugal, active/reactive, or primary/secondary.[142] Since both components must always be present, any decision as to which one to emphasize when dealing with adaptive change cannot be a function of the process but only one of the focus of inquiry. When a species adapts to an environment which is taken by an unstated assumption as essentially constant, the adaptive pressure may be described as efferent, centripetal, active, or primary. If we focus on the changes in the environment which must accompany the adaptive changes in the species, we are focusing on the afferent, centrifugal, reactive, or secondary component of the adaptive process. None of these terms is totally satisfactory; none of the pairs is sufficiently neutral. We must constantly remind ourselves that the pairs are projections of one, single, underlying evolutionary change.

The diversity of examples one may quote as those of evolutionary adaptation is certainly vast, yet it is impossible to point to an adaptive change that would not also constitute selection pressure on the environment to which it relates and/or produce some degree of reorganization in inanimate matter on which it draws. Consider, for instance, morphological adaptation. Insect mouths, which make the sucking of nectar more efficient, influence the nectar balance of the flower; the pheasant's feet, well-suited to walk on level ground, scratch that ground quite noticeably. Or, consider physiological adaptation. Sea birds, which filter out fresh water from sea water and excrete concentrated brine, modulate the salt distribution of the ocean, even if ever so slightly. The woodpecker has an intricately organized muscle system associated with peculiar nervous control that makes it possible for the bird to insert its long tongue in the tree—and exert, thereby, substantial selection pressure upon insects deep in the tree. Going on to populations, consider the dynamics of a prey-predator situation where the predator was decimated. The subsequent population explosion by the prey is an example of afferent or centrifugal adaptation by the prey. The efferent or centripetal component of this change may be a relative increase of food supply for the surviving predator. If the majority of the predators come to prefer prey of certain taste, associated, perhaps, with certain colors in the prey, and if this results in an eventual change in the phenotype of the prey which makes the predator go hungry, we again have active and reactive components to the adaptive change of both prey and predator, depending where we fix our attention. To each modification in the organism there is some corresponding modification in the environment for, in the most general sense, the roles of organism and environment must be regarded as interchangeable.

It should be clear to the reader that what is proposed here is a rudimentary and schematic formalization of ecologic interaction and of the ecosystem. An impressive amount of work of technical sophistication has been forthcoming in this field, prompted mainly by the evident needs of human ecology. The historical roots of our ecological crises have been explored but, in spite of the extensive work on ecology proper, the omission of emphasis on mutuality in adaptive change has received very little attention or none at all, as far as I can tell. Only in 1973 did the first serious proposal appear to implement, partly, the ordinary view of Darwinian selection. It is that made by L. Van Valen of a new evolutionary law that holds that each evolutionary advance made by any one species is experienced as a deterioration of the environment by its competitors.[143] Yet the change in perspective that follows from an emphasis on the mutuality of adaptation is substantial.

The vitalist-reductionist debate, for instance, is placed in a new light. To see this, let us hold first with the reductionist that life is a complex physicochemical process of matter with no unique lawfulness of its own. This perspective makes life appear both transient and insignificant. Matter gathers now into the fig tree, now into Susanna beneath the fig tree; they both come from dust and to dust they shall return. Organic evolution, then, is a manifestation of inanimate matter whose nature is essentially independent of life. Let us then hold with the vitalist that there is something like a life force. This perspective makes matter appear both transient and, from the point of view of life, insignificant. For we know that both Susanna and the fig tree metabolize and exchange their total particle complements every few years, yet they maintain their identities for decades and centuries; life itself has maintained its identity for 4,000 million years. Thus life is capable of making inanimate matter assume shapes that suit the organism. Organic evolution, then, is a manifestation of a life force whose nature is essentially independent of matter.

Clearly, both views are true simultaneously, even if evolutionary thought has been favoring the reductionist stance for historically understandable reasons. For instance, there is a clear preponderance of non-living matter, hence the environment appears unperturbable. Individuals and groups are usually judged as powerless against their environment made up of aggregates of matter and life, and our sciences were successful in dealing with the physical world long before they began to interpret biology; hence there exists a methodological priority favoring reductionism. The vitalist-reductionist views may now appear but opposing caricatures of the self-same evolutionary progress, wherein life and its environment mutually adapt. The principle of the mutuality of adaptation amounts to stressing a seldom emphasized aspect of well-known phenomena, a shift in perspective, somewhat like that from the image of an apple falling to the ground because the earth attracts it, to the image of the apple and the earth mutually attracting one another. The new outlook may make no difference for some purposes, but it is likely to become essential if we seek a universal lawfulness underlying change, whether change in the position of falling apples or in the evolution of living forms.

In my prior reasoning about the emergence of biotemporality from the eotemporal background, I invoked as illustration the image of an abstract organism; *one* organism, made up of a number of coordinated physiological clocks which originated as models of external periodicities. In this imagery the environment, though absolutely essential, remained substantially distant from the organism. But the mutuality between organism and environment in evolutionary adaptation prohibits us from believing that one single organism can possibly be meaningful in itself. Since natural selection must have been a process of mutuality from its very beginning, we must think of biotemporality not as a feature of separate identities but one diffused through life on earth. The creature present is thus a systems property of the biosphere [144] intimately joined to the pure succession of the physical world. This is not to be taken as signifying a global biotemporal "now"; Maine bees cannot collect nectar from California flowers. The creature presents are connected, instead, by physical, chemical, and biological means appropriate to different Umwelts with different temporalities.

Another consequence of the stress on the mutuality of adaptation is the relative ease with which continuity between eotemporality and biotemporality can be established. As we contemplate organisms of decreasing complexity we must think of the environment as comprising an increasingly larger percentage of inanimate matter. The operational definition of the creature present in terms of final causes (that is, purposeful actions) thus becomes hazier and gets gradually washed out as the excitation-processing-action scheme of life approaches the action-reaction scheme of physical dynamics. This continuity between deterministic and final causation is a corollary of going from the diadic forms of pure succession to the triadic form of future, past, and present. Coemergent with the creature present we also find a greatly increased domain of activity in which the expected and the encountered may differ—simply because the organism is now equipped to expect more. By the principle of the mutuality of adaptation we cannot, however, place any well-defined boundaries around the resulting stresses, because what is advantageous for one individual or group is not likely to be a neutral change for other individuals or groups; hence, we must regard the existential tension appropriate to biotemporality as a systems property of the biosphere. That is, the totality of the living must be taken as a coherent society of many forms sharing, though unevenly, the existential tensions of life.

One form of this existential tension is known as biological stress, by which is meant a tension in response to heat, cold, exertion, disease, hunger, terror, and the like; in general "any set of events which modifies steady state conditions within the organism so as to activate adaptive or homeostatic mechanisms." [145] In response to an influence named "stressor" the organism develops a general alarm reaction followed by a recovery process named "the general adaptation syndrome." [146] In our scheme of thought the stressor may be seen as any substantial conflict between conditions expected and encountered.

Biological stress plays an essential part in biological regeneration, according to A. E. Needham. Here "regeneration" is taken to include a broad spectrum of responses to biological stress: the healing of a boil, recovery from illness, or growth of the hydra out of a single tentacle. Needham also sees as regeneration (in response to stress) such phenomena as the growth into a complete organism from as little as one percent of the protozoa, or the reconstitution of the hydra from its isolated cells obtained by having forced the organism through a fine mesh. "It seems quite reasonable" he writes "to regard embryogenesis itself as a process of regeneration—of new individuals to replace casualties among the previous generation." [147]

These thoughts open up the possibility of seeing in morphogeny a strategy of conflict resolution by future-directedness, in response to stress. Certainly the closest thing to magic that life can produce is morphogeny: beginning with an insignificant-looking blob of matter, usually too small to be seen with the unaided eyes, living matter forces inanimate matter to conform to the demands of life and from amorphous elements it creates now a tulip, now a horse. The growth of an organism as the actions of a mature individual are both future-directed, though we have a spectrum of future perspectives. A chipmunk searches for a stone, then hides; a climbing vine searches for a branch then winds around it; the growing embryo develops features which, according to the genetic memory of the species, will aid him in survival. These various strategies are made possible by the multiple-choice/final-causation feature of life. Again, the dynamics of adaptation forces us to regard future-directedness as a property of the aggregate of all life.

Finally, emphasis on the mutuality of adaptation, if taken together with some prior thoughts on evolution, throws some light on the corollary features of complexification, increasing rates of evolution and apparent goal-directedness of the evolutionary process. Let us recall first our speculations about biogenesis and early evolution. We envisaged early life as comprising groups of tightly coupled physiological clocks which copied external periodicities and generated some of their own. We also found that while this copying and generating activity did improve adaptation through the creation of more accurate inner models of external regularities, on the average, it also and inevitably increased existential tension. From those beginnings on, the environmental regularities to be modeled remained forever more numerous and more complicated than the regularities actually matched. One might suppose that at first this was a matter of unsophisticated biology. But later, as the total biomass and its sophistication increased, the superior complexity of the environment to which the organisms had to adapt came to be enhanced by the very process of adaptation, because of the reciprocity of the adaptive process. As in the Tristram Shandy problem encountered earlier, each action performed so as to bring a goal closer (in this case that of complete adaptation) carried that goal farther away. A worldwide, static equilibrium of living matter could not be had, even if long periods of relatively static conditions did exist. They were metastable states, heavily weighted toward complexification and increased existential tension, for every adaptive change was a mutual transaction which, on the average, left the total system a bit more complex.

In the definition of the idea of evolution we spoke of "successive forms of existents and levels of integration." This assumes that it is possible to identify and distinguish successive forms. If we survey the features which cosmologists, evolutionists, and psychologists tend to distinguish as successive forms of their concern (such as stages in the evolution of stars; stages of speciation; stages of learning) we find that consecutive steps recognizable as more advanced are least closely spaced in the history of the universe, more closely spaced in organic evolution and follow one another quite rapidly in the many ways of learning. If, instead of surveying the trend of the integrative levels of matter, life, and mind we concentrate on the rates of organic evolution alone, we note that evolutionary development is going on faster and faster as we progress from molluscs, to mice, to monkey, and man. It is well known that in natural selection explanations using more rapidly acting agents are more likely to be correct.[148] This expression of urgency, or time-economy, is a corollary of the general trend toward complexification. Systems increasing in complexity will produce distinguishable forms at higher rates than would systems of decreasing complexity—or of generally static makeup.

The different evolutionary rates have profound consequences because, on account of them, more sophisticated creatures outgrow and outevolve less sophisticated ones. Life itself outpaces matter, and mind outpaces life. Dobzhansky noted this by stressing that "for at least 10,000 and perhaps for one million years man has been adapting his environment to his genes more often than his genes to his environment." [149] Whereas in the history of early man our interest had to be focused on man as the passive partner in adaptive change, for perhaps a million years he also became an active partner with respect to his environment. This is painfully evident today as inanimate nature is being called upon to adapt to life and all life on earth to adapt to the demands of man's mind. But because of the grossly different rates of change we encounter crises of inversion: our bodies cannot catch up with our minds, and the nonliving substance of this earth cannot adapt to life and to the mind of man.

Consider now the opinion of Aleksandr I. Oparin, that

> the universal "purposiveness" of the organization of living beings is an objective and self-evident fact which cannot be ignored by any thoughtful student of nature, [something which] pervades the whole living world from the top to the bottom, right down to the most elementary forms of life.[150]

By this purposiveness one means partly the clear evidence of final causation in the behavior of living things, and partly the apparent goal-directedness of long-term change. Pre-Darwinian evolutionists were deeply impressed by the long-term, phylogenetic components of purposiveness and saw in them expressions of teleology. A closer look at teleological arguments in evolution shows that, short of claiming revealed truth (and in addition to the curious necessity of teleological statements to be ex post facto), the fulcrum of teleological reasoning concerns the question of time available versus time necessary.[151] Certainly, the fabrication of ostriches through random mixing of particles is theoretically possible. However, considering the time that was available in evolution for the production of ostriches, if one can convince oneself of the existence of some type of directing agent that could have been responsible for the ostrich, then the animal's coming about through the work of such an agent will look so much more probable than without the agent, that ideas of making ostriches by random mixing are summarily discarded. Now, if the dynamics of adaptation were somehow sensitive to time, if, for instance, it should favor those adaptive measures which work faster, then the net effect of adaptation would be indistinguishable from goal-directedness in form of plan and overall purpose.

A brief reflection will show that we have the Fragestellung backward. Organic evolution does not take place in an abstract, evenly flowing Newtonian time, a sensitivity to which could be expressed, for instance, in adaptive policies. Instead, organic evolution creates the operational properties of, and thereby defines, a level of temporality. Biogenesis, as we learned, is an extension, a continuity of the trend of all systems to keep entropy production minimal, giving rise thereby to biotemporality. Life, it will be recalled, culled its early features from those of the inanimate world by copying periodicities, then creating some of its own, then extending the spectrum, and then splitting into hierarchical levels of different rates of change (such as the soma and the gene). With the spreading biomass there emerged the potentiality of final causation; its corollary, the creature present; and the unavoidable trend toward increasing existential tension. Each animal, including man, contains in himself the Umwelts corresponding to the prior steps in evolution. Light which excites the skin (or the eyes) is atemporal; the quantum processes of life, such as biophosphorescence (or atomic processes, such as the Brownian noise in ear) are prototemporal; movements of animals must conform to the limitations imposed by the laws of the eotemporal Umwelt. This hierarchy of rules and regulations became enlarged with the coming about of the biotemporal world. Thus, the dynamics of adaptation or, for that matter, the whole machinery of organic evolution does not *respond* to the passage of time, but *creates and extends* temporality to include such features as final causation, the creature present, and increased existential stress of types unknown in the physical world.

Let us revert now to the central dogma of evolution, namely, that living things (or groups of things) prefer to live and adapt rather than die. Spinoza acknowledged this in his concept of conatus, by which he meant the drive, force, or urge possessed by a thing toward the preservation of its own being through *amour propre*.[152] He also maintained, sounding as though he were an American psychoanalyst, that there is no conatus for self-destruction. The ideas implicit in conatus were Spinoza's way of insisting on life's superiority to death. But, whereas life and death are certainly opposites even in a unity of conflicts, adaptation and death are not opposites because death itself is a mode of adaptation. It is the most perfect and complete adjustment of life to the nonliving, under the selective pressure of the inanimate environment. Thus, the question to ask is this: Why do living things prefer to adapt by complexification and the creation of more advanced integrative levels rather than by simplification and regression to already existing integrative levels? After all, as we just said, a dead body is a completely adapted chunk of material.

The answer is to be sought in the strategy of existence. I would postulate that this consists of a preference for increased rather than decreased existential tension. As one example, in the case of the unresolvable conflicts of life (the rise out of and the return to the inorganic—also known as the two thermodynamic arrows, etc.), the solution lies not in the elimination of conflicts through mutual collapse of the conflicting opposites, that is, through catastrophe, but rather in the maintenance of those conflicts. But, as we have seen, the maintenance of the conflicts of life amounts to unavoidable complexification, and alas, an attendant increase in existential tension. This sounds as though we are speaking of Aristotelian entelechy: a form-giving cause whose essence is actuality (as opposed to temporality). Or, perhaps we may be talking about the entelechy of the vitalists, for whom it was an intensive manifoldness placed outside the physical world. I do not think that we are talking about either of these; the views here formulated are closer to those of Joseph Needham who, without agreeing either with Aristotle or with Driesch, places entelechy as a factor of true autonomy, clearly and squarely in the complexity of the colloidal constitution of the protoplasm, or even above the atomic level.[153] The philosopher may only nod a layman's agreement with the biochemist's opinion and point to the concept of entelechy in evolution as one exemplification of the strategy of existence working through a hierarchy of unresolvable conflicts.

I have maintained earlier that the environment to which organisms are to adapt is so changed by the two-way dynamics of natural selection that the gap between adaptive features possessed by the organisms and features for which selection pressures exist is ever widening. The ingenious invention of organic evolution—the periodic regression from soma to gene is helpful but not sufficient to close this gap. It is as though the biological mode of matter is on a wild goose chase: it cannot produce sufficient variations and cannot produce them sufficiently rapidly to decrease this gap. Following what we postulated as comprising the strategy of existence, we ought to seek evidence for an integrative level above that of life.

PART THREE

The Mind of
the Matter

THE observer, who has mistaken a whole, changing landscape for a static object, is examined. He is found to be a skilled maker of models and maps of the world around him, of himself, and of worlds with no prior reality. Because of these skills he cannot be considered a mere observer but must be regarded a participant in the universal enterprise of creation.

VII

THE ORGAN OF
TIME SENSE

SOME OF THE ORGANISMS whose origins and evolutions we have just considered have learned to respond to conditions which are only potentialities and are not actually imminent (such as the fact of their inevitable death), and even to conditions which could not exist, because of the various constraints of reality. They have succeeded in improving their chances of survival by substantially enlarging their capacity for memory and by extending their abilities to include the design of expectations as guides for behavior. They (and of course we are talking about man) have defined themselves as enduring identities and created a unique method of communication based on symbolic transformation of experience. These evolutionary advances have been made possible through the coordinating functions of a specific biological structure, known as the nervous system, operating under the executive powers of a portion of that system known as the brain. In this chapter I shall argue that man's central nervous system must be regarded as the biological structure responsible for our sense of time.

1. The Advent
of the Mind

THE DIFFERENCE between a dead man and another dead primate may be described by specifying their anatomies; in later stages of decomposition one may need only to state their percentage chemical compositions. However, the difference between a living man and another living primate is more difficult to describe through an inventory of chemicals, biological functions, and behavioral patterns. If attempted, any such inventory would show an ill-defined domain of residual, differential qualities which, on balance, would favor the survival of man and the growth of his society to such a great degree that the existence of an essential difference between man and other living creatures would become undeniable. In the ordinary language of current usage this difference is subsumed under the name "mental phenomena."

Bodies, Minds, and Souls

What I have just called the "residual differential qualities" between living men and living animals, have been traditionally discussed in terms of the concepts of body, soul, and mind. Ideas corresponding to these terms as a group derive mainly from three observations which appear to have been made all through recorded history. The first observation pertains to an experiential continuity in a person's life, known today as identity. The second one is this: in response to certain challenges, human actions are markedly different from those of animals. The third one is that the inevitability of death is an affrontery to the status of man with his continuous identity and distinguishable actions. Because of the changing evaluations of the significance of these observations the meanings assigned to body, mind, and soul themselves varied, ranging from expressions of fundamental importance to those of ridicule. But beneath the broad range of estimates, expressed implicitly and explicitly through the languages, there is, I believe, a rather stable syndrome; I will call it by its currently accepted name: the mind-brain problem. This syndrome is of evolutionary origin and derives, I think, from the sleep-wakefulness cycle in man.

The states which we identify as wakefulness and sleep are probably coemergent with the separation of the soma and the germ cells and are attendant to the births of heterosexuality and death. They are usually defined as hierarchical states, by characterizing sleep as a substantially reduced responsiveness to stimuli (as compared with wakefulness) which can, nevertheless, be terminated. Reduced responsiveness of this sort is not manifest in "immortal" cells, but it is in the mortal "marigold that goes to bed with the sun and with him rises weeping." Serious concern with the sleep-wakefulness cycle is limited, however, to animals and man. The most comprehensive and still accepted theory of sleep and wakefulness was first put forth in 1939 by N. Kleitman.[1] In this theory wakefulness characteristic of lower animals and of young members of higher animals is "wakefulness of necessity," which gradually evolves into the long-term wakefulness characteristic of higher animals, the "wakefulness of choice." The latter is thought to constitute adaptation to the rhythm of night and day.

The physiological inner landscape which models the outer one is familiar to us from biogenesis, as is the idea that the inner world, with its several coexisting clocks, demanded control functions and generated periodicities of its own. Certainly, whereas night and day are features of the physical world, there is nothing in that world which corresponds to sleep and wakefulness. These two states acquire meaning only as they become distinct along the hierarchy of biological complexity, reaching their most pronounced differentiation in mature man. One must assume that in a grown and healthy specimen of Homo sapiens there are two distinct though related inner landscapes corresponding to sleep and wakefulness. The inner model (of the external world) which is functional during sleep is associated with the process of dreaming; the inner model (of the external world) which is functional during wakefulness is associated with sensing.[2] It seems to me that the debate that first centered on the existence and nature of the soul or some soullike entity, then on the nature of the mind and its position in the scheme of things, derives from the dynamics of dreaming and sensing as these develop in each child, and as they coexist in a periodical fashion throughout life. I shall argue that whereas the inner landscape of dreaming is best understood as determining a bio-, eo- or even prototemporal world, the inner landscape of wakefulness in man defines a nootemporal Umwelt. The idea of "body" has generally been associated with the lower Umwelts; the concepts of "soul" and "mind" (and their many variations) with the nootemporal world. But how body is to be distinguished from soul (or mind) if at all, had been a question of changing preferences and judgements.

Philosophical Preferences

The history of the brain-mind problem in Western philosophy may be divided into three phases. The Aristotelian phase spans almost two millennia, ending only with the *Meditations on First Philosophy* of Descartes, published in 1641. The second phase is that of Cartesianism with its reactions, branches and counterreactions, leading up to the fragmentation of philosophy in phenomenology, the rise of neurophysiology, and of the naturalistic modes of knowledge.

The idea that people have souls has been an attractive one, for, whereas the body perishes, those residual qualities of man which the dead possess not, could always be thought to survive. The postmortem fate of the soul, and the placing of the soul in the scheme of things are of central concern to all the great religions of the world. Anaxagoras (fifth century B.C.) seems to have been the first philosopher to posit explicitly the power of a mind-stuff coexisting with things physical.[3] For Socrates, soul was a simple, invariable object. Plato, in *Phaedrus*, distinguishes between soul and body, and holds that the soul can exist both before and after its residence in the body, and can rule the body during that residence. That the soul and the behavior of man correlate is implied in Plato's classification of the soul into nine behavioral groups, that of the philosopher (lover of wisdom) and the *philokalos* (lover of beauty); followed by the soul of the king, the statesman, physician, priest, poet, artisan, peasant, Sophist and at the bottom, the tyrant.[4] The Platonic dualism between body and soul survives into our epoch both by explicit endorsement as well as by vehement rejection. There are two main reasons to account for this lastingness. First, it gives an uncomplicated view of reality: the self is different from all other selves and only the body but not the selfhood is under the threat of death. Second, it is conveniently associated with the duality of inward experience (expectations, memories, plans) and the outwardness perceived by the senses.

Aristotle was quite aware of the logical difficulties of Plato's dualism which arose from the connection between whatever is meant by "soul" and "body." In De Anima he put forward a theory of the soul as the form of the body. Here "form," eidos, is one of the two ultimate principles of the world; the other one is hyle, material. Matter and form are aspects of the same underlying reality, with matter being the locus of potentialities for the activity of the form. Roughly, matter is spatial while form is a potentiality, hence temporal. Soul, then, is the form of the body. While vegetative and animal souls are born and die with the body, the nous (rational soul, intellect, or mind) is imperishable. It enters the body upon conception and leaves it upon death. It has two aspects. The active soul is that portion which survives; the passive soul links man with the animal soul.

Christianity embraced the body/soul duality of Aristotle and granted it philosophical privileges until, almost two millennia after the Greeks, Descartes put the question on an epistemological footing. He struck out to save mankind through the clarity, beauty, and distinctness of mathematical thought. He placed his faith in the self-evidence of certain truths and in the (self-evident) validity of logical reasoning which lead us from these to all other truths. The basis of all was res cogitans, the reasoning self, whose existence he could not doubt. As St. Augustine had asked about time, he asked,

> But what then am I? A thing which thinks. What is a thing which thinks? It is a thing which doubts, understands (conceives), affirms, denies, wills, refuses, which also imagines and feels [and] it is so evident of itself that it is I who doubts, who understands, and who desires, that there is no reason here to add anything to explain it.[5]*

From res cogitans he could be sure only of mental events but had to doubt the existence of bodies. Matter became an inference and there opened the logical possibility of souls without bodies and bodies without souls. The first corresponded to the surviving soul of man, the second to automata, better known as animals. But while permitting mentalistic and physicalistic statements to be logically unconnected, he did assume the matter of the body and that of the mind to be substances, though with different basic natures. In the composite whole,

> there is a great difference between mind and body, inasmuch as body is by nature always divisible, and the mind is entirely indivisible. . . . I cannot

---

* As a refreshing contrast, consider Freud who agreed with Hamlet: "Doubt truth to be a liar, But never doubt I love."

distinguish in myself any parts, but apprehend myself to be clearly one and entire; and although the whole mind seems to be united to the whole body, yet if a foot, or an arm, or some other part is separated from my body, I am aware that nothing has been taken away from my mind.[6]

As the intellectual ancestor of modern science, he dealt with the problem of body by stressing the spatial.

The nature of corporeal substance consists in its being something extended and its extension is none other than is commonly ascribed to a space however "empty." [Furthermore] the idea of the extension that we conceive any given space to have is identical with the idea of corporeal substance.[7]

Body is therefore extended and unthinking, mind is unextended and thinking. By completely spatializing matter, he removed from space all features that refer to life and concentrated them in time. Bodies are physical substances independent of human thought; they are vortices in geometrical space—a description which bears more than superficial affinity to the topological metaphors used in the geometrical handling of space-time in relativity theory. The interaction between the temporal soul and spatial mind takes place in "a small gland in the brain in which the soul exercises its functions more particularly than in other parts." [8] Sensation thus remained a function of the soul influencing, but not determining, free judgement derived from free will.

Descartes' solution illustrates one of the many logically possible attitudes one may have regarding the manner in which mind and body coexist and interact. Most thinkers would agree to the practical necessity of dealing with this question in terms of such other ideas as space, time, life, and matter, but they might hold a variety of views as to the relative priority, or ontic status of these ideas. For purposes of the following discussion I shall describe any theory which is built on the conviction that final reality is physical, as materialistic; if the theory emphasizes life as the central feature of reality, I shall regard it as vitalistic; if it judges the mind as the most important feature of reality, idealistic. By monism, following common usage, we shall mean that only one of the several possible integrative levels has ontic status; by dualism that two of them do, by pluralism that many of them do. The classification here offered is rather unconventional but it is quite useful in giving some order to the proliferation of opinions.

(1) Materialistic monism

These are theories which see the laws of life and mind reducible to those of matter.

Thomas Hobbes (1588–1679), a contemporary of Descartes, held that matter and motion were the least common denominators of all phenomena, hence the only proper subject matter of philosophy. Feelings of pleasure were nothing but motion in the head and motion about the heart.[9] Karl Marx and Friedrich Engels, in no way sympathetic to Hobbes, asserted in

1846 that mind is a nonmaterial existent which, nevertheless, derives from matter. "Conceiving, thinking, the mental intercourse of man appear at the earliest state as the direct efflux of their material behavior." [10] Earlier we encountered a twentieth century example of materialistic monism when we learned from I. A. Oparin that life and, with it, mind, are "motion of matter." Marx, Engels, and Oparin admit separate lawfulness to life and mind but insist on their material origin. Theirs are, thus, less radical than the opinions maintained by the early Vienna Circle (represented by Rudolf Carnap and Otto Neurath) which held that reports on mental events are synonymous with reports on physical events but in a different language. An alternate theory of materialistic monism maintains that mentalistic reports are simply mistaken in that the subjects reported do not exist. Although one may give the benefit of doubt to such extreme views and admit that self-evident things are not necessarily so, it is difficult to understand how one can insist, for instance, that the painfulness of cancer is only a mistaken mentalistic report; or how one is to search for the physical events that correspond to a simple mentalistic statement such as "some philosophers are funny, aren't they?"

The positivist Moritz Schlick has held that the physical and the mental, once understood, will be found mutually reducible one to the other; thus, he placed the problem, though not explicitly, in the domain of epistemology. In body and mind he did not see

> two attributes of a single substance (as in Spinoza) or two kinds of mani-festations of one and the same "being" (as in Kant), but instead an episte-mological parallelism between on the one hand a psychological system of concepts and on the other a physical system of concepts. [11]

The reader will note that Schlick's insight tacitly invokes nomogenesis for, in the course of evolution, the laws of the psychological system had to appear where there were none before, albeit displaying parallelism with the laws of matter already operative.

A theory of identity between body and mind was proposed and de-fended by J. J. C. Smart. He did not hold that sensation statements may be translated into statements about brain processes (as Schlick would have it) but that "in so far as a sensation statement is a report of something, that something is in fact a brain process. Sensations are nothing over and above brain processes." [12] H. Feigl drew attention to a semantic difficulty attendant to the identity theory. He felt that mentalistic and physicalistic statements differ in connotation (in the essence of the concepts they describe) but not in denotation (exemplification of individual instances). [13] The two can be seen to coincide only if they are assumed a common essence which, in Feigl's view, will eventually be identified as physical phenomena.

The doctrine of epiphenomenalism holds that mental activities are byproducts of neural processes with no autonomy of their own. It may be classed as materialistic monism only if one also maintains that the laws of life are reducible to physical processes. This is the opinion of J. J. C.

Smart, who sees no conceivable experiment which could decide between materialism and epiphenomenalism.

Finally, under the heading of materialistic monism, we must attend to behaviorism. Methodological behaviorism demands that psychologists confine themselves to the study of behavior as measurable by an experimenter (observer) in a subject (agent). This is a very neat trick, a copy of the mathematization of physics. It removes creative processes from psychology, assuring thereby the security of law and order in the world-view of the psychologist even though it pays the price of incompleteness by definition. Figuratively speaking, it admits as valid the occurrence of sexual orgasm but not that of any associated joy. Still, the approach is useful as long as it doesn't claim to interpret temporal experience or accommodate existential stress of an agent, but confines itself to the inventory method of an observer. Strict behaviorism abandons the concept of the mind altogether. Neobehaviorism permits "self-reports" but would judge them to be reports of biological or physical events. In some ways which shall be discussed later, behaviorism remains appropriate for the study of prey/predator relationships but not for the study of the self. It is a late and refined echo of Thomas Hobbes's materialistic monism, combined with the methodology of physical science. To be self-consistent, theories of materialistic monism must provide for causal interactions among features of the world regarded as autonomous. They have the advantage of being able to point to the laws of physics as a corpus, perhaps still incomplete, but potentially valid for all known phenomena.

(2) Idealistic monism

These are theories which see the laws of matter and life reducible to those of the mind.

Bishop Berkeley represented idealistic monism when he maintained that the only existents are minds and perceptions by minds.

> It is evident to anyone who takes a survey of the objects of human knowledge, that they are either ideas actually (1) imprinted on the senses or else such as are (2) perceived by attending to the passions and operations of the mind or lastly, ideas (3) formed by help of memory and imagination.[14]

But if statements about physical objects are meaningful only when they are understood as descriptions of perceptions, the world of life and matter must exist only in the mind. In Berkeley's idealistic monism man's mind is part of the cosmic mind, of which matter and life are but constant objects.

Pure idealistic monism in the West is represented today only by the teachings of Christian Science and those of general relativity theory. While some Christian Scientists hold only that bodily events, particularly those concerning health and disease, are results of mental activity, some would maintain that all our life processes are manifestations of the mind. In this they reflect the origins of their faith in transcendentalism, believing the ideal to be immanent in the sensuous, and denying the reality of the material world. A similar progression can be found in general relativity theory. Surely, in physicalistic understanding mind must be taken as an epiphenomenon of life, life as an epiphenomenon of matter. But matter as well as energy are identified with geometry. Geometry itself is certainly not a thing in any conceivable way; it is an idea. Thus, in the abstract curvature of four-space to which all things, and life and even mind are ultimately reducible, the cosmic idealism of Berkeley reappears.

(3) Vitalistic monism

These are theories in which both mind and matter are subordinate to and/or derive from life.

G. G. Simpson and many other philosophically minded biologists hold that mind is an epiphenomenon of and reducible to life, though the laws of life are not reducible to those of matter.

> Insistence that the study of organisms requires principles additional to those of the physical sciences does not imply a dualistic or vitalistic view of nature. Life, or the particular manifestation of it that we call mind, is not thereby necessarily considered as non-physical or non-material.[15]

This statement should be completed with a telling epistemological view advocated by Simpson. He feels that since the fragmentation of science is undesirable for many reasons, there ought to be only one intellectual discipline properly called science. The unification of fragments (of current science) should be

> not through principles that apply to all phenomena but through phenomena to which all principles apply. [These are the phenomena of biology] the science that stands at the center of all science. It is the science most directly aimed at science's major goal and most definitive of that goal. And it is here, in the field where all the principles of all the sciences are embodied, that science can be truly unified.[16]

As its distinguished author himself asserts, this is not a dualistic view; clearly, it is monistic. Neither is it vitalistic by postulation of some mystical vital force, but it is certainly vitalistic in the sense that it sees all of existence and all of knowledge centered in the living process. This is also the basic philosophy of Oparin who regards mental processes as manifestations of the irritability and excitability of nervous activity.[17]

(4) Neutral monism

This is a theory of American new realism derived from the work of William James. He acknowledged a multiplicity of realities but argued that mental and physical relations can be reduced to something common to them, though neither mental or physical. Consciousness, he wrote, "is fictitious while thoughts in the concrete are fully real. But thoughts in the concrete are made of the same stuff as things are."[18] The question was left open as to what this "stuff" may be.

(5) Psychophysical dualism

These theories admit the mental and the physical as two separate integrative levels, each with its unique properties, tied together by various means with varying strength. They tend to subsume life under matter.

Double aspect theories regard the mental and the physical as full descriptions of man but under different categories. For Spinoza (1632–1677) body is "a mode which in a certain determinate way expresses the essence of God, in so far as God is extended being." [19] Extended things in the world are connected so that "the order and connection of ideas is the same as the order and connection of things." [20] A temporal congruity is thereby established between order in things and order in thought. Passion and thought are manifestations of the soul, but "the body cannot determine the soul to thought, nor can the soul determine the body to motion or rest." [21] For Spinoza the mental and the physical were, thus, double aspects of the same, single, underlying reality of God or nature; they guaranteed the unity of the dualism.

Occasionalism is a religious answer to the sources of unity between the mental and the physical. Associated with the name of Nicolas de Malebranche (1638–1715), it postulates God as the missing link. God wills "that the modifications of the mind and those of the body be reciprocal. This is the conjunction and the natural dependence of the two parts of which we are constituted." [22] Occasionalism is a theory of a continuous present; it may be symbolized by two clocks which remain in phase because God continuously synchronizes their hands.

Leibniz proposed a design which distantly resembles occasionalism but involves past and future as well. Borrowing Giordano Bruno's idea of monads (psychic and spatial atoms) he identified them with the simplest and ultimate elements of reality. Each monad was believed to reflect, according to its own nature and point of view, the totality of the world. Coherence among monads obtains not by continuous synchronization but through a harmony preestablished by God. Mind and body are monads, or groups of monads of different qualities linked by preestablished harmony. Returning to the metaphor of two clocks, they remain in phase because they were made so to remain. Monadology sounds fancy to the modern reader, but one must remember that laws of nature are peculiar expressions of preestablished harmony. The figure of an establisher need not be invoked particularly as one gets closer to physics. Leibniz himself was encouraged in his ideas of monadology by the discovery of spermatozoa by Leeuwenhoek, which suggested the existence of monads of life, as well as by his own work in the calculus of fluxions, the forerunner of differential calculus, which held the promise of possible usefulness in the abstract handling of monads.

Two other theories need to be mentioned. Interactionism derives from Descartes' idea of the mental and the corporeal, with the former temporal, the latter spatial. In man, and in man only, the two can sometimes causally interact in either or both directions. Psychological parallelism derives from

Spinoza's ideas but it was first named parallelism by Fechner. It assumes that mental and physical events are correlated in a regular way, but not through the offices of a divinity nor through ordinary causation. Both theories bring to mind the Chinese idea of synchronicity, or simultaneous resonance,[23] and with it, once again, the problem of causation. How are we to interpret causation, for instance, between the ecstasy one feels and the changes in alpha rhythm that accompany the ecstasy? How would we begin to formulate laws which express some common regularities between feelings and electrical pulses? Connecting principles other than causation are logically legitimate but epistemologically difficult to support. Noncausal connections are alien to Western thought, possibly because they describe Umwelts below the noetic—such as the probabilistic causation of the pro-totemporal world. We must leave the possibility open, however, that connecting principles appropriate to integrative levels above the noetic might also exist.

(6) Biophysical dualism

These are theories which admit the biological and the physical as two, separate, integrative levels, each with its unique properties, and hold that mind is an epiphenomenon of life. We shall encounter such views later, when we discuss "neurological preferences."

An approach to the brain-mind problem which is not classifiable under (1)–(6) above, emerged with the existential and phenomenological emphasis on intentionality as the fundamental feature of mental phenomena. The strength of this new understanding is its focus on the importance of time in the analysis of the characteristic functions of brain and mind. Intentionality here is to be distinguished from teleology by the simple means of scope. Intentionality invokes no world-plan or long-range purpose of the universe; it is, rather, the capacity of living and thinking things to relate to their environments so as to secure, if possible, a final goal. This covers a range of activities, from the intent of the cat to swallow the canary to the intent of Neville Chamberlain to have "peace in our time."

Emphasis on intentionality in modern philosophy has its origins in the writing of Franz Brentano (1838–1917) who insisted that the emphasis of the philosophy of psychology of his days be changed from psychic states to psychological processes. However, it was Edmund Husserl (1859–1938) and the phenomenologists after him who completed the reemphasis necessary for a better understanding of the body-mind problem. For the idealists who emphasize the primacy of the mind, knowledge tended to remain subjective; for the materialists who emphasize the primacy of the body, the hallmark of knowledge tended to remain something objective; phenomenologists dare not neglect either. As a reaction to the exclusive emphasis of natural science on what it regards as objective, the call of phenomenology amounts to a reinsertion of the subjective as a valid basis of knowledge. Although this is reminiscent of Descartes' absolute belief in his subjective existence, it leads to different results. The Cartesian emphasis on the nomothetic leads to quantity; phenomenology with its emphasis on the generative leads to quality.

Phenomenology insists on descriptive analysis of all subjective processes, neglecting nothing, for contingencies might turn out to be crucial. Hence its subject is that of psychology. But, whereas scientific psychology must work with strict causation, phenomenology can reject any lawfulness as burdensome, preconceived notions. For instance, our ordinary ideas of causation do not seem to be able to connect the subjective world of experience with the so-called objective world of measurements (this being one of the problems of psychophysical dualism). Phenomenology did not resolve this difficulty. But it loosened the hold of causation on the methodology of the study of the mental by insisting that the content of consciousness be examined without any preconceived notions about theories.

There appears to be no such thing as a common view of phenomenological psychology; its benefits are more likely to accrue through the works of psychologists with phenomenological bent. Such an instance is a brief but profound paper by John Cohen, entitled "Cyclopean Psychology." [24] Cohen stresses that to improve our understanding of mental phenomena we must ask two simultaneous questions: How do we represent within ourselves the world without? and, How do we represent externally what we create within? Western psychology, following the philosophy of logico-mathematical truths, sees the activity of the mind as cognition (internal representation of the world without) and defines all other mental operations in its terms. In this model the mind becomes a problem solver of pragmatic questions derived from the external world. Psychology based upon such a model remains cyclopean (one-eyed) for it is uninformed of the qualities of consciousness as this manifests itself in the external representations of the world within. Such external representations are symbolic images of experience, exemplified for instance by the metaphor-making capacity of imagination, which, far from being prelogical or prescientific, is an indispensable feature of scientific inventiveness itself. But neurological correlates of feeling and thought, so intensively sought for and so highly esteemed in modern theories of the mind, cannot be identified from the inner image of the outer landscape. They must be found, instead, through an understanding of the inner landscape—reflected in the external image. Hence, the fundamental problem of modern psychology is the challenge of recognizing a world of meaningful private experience projected by the individual upon the external world.

Neurological Preferences

For Descartes, as we mentioned, the interaction between soul and body took place "in a small gland in the brain." Textbooks at the turn of the nineteenth century often showed pictures of the human brain with various regions identified as the seats of consciousness, unconsciousness, voluntary actions, acquired reflexes, etc. Textbooks in the middle decades of the twentieth century would label a median section through the brain in anatomical terms and indicate generalized functions. The cerebrum would be associated with remembering and intelligence; the thalamus with the sensory system; the hypothalamus with the control of the heart, of breathing rates, and of digestion; the midbrain with the nonvisual input and output pathways to the cerebrum. However, mental phenomena such as learning or speaking have been found sufficiently complex to make localization in centers rather meaningless.

Since Descartes, the concept of the mind itself has undergone several gyrations among its potential positions as substance, spirit, matter, or simply, behavior. We may agree with Descartes that the loss of an arm does not usually result in losing one's mind, though it may. The same cannot be said, however, for the loss of the brain. The evidence is firm that, although consciousness has no sharply defined physiological boundaries, mental processes are tied to the function of man's nervous system but especially that portion of it known as the brain. With this realization a neurological preference enters the older debate, changing it to an inquiry into the relation between mind and brain.

Whereas most philosophers of science find the Cartesian dualism of mind and body (or brain) unacceptable [25] many scientists who work with the brain find it useful as an acceptable working postulate. A distinguished representative of this view is Sir John Eccles who in 1952, from data then available, formulated certain hypotheses in this regard. They may be summed up in four direct quotations, given without their subsidiary qualifications.

> (1) Mind liaison with the brain occurs primarily in the cerebral cortex. . . .
> (2) Only when there is a high level activity in the cortex (as revealed by electro-encephalogram) is liaison with mind possible. . . . (3) The uniqueness of each percept is attributable to a specific spatial-temporal pattern of neuronal activity in the cortex. . . . (4) Memory of any particular event is dependent on a specific reorganization of neuronal association (the engram) in a vast system of neurones widely spread over the cerebral cortex.[26]

There remains, as Eccles sees it, an apparently unbridgeable gulf between the matter-energy system of the brain and the system we identify as that of the mind. I do not believe that empirical findings alone can possibly bridge this gap. What is needed is a new view to assist in the integration of the two systems into an intelligible single unit. I shall argue in the following sections that such a program will be facilitated if we understand the brain and the mind as determining two distinct Umwelts, the biotemporal and the nootemporal, respectively; and if we understand the mind as representing a degree of resolution of the conflicts immanent in life. Instead of reasoning about mind and body we shall speak of mind and brain, for only through the finding of neurophysiology does it become possible to identify that particular feature of man's physiology, namely, the complexity of his brain, which can accommodate conscious experience.

Contrary to what might appear to be the case, scientific concern with the mind and the mental is headed not for oblivion, replaced by behaviorism, but for an uncanny success of social usefulness and danger. We may observe in the history of the idea of the mind a mutuality between its self-description and its externalization in projected functions. Without trying to marshal detailed evidence, I simply record my impression that mental processes (or earlier, spiritual processes) were usually described through whatever was judged to be the most sophisticated way of knowing man and the world. Thus, from incubi and succubi in the vocabulary of the ubiquitous Devil, we changed to feedbacks and feedforwards in the vocabulary of the ubiquitous communication engineer. Yet, the changing metaphors must have appeared through the centuries as entirely, or almost entirely, appropriate means of description to those who used them. This brings to mind the stance of phenomenological psychology mentioned earlier, namely, that the internal and the external landscapes of the mind are complementary. In its turn this suggests, for instance, that the change from the world-view of Newtonian mechanics to that of electronic communication science is a change in the self-description of the mind. But if this be the case, then our technology is headed from solid state to wet components, and from miniaturization to ultraminiaturization, imitating the neural structure. The increasing drive for the control of the mind by all political establishments, the increased importance of psychic experience as a reaction to industrialization and regimentation, the very trend of technology, and the intuitive utterances of writers suggest a convergence upon the problems of the mind as the central issue of modern society.

2. Mind as Expectation and Memory

NEITHER expectation nor memory is unique to humans. Yet the reach of man's memory and the scope of his expectations are so much vaster than those of other organisms that, in spite of the common designations, a qualitative difference is indicated. We will try to understand the mind in terms of its functions of memory and expectation as these appear to an *observer* who notes them in others, and to a subject himself, whom we shall call an *agent*.

An observer identifies an agent by the agent's morphology and history. Since members of our race often have similar features and similar histories, the observer's method can lead to confusion; identical twins, for instance, may easily be mistaken for each other. But the mature individual as an agent observing himself does not run this risk; he knows that he is not his brother. This certainty stems mainly from a fundamental methodological difference between an observer watching an agent and an agent watching himself. The technique of the observer is primarily that of classification. He prepares a list about the agent, made up of items which, if they are to be held as true, are unchanging characteristics of some sort. The items on this list must therefore be lawlike, thus, at least in principle, recordable in spatial forms. In contrast, the agent identifies himself by a process which is a continuous integration of his memories and expectations in his ever-changing mental present. The elements of this self-identification are not fixed items but processes, or segments of processes; they are connected primarily by value judgements and not by laws. The list of an observer, perhaps engraved on a slab of stone, belongs in the eotemporal Umwelt; the process of self-identification of the agent belongs to the nootemporal world. Although there must be correspondences between the items on the observer's list and the inner experiences of the agent, it would be very misleading to describe any such correspondence as being "one to one," because of the fundamental qualitative difference between the two worlds. On the following pages we will often invoke the images of an agent and of an observer, sometimes as two people playing these characters, sometimes as two aspects of the same mind.

Memories and Expectations Concerning Others

Some convincing reasons were given recently by R. L. Gregory, why in the case of visual perception one must assume that eyes, the brain, and their linkages evolved simultaneously as parts of a single, functional system.[27] Surely, transformations of the external world into a topologically equivalent inner landscape via the senses, if it is to have any adaptive value, must somehow be interpreted and translated into action; otherwise the sense impressions are both useless and meaningless. Thus, I would admit Gregory's conclusion as unavoidable and apply it to all the senses. From among these, sight permits the longest processing and translating periods between impression and action, because light travels faster than sound or moving bodies do. A stalking tiger may be seen long before it may be smelled or touched. It is not accidental, therefore, that awareness of long-term future and the attendant intentionality are most pronounced in organisms that possess sight, and that expectation (and memory) are usually thought of in terms of images now seen rather than smells or voices now sensed.

The periods of time taken up by processing vary widely. Their lower limits are those of reflex actions; their upper limits those of a man's life or that of the human race. But whether brief or very long, the processing of some excitation signal and preparation for action are always in terms of past experience (of the species or of the individual), albeit the connectivities invoked correspond to several possible levels of causation and, in the case of man, must include free will. More shall be said about this later. Presently we note that perhaps the simplest way to describe the processing between excitation and action is that there are physiological filtering mechanisms interposed between stimulus and action, forming portions of the innate releasing mechanisms which evolved through the ages.[28] A releasing mechanism, then, is a rudimentary memory derived from past experience. But this is a very tricky proposition, because the species whose experience created the releaser mechanism did not always exist but evolved out of some simpler forms of life (and, further back, from no-life), including an evolution of memory modes. Thus, we must inquire about the origins of memory.

Consider the image of an object on the retina of a cat or man. It is a peculiar light pattern at a position where there was none before. The stimulus cannot consist of the image; it must be a change from no image to image or vice versa. The origins of memory must reside therefore in changes from conditions "before" to conditions "after." In the absence of firm knowledge of how such past changes are retained as memories, most current theories accept the view that the storage consists of preferences for pathways connecting receptors and effectors. The mechanism itself may be chemical, electrical, or even mechanical. Recall and remembering, then, depend on permanent changes in the cells of neural circuits. But the final state of a change in itself, no matter how sophisticated, cannot represent a past event or condition, only a contrast of some sort can do so. "Mene, Mene, Tekel, Upharsin," on the wall carries a message (constitutes a type of memory) because the words contrast with the unannotated wall; in a cuneiform script the contrast is between the flat surface and the groove. Whatever the detailed nature of memory traces may be, they must contain some contrast. The suggestion comes to mind that the contrast itself constitutes temporal labeling of a binary type. Perhaps an engram ought to be thought of as a module of some kind with its coding of before/after but without anything that corresponds to a present. This would make the imagery of memory and the experience of remembering itself a function appropriate to the eotemporal level, with memory images belonging to a lower level of temporality than do images constructed from current sensation. This, I believe, is in fact the case.

In this respect I wish to draw attention to the curious, fleeting, static quality of remembered images and to the modular character of memory. For instance, I remember Sandy running down the hill toward the trees with the sun playing in her hair. This image appears in my mind with all its details as simultaneous, even though it does include motion. I remember her 400-foot trip all at once, in the right temporal sense; I do not remember her running out from among the trees, backwards, uphill. I can also remember what happened after she reached the trees but that is again a static, unitary module in my mind even though it includes complex motion. If I decide to connect these two modules they form a new one, with a combined content but with the same character of pure succession. If I insist on remembering the grace of her motion in the tall grass I must force myself to lift the image from the eotemporal to the nootemporal Umwelt by inserting a present which scans, as it were, the eotemporal module, and inserts into it a changing condition of future, past, and present.

The span of time covered by memory images does not alter their eotemporality. The memory of a friendship that lasted for twenty-two years, or the memory of seeing the flattened nose of a frightened soldier upon a small window, or of a performance of *Oedipus Rex* all remain eotemporal unless one lifts them into the integrative level of nootemporality by imposing upon them the experience of the mental present. With that act of will, we may force ourselves to rethink the contents of memory modules in an Aristotelian unity of time: rethinking the memory of our friendship may itself last an arbitrary twenty-two years, that of the performance of *Oedipus Rex* four hours, the flattened nose of the frightened soldier three seconds.

Stored memory is by no means a continuous record of the past, something that could be compared to a stored motion picture. Rather, it may be compared to a family album of pictures, many of them overlapping in time but otherwise helter-skelter, each with its own coding of before-after. They seem to form sets and subsets and sub-subsets with personal history being the most inclusive set. We do not subdivide past time, but only divide memory images into their segments. (When I am asked to subdivide a ten second span into five equal parts, we are then dealing with the future, not with the past!) It is known that when electrical currents are applied to certain regions of the exposed brain in fully conscious patients, one or more engrams are activated. In flashbacks elicited by electric currents, in the words of Penfield "the stream of that former consciousness moves forward again in full detail as it did in some previous period of time." [29] I would say that the eotemporal memory modules, corresponding to the activated engrams, become attached to the mental present of the conscious patient. We might also add that a very large dose of electric current applied to a fully conscious patient might stimulate his most inclusive memory module, that of his total life, and produce the panoramic review of life anecdotally attributed to the dying. [30]

There is very little doubt that engrams are unlocalizable, by which is meant that the memory storage of a module of former mental presents involves the activity of millions of neurons spread over the cerebral cortex. Furthermore, each engram has multiple representations and in converse, each neuron and probably each synaptic junction forms part of many engrams. [31] It is suspected that engrams employ frequency, phase, or pulse modulation of some sort (as versus amplitude modulation) because they are more appropriate to the binary nature of neurons. [32] The nonlocalizability and the frequency (time) sensitivity of the neurological correlates of memory have given rise recently to some interesting models of memory.

One of them involves the optical technique of holography.[33] The analogy between memory storage and holography is attractive. In both: the storage of information is distributed; the records bear no simple topological relation to the information stored and are unintelligible without the reinsertion of some type of reference information. Another quasi-holographic model suggested by Longuet-Higgins conjures up a bank of resonators so interconnected that input signals similar to signals stored would bring forth an immediate echo.[34] Temporality as an essential element of memory storage also appears in nonholographic schemes, some of them included under molecular memory. In these it is postulated that learning leads to changes in rates of chemical processes such that the impulse discharge frequencies of nerve cells become specific with respect to the coded information. A temporal pattern arriving in the brain as the result of later stimuli would then evoke a resonancelike phenomenon.

E. Roy John in his recent work on the *Mechanisms of Memory* reaffirmed the unlocalizability of memory traces and proposed that memory is based on the patterned activity of aggregates of cells.[35] He sees the formation of memory traces as assimilation of rhythms such that an increase in the probability of coherent activity involving millions of neurons constitutes the stored memory. He hypothesizes further that recall consists in the correspondence of two sets of temporal patterns. One arises from the physical characteristics of the stimulus and may be classified as exogenous or evoked activity. The second is endogenous in origin, the result of prior stimulations of the neuron network.[36] Sustained afferent stimuli modify some characteristics of certain cells, which he calls plastic cells, while others, termed stable cells, display invariant responses even to repeated afferent stimuli. Retrieval of memory comprises the assessing of the congruence between oscillations of aggregates of stable cells and those of plastic cells.

> The activity or absence of activity of any particular cell is of itself not informative. Information is the time course of coherence, or the time course of deviation from random or base line activity in extensive neural populations. In this view, information is a statistical property of the neural aggregate. [Thus, awareness cannot possibly be ascribed to or associated with any individual cell, but only with statistical processes in the brain.] Restated, subjective experience would appear to be a consequence of the emergence of order in the organization of the activity of vast neural aggregates. Subjective experience cannot be attributed to the activity of particular cells whose function is to mediate the content of consciousness, but must be a property of the organized aggregates themselves.[37]

In John's hypothesis the states of the aggregates of stable cells associated with a specific engram represent the conservative elements and may be describable as associated with what is expected. The states of the aggregates of plastic cells represent the elements of becoming and may be associated with what is encountered. Furthermore, the states of the stable cells may be labeled as "before" (for they are conservative) and those of the plastic cells as "after." If John's hypothesis corresponds to fact, we might have in his approach a way of identifying the existential tension of the eotemporality of memory in the polarity of the coexistent states of "before" and "after." The stability of such a coded arrow would, of course, be that of biological processes, subject to interference. Nootemporality, that is, the insertion of a present into these eotemporal modules, still would have to be supplied through the mental present of the person who remembers.

Three further speculative remarks may be made if one assumes that engrams are temporal processes or functions rather than spatial configurations of some sort. Consider first that nerve-impulse coding is quite unlike the stimuli upon the sense organs; in biosemiotic vocabulary, what is common to stimuli and coding is only a sign function. The transformation of this sign function, by the sense organ, to nerve impulse is the first step in the long process of symbolic transformation of experience. The predominant modes of signal transmission along the nerve paths are conductivity and pH changes. Frequencies which seem to be the easiest to handle by such means are very far below that of the visible spectrum; in fact, they are mostly in the lower audio spectrum. But then, transducing audio frequency signals into nerve-impulse coding is likely to be a simpler task than transducing light signals because oscillations appropriate to the frequency range of the air pressure variations of sound are already in the right frequency domain. This fact, if it be a fact, might account for the evolutionary primacy of hearing over sight.

Sensitivity to sound allows, on the average, a briefer preparation time than sensitivity to light. Then, if hearing is older than sight, short-term intentionality and short-term memory associated with hearing are likely to be older than long-term memory. This, of course, is the case.

Finally, it is known that sensitivity of the ear is such that it cuts off just above the level of Brownian noise. Consciousness seems to incorporate an analogous feature: frequencies characteristic of the workings of the nervous system, such as carrier frequencies and discharge rates, remain just beneath the threshold of awareness. This might account for the remarkable success of biofeedback techniques. Namely, signals just beneath the threshold of awareness might be effectively brought into awareness through relatively simple instrumentation.

A fundamental question of the brain-mind relationship, from the neurological point of view, was formulated by Lord Brain as follows: If memory corresponds to a brain state, expectation to another and sense impression again to another, and since these three coexist, just how are they told apart? [38] Assuming that our speculations are close to facts, we might begin to answer Lord Brain's question, if not from the trained viewpoint of the neurologist, at least from the viewpoint of a student of time.

Memory modules, in the theory just propounded, are essentially eotemporal; their diadic form of before/after is very likely to be reflected (whether in the way John sees it or in some other way) in the neurology of the engram. In contrast, images of the future, while also demanding succession, have no preferred direction of time. Thus, I can imagine my son walking in the snow backwards, and watch the footprints fill up with fresh snow from beneath as he removes his feet as time passes. But I must place such an image in the future (if anywhere) for I have no memory of having witnessed it. The store of images out of which expectations are selected is, somehow, lower in the eotemporal Umwelt, closer to prototemporality, having neither a preferred direction of time nor, for that matter, any restraints of the physical or biological world. Now, if one holds with any rigor that there are correspondences between brain states on the one hand and memory, expectation, and mental present on the other hand, one would tend to seek the peculiarities of brain states corresponding to expectations (contrasted with those corresponding to memory) in their unorganized and perhaps unmodular continuous nature. One must assume that these qualities ought to be identifiable in the neurological changes which are corollaries of expectation. About the third category of brain state mentioned by Lord Brain, that of the mental present, we cannot yet say very much because so far we have dealt only with the actions of an agent. We have reflected only upon memories and expectations which relate to organisms external to the observer.

Earlier in this subsection we questioned the origins of memory: How far need we go back to identify "prior experience" from which rudimentary memory may be derived, for instance, in form of a response to a releaser. Consider now that in the biogenetic situation, the environment must already have contained connections of an if-then nature. Let us say, when the sunlight vanished or decreased substantially (whether due to nightfall, or cloud, or a new environment into which the organism was carried), colder temperatures were sure to follow. We have spoken earlier about the emergence, from the strictly cyclic order, of the capacity to respond to such environmental conditions, and we identified them with the origins of linear time. Any predictable reaction to contingent excitation exhibited by early life may be regarded as the ancestor of memory. The essential feature of this ancestry is that it was coupled to some reliably consistent features of the external world, where these features were quite independent of the organism itself.

Predictions Concerning the Self

As organisms complexified and also multiplied in number, the activity of the world external to any one organism ceased to be totally independent of the behavior of the "observer." Whereas actions based entirely on the (evolutionary) memory of past conditions were useful for certain primitive conditions, they ceased to be satisfactory when predictions based on past experience had led to behavior which altered the known response of the environment. To deal with this situation it was necessary to add to the inner landscape, which is the transformed image of the external world, a new symbol for a living object not itself part of that external world. In its most sophisticated form this object is the self.

Memories and expectations about this living object came to assume certain qualities which made them fundamentally different from those concerning external objects. Functionally, this came about because the observer himself, unlike an agent, does not ordinarily appear as an unexpected image on his own retina, or as an unexpected sense impression of voice, heat, odor, or force upon his own sensory system. The consequent differentiation between the self and the nonself has such far-reaching consequences that they will occupy much of our attention through the rest of this book. Presently we shall concentrate only on one aspect of this selfhood. Namely, whereas expectations about the behavior of prey, predator, or mate based on the memory of the observer were in principle, if not always in practice, quite reliable guides for actions, expectations about the self derived from the memory of its behavior possess a great degree of undeterminism, even in principle. The undeterminism is tied to the self-referential nature of this new entity.

Starting from strictly logical precepts, Karl Popper explored why it is impossible for a system to produce accurate answers to questions concerning its own future physical states. He concluded that however complete the information that is fed into a predictor, its description of its own future will always remain partly undescribable, hence unknowable in principle. His argument runs as follows.[39]

Having been given complete data of its past activities up to the instant t_0 and programmed with the necessary laws, the machine should be able to predict all of its future states, except insofar as they involve the very activity of this determination. Thus a degree of unpredictability is introduced. Whether or not the unpredictability is negligible depends mainly on the complexity of the machine and on the length of the processing period. But shortening the processing period will not help because of the necessary increase in complexity. Popper noted further that a programmed computer corresponds to a formalized deductive system to which the Gödelian principle of undecidability may be applied. Through the use of this principle he showed that, however complete may be the computer's information about itself, there will always exist questions about its future physical behavior which it cannot answer—at least not before the events have become part of its past. Finally, he showed that it does not help to have another machine with capacity to predict the first machine's outcome and feed such information to the first machine, somewhat as the Oracle did to Oedipus or the Witches to Macbeth. The reason is that any information fed from the outside must still become self-information first.[40]

We may draw two intermediary conclusions. (1) For all sufficiently complex computers there exists a specious present, that is, a minimum time that it takes the machine to find out what has happened to it. (This is not the minimum reaction-time to an input; that time must be much shorter). (2)

The specious present introduces an element of unpredictability into the future actions of the machine, as far as the machine is concerned.

Popper's idea was expanded by D. M. MacKay into an ingenious theory on the nature and limits of conscious control of action in a totally mechanistic universe.[41] MacKay inquires into the nature of information flow between an agent with a mechanistic brain (so as to eliminate problems of quantum physics), and an observer of that brain. The agent corresponds to Popper's computer; the observer is a garden-variety behaviorist, working with the inventory method. MacKay assumes now that the observer knows the details and operation of the agent's brain in sufficient detail to be able to predict its future actions. Then he comes to the logical yet startling conclusion that as long as the observer communicates his findings to the agent, this very act necessitates a reevaluation of the observer's conclusions which did not before include the effects of a disclosure upon the agent. Unless, that is, if the observer knows that the agent does not believe him. From the point of view of the agent, he need not be ignorant of the observer's views, but must believe them to be unfounded.[42] Higher-order corrections will not help, for if I were to introduce a converging series of such corrections this would only lead, as its limiting condition, to the Popper argument of indeterminacies in the future of self-defined systems. As MacKay sees it,

> The key point is that if what a man believes affects correspondingly the state of his organizing system [which is assumed to be his brain, then] no up-to-date account of that organizing system could be believed by him without being ipso facto out of date.[43]

Thus, an observer may have a belief valid to himself but the agent could not share that belief, even if in retrospect the observer turns out to have been correct.

We may draw, then, our own next intermediary conclusion. For an agent-observer system as envisaged, the agent's time is asymmetrical by the Popper argument. That is, whereas his past is determinate and may be remembered, his future contains a degree of irreducible undeterminacy. For the observer, however, the future is totally knowable if not actually known. His memory and his expectations have the same quality of certainty; hence his Umwelt is that of pure succession.

As a schematic model of cerebral organization MacKay proposes a combination of feedback and feedforward loops connecting the internal landscape with events of the external world as reported by the senses. The information flow so represented is of the type which, he has reasons to believe, abounds in the central nervous system.[44] As a specific example of such a scheme he considers a well known perceptual phenomenon: if the gaze of the eyes changes, the optical image on the retina moves, yet the visual world remains stationary. Differing from the usual interpretation of cancellation (that the signals resulting from voluntary eye movements are canceled by some signals of the oculomotor system) he selects to interpret the static quality of the environment in terms of the expected and the encountered. Voluntary movement generates no mismatch between the two because the change was expected on the basis of the existing internal world map. Perceived motion of the environment would result "only from a significant mismatch between the signals received and those which the action was calculated to bring about," such as when the eye is gently pressed, and the visual world is seen to move. The feedback loop in this case is that from the eyes to the brain, the feedforward is from the brain to the eyes. MacKay identifies the manifest functions of these loops with the goal-seeking behavior of the organism (the expectations) and posits a metasystem whose objects of perception are the details of the internal landscape.[45] Whereas the component systems comprising the feedforward and feedback loops are adaptive organizers and thus may be said to "know" those features of the world to which the organism is adapting, the metasystem which surveys the internal field (as the cognitive systems survey the external field) may be said to know that it knows. Mackay explores the peculiarities of such a hypothetical supersystem and finds that its logic suggestively parallels the human first-person language. That is, the system seems to have an identity.

Starting with the précis of MacKay's work as here represented, we may revert once again to our own thoughts. In the coexistence of the feedback and feedforward loops it is easy to recognize the existential tension between the expected (by the internal landscape) and the encountered (by the senses); the tension itself is MacKay's "mismatch." Philosophically, "value" is a corollary to the outgoing order, "fact" to the afferent signals. Since each sense impression is partly expected and partly unexpected, it is difficult to disentangle fact from value, a well-known source of chagrin for philosophy. The feedback and feedforward activities sketched by MacKay are those of an agent; the metasystem is something that does not concern itself directly with the external world: it is of the nature of an internal observer. The two acting together (as two aspects of one mind) represent schematically some of the important features and functions of the self. For the agent, future and past are asymmetrical; for the observer they are symmetrical. The agent experiences the undetermined future of all complex systems, for it cannot make completely reliable self-predictions. Taking these two imaginary actors together, we may say that cerebration (symbolizd by the observer) changes the cerebrum (the domain of the agent). In the language of philosophy, thinking is a form of activity which alters its own content.

This model of the mind displays some peculiarities which we have earlier associated with identity. Whereas in remembering the past history of others we seem to be able to render some statistically reliable predictions, when it comes to predicting our own future actions from remembering our own past we often find ourselves with a feeling, as it were, of inadequate input. Our future *feels* more different from our past than the future of external objects (living or inanimate) from their pasts. Perhaps it was this type of feeling which made the Greek dramatist Menander write at the turn of the third century B.C., "In many ways the saying 'Know thyself' is not well said. It were more practical to say 'Know other people.'"

3. The Mind
as Strategy

SELF-PREDICTION became necessary, or at least so I have speculated, when in the course of evolution the individual came to include a model of the self among the objects that populated his inner landscape. But we have just argued that predictions based on the memory of the self go with a degree of irreducible uncertainty, very unlike predictions based on the memory of regularities in the behavior of external, selfless objects. Thus, this new phase of adaptive technique, while it enhanced the individual's chance of survival by enabling him to make more accurate predictions of what he might expect of the external world (which now included himself), it also led him into conditions of intrinsic uncertainty and increased existential tension. Namely, unlike simpler animals which could predictably respond to a releaser, complex organisms with a cognitive structure of the observer-agent type are often left with the potentiality of multiple paths of actions whose relative outcomes can not be foretold as certain, even in principle. Certainly, man shares a capacity for choice with other higher organisms; however, as in the case of memory and expectation, the scope of alternatives available to man is so much vaster than that available to other animals that a qualitative difference between animals and man must be admitted.

The Very Complex

Here we will deal with a quality of the brain which makes its functions unique among all known biological processes. This quality is its complexity.

Consider first that the very large, the very small, the very many, and the very fast all have their characteristic laws, even if the boundaries of the effectiveness of those laws cannot be exactly delineated. Thus, physical cosmology is employed to deal with problems of immense distances, though it is impossible to identify a shortest immense distance. Quantum theory deals with aggregates of particles, but there is no critical particle number where Newtonian laws suddenly enter. Special relativistic effects become important gradually as we approach the speed of light. I believe that the *very complex*, as distinguished from the very many, also has its own peculiarities, even if there is no meaningful threshold to separate it from the not-so-complex. Accordingly, a formal definition of great complexity is as difficult as defining how many is very many, or how fast is very fast.

Roughly, however, complexity does *not* have to do merely with the number of component units of a system but with the degree of their organized interconnectedness. For instance, 1 cm³ of water contains about 10^{22} molecules. This is a large number. The number of neurons in the brain is of the order of 10^{10}, also a large number in terms of macroscopic objects but not impressive as compared with 10^{22}. But, although the 10^{22} molecules may all be simultaneously acted upon by a gravitational or an electric field, there is no interaction among them that would make them into a single unit of any sort. In contrast, each of the 10^{10} neurons receives 50,000 or more synaptic contacts; therefore the number of possible points of interaction among nerve cells in a human brain is of the order of 5×10^{14}. The number of possible, even if not actual, organized or organizable paths of interaction among neurons may be metaphorically described as countably infinite, and the quality of its organized multiplicity characterized as very complex. Eccles calculated that a discharge originating in one neuron in a cubical network, with each neuron having 5 synaptic contacts, can excite perhaps 4×10^6 other neurons in 20 milliseconds.[46] Yet this is a very much simplified case, for, as just mentioned, each neuron may receive not 5 but 50,000 contacts. Thus, during periods of time meaningful in conscious experience, the brain can play a virtually unlimited number of "kinetic melodies." [47]

Let us turn to the problem of unlocalizability of the engram. K. S. Lashley demonstrated that rats retain a memory for specific maze tests and visual forms after all but one-sixtieth of the about 700,000 neurons of the cortical cells of the visual cortex of their brains have been surgically removed.[48] Evidence gathered during the last four decades makes unavoidable the conclusion that both memory and learning are distributed rather than localized functions of the brain. Yet large brains are not morphological prerequisites even for advanced mental functions, such as language learning. Nanocephalic dwarfs (dwarfs with skeletal proportions of normal adults but seldom taller than three feet) have brains whose volumes are one-third or less of the brain volume of a comparable normal adult; still they acquire rudiments of language and verbal skills. Would it be possible, asks E. H. Lenneberg, to learn to understand or to speak a natural language such as English with a brain the size of some nonspeaking animal? The answer is yes, but only if the individual is of the species Homo sapiens.[49]

One may reasonably ask, therefore, What particular feature of the brain is interfered with the least when lesions are produced experimentally, accidentally, or clinically? In our two examples, the sizes of the brains were radically decreased; in nanocephalic dwarfs by a factor of three, in Lashley's experiments with rats by a factor of two. Perhaps the feature least interfered with was the degree of complexity. Histologically the brains of nanocephalic dwarfs are not different from normal brains, nor is the residual visual cortex of the rat different from the portions excised. One must assume, therefore, that the complexity of the respective brains has not been tampered with. Thus, for instance, the number of paths within the region which a single neuron can influence, in a period comparable to some important time constant of the brain (say, alpha rhythm in man), is the same for the normal and the nanocephalic brains. Analogous arguments hold for the rat brains.

Complexity has been of fundamental interest to general systems theory,[50] and its importance to the study of the brain has been occasionally emphasized.[51] I wish to remove the concept of complexity from among features of the brain which are no more than interesting, important, or relevant and elevate it to the category of such fundamental concepts as those of motion, distance, or number. In this respect I wish to draw attention to the fact that our bodily facilities do not place us on the boundaries of the very large, very small, very many, or very fast. We can easily find examples of things larger, smaller, more numerous, or faster than ourselves. But possession of the human brain places us at the upper, limiting boundaries of complexity. I submit that the complexity of the human brain is greater than that of any other known structure, including that of the universe at large (excluding other human brains). Accordingly, I shall use the phrase "the very complex" as identically equivalent to the human brain, just as "the very small" must mean elementary particles, "the very large" the astronomical universe and "the very fast" the motion of light. It is because of its extreme complexity that the brain is capable of containing, in whatever language is peculiar to it, models of the external world, of the body of which it is part, and of things and processes which do not necessarily exist.

By way of inquiring into the functioning of the brain so conceived, let us recall once again the transcension of matter to life. Biogenesis and the evolution of the cyclic and linear order brought forth, through internal self-generation, some new cycles and biological programs which were not the models of anything that existed in the external world. Because of the mutuality of the dynamics of adaptation, such new programs came to exert selective pressures upon the environment, both living and nonliving. Now let us raise this scheme one level higher.

Let us think of the mind as comprising models of its environment, which is the brain. Such models might pertain to memories, expectations, or creative thought. The signs of the language appropriate to the mind would be such symbols as were left undetermined by the laws of life. For instance, there is nothing in biology that would prohibit the existence of a poem or a bridge, though the final products must abide by lower-level lawfulness. The contents of the mind then, comprise the next higher integrative level above that of life. It is reasonable to assume that as the programs of the mind evolve they will generate new symbols which have no prior physiological counterparts in the brain. I shall call such self-generated, new engrams autogenic imagery. "Imagery" is not to be understood as exclusively visual however; it is intended to include auditory and textual elements as well. (In very primitive forms, self-generated imagery is implicit in the non-linearity of differential equations proposed by general systems theorists as appropriate for complex systems.) [52] We may further assume that, because of the mutuality of adaptation, the mind would endeavor to impress its autogenic images upon its environment, the brain. And via the brain, the mind may be seen to exert selection pressure upon the evolution of life (and changes in matter), just as life exerts selective pressure upon the evolution of other lives (and changes in matter) via organized matter, which is the environment of life.

Autogenic imagery with no prior engrams may be identified with creativity in man: a house, a statue, or a plan of attack first exists only in the inner landscape of the mind and only later is it forced upon the external world through the functions of brain and body. The process of the mind in general, that is, the formation of symbols corresponding to brain states, but especially the formation of autogenic imagery or autogenic symbols, is ordinarily known as thinking. Among the self-generated engrams there may be some judged as corresponding to certainties; these we call laws of nature or by some similar name. There will be some which are only potentialities, that is, while not actual they are possible. Again, there will be some which correspond to impossibilities, for they would demand the removal of restraints which define the biotemporal, eotemporal, and the lower Umwelts. Pascal remarked in his *Pensées*, "The heart has its reasons which reason knows nothing of." We may add that the mind has its powers that the brain knows nothing of.

Some other characteristics of the very complex will be suggested as our reasoning progresses; again some further features will be sought through psychology. Regarding possible boundaries of the very complex, one can only guess that complexification cannot go beyond a certain level without being self-destructive or without coming up against limitations imposed by the nature of life or matter. When that condition is approached, the complexification of the brain has to yield to some other methods in the strategy of existence.

The autogenic imagery of the human brain, that is, engrams which do not correspond to past facts or sense impressions, may be identified with creativity in man. A house, a statue, or a plan of attack first exists in the inner landscape of the mind and only later is it forced upon the external world through the functions of brain and body.

Miners raising a load of ore up a shaft by means of complex machinery which first existed only in the mind. From Georgius Agricola, De re Metallica, *Basel, 1556. Courtesy, The Burndy Library, Norwalk, Connecticut.*

Conscious Experience and Free Will

It is rather an understatement that the variety and number of possibilities in the physical universe stagger the imagination. But if my reasoning is correct, then the store of images contained in and generated by the mind is immensely richer than that of the external universe, for it comprises, in addition to an inner map of that world, models of things and events which may exist, though need not; models of things and events which could not possibly exist or come about, and also a model of the self. Hence, figuratively speaking, the variety and number of possibilities in the mind should stagger the universe.

The evolutionary advantages, for the organism, of its capacity to respond to present stimuli in terms of past experience are quite clear. To the set of actual stimuli originating in the external world, the operations of the mind have added an open set of imaginary stimuli: the possibilities of conditions which may or may not ever arise but are certainly not present in the external world at the time of their genesis in the mind. Responses to such stimuli need not themselves remain symbolic but, on the average, they are likely to, because many contemplated actions cannot ordinarily be pursued simultaneously. Thus ensues conscious experience in which the self continuously acts and reacts with a variety of symbolic conditions in addition to lower functions of the biological order, such as releaser responses, reflex actions, and the like. The advantage in being able to respond to imaginative conditions with no foundations in memory is the potentiality for greatly increased foresight and hence, a substantially improved power of adaptation.

That consciousness is an emergent quality of the very complex has been suggested by W. H. Thorpe, who holds that it might be a process essential for the steering of nervous systems when they get beyond a certain degree of complexity. His work on ethology suggested to him that if the nervous system does get very complex,

> if it reaches the fantastic degree of elaboration which we find in the brain of the higher mammals, the unconscious, and even subconscious, methods of control are no longer adequate and we enter the realm of conscious control.[53]

Thorpe's insight, as well as my remarks above, are both accounts of an observer. They propose to explain how other people's minds work and give some reasons why and how they came to work that way. But when I attempt a first person account I confront the extraordinarily difficult problem of trying to reconcile the unitary quality of *my* conscious experience[54] with what I believe to be the source thereof, namely, the tensions among signs and symbols in my head. Some hints on how to get from "here" to "there" may be found in the valuable inquiry of H. A. Simon into the advantages of process versus state description of complex systems.[55]

He points to the practice in some branches of science of substituting process for state description when dealing with complexity. He then argues at length how problem-posing in various fields of knowledge amounts to the making of state descriptions, and how solving the problem amounts to the discovery of sequential processes that will produce the goal state. Human problem solving, as he sees it, is basically a means-end analysis that aims at discovering a process description of the path that leads to a desired goal. In fact, reasons Simon, this approach is not something accidental or superficial invented by science. The parsimony which attends the use of process description in lieu of state description is basic to the functioning of any adaptive organism. As one example, he points to the genetic mechanism of multicellular organisms. Apparently, natural selection found it advantageous to describe an organism through genetically encoded programs rather than by, say, genetically encoded state descriptions. Program description has been favored by natural selection for reasons which we might suspect from what we concluded about time and evolution: it is a feature which can provide faster adaptation.

Applying Simon's reasoning to the very complex, we find that the process description of the brain is not only the parsimonious one but, probably, the only one feasible. If I wish to describe my head I would ordinarily start describing how it looks, because about internal organs one can say something only if they hurt. I am unable to give a state description of my brain, but I am able to give a process description thereof, as I have been doing on these pages. This is, I believe, what is meant when someone stresses that the mind is temporal and not spatial. We may even turn this reasoning around and say that what we mean by temporality is the mode of expression of the mind.

The feature that introduces manageable order and unity into our inner world of whirling memories, expectations, and strategies is the limited nature of the psychological present within the broader scope of mental present. To elucidate this statement I wish to describe first, with the assistance of an illustration, the essence of what is known as harmonic analysis. Consider that on a phonographic record even a very complex audio signal, such as that of a symphony orchestra, is represented by one single, continuous, undulating line. The pickup needle is only at one point at any one instant and, if followed for a sufficiently brief period of time, its motion approximates that of a single synthetic tone, and not that of an orchestra. By narrowing the period of our interest we have produced a very simple image (a line) of a very complex process. Out of such a series of simple periodic signals the listener, using his memory and expectation, reconstructs the tone of the instruments, the voices of the choir and might even conclude correctly that the conductor was in a cheerful mood. A state representation of the same symphony is the musical score used by the conductor. The state and process representation of the symphony are assumed to be equivalent, and it is the task of the conductor to assure this equivalence as he interprets it.

Let us revert now to the process description of the brain: much simpler than state descriptions would be, but still very intricate. We will try to make some sense of the coexistence of the psychological and mental presents in man, and of the phenomenon of attention which makes the store of "kinetic melodies" manageable. For this purpose it will be necessary to produce a delicate *auseinandersetzung* (mutual juxtaposition and separation) of three "presents." Let us recall the idea of creature present as the (experimentally suggested) experience of simultaneity (atemporality) in animals. The capacity of higher animals for delayed gratification we may now identify as a feature of biotemporality, or, philosophically, as a phenomenon which makes final causation meaningful. Scientific study suggests that certain elements of consciousness (attention, ideation, manipulation of abstract ideas, anticipation) are also manifest in animals.[56] We also spoke earlier about the psychological present in man, which we said comprises all the lower Umwelts of temporality through biotemporality but, as compared with animals, it includes a greatly extended capacity for delayed gratification. Now, if we argue that there is some kind of machinery operative in man so as to make the process description of his brain useful, we must also extend such qualifications to higher animals. But here we are faced with what evolutionists call a divergence. Because of some changes in the evolution of the brain—changes which can only be described as explosive and which we will consider later—man's psychological present blossomed into what we have called his mental present, whereas the mental functions of animals remained in an evolutionary cul de sac.

It seems that with the substantially increased scope of expectation, the vast storage of memory, and the open store of autogenic imagery it became necessary for Homo sapiens to gather in and sharpen his temporal categories. Returning to our phonograph analogy, while in a simple, slowly changing single tune the instantaneous position of the needle is not very critical, in the case of a symphony it is; so also in the case of the struggle of symbols in the mind of man the definition of something that sharply separates memory from expectation became imperative if practical adaptive behavior was to be assured. Indeed, there seems to be a reciprocal relationship between the unity of conscious experience and the degree of focusing on a present. When one is in healthy equilibrium with the environment, aware of future and past but not overwhelmed by them, and concentrating on tasks at hand, there is also usually an unquestioned sense of unity of self. The loss of this unity is a clinical sign of disturbance and is often accompanied by the loss of the capacity to focus on the present.

Now, unlike what appears to be a fairly rigid structure of temporal hierarchies in higher animals, the mental present of man is of a continuously varying scope which broadens and narrows between extreme limits. At the one extreme is the vista of a personal identity from birth to death; at the other is the concentrated attention on the tick of the clock which separates an irrevocable past from an unknowable future. I think this plasticity is necessary to provide for unity of conscious experience in the face of the turmoil of the mind.

The domain of creative thought is immensely richer than the world of reality, for reasons already discussed. That our mental potentialities are much broader than the vital or material ones is expressed in our awareness of a free will. Thus, free will is not a property of the mind itself but rather a relationship between the broader limits of the mind and the narrower limits of the body. Since life itself also has a store of potentialities not available to inanimate matter, life may also be said to have more freedom than does matter. These hierarchical restraints fill our experiences with ambiguities. We are free to select from among potentialities represented by autogenic engrams; but we cannot do everything we can think of for we are limited by the laws of life and matter. Going one level lower, life may graze and move, hide or run at its will, but it must follow the rules of its eotemporal Umwelt. Although many of our actions are demonstrably deterministic (they are manifestations of eotemporality) there are no actions which could be scientifically proven to be absolutely free. This might have been what Bergson meant when he wrote that the mind

> feels itself free and says so; but as soon as it tries to explain its freedom to itself, it no longer perceives itself except by a kind of refraction through space. Hence a symbolism of a mechanical kind.[57]

We may now combine what we have said about the agent-observer model of self-prediction and temporal asymmetry with the idea of free will as the relative freedom of the mind over the body. For the purposes of these few paragraphs I will capitalize Agent and Observer as though they were actors in a noetic play which, after a fashion, they are. In our daily experience they may sometimes be identified with the active and reflective aspects of temperament. In some other ways they are analytical poles of a single phenomenon, that of conscious experience, with neither the Observer nor the Agent having ontic status. This unity may sometimes be seen as torn asunder, such as in certain cases of psychosis, where the patient feels himself separated from his body while still being aware of himself as a unit. Again, sometimes, such as in certain ecstatic experiences of universal harmony, the Agent and the Observer merge into an unstructured whole.

We recall that for the Observer there is no difference between memory and expectation; neither may contain the unexpected. For the Agent, however, the future contains elements which are unpredictable, due to the way the Agent functions; hence there is an absolute difference between his future and his past. Consider now consciousness as the interpenetration, or coupling, of the two actors; after all, they are but analytical projections of a single, integrated mind. Therefore, the Observer cannot be secretive; he has to keep the Agent posted. The Observer, true to his nature, will hold that the Agent's actions are predetermined, hence that the Agent is a member of same eotemporal world as is the Observer. But the Agent, because of its self-evaluating nature, must insist that the future is different from the past and that the unexpected might happen. He must also insist on his freedom of choice, for otherwise the Observer himself would be discredited. Furthermore, the Agent possesses an open store of autogenic imagery to which the Observer cannot have access, for the Observer bases all his judgement on structural description and not on process description of the Agent. In the struggle between Observer and the Agent within one single mind we may find the existential paradox of man's freedom versus necessity, expressed in history as the unresolved conflict of the two. (The reader may recognize in this view of history the one held and elaborated by Nikolai Berdyaev.) [58] In any case, the conflict of doubt and certainty is a hallmark of conscious experience.

We have already noted the historical trend to describe the operation of the mind in terms of whatever is judged the most sophisticated understanding of the world at an epoch: soul, philosophy, mechanical clockwork, or computer. In our own epoch the question is in this form: Can the functions of the mind be reduced to the functions of an electronic computer using the most up-to-date components, or components now planned in advanced research laboratories?

Answers to this question, with few exceptions, tend to lose sight of the evolutionary nature of existence and concentrate, instead, on technical details without resolving, or even acknowledging, problems of principle. There is little doubt that specific functions, directed to an increase of our store of deductive knowledge or to the improvement of our facilities to perform special tasks, could be performed by machines. This, of course, is not new. Archimedes demonstrated the power of simple machines when he invented the lever for raising weights. The peace treaty inscribed in hieroglyphics on the temple of Karnak is a memory storage device appropriate to the technology of Egypt, eleventh century B.C. Both are examples of how autogenic imagery may be forced upon the environment. Levers and glyphs came into existence only some time after they appeared as creative images in the mind of man. The requirements for a machine with artificial intelligence, then, boil down to those of approximating the brain, so that the new mind-made mind can also create levers and glyphs.

Consider the builder of such a machine developing it from first principles by analytical deductive reasoning. As he identifies an increasing number of laws and incorporates them in his machine his own mind must remove itself from the domain so defined. His self-awareness must incorporate the principles of the gadget, hence his own brain must further complexify by continuously separating the lawful from the contingent in his creative thought. Hence, he will not succeed in reproducing himself but only something less complex, even if much larger or faster. He might decide, therefore, to abandon the universal method of machine design, which is that of producing a state description, and try to construct instead a self-evolving device. He will learn in due course that his machine of artificial intelligence must be very complex, just as an artificially created particle must be very small and an artificially created light beam must be very fast. He might decide, therefore, so as to control energy expenditure and size, to use some such techniques as have already been found useful in organic evolution and abandon electronic computer techniques for biochemistry and neurological techniques. In due course, he will reinvent the brain, a soma to support it, and genes to offset cumulative deleterious change.

Could such a machine eventually possess self-awareness, free will, and identity? My answer is in the affirmative. However, the maker, if not yet devoured, is likely to be disappointed for two main reasons. First, the whole super project could have been completed more economically by engaging the services of a young couple. A man and a woman together bring to such an assignment the evolutionary experience of some three million years in making brains and, normally, they are able to deliver the final product nine months after signing the research contract. Secondly, although the mind-made mind (and its auxiliary somalike and genelike features) may be larger than man and much less subject to the influence of the weather, unless the whole artificial intelligence machine (self-reproducing and free-willed) is somehow out of this world, it would certainly be subject to the laws of the physical world, those of life and those of the very complex. If such machines were eventually to get sufficiently perfect to replace man through natural selection operating through the preservation of favored races in the struggle for life, then we would witness from our graves the emergence of a new integrative level in the hierarchy of unresolvable conflicts. We might call this state of affairs the Frankenstein syndrome.[59]

Prediction and Creativity

We shall inquire into prediction of future events as examples of the functions of the mind. Consider two predictions. (1) During tomorrow's solar eclipse you will find that the sun deflects a light beam that passes near it. (2) Flight 437 from New York to Boston will be hijacked by a kind old lady. These two predictions, assuming that they will both come true, have many features in common.

(a) The person making the prophecy and the experimenter checking its validity can be, but need not be, the same person. (b) Both predictions may be declared true only some time after they have been made. (c) Both describe unique and unrepeatable events. And (d) both derive from non-inferential knowledge. From among these points (c) and (d) need some explanation.

The story with the kind old lady would immediately be judged as unique, because it involves too many uncontrollable variables. But so is the light-bending experiment: I have in mind the 1922 expedition which confirmed Einstein's prediction made in 1916. The hijacking story would immediately be judged as noninferential. But, of course, so is the light-bending. Though the prediction was drawn by logico-deductive methods from a set of equations, the principles from which those equations them-selves derive (relativity, limiting velocity of light, equivalence of inertial and gravitational masses) are examples of the noninferential genius of the mind. So is Bohr's atomic model, or the famous Schrödinger Equations of quantum theory which can be justified only ex post facto, but not derived. Shakespeare's understanding of human nature without the benefit of psy-chology belongs in the same category, as do the predictions of George Orwell, Aldous Huxley, or Arthur C. Clarke. C. S. Peirce spoke of "natural light" and of "guessing instinct" and stressed that approximately correct theories are discovered with remarkable ease and rapidity, for man's mind has a proclivity for imagining correct theories.[60] I believe that the com-plexity of the brain, which is the source of free will, is also the source of human creativity as it is manifested in predicting the future.

But if the selection from among imaginable future conditions is not made through the restraints of remembered experience and unchanging lawfulness, on what basis is it made? The criteria, I believe, are value judgements, emotions, and feelings which operate both in the conscious and in the unconscious. We distinguish and select for attention some of our "kinetic melodies" because they appear beautiful (whether they say that energy must be quantized or that you have a blue guitar); or because they are exciting, horrible, or repulsive; or because they suggest love or hatred; or because they are frightening or funny. In any case, it is as though the syntactical relations of our autogenic imagery were on a continuous hunt for semantic realizations. Indeed, syntactically well-formed strings of symbols determine the domain of the thinkable, working as they do in our inner landscape, but they do not determine what may be found. The semantics of creative thought, that is, the relationships between the inner strings of symbols and their significance in the external world, are better known as feelings and sensations. These, as all other features of man, must be re-garded as evolutionary in their origins; they imply types of ordering alien to codified knowledge. Creative foresight, then, is the capacity of the mind to make selections from among autogenic images by means of the empirical rules of feelings, such that the selected images are significant for the future.

In the two examples in the beginning of this section I have grouped under creative foresight the socially acceptable forms of scientific inventive-ness together with the unacceptable forms known as precognition. The unacceptability of the latter does not derive from methodological objections, for the unrepeatability and unpredictability of card guessing is not different from the unrepeatability and unpredictability of creative utterances by gifted men from Anders Armstrong to Ulrich Zwingli. The mistrust might be psychological in origin and stem from the curious feeling which often accompanies the demonstration of unexplainable foresight, one which suggests the work of higher powers and thus poses a threat to man's integrity. In turn, this phenomenon (the fear of unexplainable foresight) seems to be tied to the fact that along the hierarchy of life we find that organisms are capable of controlling an increasingly longer span of the future. The span of future which is potentially under the command of a unicellular organism is determined by such features as chemical or temperature gradients in liquids. A dog with his dog senses can easily get out of the way of a flying stone, well before the stone arrives. A centipede hit by the same stone would, if it could, attribute divine foresight to the canine. Likewise, a dog would, and perhaps does, assign magic powers to its master for having got out of the way of a stationary car which began to move and scared the dog. In King Arthur's Court the Connecticut Yankee was thought to be in league with the heavens because he foretold the solar eclipse. Similar magic may be produced for the benefit of small children by any thoughtful parent. Creativity in man, either in bringing about the unpredictably new, or in foreseeing what appeared to be unpredictable, seems also something magical, implying certain advanced capacities of the mind which are unintelligible to a mind with less advanced capacities.

I have reasoned in the preceding subsection that a machine model of the mind can be only of limited use unless it becomes very complex (in which case the machine would act as the brain, rather than the brain as a machine). One might, however, obtain some interesting hints from computer models—as one could also obtain from the soul-models, or from the clock-models. Specifically, I am thinking of von Neumann's interesting point that one need not assume that the brain operates as do present day computers, in successive operational steps. He agreed that the main contrast between the best of computers and the brain is complexity; he expressed this in terms of the number of actions that can be performed by a computer and a brain of the same total size (defined by volume or by energy dissipation) in the same interval. It is known that reaction times of neurons are of the order of 10^{-2} seconds, whereas transistors can recover in 10^{-7} seconds. It would seem therefore, that while computers can well afford to be sequential, the brain is likely to favor simultaneous processing. Hence, the logical approach and structure of the brain may be expected to differ from those of artificial automata.[61] It seems to me what whereas processes of reasoning in the brain must operate serially, as assumed in the Popper-MacKay analysis, essentially because actions can be taken only in a serial order, other mental phenomena, such as feelings or unconscious judgements might derive from modes of operation which are nonsequential and follow a peculiar, and for the moment unknown, logic. Selections from the repertory of autogenic imagery might well be guided by feelings and judgement which are not themselves expressible through the necessary logical sequencing of discursive cognition.

The history of man can certainly boast of fateful utterances about individual and communal destinies which do not seem to have been derived by logico-deductive reasoning. They form a broad spectrum, from the many "announcements from the mount" to the corporate sensitivity of poets and writers. It is precisely these prophetic capacities, together with knowledge of the inevitable, which inform all levels of human civilization with a tragic sense of freedom and fate. The seventeenth century English physician and author, Sir Thomas Brown, likened the elusive logic of creativity to the "mystical mathematics of the city of heaven." I would like to suggest that the "mystical mathematics" of feelings and sensitivities, unlike the logical processes of thought, need not necessarily favor the lives and loves of the individual but favor, instead, some integrative level higher than that which corresponds to the individual. This idea is an extended version of one suggested by Jule Eisenbud as a possible way to accommodate precognitive instances within the conceptual framework of the psychoanalytical situation. He writes that if precognition exists

> then the goals it serves are not only not necessarily those which the individual might consciously hold important but almost surely, rather, those resulting from the interplay of his unconscious needs, determined by his genetic and psychosocial history, with the overall requirements of a probabilistically geared universe. In this view, finally, man is no more a puppet than he would be otherwise. There is merely a slight shift in bookkeeping methods in what in any case adds up to a highly complex probability calculus.[62]

If the reader is about to conclude that this sounds too mystical, he should remember that the operation of a higher integrative level always appears unintelligible in terms familiar on the lower level.

A brief review and preview is in order here. The brain-mind theory here proposed differs from other theories reviewed, although it resembles many of them. It represents a neurological rather than philosophical preference and sees the peculiar properties of the human mind in the very complex quality of the human brain. It is monistic in that it regards matter, life, and mind, and whatever might emerge from mind, as manifestations of a single natural order. But it is pluralistic in that it holds that the nomothetic qualities of matter, life, and mind are unpredictable from and irreducible to the laws of the levels respectively beneath them. Each level seems to operate in the region left undetermined by those below it but abides by the restrictions (and displays the indeterminacies) of those lower levels. Thus, there are no physical laws against the formation of tulips: organic evolution could produce a tulip; there are no laws in biology against the formation of symbolic images, hence the brain could produce the mind.

I speculated that what we recognize as thoughts are process representations of brain states, or dynamic models of brain states, in analogy for instance, to physiological clocks which are dynamic models of processes in the inanimate world. Although there is a degree of analogy between the two examples of transcendence, there is rather a significant dissimilarity between the two as adaptive techniques. Genetic symbolism, representing the learning of the species through organic evolution, must be carried by a living thing; it must be engraved, as it were, on living matter In contrast, thought may be recorded on inanimate matter in ways which make possible the passing on of learned knowledge (acquired characteristics) to other members of the species not present or not yet existent. The mind can thereby reduce experience appropriate to biotemporality to modes of expression (such as signs engraved on a stone) appropriate to eotemporality. The mind as here understood is disembodied in that it is a function, rather than a thing, but it bears little if any relation to the idea of mind as soul or as world spirit. Still, the very complex possesses some yet unknown laws and undeterminacies sufficiently large to permit the emergence of geniuses for good or evil, to accommodate interpersonal relations of great depth, to provide for interspecies communication and even to allow some things to go "bump" in the night.

4. The Mind
as Communication

IF I were to regard my horse as unnamed and unnameable, as indeed it must regard me, we could still go riding together when the fall leaves are down. But I doubt that we would.

The Gift of Tongues

In our first look at the problem of language in chapter 3 we found that language and time perception are likely to have emerged simultaneously. We also noted that the deep structure of language, which is regarded by Chomsky, Lenneberg, and others as innate in man yet seemingly absent in nonhuman primates, is perhaps forty thousand years old. The age of Sanskrit is possibly just over thirty-five hundred years; the age of Western languages which have special time distinction built into their verbal systems may be less than twenty-five hundred years. Compared to the age of life on earth (3,000 million years) and even the age of Homo sapiens as a distinguishable species (10 million years) the age of the deep structure, or that of Sanskrit, is very, very brief. How did language come about? Perhaps the physiology of the brain changed by natural selection until it was able to produce that internal cosmology of plastic symbols which we know as language. This would have had to go with a gradual development of all the perceptive faculties which, together, make language possible. Or, perhaps our ancestors discovered some novel uses for their already existing perceptive faculties. The extreme rapidity with which man himself emerged, and the assumed rapid emergence of spoken language, suggest that the evolution of the mind is likely to have been of this latter type. Our nonhuman ancestors might have discovered something which catapulted them into new modes of behavior: perhaps the use of tools made it possible for them to speak and to plan. Then again, in the evolution of the language, they might have discovered some new uses for already existing faculties which again catapulted early man into his modern mode of behavior: complicated languages, social and individual planning, sophisticated control of the environment. In each case, a few judicious (or chance) selections from the vast store of autogenic imagery might have done the trick, provided it received communal reinforcement; echoes, as it were, from the brains of humans ready to make similar selections. The continued adaptive pressure exerted by the mind upon the brain would have forced the latter to evolve biological modes to accommodate the demands of the new Umwelt.

Virtually all who study human descent link the rapid phylogenetic development of modern man's brain to the birth of our capacity for language, possibly associated with certain neurological changes. The English neurologist William Gooddy drew attention to the fact that the function of language in a right-handed person is associated with the left half-brain.[63] Since the two hemispheres controlling, respectively, the opposite-handedness are not anatomically differentiable, the decision to use the right (or left) hand for certain tasks leads to a competition between the hemispheres. He concluded that the decision to use one side rather than the other, implies special training of the chosen side and rejection of the other. Gooddy suspects that this selection was contemporaneous with the earliest production of enduring symbolism, datable in the Sumerian era no more than two or three millenia B.C. Regarding its details, this suggestion must remain spectulative; the causes might not have been any one, but many correlated decisions. But Gooddy's remarks have the pleasing feature that they associate a neurological event in organic evolution with the emergence of a new form of behavior: that of communication by written symbols.

Some would place the crucial neurological events much further back in the past. John Cohen believes that the apprehension of a future might have been the reason why man began to make tools,[64] hence the source of all further developments that employ tools. But this is paleolithic history, dating back almost two million years. Cohen also points to the likelihood that the separation between the intelligible and the emotional portions of language was cotemporal with the first tools—hence also a paleolithic event. This is a provocative thought, for experimental evidence suggests that the intelligible aspects of langauge tend to point to the future and to the past, while the emotive aspects point mostly to an (unanalyzed) present. These and other speculations, when put together, point to mental phenomena as a form of behavior which, whether available only through privileged access or observable by others, pertains to the emergence of nootemporality. From the point of view of society, it makes possible the creation, accumulation, and the passing on of acquired knowledge to other members of the species. The major tool of this activity is language.

The capacity of language is species specific. It is genetic to the extent that it demands peculiar ontogenetic phases and certain rates of maturation [65] and in that it also has a number of anatomical and physiological correlates.[66] The study of its evolution favors a theory of discontinuity, or at least a very sharp separation between animal communication and human language.[67] Insisting on discontinuity is not a view of special creation but only one of evolutionary uniqueness, such as biogenesis. Since the necessary neurological changes took place in the soft tissue of the brain, they cannot be reconstructed from the fossil records. Alas, we could not reconstruct early languages even if we were to find the brain of the Cro-Magnon man preserved in alcohol. Neither is comparison of language with animal communication much help in identifying the origins of language, because language is characterized by the simultaneous functioning of many traits which either do not exist in animals at all,[68] or which exist in widely scattered examples among otherwise unrelated members of the animal kingdom. Among other investigations, Dobzhansky, with reference to the earlier work of Hockett, Ascher, and others compared animal calls or signals with language. He saw the essential difference in that whereas animal calls are in response to a situation and are selected from a repertory of calls, human beings can speak of things that are out of sight or in the past, in the future, or are imaginary.

In the ontogeny of language, the assigning of names to things is one expression of the self-definition of the child. We can find interesting illustrations for this point from two sources as diverse as philosophy and folklore. Cassirer holds that

> by learning to name things the child does not simply add a list of artificial signs to his previous knowledge of ready-made empirical objects. He learns rather to form concepts of those objects. . . . Eagerness and enthusiasm to talk do not originate in a mere desire for learning or using names; they mark desire for the detection of an objective world.[69]

In folk literature the ability to speak is often equated to the power of magic, or even to life; the loss of speech is a source of panic and is often interpreted as a sign of punishment and/or the loss of the self.[70]

I would like to point to some aspects of the spoken language which might elucidate its relationship with identity and the sense of time. Specifically, I wish to stress that the auditory holds a unique position among the senses. Consider first that an individual cannot normally see his total body; if he wishes to do so he must construct that image by means of mirrors, photographs or other artifacts. Neither can an individual tactually explore himself very fast and in toto, neither can he ascertain with any rapidity the total spectrum of his bodily odors. In spite of these difficulties we nevertheless have body language (to be explored by sight) and communication (mostly with others) by touch and odor. In contrast to these faculties, however, our auditory system offers a certain immediacy regarding even our inarticulate cries; and when it comes to spoken language, the loop made up of our nervous system, our brain, our faculties of speech and hearing, offers a sense of intimacy second to none.[71] I would like to postulate that because of this privileged position of the speaking self, articulate vocalization was indispensable for the process of inserting in the inner landscape a symbol of something which is not in the outer landscape. To my faculties of visual, tactile, and olfactory explorations, my body bears some characteristics which it shares with the external world; my speaking self, as it were, is immediately and intimately *not* in the external world.

None of these claims should be construed as neglecting nonauditory languages, or the significance of nonauditory means in the development of language. They only point to the privileged position of hearing among the senses (in the context explained) and suggest some reasons why language has been the major instrument in differentiation of the self from the nonself. Similar arguments, applied to the formation of inner models of things and processes in the external world might explain the origins of language thinking. Furthermore, if the mind is a process description of the physiological states of the brain, as I have argued it is, then the temporal character of sound modulation is there for a new use: a way of transposing that description into the physical variables of the eotemporal Umwelt.

It is well known that children vocalize before they learn to read, then for a period they vocalize while they transfer their perceptive modality from the audio-mind channel to the visual-mind channel. In Piaget's keen observations the small child babbles and "thinks out loud" and does not care whether anyone is listening. The child's development of egocentric speech goes through the stages of repetition, monologue, and then the collective monologue.[72] Only at the end of this process of self-definition does the child enter the phase of socialized speech, and even then, not to communicate but to play. It is at this point that magic causation is replaced by more reality-directed connecting principles. It is true enough that nonverbal communication is older than verbal communication and it is very much around and alive, both in animal and man.[73] But the sudden evolutionary emergence of language suggests the superiority of verbal over nonverbal communications. The realization of this superiority might have been the crucial discovery that constituted a novel use for already existing perceptive mechanisms. These remarks emphasize the role of language in the formation of self-identity which, in its turn, must be coemergent with the advanced uses of language in social intercourse. Thus, language not only connects the individual with his fellows but, simultaneously, it links the individual as a biological and physical source of language with the individual as the source of yet unnamed thoughts, dreams, and feelings.

> . . . as imagination bodies forth
> The forms of things unknown, the poet's pen
> Turns them to shapes, and gives to airy nothing
> A local habitation and a name.
>
> A *Midsummer Night's Dream* (V, i, 14)

Communication by spoken language presumes a source and a sink equally adept in temporal structuring of sound.[74] The carrier wave produced by the vocal fold constitutes one way in which the cyclic order in biology has been adapted to subserve speech. But the information is not in the carrier but in the temporal variations imposed on it by the articulatory mechanism. The peculiarly human feature of language, therefore, is not in the cyclic order but in the modulation thereof, demanding memory and expectation. Opinions as to what the study of language ought and ought not include have varied since 1821, when Wilhelm von Humboldt first called language an intellectual instinct of the mind. Current understanding may be summed up by the statement that all natural languages are variations of the species-specific ability of man that constitutes a faculty of language. Accordingly, a search for linguistic universals must be regarded as a scientifically legitimate activity, although disagreement about details might be wide. I wish to reflect briefly on attitudes to time as a guide to linguistic universals.

In chapter 3 we left Haldane's bees conveying, through their dance, coordinate information about flowery fields. To the bee language we must add several unbeelike capacities, such as communicating about "the forms of things unknown," about possibilities and impossibilities, and about a privileged internal world. The self is said to be distinguished by privileged disclosures because, although the self is visible to all nonself, its selfness can only be adduced by others from how it behaves. But the self is equally mysterious to the body so designated because it is a symbolic reality, only partially subject to sense data and known mainly by what it thinks—and that gets us back to language. It should not be surprising, therefore, that languages of successive epochs and/or levels of civilizations (measured by some such parameters as the use of tools, or complexity of social organization) display an increasing polarization of beinglike and becominglike forms and concepts, to which the permanence of the self and the contingency of the nonself stand as paradigms.[75] We already noted, for instance, the ancient Sanskrit preferences expressed in the use of nouns instead of verbs. Thus Sanskrit tends to timelessness by concentrating on the unchanging; Hopi and some other languages tend to timelessness by concentrating on the present. What in English would be nouns such as "summer" or "morning," the Hopi expresses in "phase terms" modifying present conditions such as "while morning-phase is occurring" or "summer now."[76] The Hopi would not say, "I stayed five days," but would concentrate on the present, "I left on the fifth day."[77]

Benjamin Lee Whorf in his pioneering work in the comparative study

of languages placed the origins of our division of nature into the permanent
and the contingent in the domain of language. What is a verb, therefore
an event, and what is a noun, therefore something lasting, is determined
by what the particular language classes as a verb or a noun. This itself is
interesting enough. Revolutionary, however, is his emphasis on the fact
that different languages slice reality differently. Hence no status of absolute
reality can possibly be claimed, from any language, regarding the specific
division of temporality into beinglike and becominglike components. But
certain types of separation have certain advantages or disadvantages. For
instance, Hopi expressions for the present are unambiguous because of the
absence of emphasis on other tenses. Compare this with English, which
is highly sensitive to temporal categories; in English the use of the present
is difficult and inconsistent, as any careful copy editor knows. On the
other hand, the three-tense system of Western languages makes them very
well suited for geometrical, linelike representations of time, for on such a
line the self (well defined in English) may be thought of as a moving point
along a line of nonselves. Perhaps certain linguistic habits were necessary
prerequisites for the scientific revolution of the Renaissance.

In one of his last paper, Whorf reflected on the epistemological and
ontological significance of his findings about language.

> . . . one of the important coming steps for Western knowledge is a re-
> examination of the linguistic background of its thinking and for that matter
> of all thinking. [Such an inquiry will reveal] the *premonition in language*
> of the unknown, vaster world—that world of which the physical is but a
> surface or skin, and yet which we *are in,* and *belong to.* For the approach
> to reality through mathematics, [as one example] which modern knowledge
> is beginning to make, is merely the approach through one special case of
> this relation to language.[78] [italics his]

Whorf's suspicion is equivalent to Shakespeare's outright claim that
language gives "to airy nothing local habitation and a name." In terms of
this chapter, what Whorf and the Bard are talking about is the autogenic
functions of the mind. Lord Brain, reviewing a symposium on brain
mechanism and consciousness, remarked that from the semantic standpoint
the symposium might be called "a symposium upon a hyphen, the hyphen in
the brain-mind relationship."[79] Language could probably be identified with
that hyphen.

Linguistics, following Whorf, has been dominated by the work of
Noam Chomsky. His theory of syntax paves the way for tracing the origins
of language back to the mind and the brain. He places linguistics squarely
in the domain of cognitive psychology,[80] thereby removing it from the
behavioral sciences which he believes merely mimic the surface features
of natural science and achieve scientific character only by concentrating
on peripheral issues. Countering the trend which pointed away from
deeper-lying issues and favored concentration on evidence per se, he went
back to the classical task of searching for the universal. In his basic state-

ment on *Aspects of the Theory of Syntax,* he stresses that language learning would be impossible unless the child brought with him, to language learning, a specific capacity for approaching language data and selecting from it the rules of a grammar.[81] Hence he must possess some initial knowledge concerning the nature of language; not any specific language, for he cannot know what language he is going to speak, but some truly species-specific universals. (There is an element in this Socratic epistemology which must sound convincing to everyone who has acquired a complete mastery of a language other than his native tongue. To those who have done so it is usually abundantly clear that the new language, hence, presumably the old one too, is governed by an unwritten and, possibly, unarticulable logic which one does not learn but feels and knows.)[82] Thus, central to Chomsky's message is that linguistic ability stems from a mental universal which he calls generative grammar. He regards

> linguistic competence—knowledge of a language—as an abstract system underlying behavior, a system constituted by rules that interact to determine the form and intrinsic meaning of a potentially infinite number of sentences.[83]

Generative grammar places linguistics in the rationalist camp with Descartes, Leibniz, and Humboldt; knowledge, including the knowledge of language, is derived from internal givens. Opposite views are held by the empiricist Locke, and by learning theorists of behavior who believe that all knowledge derives from individual experience. In Chomsky's linguistics man is a syntactical creature with the rules of syntax determined by the structure of his brain. This is what I have argued earlier, starting from different and much broader precepts than those pertaining to language alone. But if such a broader base is possible even in principle, one may then question whether Chomsky's revolutionary stance is radical enough.

Search for a universal grammar, following the reasoning of Chomsky, permits the application to linguistics, of the tools and methods of the computer engineer. While this is useful in many ways if it is not taken very seriously, it amounts to the taking of the lure of misplaced precision, if it leads one to conclude that language is primarily a nomothetic or stationary rather than a creative process. Interestingly, both the rationalists and the empiricists tend to remain curiously preevolutionary in their perspectives. The rationalists focus on a completed stage of the species as though that stage had always been there, overlooking that it is through organic evolution that linguistic universals came about. The empiricist stand focuses on ontogeny—a growth process quite meaningless unless understood as the source of phylogeny. In an uncharacteristically conservative stance, Konrad Lorenz once maintained the validity of Kantian apriorism with respect to ethology.[84] But linguistic universals cannot be Kantian a priori modes of perception, unless one alters Kant so as to see in his idea of the innate form of human thought not the cosmic universals which he himself had seen in it, but some transient human universals valid only for a particular period in organic evolution. While this might well be the case, it ceases to be a Kantian view.

Linguistic capacity displays some of the time-binding methods of the strategy of existence. At any time, the nomothetic portion of language is a generative grammar, or some more basic lawfulness; the creative portion is the individual's contribution to the corpus of vocal communication. I suspect that by the time that linguistic universals common to such distinct languages as, for instance, Hopi, Sanskrit, Hebrew, and English are determined, they will not amount to a common grammar, but attest only to a hierarchy of common Umwelts, that is, common universes of perception among the speakers.

The Mind in Its Many Settings

The task of the mind, according to the particular understanding of the mind I have put forth, is seen as the creation, gathering, and coordination of expectations and memories. From among the categories of future, past, and present, as seen in the future-directedness of animals and judged by the archeological records, the oldest and simplest is that of the future. A relevant argument concerns a feature which made Paleolithic man distinguishable from his nonhominid contemporaries, that is, the reverential care with which he buried the dead. There is evidence, as S. G. F. Brandon put it, that

> the savage creatures which lived in the vicinity of the place where some 300,000 years later the city of Pekin would stand, and whose remains were found in a rock shelter at Chou-K'ou-Tien, may already have carefully, perhaps ritually, disposed of their dead, since the skulls discovered there lacked the neck vertebra, which fact indicates that decapitation took place after death and suggests that the bodies had previously been buried until decomposition and then the skulls recovered and carefully preserved—a practice for which later, elsewhere, there is abundant archeological and anthropological evidence.[85]

Burying could not have been a hygienic measure, wild dogs and the midday sun can decompose a body faster than burial does. The ritual burial of the dead is witness, instead, to an awareness of an individual future which includes one's own death; hence it implies personal identity as well as a belief in social continuity. It seems to stand for the thought, "I do unto others now as I want others to do unto me after I die." But if this be so, it implies a clear categorization of events into future and past, where the future does contain "my death" but the past does not.

Whereas all animals show mortal terror in face of immediate danger, the knowledge of death as an ever-present goad is unique to man. It belongs to the family of potential stimuli which are images in the very complex, that bring forth responses even though they do not correspond to anything present. From among them, death is one image which is intractable. Its selective advantages are great. It is something of a distributed and diluted mortal terror, a mere symbolic entity which can command the mind to subdue all the nonself in response to its distant and ill-defined possibility. The knowledge of death in time is itself a selective pressure. The individual's discovery of his inevitable death, made at a stage when, no doubt, some interest in the past was already dimly emerging, might have been the trigger that sent man off in search of immortality. The unresolvable conflict between growth and decay thus became the hallmark of earthly existence. The mind could now have a goal with respect to the past: it could search the past in service of the future. With each expansion of the memory store the number of thinkable future possibilities would increase many times because of the large number of new possible combinations with data already stored. This sounds familiar: it is a manifestation of self-generated imagery.

When in connection with biogenesis we discussed the coming about of multicellular organisms, I pointed to the advantages that coordinated actions could offer in the control of the future, as compared to what individuals could do alone. In multicellular organisms the individual cell relinquished its autonomy and became redefined in terms of its ability to contribute to the integrated functions of the whole. During the long stretches of evolutionary history the integrating function itself remained one of biology, as exemplified by the developmental process which accompanied the separation of germ cells from the soma and reached its most advanced embodiment in the nervous system. Apparently, the nervous system, including the brain and its autonomous model, the mind, became so well suited to function as the integrating and executive organ of the individual that it did not stop expanding its role; it extended itself to coordinate groups of individuals in society. Perhaps the richest interpretation of the function of the mind in the strategy of existence obtains in the context of this extended role. The mind is the agent which accomplishes, through nonbiological means, the same type of coordination among members of a society which the central nervous system accomplishes by biological means for the individual. This extension of effectiveness amounts to a new, creative employment of an already existing biological structure, just as the emergence of living organisms amounted to a new use for already existing physical structures. The role of nerve impulses sent along nerve paths carrying efferent and afferent coded messages is accomplished, on this new integrative level, by the symbolism of languages (spoken, written, painted or sculptured, etc.) carried along many and diverse channels.

During the evolution of the individual the self remained of a privileged nature. Certainly, to the observer, the agent's identity must remain an inference; but now we may also see that even for the agent, his own identity must remain symbolic. The self is thereby separated from the nonself not only by the physical boundaries of the body but, perhaps more importantly, by the fact that the self is a mental construct. As such, it is as alien to its own body as it is to the bodies of others. To this self the outer landscape of the senses is as private as is the inner landscape. The difference between the two resides in communally approved reinforcement of the outer landscape. For instance, the content of space, which is the world of kinesthetic experience, is reaffirmed, as it were, by a majority vote.[86] Some would even argue that not only the features of the external world but, in fact, all mental conceptions are communal and not individual in character.[87] Be that as it may, there is a warning here against trying to understand nootemporality in terms of the functions of the individual mind alone, without regard to the larger unit of which it is part.

The humanoid features of the brain developed from approximately 500 cc to 1,400 cc within about half a million years in the frame of the evolution of living matter that spans some 4,000 million years. In a world which may be 15,000 million years old the very complex came to challenge the limits of the universe: the very large, the very small, the very many, and the very fast. This explosive rapidity might perhaps be accounted for if it were the manifestation of the coemergence of many, mutually reinforcing processes: the discovery of death; articulate language (the separation of the emotional from the intelligible); writing in its many forms (including primitive art); toolmaking (and preference for handedness); long-term memory (with sudden enlargement of the store of autogenic imagery) and personal identity (as the symbolic executive of strategy). Since mature awareness of temporality was necessary for all these features, we could say that the single discovery that mattered was the discovery of time which, then, made all these skills possible. Or, we might turn around the argument and say that the simultaneous appearance of a constellation of traits may be diagnosed as the emergence of a new level of integration, that of the nootemporal mind or, as we called it earlier, the sense of time.

Recapitulation theories tend to turn out to be only approximations at best, yet they are tempting; it is much easier to observe the ontogeny of time sense than to reconstruct its phylogeny. In the growing infant, clinical observation shows a reciprocity between a sense of time on the one hand and, on the other hand, deferred gratification and separation anxiety.[88] Readiness for, and initiation of, separation in the service of self-identification are functions of maturity; but they are also forces which urge and promote maturation. Thus, in a way, the drama of the discovery of time, and self, and death are reenacted in all stages of alienation, but especially in that of the adolescent from the parent. The infant seems to know a continuous present at twenty-four months; he will master behavior corresponding to an idea of "tomorrow" at thirty months and to "yesterday" at thirty-six months.[89] Eventually, at the age of eight, ten, or twelve years the child will develop an intellectual and behavioral command of "today" in the sense of mental present. This mature "today" is to be distinguished from the "today" of the infant. Whereas the latter corresponds to an atemporal

Umwelt, the former is that of a nootemporal Umwelt. Thus, the developmental sequence is not truly that of present, future, and past but rather atemporality, futurity, pastness and only then, the complex mode of the mental present.[90]

In the spirit of the heavily evolutionary approach which I have taken, this question easily comes to mind: if time was discovered through consecutive steps of futurity, pastness, and then the mental present, and if evolution is indeed open-ended, what comes next? let me first rephrase the stages more precisely as they emerged, with futurity leading as the longest organ pipe. The stages are: (1) futurity, (2) future and past, (3) future, past, and present. Since the physical (eotemporal) Umwelt comprises only pure succession, these stages start with biogenesis and continue with organic evolution. It will be recalled that first there developed an existential tension between matter in its organic and in its inorganic form. With the integration demanded by complexifying life, the category of a present was born into a background of pure succession, and a continuously increasing existential tension ensued. From then on we can understand the growing polarization between pastness and futureness and the growing definition of a present, with ever-broadening horizons. We can understand it because we can share it. Below the biotemporal we can understand temporality but can share it only with difficulty. But when we attempt to make some assertions about nootemporality with the same authority as we feel we can make assertions about simpler Umwelts, we can neither understand it nor share it, and we find ourselves wanting a cliff whence to behold the vistas of this highest level of temporality.

The difficulty of dealing with nootemporality resembles, at least partly, the difficulties of scientific cosmologies. There the universe is the limiting size of all of space; hence we cannot compare it with other universes. Yet we may meaningfully discuss its properties because of the privileges of our brain, itself more complex than the universe. We can have the universe rebuilt, as it were, within our brains. This is what I meant when I claimed in chapter 4 that even the astral geometry of Gauss reverts to questions about man, the formulator of that geometry. Likewise, with the assistance of the mind (expressing itself in symbols such as those in this book) we were able to categorize and describe the atemporal, prototemporal, and the eotemporal Umwelts, even the emergence of life and the biotemporal Umwelt corresponding to life. We could even make guesses about the development of the sense of time in the species and in the individual. We could do so because of our position on the top of the hierarchy, incorporating within ourselves samples of all the Umwelts. We were able to imply the outlines of lower Umwelts because we know them as portions of more advanced Umwelts.

If these regularities have any predictive value, they suggest that our only hope for further elucidation of nootemporality is an eventual look backwards from an integrative level higher than that of the individual mind.

We might, from there, discern the salient features of the mind. To put it quite crudely, we shall not be able to understand totally (even in principle) what time does represent in the existence of man until we can tell, in terms of a temporality higher than the noetic, what regions of nootemporality are undetermined. Hence, there must remain a residual mental anguish or, at best, a mystery, after all noetic analysis of time.

In earlier cases of transcendence, and even along steps of evolution, each improved adaptive function also resulted in an increased existential tension. It seems, for instance, that life reached the limits of the living mode of matter in that it could not keep up with a rapidly changing, living environment. Death and sexual reproduction helped, but were not sufficient. Thus a new mode of matter emerged in the complexity of the brain, capable of controlling the future and the environment through cumulative, symbolic transformation of experience, in forms which could be passed on to others not present or not yet born. But here again, having a mind was not an unmixed blessing. It resulted in increased existential tension because the region of conflict between the expected and the encountered increased tremendously, as compared with the experience of man's living cousins with less complex brains. Presently, we shall attend to some problems of the increased existential tension attendant on having a mind.

VIII

OUT OF THE DEPTHS

WE HAVE POSTULATED that the brain is the most complex of all known systems, far more so than anything else we know of, and that for this reason it is likely to have its own, peculiar lawfulness. Then we considered an understanding of the mind as a process description of the brain. Such a perspective permits, at least in principle, the development of a body of knowledge dealing with the very complex. The amazing wealth and variety displayed by human thought and action suggest that a systematic study of the mind following a phenomenological approach might, however, be of more immediate use than one built up from known first principles, even though the latter is the potentially more complete approach. Studying the mind as an accomplished fact and having relatively little concern with its evolution has been the road taken by the pioneers of modern psychology.

On the preceding pages I often stressed that the various conflicting manifestations of existential tensions ought not be granted separate ontic status, but we were not able to deal directly with such tensions. Here we will attempt a direct approach as far as man is concerned. We place our hope in the fact that whereas in the case of ponderable matter emerging from the atemporal world, or life emerging from nonlife, or living matter complexifying, we can only observe behavior and put questions to nature, in our own case we can be both observer and agent.

1. A Region of Functions
between Life and Mind

ACCORDING to the Old Testament, Moses was hesitant to address the Pharaoh on behalf of the Sons of Israel because, he said "I am slow of speech, and of a slow tongue." (Exodus 4:10). But God reassured him saying, "Go and I will be with thy mouth, and teach thee what thou shalt say." (Exodus 4:12). Likewise, in the Matthean Gospel Christ allays the fears of His disciples: "But when they deliver you up, take no thought how or what ye shall speak. . . . For it is not ye that speak but the Spirit of our Father which speaketh in you." (Matt. 10:19–20). Thus three and two millennia before Freud's "unconscious," Jung's "collective unconscious," or the behaviorists' "operant conditioning" the authors of Exodus and the Gospel knew that people sometimes rely on a store of knowledge, or display a capacity of which they are not aware before it actually manifests itself.

What in post-Freudian terms are understood as manifestations of the unconscious (Ucs.) were attributed at various times to god and devil; to the superiority of the mind to matter; to the power of knowledge; wisdom of nature; vastness of memory and sensations; guidance of the muse; blindness of passion, or simply the irrationality or genius of man. Certainly, the ancient Yahwist writer is not alone. A list of those who have reflected upon, marveled at, or been frightened by the power of unintelligible and unpredictable forces that seem to be directing the actions of man, would read like a "Who's Who in Humanity." Therefore, our interest must be limited to no more than giving a hint of the continuity of that awareness.

Much of the unpredictability and/or unintelligibility of man's actions, feelings, and creativity (in which we must include destructivity) have been traditionally attributed to the soul. For what reason would one seek, classify, and study the invisible and intangible if it were not that knowledge of the visible and tangible does not supply enough clues to the puzzles of life, death, suffering, and predictability? The idea of the soul has been a safety valve in religion and in philosophy. Its presumed existence permits man to be himself and yet become something much more, or much less, than he himself or others around him would expect him to be. St. Augustine tried to fathom his soul and found it full of memory. The powers of his memory, the residents of his soul, were in his command—but not altogether. Hence, he wrote, "I myself comprehend not all that I am." [1] Paracelsus knew that some "diseases that deprive man of his reason" come quite unnoticed by the patient (and inveterate physician, chemist, and astrologer that he was, he began to seek the etiology of the "suffocation of the intellect" in humors, airs, and the influence of the moon). [2]

Shakespeare's understanding of man created many characters whose motivations are derived from sources unknown to the characters themselves. Thus, in *The Merchant of Venice* Antonio enters declaring,

> In sooth, I know not why I am so sad
> (I, i, 1)

sounding like Heinrich Heine three centuries after Shakespeare

> Ich weiss nicht was soll es bedeuten
> Das ich so traurig bin.

In *Troilus and Cressida* Achilles has presentiments of both ruin and victory, and that of the "ugly night" that will come breathing upon the heels of the day.

> My mind is troubled, like a fountain stirr'd;
> And I myself see not the bottom of it.
>
> (III, iii, 314–5)

And Rosalind describes her love for Orlando, in *As You Like It*, as an affection which "hath an unknown bottom, like the bay of Portugal." (VI, i. 219). Much more important, however, than these pebbles randomly picked is the uncanny humanity of the Bard's characters. The darkness and light of the mind follow their inarticulate laws, as action and utterance upon action and utterance unfold; the inner secrets of men, women, kings, and fools are disclosed, clear to the reader but unknown to the characters themselves.

The first major apologist of the Ucs. was Eduard von Hartmann (1842–1906). His *Philosophy of the Unconscious* (1868) was influential and very much in vogue at the end of the nineteenth century.[3] The work freely and eruditely draws on biology, anthropology, linguistics, aesthetics, poetry, philosophy, and on Hartmann's fertile mind. He laid great emphasis on instinctive sexual behavior when dealing with hypnotism and hysteria, and assessed love and religion as illusions based on sex and the Ucs. He also stressed a quality of the Ucs. which was later appreciated by Freud: its essential conservatism. Tracing the historical influences in Freud's work is not our purpose; however, those influences did exist.[4] Even as regards Hartmann's forerunners, according to *Brett's History of Psychology*, at the time of the publication of his major works the doctrine of the Ucs. as a fundamental active principle of the universe (not only of man) was widely known.[5] It is so much more remarkable, therefore, that except for a slender volume by L. L. Whyte,[6] an extensive and definitive study of the history of the idea of the Ucs. before Freud does not exist. The reason for it might have been expressed a century ago in the introduction to Hartmann's massive volume: "Mankind very naturally began its researches in Philosophy with the examination of what was immediately given in Consciousness."

If there is any trend to be discerned in the history of the idea of the Ucs. it is likely to be that of man's maturing self-identification expressed in changing terms, intelligible to people at various places and epochs. The modern history of the idea of the Ucs. may be summed up in four precepts: (a) that the irrational, the destructive, and the creative in man belong to the natural order; (b) that the potential for strange and uncommon behavior exists in all members of our species; (c) that the life of the sources of our motives of which we are not aware displays lawfulness subject to test; and (d) that understanding Ucs. motivation is essential to a humanistically and scientifically acceptable evaluation of the nature and destiny of man.

It is said that there are no less than 39 distinct meanings of *unconscious* as an adjective.[7] This does not hinder serious analytical work, but it does lead to easy abuse of the concept and it offers a cheap entry to criticism against it. Yet, while the concept is intricate, it is not mysterious. Taking one's total capacity for granted and unsurprising is the rule, rather than the exception, among the living. A horse, for instance, can get around many unexpected obstacles in many ways, as do cats, dogs, and rats, but never with any expression that would indicate that its actions were anything unusual. Before an organism can be capable of being unaware of some of its own motivations it must possess awareness of some others: only by possessing conscious experience can we thus begin to think of unconscious motivation. The pitfalls of the idea of the Ucs. exist because of the difficulties of the idea of consciousness. Therefore, we must consider these two in contrast.

Earlier, I identified the self as one symbolic continuity among the many symbols of the inner landscape but distinguished from all other symbols by not having an external referent. Let us ask what may be meant by the self in the context of death. The "I" in, "I don't want to die," coming from someone before a firing squad is not difficult to interpret: the prey faces the predator and opposes it; the "I" happens to be the prey. How about the same statement coming from someone dying of metastasized carcinoma? There the spirit is unwilling to perish but the body relentlessly seeks death. In the freely willed death of an otherwise healthy body the roles get exchanged, "the spirit indeed is willing, but the flesh is weak." (Matt. 26:41). In both cases the "I" insists on remaining the executive faculty; it is its will that ought to be done, not that of the nonself. It would seem, then, that even the physical and biological functions of that very body whose brain is the environment of the mind must appear to the "I" as functions of a nonself.

If the self is to maintain its level of integrity over life as life maintains its level of integrity over matter, then the self would also have to classify not only the living body but also some lower mental functions as activities of the nonself. Indeed, short-term recall is something I can do without involving my feeling of personal identity, and the same is true for such motor actions as the routine playing of the organ. Some such tasks feel as though they were done by other people because the "I" can direct its attention to other matters while still making decisions about organizing details. Nevertheless, I can make myself become aware of what I am doing if the "I" so desires, precisely because the tasks are still sufficiently far from the strategist, the self. Because of the hierarchical continuity of man as a unitary organism there are likely to exist, however, functions which are not purely biological or motor functions and thus classifiable as those of the nonself; neither do they demand the type of symbolism that would make them part of the self. Functions of this type, belonging to a no-man's-land between body and mind, are confusing and thereby threatening to the executive powers of the mind.

We have already encountered another very "unpopular" state of matter, that between life and no life. Although biogenesis is the only generally acceptable and accepted explanation of life on earth, things which would be just emerging on the boundary line between life and no life are nonexistent at this time. As we have seen, the usual and convincing argument is that such ill-defined life would either be devoured in support of higher life forms, or would return to the inorganic. I think that a similar process of banishment is operative in the region of transcendence between life and mind. By the Ucs. portions of the mind I would understand those functions which can neither be completely identified as spatial biological processes (hence, be considered as nonself), nor can they be identified as totally symbolic. In the case of transcendence from matter to life, protolife might indeed have disintegrated or become absorbed; I can think of no necessary and lasting tasks for such organic forms. In the case of transcendence from life to mind the situation is quite different, for the brain must remain an integral part of the body. Hence there remains a large region, or interface, which can only be disowned by the self but not eliminated. This buffer zone between the mind and the body thus becomes the depository of functions higher than the biological but, for various and sundry reasons, not altogether mental, hence dangerous to the autonomy of the mind. Motivated by the content of this region of his psychobiological structure a man, unlike a horse, can most definitely act in ways surprising to himself—as though reacting to the unexpected behavior of someone else.

2. The Devil
of Vienna

FREUD'S determination to explore the unknown depths of the Ucs. exudes a fierce defiance of the then established morality, logic, scientific methodology, and even common sense. His *Interpretation of Dreams* (1899) bears as its inscription the words of Juno, representative of sex and of the female principle, quoted from *The Aeneid* of Virgil: "Electere si Nequeo Superos, Acheronta Movero," "If unable to change the world above, I will move hell itself." This is precisely what the gentle Viennese doctor had done at the birth of the twentieth century. In spirit and temperament he might have felt like the imprisoned King Richard II in the dungeon of Flint Castle.

> I have been studying how I may compare
> This prison where I live unto the world:
> And for because the world is populous,
> And here is not a creature but myself,
> I cannot do it; yet I'll hammer it out.
> My brain I'll prove the female to my soul;
> My soul the father; and these two beget
> A generation of still-breeding thoughts,
> And these same thoughts people this little world. . . .
> *Richard II* (V, v, 1)

From the vantage point of eight decades Freud appears a philosophical and moralistic genius first and a medical man second, who generated enough thoughts to people our world and who perceived man as roaming the earth to which his destiny is forever bound.

Freud's principal idea and the mainstay of psychoanalytic theory is the assumption that the manifest contents of dreams may be interpreted so as to disclose the contents of the Ucs. The precept upon which this idea is based is that dreams constitute an archaic and, primarily, visual language. The images of this language are its words; they are mostly objects, actions, and fantasies of the waking state. However, both the syntax and the semantics of the language are unique. The rules of its syntax are connectivities such as one would tend to ascribe to the coordination of sense impression by children, primitive man, or even animals. The semantics of the language is the Rosetta stone: the words and groups of words do not normally correspond to the objects of reality which they represent. Freud found it possible to identify Ucs. drives and desires by using certain intricate but rational methods of dream interpretation, such as free association, with respect to the manifest contents of dreams. As invented and foreseen by Freud, the exploration of dreams became the royal road to the exploration of the Ucs.[8] Freudian dream analysis is a working scheme which does not depend on any specific hypothesis regarding the evolution of conscious experience, although its success does encourage certain views. To appreciate what these are we turn to the Freudian idea that the totality of the mental structure or psyche comprises three dynamically connected systems: the ego, the id, and the superego. Although delineation of their exact relationships and roles are subject to debate, the significance of this division within the complete personality has been established beyond any reasonable doubt. The ego stands for the person, the "I" or the "self" and is characterized by conscious experience. It is to the ego that psychoanalysis assigns the powers of reality-regulated striving which it is said to pursue by means of perception and thought, in contact with the outside world. It follows that the Umwelt of the ego is defined by the perceptual apparatus of man. In this scheme, then, reality may be defined as the Umwelt of the ego.

The superego is developed by incorporating in the self the parental and social standards as perceived by the ego. When we discussed creativity we wondered by what methods selection could be made from among all the strategies of autogenic imagery. I held, then, that such selection rules would have to be of earlier evolutionary origin than deductive logic: perhaps aesthetic or ethical in nature. The Freudian superego comprises one such set of rules; it is a portable cage that limits the strategist in its choice. It is a corpus of selection rules with origins in the body social but transformed, sometimes very radically, by the process of introjection which assimilates it into the internal landscape.

This transformation is profound because it is influenced by the deepest part of the psyche, the id. As Juno's hell, the id is thought of as the depository of instinctual impulses demanding immediate gratification of primitive needs. It may be described as chaotic (by the organizing standards of the ego and superego), devoid of morality (compared to the demands of the superego) and indifferent to the restraints of matter and life. Freud at first was chiefly aware of the sexual strivings of the id and only later did he discover in it the presence of inborn destructive impulses. He lumped together the sexual and constructive strivings under the term Eros, for the Greek god of love, and the destructive ones under Thanatos, the god of death. He conceived of the id as the true Ucs. and believed that it is dominated by the pleasure principle which demands that instinctual needs be immediately gratified. The id is said to be in contact not with the external world but only with the body. Clearly, then, the id whose content is banished from awareness so that its effects appear as Ucs. motivation, is the no-man's-land between the integrative levels of life and that of mind.

Freud's direct comments on time are few, but the significance of his contribution is great, even though indirect. In the *New Introductory Lectures to Psychoanalysis*, Freud remarked that

> the laws of logic—above all, the law of contradiction—do not hold for processes in the id. Contradictory impulses exist side by side without neutralizing each other or drawing apart; at most they combine in compromise formations under the overpowering economic pressure towards discharging their energy. There is nothing in the id which can be compared to negation, and we are astonished to find in it an exception to the philosophers' assertion that space and time are necessary forms of our mental acts. In the id there is nothing corresponding to the idea of time, no recognition of the passage of time and (a thing which is very remarkable and awaits adequate attention in philosophic thought) no alteration of mental processes by the passage of time.[9]

I would like to take several of these ideas and refine them in terms of the theory of time as conflict.

That contrary impulses can exist side by side without neutralizing each other or drawing apart amounts to claiming that from the point of view of the id, they are not really contradictory. The law of contradiction is irrelevant to them, and there is no existential tension between them. But we can expect the id to determine only an Umwelt whose temporality corresponds to the conflicts which the id *does* acknowledge. If the id is closer to life than to the mind, and its concern is that of the soma, then its Umwelt is biotemporal or even eotemporal. That this may be the case is suggested by the nature of the pleasure principle on which the id operates. The demand for immediate satisfaction that characterizes the pleasure principle reminds one of the creature present which displays no long-term expectation or memory. In contrast to the id, the ego does admit contradiction between certain impulses which, to the id, do not appear contradictory. Hence, the ego generates an existential stress corresponding to those conflicts and thereby determines an Umwelt of higher temporality than that determined by the id. The temporality of the ego would have to be the most comprehensive one known to man, that of nootemporality. The id, then, which for our purposes may be identified with the Ucs., is seen to determine a buffer zone between biotemporality and nootemporality, somewhat closer to the former.

Freud's contention, that the mental processes of the id do not change, was experimentally tested by Calvin S. Hall on a sample of 770 individuals. He asked whether there is any significant change with age in the number of dreams that evoke anxiety, in the number of danger situations which evoke anxiety in dreams, and in the dreamer's capacity for dealing with such situations. He found that there was no significant change and concluded that the "study supports Freud's concept of a timeless unconscious. The far reaching implication of this concept is that motives do not change with age once the unconscious has reached its full development at the age of five." [10] As I see them, Hall's findings suggest that the Ucs. is part of the phenotype; it reaches its final form at some point in life, remaining thereafter substantially constant. However, it cannot possibly remain unaffected by time, for if it were it would survive the body. All one can say is that it is remarkably stable, that is, it changes very slowly. Beneath this phenotypal quality I would envisage the genotypal quality of the Ucs. evolving under selection pressure. This is a necessary assumption, or else we would have to postulate that all our ancestors, reaching back to the primordial broth, had the same mental structure as ours. In any case, it is a category mistake to call something timeless when in fact it is only very stable.

Freud's contention that the id does not recognize the passage of time came under scrutiny by Marie Bonaparte and others. Bonaparte argued that the Ucs. has no knowledge of time as an intellectual concept because it is a primitive reservoir of our instincts and our will to live and it knows nothing about concepts.[11] Bonaparte's stress on the concern of the id with life, if taken as representing psychoanalytic findings, reinforces our prior suspicion that the id, though part of our mental structure, is rather close to the bio-temporal Umwelt. In the structural theory of modern psychoanalysis, time is believed to operate both in the conscious and in the unconscious as a mature ego function.[12] Perhaps a careful statement would rather hold that temporality operates both in the Cs. and in the Ucs. but always in a manner which corresponds to the Umwelt of the appropriate structural level; for the ego it is nootemporality, for the id it is a region close to biotemporality. Based on another opinion expressed by Freud in *Beyond the Pleasure Principle* we might even suspect that the lowest regions of the id are proto-temporal, that is, countable but not ordered. He wrote that "we have found by experience that unconscious mental processes . . . are not arranged chronologically. . . ." [13]

I would like to draw attention to three features of the manifest contents of dreams which identify the dream Umwelts, hence, presumably, those of the Ucs. as bio- , eo- , and prototemporal. These features concern (a) the topology or, we might say, the syntax of dream imagery; (b) the emphatic presentness of dreams; and (c) the curious fore- and hindknowledge in dreams.

The topological element is the connectivity among dream images; this is the syntax of dream language. Events in dreams hang together by what is sometimes described as magic causation. Whereas the peculiarities of the ways dream images connect would strike us dumb if witnessed in the waking state, in dreams they are taken as a matter of course. It will be useful to remember our earlier conclusions that causation is an emergent quality and that it displays, both ontogenetically and, possibly, in the history of the universe, successive stages of differentiation. In the early universe, as in the Umwelt of the young child, connectivity and intentionality are yet undifferentiated. This indeed is the case in the manifest contents of dreams where my wish (or fear) becomes the unquestioned law that binds events together. On this account, the contents of the Ucs. itself must correspond to Umwelts below the nootemporal.

That the predominant mode of dream is that of the present may be rapidly revealed by a bit of introspection, though readers who prefer to trust others may consult Doob's summary of this issue.[14] Although not everything happens at once, whatever does happen and unfold has an overwhelming aura of presentness. Fear, hope, anticipation, and memory transform to present emotions and actions. The difficulties of regressive sharing inhibit us from experiencing this type of present in the waking state; it can only be approximated through art or ecstasy. It is *not* the mental present of nootemporality but rather the creature present of the eotemporal Umwelt with its pure succession and without sharply distinguished future/past/present. Interestingly, this dream present is sometimes informed of past and future. We may see in a dream something motionless, yet we know that it is going to crawl, or grow, or shrink—and it does; or, we see someone move and we know where he came from and where he is going. But these futures and pasts have very little of the experiential qualities of waking futures and pasts. The situation rather resembles the second reading of a story, or the assumed omnipotence of a divinity: we are simultaneously surprised, and not surprised, by knowing the future. While we might be terribly frightened or very pleased, we still do not experience the complete tension that characterizes the future becoming past in the nootemporal Umwelt until we are actually awake. When we spoke earlier about the wakefulness-of-necessity characteristic of lower animals and of the young of higher animals and the wakefulness-of-choice characteristic of the adult forms of higher animals, we already implied the difference in Umwelts between that which in man is unconscious and that which in man is conscious.

Once wakefulness is achieved and the nootemporal Umwelt determined, the mind guards itself against being ruled by the laws of life as zealously as life guards itself against being ruled by the lawfulness of matter. In both cases of transcendence, however, the success can only be partial and transient. Eventually "the grass withereth, the flower fadeth." Mind also struggles to keep the powers of light (of seeing and foreseeing) independent of the powers of darkness (of feeling, eating, copulating). In psychology this struggle is identified as that of the ego against threats to its identity emerging from the Ucs. For Luther the struggle was that of man against the Devil.

> For still our ancient foe
> Doth seek to work us woe;
> His craft and power are great
> And armed with crucial hate,
> On earth is not his equal.
>   "A Mighty Fortress Is Our God"

If ego integrity is lost, so is the sense of time. As is well known in psycho-analysis, the condition brought about when the instincts of the id begin to control behavior is often that of panic and distressing anxiety. Often in such cases the concept of time, by which is meant the concept we have been calling nootemporality, is called upon to drive back the instincts. This may be done by arraying object attachments against the instincts, such as by promoting interest in shining machines instead of shining eyes, and thus diverting attention from the clamor and demands of life.[15] Clinical mani-festations of the fear of losing the sense of time, which is tantamount to losing the self, are numerous; they range from frequent female resistance to sexual orgasm because of the fear of the timelessness of ecstasy, to the rejection of the id through complete loss of contact with reality in some cases of petit mal.[16] We have a beautiful, if not very precise statement on time as a protector of the ego in T. S. Eliot's *Burnt Norton.*

> Yet the enchainment of past and future
> Woven in the weakness of the changing body,
> Protects mankind from heaven and damnation
> Which flesh cannot endure.

From the time when the self is first defined in adolescence we find an incessant conflict between the forces which tend to maintain the ego and those which work against it. This I would regard as the unresolvable con-flict of the mind, a struggle that manifests itself in many ways, and one which is to be distinguished from the unresolvable conflict of life. In Freudian literature both struggles are sometimes symbolically described as those between Eros and Thanatos.[17] In terms of the theory of time as con-flict the unresolvable conflict of the mind is a determinant of the noo-temporal Umwelt. Looking back, as it were, from the level of noetic tem-porality the bio- and eotemporalities of the Ucs. should appear timeless, as I believe they do. The eotemporal features of the Ucs. appear timeless for they are those of an endless world: pure succession admits no present with respect to which spans of time could be delimited. This state of affairs should be familiar to us from our earlier deliberations about eotemporality. The biotemporality of the Ucs. appears timeless, for its Umwelt is that of the creature present which knows nothing about death. When Freudian psychologists hold that self-awareness stands for death, they are saying that only an Umwelt with the attributes of conscious experience, which are those of nootemporality, can accommodate the knowledge of death.

There are many ways of escaping from this Umwelt. When Freud postulates the existence of a death instinct and sees its evidence in such behavior as the repetition compulsion,[18] he describes the regression from the future/past/present of nootemporality to the world of cycles which are, as we know, eotemporal. The forces which cause man's mind to negate itself by returning to the level of organic existence are great; their roles among the functions of the mind parallel those of aging among the functions of life.

Examples of the return toward the level of life, and thus of the resolution of conflict of the mind by regression, include the giving up of one's individuality by becoming, for instance, an undifferentiated member of a group. With the disappearance of individuality the Eros-Thanatos struggle vanishes and so does, of course, the sense of time. This type of regression, and the resultant feeling of unity (absence of the conflict of the mind) is known to children, archaic man, saints and, occasionally, to most of us. In primitive society, as Gunnell noted, this

> lack of temporal awareness is grounded in the participative character of the mythic vision in which the individual, society, and nature are experienced as a compact and consubstantial unity. [In the reciprocally sustained unity between society and cosmos] all events and activities of thought and action converge in a timeless paradigm in which collective memory as the awareness of individuality is absorbed in the uniform rhythm of cosmos and society which abrogates time and history.[19]

Fearless death by voluntary regression into the biotemporal and eotemporal Umwelts is one way to resolve the conflicts of the mind. Others are the many ways of ecstasy which lessen the existential tension of the mind. In Greek *enthousiasmos* from *en-theos* means, "having the god within"; it is the identification of death-knowing mind with immortal god. When, for some reason, the attempt of the mind directed to the resolution of its unresolvable conflicts through the control of its immediate environment fails, the mind then tends to become the controlling agent of the extended, collective self. Often this takes the form of forcing other members of the species to act out the desired but unsuccessful methods of conflict resolution in the individual, a process which often takes the form of war. "Time," wrote Kierkegaard, "does not exist for dumb animals who are absolutely without anxiety." [20] Indeed, though animals are instinctively aggressive (in my opinion), they do not have social conflicts comparable in scope to human wars (in the opinion of many people).

The world of the child resembles the creature present of other animals with nervous systems. He combines the cyclically lawful and the unpredictable elements of his experience in the unity of his psychological present. As he reaches adolescence and differentiates his psyche into the id, ego, and superego, with all their complex dynamics, he has to learn to live with the continuous stress of the unresolvable conflicts of his mind. The story of man's attempts to resolve these conflicts through biological and intellectual means is one possible interpretation of the origin of civilization.

3. The Sage of Küsnacht

IF THE PSYCHOLOGY of Freud is described as that of man roaming the earth, the psychology of Jung is that of man roaming the universe. Freudian concerns with guilt, repression, and the autonomy of the individual naturally lead to an interpretation of civilization as that of psychic discontent. An appropriate emblem for such an outlook would be that of Yahweh, the God of Revenge. The concerns of Jung were with the psychic unity of man and the unfathomable depth of the universe of which he is a part. The appropriate emblem for this outlook would show the Christian godhead of Agape and the hilltop with a lonely ash tree, under which the Romansh inhabitants of Canton Schwyz were said to have worshiped. The two imaginary emblems may assist us in understanding the different philosophies of these two men stemming from two very different social settings. Freud lived the bitter-happy life of the Jewish Viennese intelligentsia where racial hatred was just beneath the surface of bourgeois security. Jung lived the life of the aloof Swiss in a country where God resided in everything but spoke through Calvin and Luther. Freud's reactions to the events of 1914 resembled those of most of his countrymen: he hoped for a quick German victory and a lasting peace thereafter. To the extent that the events interfered with his work, he was annoyed.[21] For Jung the events amounted to a gigantic experiment, laying bare some of the forces that move mankind:

> on August 1 the world war broke out. Now my task was clear: I had to try to understand what had happened and to what extent my own experience coincided with that of mankind in general. Therefore my first obligation was to probe the depth of my psyche. I made a beginning by writing down the fantasies which had come to me during my building game. This work took precedence over everything else.[22]

The two major working concepts which Jung developed for the purposes of identifying the way individual experience coincides with that of mankind are that of the collective Ucs. and that of the archetype. In this section we shall turn our attention to the significance of these two concepts for the study of time, rather than to a discussion of Jungian psychology in general.

The reasoning that leads to the postulate of a collective Ucs. is straightforward though seldom clearly stated. It may be put forth by means of a biological analogy. Although each individual has many morphological features which distinguish him from others of the same species, he nevertheless shares some patterns of biological organization with all members of that species. Likewise, if the psyche is assumed to be part of the natural order and a product of evolution, there must then exist certain patterns in the psychological makeup of every individual which he shares with other members of our species. In post-Jungian jargon, then, communication among individuals along the channels of shared, common features of the mind is likely to have distinguishable qualities. Accordingly, in Jungian thought, the collective Ucs. is a continuous substratum of man's world in space and in time. Jung held that as one probes the deeper layers of the psyche these layers lose their individual uniqueness and become "increasingly collective until they are universalized and extinguished in the body's materiality, i.e., in chemical substance." [23]

The elements of the language of the collective Ucs. are certain primordial images called archetypes. In Jung's own language an archetype constitutes a facultas praeformandi with no material existence of its own, a possibility of representation given a priori.[24] This is a medieval way of describing what today we would call modes of perception; in a more general sense, it is an epistemological claim, because all scientific laws are but faculties of preformation applicable to specific segments of reality.[25] Jung defined the archetype as "an irrepresentable, unconscious, preexistent form." [26] By "irrepresentable" one could not mean completely unmanifest, but only as not representable by means of a spatial model or in the words of a natural language. This is a condition familiar to physicists in such concepts as, for instance, the wave function of quantum theory or the geometry of space-time. Both are formalisms employed because they make pragmatic sense, but they are not spatial models of the world, and they are certainly not in a natural language. By "preexistent" one could not mean sempiternal, with no beginning, but only that it is genetic. Philosophically, an archetype is an a priori in biology, a matter on which Konrad Lorenz commented interestingly in a little known paper:

> The a priori is due to hereditary differentiation of the central nervous system which have become characteristic of the species, producing hereditary dispositions to think in certain forms. [As the fin of a fish is given before individual negotiation of the water and, on the contrary, it is this form that makes negotiation of the water possible] so it is also the case with our forms of perception and categories in their relationship to our negotiation with the real external world through experience.[27]

Jungian archetypes, if taken seriously, must then be regarded as products of organic evolution as are toes or facial expressions. They are primordial images which evolved as potential forms of mental constructs and survive in the collective Ucs.

If archetypes so understood are to be made a part of our generally accepted understanding of man they must have a place in the evolutionary progress of life. Thus, for their origins we ought to look to experiences that have been common to all men, such as life and death, the transitional stages of growth, the daily journey of the sun, darkness and fear, hunger, or the forces of reproduction. One would expect them to be largely independent of historical epochs, ethnic background, geographical location, or cultural conditioning. These are, in fact, the hallmarks of archetypal forms as seen by Jungian psychologists. They see them appear as purported explanations and preferred analogies in the description of such primordial experiences as I have listed above. They also find them expressed in dreams, in adult fantasies, in fairy tales and in the delusions of the insane. Also, they believe to have identified the archetypal origins of certain universal feelings and of certain expressions of veneration directed to some shapes and forms. It is this last group of archetypes which suggested to Jungian psychologists that number is a form of primordial image.

Number so understood may be interpreted as built up by qualitative generation. Thus, oneness is unity, *the* unity of philosophy, monism; twoness is the simplest possible symmetry; threeness is the smallest number through which rhythm may be expressed, etc. This has been worked out in any detail only for the first four numerals.[28] The point they are making is that natural numbers represent preconscious (i.e., not present in the conscious but recallable) patterns of thought common to all man and, therefore, a suitable basis for transmitting knowledge; more so than mythological images (which they resemble) but which are socially and ethnically conditioned.

To the scientifically trained reader this will appear as an inverted direction of reasoning: the picking, from nature, of examples of what we know as "oneness," "twoness," etc. Reflection will reveal, however, that if we assume the natural origin of numbers, though there is no obligation to follow Jung, this is the only direction we can proceed: from perceptive faculty to number. The leading proponents of the literalistic philosophies of number—Peano, Kronecker, Cantor, and Bertrand Russell—all began by taking for granted our capacity of forming sets of real objects. Then they would say, for instance, that "one" is (1) the name of a natural number, (2) the name of a rational number (which is a set of ordered pairs of natural numbers), specifically, the name of a set containing all the sets of ordered pairs 1;1, 2;2, 3;3 . . . , and (3) the name of a real number, specifically that of the innumerably many rationals which are smaller than the rational number "one." Both Jung and the set theorists begin from considerations of our perceptive faculties but because of their different focus, they differ in unstated assumptions. Jung did not dare to neglect man the perceiver and assumed that a science of mathematics could eventually be built up from psychological precepts. The philosophers of mathematics did not dare to include man but did tacitly assume the self-consistency of our perceptive faculties as well as the independence of these faculties from cultural conditioning. Otherwise Peano's axioms could not hold equally for mathematicians with different ethnic backgrounds.

Further assistance in this fascinating matter comes from Piaget's genetic epistemology of number, which threw a narrow beam of light upon our ignorance concerning the path the child takes from his feelings of "oneness" or "twoness" to being able to count.[29] He drew attention to the consecutive emergence in the mind of the child of cardinal and ordinal numbers. That children learn cardinal numbers long before they comprehend their sequential feature of ordinality is a matter of working knowledge for all elementary school teachers and all parents who watch their children. From the operational point of view, Piaget identified these two properties of number with "class" and "asymmetry." Class is the numerical value of a set. As the child grows, he enriches his knowledge of class by acquiring an awareness of the conservation of number, that is, that the spatial rearrangement of a set of objects does not alter the numerical value of the set.[30] In producing a class, that is a cardinal number, all members of the class are taken as identical; they are indistinguishable. Ordinality is discovered when the child is able to assimilate asymmetry of the "more than" or "less than" type.[31] As he learns the various simple rules that derive from ordinality, he practices "progressive arithmetization" of number series: he applies his rules to smaller numbers first, then progressively to larger numbers.[32] What he learns is an asymmetry, but one without preference for cardinality: natural numbers ordered in an increasing sequence are also thereby ordered in a decreasing sequence.

Earlier we followed the emerging self-identification of the child and learned how the continuity of the self becomes the paradigm of permanence, which is oneness. Subsequently his world becomes populated by other permanencies independent of him and he begins to grow out of his infantile monism. It is at this level that he first comes to recognize cardinality, followed by ordinality; later he learns about the number zero and about negative and positive integers and multitudes of identities come to populate the inner landscape of his mind. These have their own life and generate new symbols. How to select from among the new autogenic images those which do have external status must be learned as the child develops. Some of the selection rules regarding the world of numbers ought to sound familiar to the reader: cardinality, for instance, is the countability of indiscernibles which cannot be ordered; it determines a prototemporal Umwelt. Ordinality is the orderability of numerals in a series of pure succession with no preferred direction; it determines an eotemporal Umwelt. The introduction of "zero" (a natural number in Peano's axioms but physically a symbol of negation) provides an absolute separation between two semi-infinite rays of pure asymmetry but still with no preferred direction; this corresponds to an early biotemporal Umwelt. We may now be able to answer, from the theory of time as conflict, the profound question that Eugene Wigner had posed: Why is mathematics so unreasonably effective in the description of the world by physics? [33]

It will be recalled that the autonomous unity of man as an organism leaves him, nevertheless, subject to the lawfulness of all integrative levels. Each of his Umwelts is shared with other beings: the nootemporal with other men, the biotemporal with all of life, the lower ones with all of the physical world. It follows that the mind, that is, a process description of the brain (which comprises matter in its biological mode of existence) will con-

tain statements about the biological as well as the physical Umwelts. Some
of the imagery of the mind must have been incessantly reinforced since
the appearance of Homo sapiens. Some other images may concern features
which the brain shares with all of life and which have been reinforced since
life emerged from the primordial soup. Some further features of the brain
must have an even longer history, going back to geological antiquity; after
all, the brain is made of ponderable aggregates of matter. One consequence
is that the deeper we probe into the imagery of the mind the more stable
and universal those images are likely to be. We can say even more than
this. As we go deeper we ought to discover—and we do—beneath biotemporal features such as goal-directedness, some eotemporal features, such
as ordinality. Whether the primordial images are conscious or unconscious
is a structural problem; it is interesting but not as significant as the fact
that they have their roots in primordial Umwelts. Descriptions of increasing
sophistication, such as those in terms of set theory, seem to stem from those
early conditions. The ontogenesis of number itself parallels the ontogenesis
of the sense of time: we progress from atemporality with no numbers, to
prototemporality with cardinal numbers, eotemporality with ordinal numbers and biotemporality with the introduction of the triadic division of
numbering systems, to wit: negative numbers, positive numbers, and zero.

Much has been made, in the philosophy of science, of the correspondence between counting and time. We even witnessed the appearance of a
modern mysticism of number and time in physics. But what we have just
learned suggests that the interpretation of time by number series, no matter
how sophisticated, can speak only about prototemporality and eotemporality. The "unreasonable effectiveness" of mathematics in physics
remains true. It derives from the fact that the brain is a part of the physical
world; equations of conic sections selected from among the autogenic images
of the mind stem from the same Umwelt whose laws control the motion
of the planets. It is precisely the primitiveness of the roots of mathematics
which gives it its awesome universality, power, and beauty; it is the same
primitiveness which makes any specific mathematical tool increasingly useless as we try to deal with time in its biological and mental aspects. The
great sophistication of the queen of the sciences is no counterargument; she
does remain queen but only in her own realms.

Reverting now to the work of the man whose psychology permitted us
to make an epistemological detour, we must mention his total commitment
to a principle of connectivity which he called *Koinzidenzprinzip* or synchronicity. By this he meant the coincidence in time of two or more
causally unrelated events or series of events which are connected by some
significance or meaning to the person observing them. Anecdotal reports
abound in pronouncements about coincidences which have profoundly influenced life histories. Such events or series of events, because of their
great importance to the individual, tend to appear as though specifically
arranged by a superhuman agent for harm or for help.[34] The sources of the

feeling of superhuman origins stem from the impossibility of thinking of such events as having been causally produced. For instance, my arrival in Norway from New York at the same hour as a friend's arrival in Norway from Scotland profoundly influenced the course of my life, yet, except for an assumption of rigid and complete determinism, I cannot imagine these two events to have been causally determined. Jung would say that a connecting principle nevertheless exists.[35] He specified the emotionally loaded character of such coincidences as qualifying them to belong to this category. He admits that synchronicity is a difficult concept to handle because of the ingrained belief of the Western mind in the sovereign power of causality.[36]

The idea of acausal connections was not invented by Jung. Ancient Chinese philosophy could accommodate causation working backwards; that is, it could permit cause to follow its effect. Or, a cause could bring about an effect by a type of simultaneous resonance. At first sight such ideas seem to play havoc with our concepts of the natural order. Yet they claim no more than that of connectivity below the level of deterministic causation. Thus the Jungian idea of synchronicity, whatever merits its specific exemplifications may have, deals with connectivities of the type we find for instance in the probabilistic causation of the prototemporal world. Whether and to what exent such lower-level causations can enter our conscious experience is not known.

However, the point has been made, that Jungian psychology places man in the perspective of his cosmic origins.

4. The Evolution of Conscious Experience

IN MANY FIELDS of knowledge, such as those which deal with life, society, or the earth, the evolutionary character of the subject matter must be appreciated and, in lieu of first principles, its details known before satisfactory theories of present phenomena can be advanced. Assuming that epistemological rules apply with equal strength to the study of the mind, it is likely that the evolution of self-awareness will have to be understood quite well before we can deal satisfactorily with its present problems. In this section I will speculate about some evidence that bears upon the history of conscious experience of man and will try to draw some conclusions from these speculations regarding the structure of knowledge.

Future, Past, and Present

In the physical sciences experimentation is based on the metaphysical confidence in unchangeable laws. The results of today's tests today are assumed to have been true in the past and to remain true in the future whenever the conditions of the experiment can be thought to exist. In physiology the situation is more difficult because brains available today are unlikely to be the same as the brains of man two or ten million years ago. Furthermore, having an ancient brain in vitro, or even in vivo, could supply very little information about the range of mental states of its owner. The only avenue left is to assume, as indeed most students interested in the history of Cs. experience have done, that (1) there exists some type of recapitulation between ontogeny and phylogeny, even if the recapitulation itself has been modified by evolution, and that (2) the mind of man examined in its mature and contemporary expressions retains vestiges of its evolutionary history.

The Jungian understanding of the origins and history of consciousness as put forth by E. Neumann is grounded on the belief that in the course of its ontogenetic development, the individual consciousness has to pass through the same stages which determined the evolution of consciousness in the life of humanity.[37] This is a rather strong recapitulation principle whose specifics were worked out by Neumann in fascinating detail in terms of the Jungian elements of the collective Ucs. and archetypes, reflected in the mythological stages of the evolution of consciousness, paralleled by the ontogeny of the psychological stages in the development of the self. Freud implied an equally strong recapitulation principle—with some caution. He spoke, by analogy, of a visitor to Rome who may be able to trace out in the plan of the Eternal City the outlines of the ancient Roma Quadrata and point to sites where buildings of various epochs stood.[38] He cautions, however, that in the body and psyche of man, as in the ruins of Rome, the earlier phases of development have been absorbed into the later phases for which they have supplied material. The embryo, for instance, cannot be discovered in the adult. But in the mind, unlike in the body, the preservation of the past is more complete than in biology and, in fact, as he saw it, "only in the mind is such a preservation of all the earlier stages alongside the final form possible."[39] Presently I shall attempt to provide some insight into the evolution of nootemporality and conscious experience in man, with elements of the Jungian stance, but growing out of the analytical work of Freud. Specifically, I wish to examine the curious phenomenon of the uncanny with respect to temporality.

The uncanny is a feeling as is love, terror, or ecstasy. Behavior corresponding to such feelings might perhaps be described with scientific precision, but such descriptions remain annoyingly vacuous compared with the power of the inner experiences themselves which, however, resist articulate description. In the following, therefore, I will try to circumscribe rather than describe the uncanny.

Conditions and events tend to bring forth a feeling of the eerie if they imply the workings of forces which are not only beyond our control but, more importantly, appear to be beyond our comprehension. That is, if they seem to defy some of the basic features of our rational categories such as space, time, finality of death, or causation. Uncanny things imply existence of powers "below," "above," "beyond," "far away," or "long ago." Employing the uncanny in art and literature has been popular. Manipulation of time, space, life, death, and causation in plays and stories is an easy way to hold the attention of the audience. If a plot involves noninferential knowledge of future, past, or distant events a reading or viewing audience is more or less guaranteed.[40]

In a paper first published in 1919 Freud asked the question: What kinds of things appear uncanny?[41] Through comparative word analysis in Greek, English, French, Spanish, and German he first established that the uncanny (1) is associated with such other ideas as strange, dismal, foreign, sinister, demonic, gloomy, concealed, and secret. Yet (2) it often displays origins in concept which stand for "familiar," "well known." For instance in German *heimlich* (homelike, warm) is a word the meaning of which develops toward an ambivalence, until finally it coincides with its opposite, *unheimlich* (uncanny).[42] Taking his cue from this linguistic ambivalence, Freud reviews literary pieces which are known to bring forth a feeling of the uncanny. Common to them is an uncertainty about the position of the event or thing on the scale of reality. Thus a creature which may be a doll or a real person appears uncanny. So does a person who looks exactly like another, or like oneself but is not the other and is not oneself; an organism which may be an animal or a human; a man who may be alive or dead; a house which looks thoroughly familiar yet we know we have never been in it.

Recent work in linguistic anthropology enlarged this store of examples. Edmund Leach, in a study on the relationship between animal categories and verbal abuse has shown that animals whose position is ambiguous in the edible/inedible or hostile/friendly classification tend to be regarded with special anxiety and shame and are subject to verbal abuse.[43] Pigs, for instance, were until quite recently members of the household, just as much as cats and dogs; they still are in primitive environments. They were friends, therefore inedible. But, by and by, the child learned that pigs are walking sausages; the blame for the abominable fate for the animal attached itself to the pig; the former friend became a detestable creature, a pig, while his meat changed into pork. For we ought not eat friends.[44]

In linguistic usage Leach found support for the theory that in naming animals we tend to create binary distinctions in their categories, for instance, as "man, not-animal, tame, friendly" versus "not-man, animal, wild, hostile" and then mediate the distinction by ambiguous beings such as man-animals, pets, and games. These animals of ambiguous standing tend to become taboo and are either highly esteemed or highly abused. But the highly charged emotional reactions to creatures whose positions are ill-defined between two poles or levels is not limited to animals. Things may also be ambiguous. John Cohen has examined, in the framework of the history of ideas, the ambiguity of things which are perhaps man perhaps machine. In a delightful book entitled *Human Robots in Myth and Science*, he traced the awe and fascination attached to the uncanny motive of mind-made-man from Egyptian burial rituals forward to our own epoch.[45] Feces, urine, semen, menstrual blood, hair, nail clippings are also universally objects of taboo. These are ambiguous substances for it is unclear to the child, and to the child in the adult, whether such things are the self or the nonself. On the integrative level of the mind, in the words of Edmund Leach

> the gap between two logically distinct categories, this world/other world is filled with tabooed ambiguity. The gap is filled by supernatural beings of a highly ambiguous kind—incarnate deities, virgin mothers, supernatural monsters which are half man, half beast. These marginal ambiguous creatures are specifically credited with the power of mediating between gods and man. They are the objects of the most intense taboos, more sacred than the gods themselves.[46]

Freud concluded his examples of the eerie with reference to the feeling of powerlessness that accompanies odd repetition of identical facts: names, numbers, or events.[47] In these he believed he had detected the source of the uncanny; encountering unlikely repetitions reminds a person of his infantile repetition compulsion, something once well known, then repressed. Going back to the earlier examples, suggested by Freud and by others, we may now see that in each of them the ambiguity is likely to have arisen because the childhood status of the event or thing was in some ways the opposite of its later status. Thus, a child does not distinguish between a doll and a living person: the loss of a doll or a person may be equally traumatic, their well-being equally rewarding. The delineation of difference between dolls and people does not take place until after the ego of the child begins to form. The man-doll identity was once familiar, then repressed. The uncanniness of one's own double plays on long-repressed childhood memories. When every person was judged identical with the child himself because the self had not yet been established, excretum was also part of his body, but now it is banished to the sewer. The conclusion of Freud was that "the 'uncanny' is that class of the terrifying which leads back to something long known to us, once very familiar." [48]

Tampering with matters that were once familiar but are now unfamiliar is often discouraged by forceful taboos. We are forbidden to regard dolls as humans, feces as parts of the self, or copulate with a formerly familiar corpse. There also seems to be a very strong taboo against tampering with temporal categories—as though the mind would want to defend zealously its greatest discovery and asset, and perhaps its own reason for being. Should any of these taboos be violated, the resulting feelings are likely to be those of the uncanny. Thus, for instance, in a psychoanalytic interpretation of artistic vision and hearing, Anton Ehrenzweig has asked why tampering with temporal order in music and films so often leads to laughter or a feeling of the uncanny.[49] By the Freudian argument, that the source of the uncanny (and often of laughter) is related to the recognition of the formerly familiar in the now unfamiliar, he theorizes that some level of our mind identifies events without regard to temporal order. When we hear or see a record or a film played backwards, the time-ignorant imagery returns to the threshold of consciousness and is there recognized as formerly familiar. In terms of our own concepts we would say that eotemporality (pure succession, as represented, for instance, by a film shown backwards which, thereby, removes a preferred direction of time) appears uncanny to consciousness because at one time eotemporality was very familiar. We may add that prototemporality expressed, for instance, in probabilistic causation also appears uncanny for similar reasons.

There are good reasons to believe that futurity is the earliest level of nootemporality in man, and also the most manifest mode of time in the higher animals. As we ascend on the evolutionary ladder, the first instinctual activity which may be considered biotemporal is that of goal directedness: making provision for anticipated environmental changes. Many organisms which give no hint of any but very short-range memory do, instinctively, provide for anticipated distant necessities, although only for periodically predictable events. The earliest developing ideas in children are not memories but expectations,[50] and references to future are grasped earlier than references to the past. We have already discussed some aspects of this sequential emergence in the context of cosmologies, the predominantly intellectual, hence secondary, interest in the creation of the world as contrasted with the predominantly emotional, hence primary, interest in apocalypse may be adduced as suggesting the phylogenetic primacy of future over past. John Cohen in his survey of the history of divination lists over one hundred methods of fortune telling, witnesses to the keen interest in the future, in epochs when peoples' interests in the past was only of a washed-out, generalized nature, if manifested at all.[51]

Granting, then, that concern with the future is the first form of nootemporality: what is it in the future that threatens the ego and therefore must be excluded from awareness by repression? It is, I believe, the inevitability of the death of the individual, the only absolute challenge to the executive powers of the mind. Curiously, the inevitability of death possesses an air of timelessness, for it removes with unarguable authority the element which characterizes the future, namely, that of ignorance. From the point of view of adaptation, it would hardly be helpful if one were to know exactly the date, place, and mode of one's death. Though for some unusual conditions a case might be made for the contrary, generally such a foreknowledge would be a crippling handicap. Thus, while we strive to know about the future, we cannot permit ourselves to be continuously reminded of the regression to lower temporalities implicit in death. We have, therefore, keen interest in the future, combined with a repressed fear of it, a human condition of existential tension known as dread. I believe that the repression of reminders of our future death, which would continuously tell us of temporalities below that of nootemporality, is the most ancient element of self-awareness.

Concern with the past, as we have seen, is not prehistoric; it emerged in historical times. Granting that concern with the past is the next emergent form of nootemporality, we may then ask: What is in the past that threatens the ego and therefore must be excluded from awareness by repression? It is, I believe, the set of all those memories which remind it of its fallibility, thereby threatening its executive powers. They may be subsumed under the concept of guilt. Without getting into the problem of guilt, conscious, unconscious, or imagined, I wish to stress that the sense of guilt exudes an air of timelessness. For reasons that are rather complex, guilt brings forth a desire to "turn back the clock" by magical means (such as by descent to the eo- or prototemporal); also, it often carries an atmosphere of predestined doom, another feeling of timelessness. Thus, guilt is a reminder of earlier levels of temporality. From the point of view of adaptation, it would hardly be helpful in the struggle for survival if one were continuously to be reminded of past misdeeds (determined by some prevailing norms) and thereby told of the impotence of one's executive faculties. While we strive to learn about the past and to remember it, we cannot permit ourselves to be reminded of the lower-level temporalities which attach themselves to the sense of guilt. Here, too, we have the human form of existential tension, or dread: the need to be reminded of, and to forget the past, simultaneously. I believe that this repression of certain reminders of past failures that continuously tell us of lower temporalities is the second oldest element of conscious experience.

The mental present, as we have reasoned earlier, is the latest and most complex mode of nootemporality. We may then ask: What transpires in the mental present that threatens the ego and therefore must be excluded by repression? What is it that later appears as "once familiar" when brought back from the Ucs. in response to alleged feats of telepathy or clairvoyance? This is, I believe, that component of mental present which might be described as a creature present; one which takes telepathic and clairvoyant knowledge of distant events for granted. For the child listening to a fairy tale, there is nothing unusual if the Queen immediately knows that her distant daughter is in danger; she might even know all the details. Subsequently, however, communications of this type are labeled magic, belief in them is repressed, and new modes of thought involving deterministic causation and temporal order are established. When telepathy is encountered under conditions which assign to these communications the status of reality, they will appear uncanny. But from the point of view of adaptation it would hardly be helpful in the struggle for survival if we were to try to operate with magical causation. I believe that the repression of ideas about connectivities which were hallmarks of lower levels of temporalities is the latest emergent element of conscious experience.

With poetic latitude, T. S. Eliot has described futurity, pastness, and presentness in the mind of man.

Men's curiosity searches past and future
And clings to that dimension. But to apprehend
The point of intersection of the timeless
With time, is an occupation for the saint.
No occupation either, but something given
And taken, in a lifetime's death in love,
Ardour and selflessness and self-surrender.
T. S. Eliot, *Burnt Norton*

Let us return now to the feelings of the uncanny which, I claimed, accompany alleged noninferential knowledge of the future, the past, and the present and which may now be seen as originating from various stages of the evolution of consciousness. They are features of the coexistence of knowledge felt and knowledge understood, and refer to such matters as the inevitability of death, the sense of guilt, and connectivity among events. For various reasons these old acquaintances are forgot or, more precisely, repressed, and become the sources of unresolvable, continued tensions. These tensions must be seen, therefore, not as incidental to our sense of time but as being, in fact, at its very roots.

The Experience of Timelessness

For the mature mind all but the noetic level of temporality appears as somewhat incomplete. This is reflected in the ordinary use of "time" versus "timelessness." When we speak about time we usually mean something (whatever it may be) as experienced by people in full awareness of themselves. Most other experiences to which temporality may relate directly or indirectly, are lumped under "timelessness." In this subsection we shall inquire into experiences usually described as those of timelessnes.

The rhythm of a marching band is a powerful device of attraction; where bands are out of vogue, such as in the United States of the nineteen seventies, the preoccupation of music with structure and rhythm does more than make up for the loss.[52] The contagious nature of rhythm can be observed throughout the animal kingdom, and we may assert with very little reservation that everything that moves also dances, from gyrating fish and mating dragonflies to swinging men and women. From among the many reasons that attract people to dancing I wish to single out one which I will call the *ecstasy of the dance*. By this I mean an experience of timelessness, a radical decrease in existential tension that accompanies the increased emphasis on becoming (through continuous, rhythmic change) and decreased emphasis on permanence.

Aldous Huxley expressed this admirably in *The Devils of Loudon*.

Consider for example the way [in which] rhythmic movement [is used] to escape from insulated selfhood into a state in which there are no responsibilities, no guiltladen past or haunting future, but only the present blissful

consciousness of being someone else. . . . It would be interesting to take a
group of the most eminent philosophers from the best universities, shut
them up in a hot room with Moroccan dervishes or Haitian voodooists, and
measure with a stop watch the strength of their psychological resistance to
the effects of rhythmic sound. Would the Logical Positivists be able to
hold out longer than the Subjective Idealists? Meanwhile, all we can safely
predict is that if exposed long enough to the tom-toms and the singing,
every one of our philosophers would end by capering and howling with the
savages.[53]

The predominance of the cyclic order in behavior, such as in intense and
protracted rhythm, dissolves the boundaries of the ego and facilitates
regression to a relatively undifferentiated, timeless state. Happiness, one
might muse, is sometimes a periodic time-rate of change; a return to an
eotemporal Umwelt.

In contrast to the ecstasy of the dance, emphasis on permanence and
deemphasis on becoming can lead to states of mind which I would describe
metaphorically as those of the *ecstasy of the forest*. Instead of perpetual
motion, this feeling derives from perpetual no-motion. Ideas of eternity, or
the sense of awe that accompanies scientific or religious insights often tend
to induce such states of mind. Goethe's youthful reflection expresses this
briefly and beautifully:

> Über allen Gipfeln
> Ist Ruh
> In allen Wipfeln
> Spürest du
> Kaum einen Hauch;
> Die Vögelein schweigen im Walde

> (Over all the mountain tops,
> Rest;
> Among the tall trees
> Not even a whisp;
> The birds are silent
> In the forest.)

•••

*From among the reasons that attract people to dancing I wish to single out
one; it may be called the ecstasy of the dance, an experience of timelessness. The continuous, repetitive change of the dance amounts to an
increased emphasis on becoming, which, then, can lead to a radical
decrease in existential tension. It is this decrease in tension which we
experience as timelessness.*

*The Shakers believed that the rhythmic agitation of the body produced gifts
of prophecy. The illustration is a lithograph of uncertain origin, made
by an unidentified artist. Shakers near Lebanon, State of New York,
some time during the period 1835–1850. Courtesy, The Henry Francis
du Pont Winterthur Museum, Winterthur, Delaware.*

•••

The same sentiments are found in the haiku of Chiyo-Ni (1701–1775):

> Full moon tonight, when even birds that bide
> In nests will leave their doors and windows wide.
> *The Chime of Windbells* †

There is no implication that the two types of timeless ecstasies appear in life with pristine purity. They are normally mixed, even in the most commonly available ecstasy, that of sexual union. A complex adventure of mind and body, sexual intercourse displays at one fell swoop the timelessness of the dance followed by that of the forest. Its universal power of attraction derives from these two connected experiences of timelessness.

States of timeless ecstasy are very poorly understood and are scientifically almost totally unexplored. One of the few exceptions is the work of Roland Fischer who proposed to map states of consciousness along a perception-hallucination continuum.[54] Along this scale he perceives each conscious state as made up of two complementary components: one of ergotropic arousal (activated psychic state) and one of trophotropic arousal (states tending to restore energy). With increasing ergotropic and decreasing trophotropic arousal the extreme of the self is reached in a mystical rapture which corresponds, in our terms, to the predominance of becoming, the ecstasy of the dance. Proceeding in the opposite direction, the extreme of the self is reached in meditative ecstasy corresponding to predominance of being, the ecstasy of the forest. Time, as Fischer clearly sees, disappears at both ends of the perception-hallucination continuum. He speaks of the "self" as the knower and image maker (this is our Observer) and of the "I" as the known and the imaged (our Agent). The "self" is the mental dimension of the "I" which is its space-time embodiment located midway, as it were, between the timeless selves. This concept of the mind differs from the one discussed earlier in that we have taken the observer and the agent as projections of a single, underlying conflict and interpreted timelessness as the lessening of that conflict. The timelessness of the ecstasies we just described is, in fact, temporality, but of lower order than that of conscious experience.

Ingestion of certain chemicals can lead to states of mind which may be described sometimes metaphorically, sometimes accurately, as those of the *ecstasy of the mushroom*. From Aldous Huxley's experimentation with mescaline[55] to the religious cult that grew up around Timothy Leary's "League for Spiritual Discovery," the literature of drug-induced ecstasy has been increasing. This is one symptom of an underlying fear of, and flight from, time. That an ancient escape from the anxiety of life has been rediscovered and energetically advertised, is quite clear.

An eminently reserved and scholarly survey of established knowledge of drugs and their effects on the sense of time is given by L. Doob.[56] A summary of his summary would say that there is no solid, repeatable

---

† Translation and explanatory notes by H. Stewart, Tuttle, Vt., 1969, p. 87.

knowledge available on the effects of drugs on the perception of time, though most drugs do alter time perception as well as the sense of time. These alterations are subject to many, often uncontrollable variables: personality, somatotype, education, occupation, age, health, premedication, time of day, food in the stomach, and attitude to the experiment. Perhaps the problem with controlled studies is that they tend to lose sight of the forest because of the trees: in this case, it is the intention of the drug user which is lost. Plainly put, this intention is always a desire to escape from existential tension.[57] The willful ingestion of drugs to achieve whatever "high" they are promised to offer, represents a desire to return to the timeless. The experience sought has been acknowledged and described by Freud as an "oceanic feeling" and "oneness with the universe."[58] In Norman O. Brown's epic, *Life Against Death*, it is childhood lost; in Jungian terms, it is the reachievement of a state which individually one might never have experienced. In all cases, it is an escape from the unresolvable conflicts of individuation.

In a curious juxtaposition of concepts one may say that the most precise description of experiences of timelessness is that they are mystical. Perhaps all authoritative descriptions of mystical experience insist that they are ineffable, that in them the separation of the self and the nonself, the I and the Thou vanish and is replaced by a unity with the universe and by a prevailing atmosphere of timelessness.* For information on these states of mind we may look to the great mystic saints of Roman Catholicism, such as St. Theresa, St. John of the Cross, or the Dominican, Eckhart von Hochheim, better known as Meister Eckhart. We cannot expect analytical statements but only poetic, theological, and artistic expressions which imply timeless states of unity.

Meister Eckhart (d. 1327) was an intellectual mystic. In many a stern sermon he insisted on the importance of stillness and spiritual passivity, on the giving up of one's identity to God. He perceived the goal of selflessness as the reaching of the fullness of time, "when time is no more." The search for the fullness of time consists, he held, in setting one's heart on eternity in which all temporal things are dead.[59] In psychological terms he did what oriental mystics do: promote through their behavior a return of their state of consciousness to the undifferentiated ego of the child. But, importantly, the posttemporal experience of regression is not exactly the same as that of the infant yet ungrown. The regression to timelessness of a mature ego is the journey back, as it were, of someone who has been away; it is a return. The child has not yet been away.

Meister Eckhart's medium was that of Christianity. The medium of William Wordsworth was that of romantic poetry and belief in nature religion. Complying with his personality, social status, and place in life he

---

* Mysticism in its noble sense must be carefully distinguished from the perverted use of that concept to describe things that are unknown, or not yet discovered, or simply uncanny.

called for the same stillness as did Meister Eckhart. But Wordsworth
wanted to unite with nature, not with Christ.

> Once again
> Do I behold these steep and lofty cliffs,
> Which on a wild secluded scene impress
> Thoughts of more deep seclusion; and connect
> The landscape with the quiet of the sky.
> "Lines Composed a Few Miles Above
> Tintern Abbey. . . ."

For St. John of the Cross (1542–1591), mystical theologian and great
lyric poet, the flames of love for God, man, woman, and all creatures are
brought together in a holy union:

> O lamps of fire!
> In whose splendors
> The deep caverns of feeling
> That were dark and blind
> With strange new beauty
> Give warmth and light to their beloved.[60]

His lyric, like that of the Song of Songs, sounds like Eros and not Agape,
though they were not so interpreted by the mystics. He tried to express
the inexpressible through the use of antithesis implying, philosophically, a
unity of the opposites: *soledad sonora* (resounding solitude); *musica callada*
(tacit music) or *saber no sabiendo* (unknowing knowledge). The crucial
moment of his life was the experience of blinding light in his cell in
Toledo, so beautifully depicted in the painting of El Greco. Though his
prose is silent about it, perhaps it is that light (the one that connects the
internal landscape of man to the external, the one that makes foresight
possible) which he speaks of in a "Song Written about an Ecstasy of High
Contemplation."

> The higher one climbs
> The less one understands
> Since it is the darkest cloud
> That gives light to the darkness of the night;
> And so whoever knows this
> Remains always knowing nothing, where
> All science was transcended.[61]

Reports that suggest experiences similar to those of St. John of the
Cross seem to come from those drug users who are articulate enough, and
interested enough, to try to communicate their experiences and fortunate
enough to be able to do so. Only feeble efforts have been made to admit
states of ecstasy and mystical union into the fold of legitimate knowledge.[62]
For numerous reasons, most importantly perhaps because altered states of
consciousness have been subject to social taboos, there exists no scientific
survey of the "ecstasy of the mushroom." We will have to rest the argument
with Aldous Huxley. But, if there are such basic similarities, perhaps even

identities, between the religious ecstasy of selflessness and timelessness on the one hand, and, on the other hand, the ecstasy that can sometimes be produced by drugs, where does theology stand regarding the nature of mystical experience?

There are no a priori reasons why there should be a difference between certain ecstasies produced by earthly and by heavenly mushrooms. The former are drugs manufactured by the nonself; the latter are drugs which are, as likely as not, manufactured by the soma under the selective pressure of the mind. It has been partly demonstrated, partly plausibly argued that one product of metabolism of alcohol in the brain might be hallucinogens related to those found in the cacti that yield mescaline. Paul Tillich believed that in theology there need be no dichotomy between eros and agape—natural and divine love, thereby putting into modern terms the poetry of St. John of the Cross.[63] From the pantheistic view of this author, it would seem that there need be no difference in principle between the ecstasy of the mushroom made for my body by my mind or made by other minds, although there may be some pragmatic differences. We have learned that the mind protects its integrity by separating itself from its lower strata. When descent into these lower strata is the consequence of external "mushrooms," the results, though they may appear divine, can nevertheless be attributed to identifiable, earthly nonselves. When the chemicals are manufactured under the command of the mind, or by some function or dysfunction of the somatic system, the origins of the ecstasy are internal, unassociable with external causes. We have reasoned that for each integrative level the capacities of more advanced levels tend to appear methodologically unintelligible, hence, mystical. It is not surprising, therefore, if certain experiences consequent to internally generated conditions have been traditionally diagnosed as being divine in their origins.

A rather pragmatic difference emerges however, between ecstasies induced by the self versus those induced by nonselves. As images of creative men and women march before me the impression is left that most of them employed whatever methods were known and available to them to enlarge and enhance some preexistent seeds of internally generated capacity for timeless ecstasy. To remain creative however, they had to keep the external aids (sex, work, power, song, or wine) under their control; otherwise they ruined themselves in the fashion of Dr. Jekyll and Mr. Hyde.[64] That is, they had to retain a link with reality for a climb back into nootemporality.

The Origins of Being and Becoming

Creative people who journey into the lower Umwelts of temporality can usually return to the noetic world which they share with other people, express their visions, and thereby make them the communal property of man. This is consistent with one of the roles of the mind in organic evolution: the sharing of ideas which makes communal actions possible. Regression to lower states of consciousness appears to have an effect on the mind best described as that of refreshment, or new insight; but it may also be fatal if, for some reason, it turns out to be irreversible. The socially approved avenues to timeless ecstasies—those of the search for the true, the good, and the beautiful—are presumably also the avenues which provide safe returns to nootemporality.

The psychodynamics of regression to timelessness (more precisely, to bio-, eo-, or prototemporal experiences) is that of the lessening of that degree of existential tension which is the corollary of the noetic experience of time. I wish to postulate that there is an instinctual drive towards the reestablishment, in the mind, of those early Umwelts which were characterized by permanent certainties. That analytical component of existential tension which we describe as "being" corresponds to such permanent certainties. Our instinctual drives in this regard find expressions in scientific and religious laws and in philosophies of being which see the strategy of existence in unchanging continuities. The postulate of an instinctual drive for the identification of time with permanence is not to be confused, however, with the question of whether or not such continuities have ontic status. The postulate only claims that we instinctively seek permanent relationships and that we are fulfilled when we believe to have identified them.

Regressing to the lower Umwelts of temporality under the control of the ego also leads to worlds where connections among events are pre-deterministic, such as magic causation. I wish to postulate further that there is an instinctual drive toward the reestablishment, in the mind, of those early Umwelts which were characterized by unpredictable connections among events, interpreted as being in the service of some overwhelming and unintelligible power: deity, father, mother, or nature. That analytical component of the existential tension which we call "becoming" satisfies the demands of intrinsic unpredictability. In the broadest context, the play of the unpredictable is expressed in the contingencies of science, religion, and the arts, and in philosophies of becoming which maintain that the strategy of existence resides in unpredictable qualities. The postulate of an instinctual drive for the identification of time with the unpredictable elements of experience is not to be confused with the question of whether or not such intrinsically unpredictable elements have ontic status. The postulate only claims that we instinctively seek the unexpected and have a sense of completeness when we believe we have identified it.

Thus the specific ways in which we slice temporality into being and becoming, or its corollary, the way we see the world as made up of stationary and creative processes, are characteristics of the human mind. Yet, metaphysical and methodological solipsism is removed if we remember that the existential tension of the mind, of which the nomothetic and generative aspects of time are projections, is only one level, albeit the most advanced one, in an open-ended hierarchy of unresolvable conflicts immanent in nature.

5. Some Implications of the Deep Structure of Time

THERE ARE three classical problems which have bedeviled the study of time: (1) the beginning and the end of time; (2) atomicity versus continuity of time; and (3) motion and rest. It is very easy to show that beneath the usual understanding of these problems lurks the assumption that the time/timeless relationship is a meaningful one. As we have been arguing in extenso, there is nothing in nature that corresponds to such a sharp division. Accordingly, the theory of time as conflict does not attempt to answer these problems as ordinarily envisaged. Rather, it suggests a different Fragestellung, that involving the hierarchical structuring of time, in terms of which one can at least hope to arrive at satisfactory answers.

(1) The beginning and end of time

In the physical world we have four potential manifestations of this problem: going from the timeless to time in the beginning of the universe and in the end of the chronon; going from time to the timeless at the end of the universe and in the beginning of the chronon. Let us look at the time/timeless interface question in terms of transitions between temporalities.

In the atemporal world of electromagnetic radiation questions of time and the timeless have no meaning, hence we cannot ask about beginnings or ends. Out of the atemporal substratum arose, so we reasoned, the prototemporal world of countable but unorderable, indistinguishable particles. Presumably, the prototemporal world could also return to the atemporal one. But would a beginning or an end of the prototemporal world correspond to a beginning or end of time? After a fashion, yes. But its coming into being and going out of being has not even a faint resemblance to our fantasies of a beginning and end of time. We already mentioned at the end of chapter 5, section 4, that such a beginning is something we may infer but not locate in time, even in principle, for events and things in the prototemporal world are unorderable. In its turn the prototemporal world gave rise to (and conceivably could reabsorb) the eotemporal world of ponderable matter. Such transitions are also a bit strange. We have already reasoned that pure succession determines an endless world because it has no present with respect to which a limit of time may be set. Also, it has no preferred direction. Hence, neither an end nor a beginning may be properly determined (at least in terms of our usual ideas) nor can a change be identified as belonging to the future or to the past. Yet, the emergence or submergence of the eotemporal in the prototemporal world still exemplifies a type of beginning and a type of ending, though these do not resemble the speculations of cosmologies—scientific or narrative. We do not even have a suitable word to describe an event or a type of change which is both a beginning and an ending.

With biotemporality we witness the emergence of a present (meaningful in terms of the purpose of the organism) and with it beginnings and ends of sorts. They are known as births and deaths. If all life were to vanish from earth, this would still not be an end of time, however, but only a reabsorption of the biotemporal into the eotemporal. Going on to the nootemporal world, we can argue for a personal cosmology with its beginning and end defined in terms of the identity of the individual. A type of beginning of time, then, is perhaps the coming of age of the child, perhaps his birth, perhaps the evolutionary emergence of the brain. The death of a man is, then, a type of an ending. Now, stepping back to behold these frighteningly complex matters we see that the various features here mentioned do fit well into a scheme of temporal hierarchy, but none of the boundaries mentioned do suggest themselves as hopeful candidates for the time/timeless interface sought by philosophical, religious, or scientific understanding.

Let us look at the same reasoning in terms of finitude versus infinity of time. Life, we would say, is finite and so is the life of the mind. But neither nootemporality nor biotemporality alone may be identified with the totality of time, hence statements about time and mind, or time and life cannot be sufficiently broad to answer questions about the finitude/infinity of time. The world of pure succession is infinite by default, as it were, because of the absence of a present (supra). We are back at the prototemporal world which is unorderable and the atemporal world where no idea of time can be applicable.

From the macroscopic world, let us turn to the microscopic one. The chronon may be thought of as an inside-out universe: it begins where time ends and ends where time begins. Since all arguments which seem to support the idea of a chronon come from physics,* we must be talking about some features of the lower Umwelts. We do not have enough knowledge to be able to reveal the precise position of the hypothetical chronon even within the physical world. But if we assume it to be a remnant of the primordial, atemporal world then the limits of chronons must be determined by the properties of pure succession. It would follow that chronons (if they exist) cannot be placed end to end, for their limits are ambiguous: we cannot tell their beginnings from their ends. Thus, no number of chronons can add up to duration.

* The chronon must be distinguished from minimal processing times in physiology or psychology.

The summary point of these thoughts is that nowhere in nature do we find anything that corresponds to the harsh contrast we usually *feel as being meant by* an interface between time and the timeless, or between finite and infinite time. What we find in both cases is a hierarchical structure of different temporalities which form a continuous spectrum between our noetic experience of time and atemporality, and also between the finitude of our lives and the infinity of the atemporal world.

(2) Atomicity versus continuity of time

Implicit in all systems of atomism is the working assumption that the nature of the atoms themselves is not questioned. For if it were, then the former atoms themselves would become composites of more elementary constituents. The history of atomism as it pertains to matter, has been that of successive discoveries of levels of organization which pushed the limits of atomism to more and more fundamental constituents. Each new step involved not only a change from something physically larger to something smaller, but, more importantly, a change in the nature of the atomic constituent itself. We can also identify a hierarchy of temporal atoms in the structure of temporality.

The clockwatcher's own identity is his primary reference. From the point of view of some higher integrative level, the unity of a man's mind might be taken as an atom of that higher temporality. Our nootemporality itself is made up of atoms of biotemporality. They are those aspects of life phenomena which remain identical with themselves. In my internal world, examples are one heartbeat between two maxima of the *P* wave in an electrocardiogram, or the minimal processing time necessary to separate two sense impressions (the physiological present). In the external world we have other biotemporal atoms: one dog named Frieda, or the life span (reconstructed) for an entire species of now extinct animals. As we "timesect" our world further, we find that the biotemporal atoms themselves are made up of components which belong to the eotemporal world of physics. Thus, the temporal atoms of a simple living structure, such as the DNA molecule, are defined by the properties of the physical world; they are eotemporal. In their turn, the temporal atoms of the eotemporal world are the identities of the prototemporal world. These, again, are made up of atemporal atoms of electromagnetic waves, also known as wave pockets.

The summary point of these reflections is that there is nothing in nature that corresponds to what *we feel is meant by* timeless elements which would make up our experience of duration as hydrogen atoms make up hydrogen gas. What we do have, instead, is a complex hierarchy of temporal atoms. As we trace the identities of these temporal atoms from the continuity of the self in the nootemporal world, we journey into the deep structure of time. When we arrive at the last, fundamental time atom we find ourselves reabsorbed in an Umwelt which is atemporal.

(3) Motion and rest

In kinematics it is possible to generate relative rest by superposing states of relative motion, but it is not possible to generate relative motion by combining states of relative rest. In a more general way, it is possible to reproduce an earlier condition through steps of change but it is not possible to produce a new condition by steps of no change. The asymmetry suggests that motion and change are somehow more fundamental than rest and no change.

Separating in the mind the changing from the permanent is not a simple task. The process begins, or grows out of, the capacity of the mind to create autogenic images and select from them some which, in course of time, will correspond to some external conditions. Then out of such "truths" we abstract those which have this in common with our identity: they endure. Such features we call permanence (as opposed to change) or rest (as opposed to motion).

How would motion appear to an organism whose Umwelt is no higher than biotemporal? We can only speculate. I drew attention earlier to the dream images of motion and described them as static tableaux of change. We know that these images are inspired and guided by the Ucs., which is the depository of functions that fall between those of the mind and those of life. Thus the imagery of the manifest content of dreams, as well as the connectivities among these images, are culled from many levels. Perhaps, by bravely assuming a degree of recapitulation between, on the one hand, the ontogeny of the Umwelts of man which are retained in his dreams and, on the other, the phylogeny of Umwelts in higher animals in the waking state, we might see in dream imagery a type of waking biotemporal Umwelt of some animals. In such an Umwelt, motion would be characterized by instantaneity and immediacy, by a quality of responsiveness rather than creativeness, and would have no long-term implications and potentialities. Rest might correspond to the absence of necessity to respond, which is, then, permanence.

How can the idea of motion be accommodated in an eotemporal world? Since there is no present but only pure succession, and no preferred direction to that succession, motion can only be described as unanchored change. Rest might be associated with deterministic causation, albeit without preferred direction, thus without criteria whereby cause and effect may be separated. For a conscious subject, such conditions might perhaps appear as the temporally disoriented worlds of schizophrenics; in one word, hellish. Change and permanence in the prototemporal world is even stranger. Happenings are connected by probabilistic causation, hence the laws themselves (which on higher levels are embodiments of certainties) contain a degree of unpredictability. The inhabitants are indistinguishable and not orderable, and the polarization between the predictable (permanent) and the unpredictable (change) is rather weak.

Finally, in the atemporal world, all changes take place at once. Since the changes cannot be organized or understood according to lawfulness (which would be permanent), the atemporal Umwelt must be thought of as that of pure change and no rest. The primordial unpredictability is, then, ontologically prior to predictability. This priority is consistent with the cosmological view that the universe itself, if understood as having evolved from a world of pure radiation, is a contingency and not a necessity. Necessities, in the form of levels of regularities, subsequently evolve in the course of nomogenesis, leading to the universal polarization of temporalities into being- and becoming-type components. But in the atemporal world of the original fireball and its surviving remnants of atemporal phenomena, there is nothing that would correspond to our ideas of rest.

The motion/rest relationship must, then, be interpreted on all levels of integration in ways appropriate to the temporality of that level. At the upper end, in the noetic world, we have our ideas of motion and rest derived from sense experience by the sophisticated machinery of the mind. As we descend into the deep structure of time, the polarization motion/rest becomes fuzzy and finally disappears in a world which appears to us as that of permanent motion. Meanwhile, in the world of Eleatic and other clock-watchers Zeno's arrow keeps on flying as well as holding its fixed length. The intricacies of this classical puzzle are usually examined in terms of the ideas of rest and motion, that is, in terms of "timeless" and "temporal" features. Our deliberations suggest, however, that we are not faced here with such simple pairs of disjunctions but with a hierarchy of temporalities, each with its specific Umwelts and corresponding "rests" and "motions." I do not believe that any solution of Zeno's paradox of the flying arrow (or the other paradoxes of motion mentioned in chapter 1, section 1) may be satisfactory without an appeal to that hierarchy.

Assuming now, as we must, the open-endedness of evolution, we have to look at the deep structure of time as a record of history, with other levels of temporalities yet to come. Nootemporality might appear to future individuals or to new communal entities as inadequate for the understanding of nature, man, or society. I have already implied this much when I speculated that some puzzling aspects of nootemporality may be intelligible only when inspected from an integrative level higher than the noetic. We have witnessed how mind emerged as a solution to the unresolvable conflicts of life and has led to some unresolvable conflicts of its own, as well as to an increased rather than decreased existential tension. In the following three chapters we shall examine the broad outlines of some practices which seem to offer resolutions of the unresolvable conflicts of the mind. They are the search for the true, the good, and the beautiful. These great continuities of civilizations are not corollaries of further complexification of the brain (the "very complex"), but rather methods of learning to use its potentialities.

PART FOUR

Collective Greatness

FOR MAN, the most significant, though certainly not the only interesting features of temporality are those which manifest themselves in the context of his social functions. Just as life is born when the new growth of the plant introduces organization into matter, so through his individual and collective search for the true, his aspirations for the good, and his demands for the beautiful man seeks to identify and strives to extend the temporal order.

IX

EPISTEMOLOGY AND
THE TRUE

IN THIS CHAPTER I suggest that truth is that class of knowledge which individual and communal perception judge to be timeless. In our epoch the foremost representative of that class of knowledge is called science, hence, there now prevails a common cause, as it were, between the idea of the true and that of the scientific.

The many departments of knowledge, both within and without science, are products of many individual minds. Since each mind displays constellations of preferences in its rational transactions, and these are usually associated with different personality traits, the departments of knowledge themselves ought to display what may be called their "personalities." It would follow that epistemology, which today has come to mean the study of the limits and nature of scientific truths, might be guided by an understanding of preferred individual and personality traits. These, in turn, may be conveniently associated with distinguishable estimates of, and attitudes to, temporality.

To investigate the validity of these suggestions, we shall examine a few of the many and changing ways in which various branches of knowledge prefer to separate the permanent from the contingent in the structure of their evidence. As in some prior chapters, we again begin by setting the historical, developmental background through a brief, critical review of the subject matter.

1. Epistemologies

Theories of Knowledge as Philosophy

Since their remarkable origins in Greek thought, theories of knowledge have been regarded as inquiries into the nature of knowledge and not as attempts to gain knowledge. However, the judgements have varied widely as to what specific discipline ought best to incorporate theories of knowledge. There was little doubt in the minds of Plato, Aristotle, or Kant, however, that the appropriate discipline was philosophy.

We have already considered certain theories of knowledge when we discussed representative ideas of time in Western thought. Thus, Plato's distaste for the temporal was consistent with his position about knowledge as summarized in his famous symbol of the Divided Line. This geometrical metaphor may best be described as a totem pole comprising all conceivable things and functions.[1] The total body of the knowable derives from the Good, which is the uppermost portion of the totem pole. Fundamental things are timeless; all other things are more or less prostituted images of that reality—having lengths proportionate to their intelligibility and value. Aristotle's treatment of knowledge, as that of time, differs little from that of Plato. With a slight emphasis on empiricism, he assigns minor roles, but no more, to sense perception.

During the two millennia between the Schools of Athens in the third and fourth centuries B.C. and the emergence of natural science in the seventeenth century, interest in theories of knowledge was slight, although with the privilege of hindsight one may detect the slow maturing of the idea of universals, which became, eventually, the paradigm of natural law. Articulation of this paradigm as a feature of nature did not come about, however, until after Descartes had liberated Western thought from the doctrines of the Church of Rome. Underlying the Cartesian epistemology is the metaphysical dualism that regards all ideas as innate, sensations themselves being but functions of the soul.

It is this very strong rationalism that was attacked by the empiricists, foremost among them John Locke (1632–1714) who denied in toto the existence of innate ideas, categories, or moral principles. In his philosophy, the sources of all knowledge are sense experience and reflections; these are not themselves knowledge, but provide the mind with substance for ideas. Ideas, thus, have two sources. One, depending upon our senses and transmitted by them to the understanding, he identified with sensation.

> The other fountain from which experience furnisheth the understanding with ideas is the perception of the operation of our mind within us, as it is employed about the ideas it has got; which operations, when the soul comes to reflect on and consider, do furnish the understanding with another set of ideas, which could not be had from things without.[2]

Whether reflection itself derives from sensation is not very clear in Locke, though the answer is probably in the affirmative. Reflection, though it has nothing to do with external objects, "is not a sense yet it is very like it, and might properly enough be called internal sense."[3] In A. C. Fraser's scholarly interpretation this is understood to mean that reflection is empirical apprehension of mental states.[4]

From the purely empiricist theory of knowledge, an empiricist view of temporality must follow. Locke paraphrases St. Augustine: "time which reveals all other things, is itself not to be discovered"—and disagrees with him. Duration, time, eternity

> however remote they may seem from our comprehension, yet if we trace them right to their originals, I doubt not but one of those sources of all knowledge, viz. sensation and reflection, will be able to furnish us with these ideas, as clear and distinct as many others which are thought much less obscure.[5]

Locke's pure empiricism could have been made more convincing had he lived in our post-Darwinian epoch. Today he might have held that we must distinguish between the sensation experienced by the individual and the cumulative effects of sensations experienced by the evolving species. What he calls reflection, with certain modifications, could then have been attributed to the working of organic evolution.

We have already dealt with the epistemologies of Hume and Kant in the context of their theories of time. Thus, we may now turn to some twentieth century developments in epistemology characterized by the narrowing of the field in the works of logicians and by its broadening in the understandings of psychologists and sociologists.

The empiricist view came under the keen criticism of Bertrand Russell who held that knowledge is a subclass of belief and it is a matter of degree and not one of absolute status. The question, "What do we mean by 'knowledge'?" has no unambiguous answer.[6] Truth of knowledge is obtained by inference but the process of inference is ill-understood. Deduction is not as powerful as it was supposed to have been; it does not give new knowledge, but only new forms of words for stating truths in some sense already known. Methods of induction have never been satisfactorily formulated and when formulated at all, they give only probabilistic conclusions. In Russell's reasoning the next crucial step seems to be missing. Namely, deduction is a nomothetic process, induction is a creative one; deduction can formulate only the predictable, whereas induction pertains to the logically unpredictable.

Russell does stress however that whereas "existence propositions" may be obtained empirically, two other, equally essential, ways of knowing are unobtainable through empiricism; they are "universal propositions" and "unexemplified existence propositions." The difficulty with the first one is that, strictly speaking, they are valid only for periods of observation, they cannot tell us anything about what happens at other times. In particular, they can tell us nothing about the future.[7] The problem with unexemplified propositions is that they must be inferred from experience via the use of other concepts, such as causality. But then, causality must be assumed as prior to experience. Thus, empiricism has its limits in that we must of necessity depend on sources other than our own experience. It is only at the very end of his long work that Russell remarks that "the forming of inferential habits which lead to true expectations is part of the adaptation to the environment upon which biological survival depends."[8]

The validity of empirico-scientific detachment was rejected on the basis of disciplined introspection by Michael Polanyi in his penetrating study of *Personal Knowledge*. The art of knowing, as Polanyi sees it, requires commitment and active participation by the knower; and rests on the belief that there is discoverable order beyond what is immediately evident.[9] The tacit assent of the discoverer himself cannot be disregarded in epistemology. If it is, this amounts to the assumption that scientific truth is absolute in that it is independent of man. But this makes human judgement and emotions superficial, a stance whose social consequences are likely to be destructive. Similar thoughts must have occurred to George Sarton when he suggested in his utopian Culver lectures on "the new humanism" that "The New Humanism will not exclude science: it will include it, and so to say it will be built around it. . . . It is the source of our intellectual strength and health, but not the only source. However essential, it is utterly insufficient. We cannot live by truth alone."[10] But the search for truth still goes on, with no less thrust than the labor for bread, possibly because, as Spinoza put it, "Only love towards things eternal and infinite feeds the mind wholly with joy, and is unmingled with any sadness."

Theories of Knowledge as Biology, Psychology, or Sociology

What is an almost apologetic aside in Russell's comments on the evolutionary origins of our inferential habits has become the substance of a number of current scientific inquiries. I shall try to pull some of them together as a background to a theory of knowledge within the theory of time as conflict. Specifically, I wish to consider the epistemic message of Piaget's vitalistic reductionism and one example of epistemology derived from psycholinguistics.

A fair, though not a complete, description of Piaget's psychology would say that it concentrates on the gradual development in the child of the formation of such concepts as motion, substance, number, time, and space.[11] Of interest to us is that he demonstrated, among other features, how the child learns of visible and nonvisible connectedness, leading to his capacity to recognize identities in objects and in the self. Piaget takes the position that at any stage of development abilities can be fully understood only by reference to their growth out of behavior at earlier stages. For instance, perception and conception of velocity, according to Piaget, precedes the perception and conception of time.[12] This fact of evolution, then, seems to invalidate all idealistic epistemologies on empirical grounds.

In the Piagetian system the child, even an infant, is continuously constructing and testing theories of the world; at each level his knowledge depends on and is limited by his sensory-motor endowment. In our own earlier terms, the growing child constructs and tests cosmologies which, because of his maturing psychobiological organization, form a hierarchy of Umwelts. On each level we can observe a logic characteristic of that level, but only in retrospect. At each stage the mind believes itself to be apprehending an external world independent of the thinking subject.[13] The communal equivalent of this feature is that the "common sense" of an epoch often appears to later epochs as a distorted, partial, and prejudiced understanding.

According to the witness of his autobiography, Piaget's interest in the development of cognitive functions in the child came from his desire to combine biology and epistemology; he suspected that by considering development as a kind of mental embryogenesis, a biological theory of knowledge could be constructed. Not surprisingly, this turned out to be one of vitalistic monism. It sees the cognitive function as a specialized organ which regulates the interaction of the organism with the environment.[14] Since the knower and the known are inextricably intermeshed, so he reasons, one ought to understand knowledge by its own construction; since knowledge is truth and truth is the organization of the real (again, following his words) we must ask how this organization is organized, and that, as he sees it, is a biological question. Moreover, since society is also a product of life, questions about society are also biological ones.[15] Cognitive regulations are seen as having retained the "deficiencies" of organisms, such as their limited range of reactions to the environment. Hence, in the "superior states of evolution" there is a "final breakup" that dissociates instinct into logico-mathematical structures and experiential knowledge. But the building up of intelligence (by which he means all "coordination" of a "structure") is so complete in its "convergent reconstructions with further evolving" that "hardly one theoretician of logico-mathematical knowledge has thought to search for an explanation in the indispensible framework of biological organization." [16]

The evaluation of Piaget's epistemology is hampered by the poverty of his exposition, which makes his readers scurry to exegeses. Without reference to secondary sources, we can still note the following. That life is indispensable for human knowledge, means only that it is necessary but not that it is sufficient; insistence on the exclusive sufficiency of life for a theory of knowledge leads to certain difficulties. Thus, one cannot account for the curious quality of the mind which makes it an expression of life as well as an existent in its own right. Furthermore, since vitalistic monism must (properly) insist on the "vital deficiencies" of the mind, it prevents us from recognizing how mind offers a partial resolution of the unresolvable conflict between growth and decay of the body. The postulate that instinct breaks up into logico-mathematical structures *and* experiential knowledge removes logico-mathematical structures from experience and endows them with an idealistic aura, defeating the purpose of a developmental epistemology. It is only one step, but a crucial one, to see that what we regard as logico-mathematical structures are examples of the time-binding functions of the mind. These structures distinguish the permanent aspects of the Umwelt not from experience but from the contingent portion of experience.

However, Piaget does point correctly to one difficulty in understanding knowledge as a mental function: if our mental images are copies of the external world, then we have nothing against which such copies may be checked for validity except other copies. Yet what happens is evident: the mental images are checked against other mental images distributed in time; it is temporality which breaks the otherwise circular problem of biologically determined response. The inner landscape of the mind need not and cannot measure up to some abstract perfection: all that it need do is predict future events correctly (in aid of maintaining life, etc.).

In principle, any theory of man, if sufficiently complete, should either offer an epistemology or be suitable for the construction of one. It is not surprising, therefore, that a theory of knowledge may be constructed on the Freudian understanding of man. Piaget's epistemology is Apollonian in that it is written in the framework of reasonableness in man's life. Accordingly, in methodology it tends toward the timeless security of exact science. In contrast Freudian epistemology is Dionysian in that it emphasizes the emotional and the instinctual in man. The two approaches are themselves demonstrations of two different personalities of knowledge. Piaget is aloof and disorganized with a French accent; Freud is all emotional with German precision.[17]

As far as I know, Freud himself never constructed an epistemology on the basis of his theories, but there have been attempts by others to do so.[18] I shall limit myself to the fascinating but little-known work of Thass-Thienemann,[19] which is built on psychoanalytic insight into the psychodynamics of language, and on the tenet that language is continuously shaped and transformed by repressive anxieties. We would like to find out what he says about that portion of language which deals with the regions between existence and nonexistence, because it is there that we might find topics of transcendence which pierce the boundaries between time and the timeless.* This, then, might assist us in understanding how individual and communal perception separates the timeless (truth) from the temporal (contingency). The foremost items in that region pertain to birth, sexual union, and death. Original references to these, according to the postulates of Thass-Thienemann, tend to be purged from the language or at least substantially modified in ways that repress anxiety-provoking meanings in favor of meanings with intellectual content. We can immediately see the time-relatedness here. Birth and death, while most intimate, also suggest inquiries into the beginning and end of time. We have discussed this earlier. We have also seen that heterosexuality is a biological method of overcoming death by aging and thus making life eternal, or at least longer lasting. The following conclusions are derived from the examination of over twenty languages in their handling of the emotionally loaded issues of birth, sexual union, and death.[20]

Our languages give evidence of three basic and different ways of

---

* Time and the "timeless" are used here in their ordinary, nonhierarchical meanings.

grasping knowledge: through the mouth, through the genitals and through the eyes; accordingly, he distingiushes oral, genital, and ocular knowledge, each with well-delineable qualities. The mouth of the infant is the primary and oldest organ of knowledge. According to the rules of subconscious language, since sense perception of taste, smell, and touch become charged with emotions, the meanings of words in this category drift from the sensual toward the spiritual.[21] As an example, such changes can be traced from "taste" (by mouth) to "taste" concerning things and behavior as ethically or aesthetically good or bad. In this epistemology, knowledge whose linguistic symbolism is oral is likely to be rudimentary and primitive.

Genital knowledge is of a different texture. Instead of being totally self-directed it involves the most intimate knowledge of another person; its paradigm is the "carnal knowledge" of woman. The impulse that appeared on the oral level as introjection in the mouth, becomes differentiated. In the male it turns into aggressive penetration of the unknown, whether the unknown is a virgin woman, virgin forest, the structure of the atom or that of economics. In the female it assumes a new form of reception: the desire to internalize the male. The history of words that describe the sexual act shows a strong tendency to develop into concepts of intellectual knowledge; the changes of meanings from sexual to intellectual reveals that yearning for carnal knowledge is the generative agent of intellectual knowing . It further suggests a subconscious awareness of this process. The genital act has been perceived primarily as a cognitive act since prehistoric times: "Adam knew his wife; and she conceived" (Gen. 4:1), and the verbal forms of many languages understand the sexual union as an epistemological function. Vice versa, the perception of knowledge as a genital act is an old and genuine part of our ancestral heritage; "grasping" and "conceiving" are necessary for the production of new people as well as new ideas.

The distinguished experimental psychologist, John Cohen, has drawn attention to sexual experience as displaying a continuum between work and play.[22] When the meaning of work (which in terms of this chapter is the gathering and/or creating of knowledge) becomes particularly intense it tends to acquire sexual significance and is so represented in symbolic forms. From among European writers, Nietzsche, Robert Southey, Galsworthy, Milton, and Madame de Staël have described their creative experience in terms which have a markedly erotic flavor. W. B. Yeats put it rather straightforwardly.

> Bald heads forgetful of their sins . . .
> Edit and annotate the lines
> That young men, tossing on their beds,
> Rhymed out in love's despair
> To flatter beauty's ignorant ear.
>
> "The Scholars"

Creative writing as a way of seeking truth certainly ranks among the three or four most intense occupations. It is an endless striving to imitate and maintain those states of mind and emotion which the writer learned to know in moments of full embrace. It is a continuous search for something hidden in the object of knowledge, for a reality which is more than what is at hand, something which may enter conscious experience at any moment or be revealed by being entered into. Since the desire to know thus precedes whatever is discovered, the yearning to create precedes whatever is created, the product of creative thought, such as this book, belongs not to its author alone but to those who in rare moments have reached with him for the timeless horizon.

Whereas oral and genital knowledge are Dionysian, the third major mode of acquiring knowledge, that by vision, is Apollonian. Light has been perceived in recorded history as a symbol and source of reason; darkness has stood for fear and ignorance. "Enlightened" work is not by trial and error ("touch and go") but by "foresight." Unlike genital knowledge, ocular knowledge has no external partner detectable by the senses but only an internalized one, an "inner eye" that belongs to the self.

In terms of our earlier arguments, there is an interesting confluence here among the oral, genital, and ocular modes of knowledge. Phylogenetically, oral knowledge corresponds to an Umwelt which the organism can explore by touch; its capacity to learn about and control the future is limited to the content of space within its tactile reach. If plants and primitive, deathless organisms could write epistemologies, their knowledge would be of the oral type. We have reasoned that as life began to extend and expand in the evolutionary process, heterosexuality, senescence, and the death of the soma evolved together. It should thus not be surprising that heterosexuality, among its many other functions, also determines a distinct way of knowing. What began as a biological method to secure the continuation of life, produced a type of appreciation of the world unknown and unavailable to organisms which do not reproduce heterosexually. The search for truth hidden in the flesh, while retaining its original purpose and character, also came to serve as a paradigm of search for truth hidden in nature, in general. The Book of Genesis tells us about the multiple effects of the apple that exchanged hands in the Garden of Eden. The importance of this symbolic action, the acquisition of heterosexuality, carnal knowledge, nootemporality, and genital epistemology probably account for the lasting qualities of this story. In the words of an unknown chronicler of the fourteenth century:

> And all was for an appil,
> An appil that he tok,
> As clerkés finden
> Written in their book.
> Deo gracias! Deo gracias!

Ocular knowledge, which is the third way of knowing, is the paradigm of knowledge by reason, such as truth by science. Judged by its success in controlling human destiny, its significance is great but so are its dangers.

In our model of the mind, the inner eye which replaces the carnal partner is that of the observer watching the agent. Having no external partner to oppose its excesses it is capable of detaching itself from life and becoming an uncontrollable tyrant in a world of images.

The necessity of referring a theory of knowledge to evolution, that is, to something organic, unpredictable, and changing would have seemed to Plato to be ungodly and preposterous. For contemporary thought it is a self-evident necessity. But evolution is a communal enterprise. Therefore, a theory of knowledge cannot be complete without reference to those communal features which suggested as well as permitted its formulation. At the turn of the sixteenth century, Francis Bacon had already recognized the connection between knowledge and society when he identified the four idols which, in his judgement, determine the thought of man: those of the Tribe, the Den, the Market Place and the Theater. During the first half of the twentieth century Karl Mannheim drew attention to the effects of social control on the forms and substance of knowledge. He noted that theoretical attitudes (even those of science)

> are by no means merely of an individual nature, i.e., they do not have their origin in the first place in the individual's becoming aware of his interests in the course of his thinking. Rather, they arise out of the collective purpose of a group which underlie the thought of the individual, and in the prescribed outlook in which he merely participates.[23]

Mannheim felt the need for a science—the sociology of knowledge—to study the working criteria which determine the interrelations between thought and action. His philosophy seems to derive from a criticism of the individual's attitude toward his own thought: "Individual thinkers, and still more the dominant outlook of a given epoch, far from according privacy to thought, conceive of thought as something subordinate to other, more comprehensive factors."[24] As Mannheim saw it, human thought reorganizes itself around such more comprehensive factors in every epoch. It is these changing focuses of interest which demand that epistemologies take into account the social determinants of knowledge.[25]

After Mannheim the sociology of knowledge became a lively field searching for first principles.[26] Some such principles have been implied by the French social anthropologist, C. Lévi-Strauss. He advocates a contemporary form of Vico's "New Science." What for Vico were the laws of inscrutable Providence are, for Levi-Strauss, the timeless structures that underlie the reality of man's world. Unlike Piaget, whose epistemological center is the study of life, Lévi-Strauss writes, "I regard anthropology the principle of all research."[27] He uses language as his model of man's essential gift as a patternmaker. Art, music, ritual, myth, religion, cooking, literature, tatooing, the barter of goods, and the kinship system are arbitrary forms but they are all means of communication whose roots are in the permanent nature of the mind. While neolithic man worked out his theory of the sensible order, modern man works on his theories of the physical order.

The final products of both are the unchanging structures rooted in man. The entire process of human knowledge thus assumes the character of a closed system. As Lévi-Strauss sees it, the scientific spirit of contemporary society only legitimizes the principles and the structure of savage thought and reestablishes them in their rightful place.[28] It is not difficult to recognize in the structure of the structuralists, in the syntactic rules of language sought by Chomsky, or in the sociology of science, the Parmenidean search for the timeless.

Knowledge, Truth, and Time

In the beginning of this chapter I suggested that truth be regarded as that class of knowledge which individual and communal judgement regard as timeless. It will be recalled that the ordinary use of the idea of timelessness subsumes many levels of temporality which are not atemporal but belong only to levels of time beneath the noetic. Thus, with the several loose and ill-defined concepts of "timeless," or "permanent" we might expect to find truth appropriate to the biotemporal, eotemporal, proto- or atemporal world, with a hierarchy of rules of validation and evidence. I think this is, and has been, the case, for truths have always comprised many degrees and forms of beliefs and opinions.

It is clear that knowledge is not a matter of exploring a fixed and given world, but rather a process of growth involving the knower and the known. A world with planetary orbits and organic evolution was unthinkable until Copernicus and Darwin gave an account of them. Today the world is unthinkable without them and will remain so until a genius, following the alchemy of creative thought, proposes some more convincing alternatives. At that time the truths of planetary motions and those of life will again look different. I do not think that any specific discipline alone, be it philosophy, biology, psychology, or sociology, is capable of leading to a truly satisfactory epistemology. But they do appear to be convergent in the larger context of organic evolution. In this enlarged framework truth may be seen as that knowledge (or, if you wish, those aspects of behavior) which the organism may successfully employ in the control of its destiny. Accordingly, questions of knowledge and truth become functions of the complex implications of many temporalities. This view is to be distinguished from the pragmatism of Peirce and James. For, whereas in pragmatism testing for pragmatic consequences is conceived of as a method of philosophy, in the relativity of truth to temporal levels the search for "timelessness" is seen as a working method of natural selection.

In man, truth as timeless knowledge ought to be understood as one of the means of selecting from among the practically unlimited autogenic images, those which have high probabilities of corresponding to future conditions. I have stressed that language, the foremost representative of the symbols of the inner landscape, changes under psychobiological pressures.

It would seem that ways of knowledge, being themselves aggregates of symbols, would also change according to the pleasures and dissatisfactions of the psyche and of society. But if, on the one hand, knowledge depends on what the knower regards as timeless, while views of time are functions of individual and communal personalities, then the symbolic structures known as fields of knowledge ought themselves to reflect certain preferred attitudes to time—according to individual and collective personalities. This thought was already put forth as speculation in the beginning of the chapter.

2. Personality and Attitudes to Time

THROUGHOUT the history of psychology those interested in the systematic study of man have been writing theories on personality. This expresses a continued and strong desire for the unification of our knowledge of man under a single rubric.

Typology based on personality goes back to Galen (circa 130–200), who recognized four humors: blood, phlegm, choler (yellow bile), and melancholy (black bile). From these he saw derived four basic temperaments: sanguine, phlegmatic, choleric, and melancholic. This ancient classification already implied differences in attitudes to time. Sanguine temperament is characterized by hopefulness, which is faith in a better future, in exclusion of the evils of past and present. A phlegmatic person is apathetic, his interest in the passing of time is very small. Choleric and melancholic temperaments imply an inability or unwillingness to place things in their temporal perspectives of future and past; also a feeling that the present is the most important category of time. But, whereas this evaluation of events by a choleric person produces a desire to overpower the present (by a quick temper), in a melancholy person it manifests itself in apathy and in being overpowered by the present. The classification of personalities into humoral types has been superseded during the last centuries, at first by typologies based on physical features, then by the Jungian system of psychological types, and finally by the analytical approach of traits as psychophysical dispositions with likely but not necessary attachments to personality types.

There exist many definitions of personality. Gordon Allport, distinguished pioneer of the scientific study of personality, is said to have listed over fifty meanings—and is likely to have missed a few. H. G. McCurdy, at the end of an almost six hundred page survey of the psychology of personality, concludes that personality ought to be understood as "the whole universe from the point of view of one individual" and that nothing less would really do.[29] L. W. Doob defines personality as "the actual, unique organization of traits (characteristic way of behaving), cultural beliefs, and attitudes within the individual; included here is intelligence or skill as well as general or specific temporal perspective." [30] His review of personality and attitudes to time is one of the two most comprehensive works available; the other one is in J. E. Orme's, *Time, Experience and Behavior*.[31] I shall follow Doob because, unlike Orme, he does strike out, though very guardedly, beyond the personality versus time perception to the domain of personality versus the sense of time.

Doob starts by noting that a person's "temporal potential" (all those factors that influence his temporal behavior) and his "temporal orientation" (transient or permanent preferences for future, past, or present) are fundamentally affected by his personality traits.[32] The abundance of papers in this category show that the experimenters share the common impression that a person's view of time relates to his personality. The reports are, unfortunately, incommensurable because they lack common assumptions, common terminology, compatible experimental conditions, and comparable subjects.[33] Even more importantly, however, up to the last few years experimenters have concentrated overwhelmingly on time perception as a variable, rather than on the sense of time, a practice which might be attributed to the distaste of the scientific way of thought for phenomena which are difficult to quantify. But, sometimes it is more advisable to consort with dreamers who see castles than with masons who know how to build walls. Accordingly, I shall pick out what I regard as early representative examples of major trends yet to be supported by experimental evidence, in the study of the relationship between personality and attitudes toward time.

The reading of the original papers concerned with the finding of correlations between personality and attitudes to time reinforces the impression that a correlation exists between personality on the one hand, and on the other hand (1) attitudes to change and permanence, (2) preferences in the interpretation of history, (3) preferences for personal tempo, (4) interpretation of ambiguous scientific data, and (5) need for achievement. "Personality" is taken to reflect both hereditary and environmental factors such as age, sex, education, social pressures, stresses, and states of health.

Foremost among the early papers are the metaphor tests which bring into play the total conscious and unconscious machinery of the mind as expressed in language: they delineate orientation to time by recognizing preferences in the metaphorical identification of time. The time metaphor tests were suggested by the general usefulness of metaphor tests to measure attitudes.[34] Similar metaphorical figures occur in widely separate languages suggesting that the archaic logic of primary thought processes (those of the id) may be common to all men, and that they change very slowly. In contrast to the collective character of the primary processes, secondary thought processes (those of conscious experience) vary greatly with personality. Hence, a study of the imagery which connects primary and secondary processes ought to give some guidance in determining whatever separates the individual from other individuals; that is, it ought to relate to his personality.

R. H. Knapp examined for correlation six metaphor scales dealing with time, love, death, success, conscience, and self-image. He and his associates enlarged the time metaphor sample, breaking them out into three main groups described in different papers as dynamic-hasty or vectorial (a dashing waterfall, a racing locomotive), naturalistic-passive or oceanic (a quiet ocean, drifting clouds), and humanistic, involving humans, human surrogates, or human artifacts (an old man with a staff, a tedious song, a burning candle). Preferences for such groupings as these were then tested directly or indirectly against pencil and paper personality tests on individual traits, such as the achievement motive, vocational interest, aesthetic choice, degrees of introversion and trends in anticipation. The degree of need for achievement (as measured from Thematic Apperception Tests) was found positively correlated with vectorial descriptions of time and negatively with preference for oceanic and humanistic metaphors.[35] In this context Knapp drew attention to the prevalence of vectorial and dynamic imagery of time in Western civilization, the prevalence of the oceanic imagery in Oriental philosophies, and the importance of humanistic interpretations of reality in the Protagorean (Mediterranean) world-view of man as the measure of all things.[36] Persons who scored high on a test of introversion scale were also found to prefer oceanic images of time. Vectorial images of time were found prevailing among scientists who also preferred somber "cold-open" designs to gay "warm-closed" designs from among thirty Scottish tartans.* People who generally followed the principle of parsimony of time and thought (generally associated with the practice of Protestant ethics as understood by Max Weber) tended to underestimate the time taken by short tasks.

These and related studies [37] are diverse and each one, alone, is a mostly unattached bit of interesting information. Even in the records of single investigators the work at large misses direction. The absence of direction is the result of the absence of sufficiently broad and convincing unifying perspectives. One such perspective is suggested by a reconsideration of the Augustinian dilemma: If no one asks me what time is, I know what it is; if I wish to explain it, I know not. This is an "indeterminacy principle" according to which time felt and time understood are, at any one instant, complementary and coexistent qualities which become mutually exclusive at either extreme. As I increase the keenness of my understanding of time, I end up with some imposing intellectual structure of — timelessness, such as, the theory of relativity, or quantum theory. (Our inquiries revealed that their "timelessness" is a mixture of temporalities below the noetic, but that is irrelevant in the present context.) If, by going the other way, I attend

---

* The interesting change in attire among scientists and engineers which has taken place since the Knapp tests in 1960 offers rich material for research in aesthetic judgement versus world-view.

more and more to my feelings about the passing of time, I am likely to end up by hearing unrepeatable voices and experiencing undescribable ecstasies, all of a timeless nature. (Our inquiries again revealed that their "timelessness" is a mixture of temporalities below the noetic.) The complete experience of temporality obtains only in the middle region where the unresolvable conflict in the mind between the phylogenetically older portions of feeling and the newer portions of knowing is fully active. Either extreme tends to decrease the existential tension which determines the nootemporality.

In terms of this gamble of balance, and in a skeletal way, personality then may be thought of (1) as the peculiar manner in which an individual prefers to play, as it were, his archaic logic against his discursive logic, that is, balance the lower Umwelts against his noetic Umwelt; (2) as a preference for the direction taken along the feeling-understanding axis to lessen the existential tension between the two; and (3) as the degree to which a person is willing and able to pursue the retreat from that tension and still retain his actions under ego control.

When called upon to make a decision, preferences for one or the other personality type must prevail (for only one action is possible at one time) even though both are continuously operative in the self. But as personality becomes defined and refined in response to the challenges of the day, one particular weighted mixture of attitudes is likely to become established in the normal course of maturation. Although a person's attitude to time, as expressed in the balancing act just described, may vary from instant to instant, it remains, I believe, a reliable index of his personality, subject ordinarily only to slow change, if any.

3. Personality and Preferred Ways of Knowing

THERE IS a jarring obviousness to the claim that, given more or less free choice, people with certain personalities would select certain occupations. What is also obvious yet seldom stressed is that the various occupations, or, in the intellectual domain, fields of knowledge, were created by people who selected to understand ideas and do things in ways they found compatible with their personalities. Thus, differentiation among personality characteristics must have preceded the differentiation of labor. For, unless there were people who were capable of, and perhaps even enjoyed, making arrows, there would have been no arrowmakers and no arrows.

Some Individual Preferences

We read in Aristotle's *Politics* that Archytas invented a new baby-rattle. To Plato, unmarried and childless, this appeared as a welcome aid in teaching the child rhythm at an early age and, thereby, mathematical readiness. To Aristotle, man of affairs, husband, and forever practical, this was a "good invention which people give to children in order that while occupied with this they may not break any of the furniture; for young things cannot keep still." [38] This twofold evaluation of one baby-rattle shows a division of knowledge into the abstract and the applied. Such a division probably goes back to prehistoric times and was certainly not new with the Greeks. But it is since the age of the Pythagorean Parmenides and the aristocratic Plato that the paradigm of knowledge in the West became not the functioning of living matter, but the rules of geometry and those of the multiplication table. These rules are superb means of emphasizing the lasting in the experience of man, but they are not the only means. Let us consider some others.

In Hebrew as well as in Greek tradition, as witnessed by language, the seat of the recreative, hence time-binding power of man and woman resided in the bone marrow. The Greek *aion* meant "spinal marrow," "eternity," and also "human lifetime." The word relates to the Latin *aevum* (age, always, lifetime) and survives in the English "aeon." Shakespeare also informs us that virility resides in the marrow and lets Parolles, in *All's Well that Ends Well*, berate conscientious objectors:

> He wears his honor in a box, unseen
> That hugs his kicky-wicky here at home,
> Spending his manly marrow in her arms,
> Which should sustain the bound and high curvet
> Of Mars's fiery steed. . . .   (II, 3, 298)

For Yeats:

> He that sings a lasting song
> Thinks in a marrow bone.

In retrospect these ideas may be berated as fantasies because we know that the generative power of man resides not in his bones but in his genes; this is *our* theory. But for those who created the early Hebrew and Greek words, the bone marrow interpretation must have been a theory, not a fantasy. What we see here illustrated is a flow of symbols from their origins as autogenic images, checked against evidence according to changing rules and, if proved, turned into theories. That the substance of fantasies share collective imagery but are strongly dependent on personalities for their significance, is well known. But, if theories accepted at an epoch can appear as fantasies in later epochs, and if the character of fantasies are functions of personalities, then the theories themselves are but reflections of personalities, individual or communal.

Without any doubt the predominant trend of contemporary knowledge has been a quantitative thrust. Yet in the minds of creative scientists, though not in the minds of engineers and production managers, the quantitative feature is only superficial. Newton had willfully and by design manipulated his numerical calculations of the acceleration of gravity, velocity of sound, and the precession of the equinoxes so as to make his theories appear more acceptable to the demands of precision set by the very scheme he created.[39] Newton must have been convinced (as the run-of-the-mill scientist of our days is not) that science and accuracy are not synonymous. He also must have been convinced of the truth of his universal law of gravitation and feared no ridicule if found out, for, I would think, he trusted in the confirmation of his thought in the long run. Gauss measured the sum total of the three angles of a large triangle and found them very slightly over 180° rather than under, as he expected them.[40] Yet he did not give up his "astral geometry" but declared the experiment to have been insufficiently accurate.

Then, whence these convictions as to what is permanent in nature and what is contingent? They could not have come from measurements, for the measurements had either to be fudged or discarded. They had to emerge from personal knowledge which, in the case of Newton, was probably his faith in the existence of permanent structures which defy time as do the laws of God. Yet the Newtonian law of gravity was proved to be a fantasy when Einstein replaced action at a distance by applying variational principles to local regions of a purely geometrical structure known as space-time.

Leading scientific theories in their own times are recognized by those in the profession by an atmosphere of truthfulness. The sudden revelation or "satori" afforded by a new scientific law exudes mental security. From the instant of its discovery, within the purview of that law, the unexpected cannot happen. Psychological interpretation of this truth-feeling seeks its sources in the desire to control the free, hence frightening play of impressions, resembling the work of the painter who fixes the fleeting moment in a structure of color. Fixing nature in scientific laws and in permanent images are methods of separating the permanent from the contingent. Both, in different ways, please their creators.

Highly original men in all fields have a tendency toward femininity, highly original women toward masculinity, implying a hermaphroditic scope of personality which creative people have learned to contain and direct. Creative scientists (as judged by their influence upon their fields of specialization) display, in addition, certain well-recognizable traits. They like disorder, provided it appears resolvable without reference to human affairs; their interpersonal relations are generally of low intensity and they dislike interpersonal controversy; most importantly, however, they tend to be preoccupied with things and ideas rather than with people.[41] The latter is recognized in psychology as a defense against instinctual drives.[42] It is a method of repression which in Western civilization is associated with Christianity in general but became a powerful syndrome only with the rise of Protestantism. The continuous conflict between instinctual drives clamoring for recognition and thing-orientedness was recognized by D. W. McKinnon and his coworkers as evidence of psychopathology which creative scientists learn to control.[43] One study concludes thus: "That a man chooses to become a scientist and succeeds means that he has the temperament and personality as well as the ability and opportunity to do so. The branch of science he chooses and even the specific problems he chooses and the way he works on them are intimately related to what he is and to his deepest needs." [44]

While creative scientists and artists, qua creative humans, have many common personality traits, the deeper strata of their personalities are quite different, as suggested by a study of dream recall. In this study the dreams of "convergers" and "divergers" were examined. Convergers are people who score high on conventional IQ tests but score low on open-ended tests which require imaginativeness and mental fluency; divergers score high on the latter and low on the former. It was found that convergers are more likely to specialize in the physical sciences, divergers in the arts. In reporting their dreams convergers showed significant inhibition about emotions and about the nonrational, in contrast to divergers. This was taken as supporting the experimenters' assumption that convergers prefer the security of analytic reasoning because it protects them from dealing with the id. The inhibitions, in turn, reinforce their capacity for "logical constructions" at the expense of "combinatory play." [45]

British and American schoolboys seem to recognize a difference in personalities between those who specialize in the sciences and in the humanities. Several hundred American and British students were asked to rate typical figures (mathematician, poet, etc.) on a semantic differential scale (relative warmth, dependability, etc.). The findings were remarkably similar: "novelist"-type figures were rated more imaginative and warm versus "physicist"-types who were thought more dependable and hard working.[46] The important point here for our personality-based epistemology is that since "the boy is the father of the man," as the boy selects his occupation to suit his temperament, so the various ways of seeking truth will bear the hallmarks of the man's personality.

Instead of asking schoolboys about typical figures, we can inquire with social scientists about the politics of academic scientists and engineers. From a total sampling of about sixty thousand people, those in the social sciences and in the humanities were found to be politically liberal; those in agriculture, civil engineering, and business, conservative.[47] I seriously suspect that the true direction of causation is not from profession to worldview, but vice versa: world-views and personalities determine scientific views. In nineteenth century Britain, liberals tended to be hereditarians and conservatives talked like environmentalists; by the early twentieth century the roles were reversed. In mid-twentieth century America, in the nature-nurture controversy political conservatives emphasized heredity as the prime determinant of human nature, whereas liberals and radicals emphasized the primacy of the environment in determining human nature.[48] In the current controversy in California on whether or not students should be taught Darwinian evolution, conservatives champion the cause of special creation, progressives the cause of natural selection.[49] I have already stressed that those who used to endorse the steady-state cosmology came to embrace, generally, the oscillating universe model, after the steady-state model became untenable on experimental grounds. Clearly, in the long run, an oscillating universe is as timeless as a steady-state one, and the elimination of time is an essential feature of certain world-views.

Almost a century ago James Clerk Maxwell wrote that there are some minds which prefer to contemplate pure quantities presented as symbols; others prefer geometrical shapes built in empty space. Again others calculate the forces upon the heavenly bodies, and to them momentum, energy, mass, "are words of power which stir their souls like memories of childhood." Therefore, says Maxwell, for the sake of these different personalities,

> scientific truth should be presented in different forms and should be regarded as equally scientific whether it appears in the robust form and vivid coloring of physical illustrations, or in the tenuity and paleness of a symbolic expression.[50]

In Maxwell's time, science was still natural philosophy, hence a depository of practical and philosophical truths. A century later the polarization between knowledge felt and knowledge understood became too deep, and the personality of the natural philosopher broke into two new personalities: the scientist and the humanist. It is surely not by chance that G. J. Whitrow's classic treatise is called, *The Natural Philosophy of Time*, and not "the science of time." It is, as I see it, a call for a new natural philosophy.

Reviewing this subsection we note that, as we have progressed, our considerations about personalities and ways of knowing have come to imply more heavily the communal origins of preferences. This is not surprising. The concept of personality can only be understood as a comparative quality among several people. In a world which contained only one, single human being, or in a world of indistinguishable people, personality could have no meaning. Thus, we ought to turn now to the social origins of the "personalities" of ways of knowing.

Collective Perceptions of Science as Truth

(1) Perspectives

The relationship between individual dispositions and social structures has been of interest to anthropology, psychology, and sociology, and there are a number of theoretical formulations linking personality and culture.[51] While keeping in mind the intricacies of this relationship and the difficulties of obtaining quantifiable data, we again begin by seeking a hierarchical structure.

The evolution of multicellular organisms made it possible for a new, more complex organism to secure a more favorable future for its corporate self than was possible for its component cells acting alone. Under the pressure of corporate demands the cells relinquished their original autonomy and became identifiable only as parts of a larger whole. The mind, as an expression of the strategy of existence, overcoming some of the intrinsic limitations of life, also shows a trend of bunching together into larger corporate entities for improved collective control of its collective future. Under the pressure of the communal demands, in the course of social evolution * the individuals learned to give up some of their autonomy in favor of a presumably securer future. Social theorists have argued this matter from well before to well after Jean Jacques Rousseau. With a continual flow of refinements, modifications, corrections, ifs and buts, this theory of freedom exchanged for security remains the basis of all theories that attempt to account for the emergence of social organization in the singular history of the family of man.

In a multicellular organism, the specialized tasks are performed by identifiable organs; within the organs by organelles, etc. In each case the function of the organ (or organelle) can be understood only within the framework of interconnected functions of the higher unit. In society the tasks are performed through the division of labor, whose individual features make sense only in terms of the purposes of the society. These individual features are created under the guidance of socially gathered knowledge. The individual's contribution to that corpus of knowledge is, in its turn, guided by a consensus, or communal perception of what is true, that is, unchanging in time. But just as in organic evolution, selection pressures can assist differentiation only because there is a spread of capacities among the units from which the selection is made. There is indeed a natural—and very wide—spread of capacities among individuals, upon whom the pressures of social selection may work. As C. D. Darlington put it in the conclusion of *The Evolution of Man and Society*, in selecting an occupation,

> the man of destiny follows his own judgement. He seeks out and creates his own environment and even his own posterity. At the other extreme, the weak man is compelled to heed the opinion of others and [do whatever] work they will give him. But between these two extremes lie most of the interactions of individuals with society and with the world as a whole in which they find themselves.[52]

Perhaps the greatest of all advantages human society has over animal groups is precisely the enormous variety of individual predispositions, especially those of the "man of destiny," from which society may select.

Cultural and social determinants enter into scientific delineations of truth not only through the obvious pressures for social acceptability of subject matters, and alas, even of results, but through such subtle means as the preferred conceptual statement of a problem. It has been con-

---

* There is no implication intended of any easily depictable pattern of social "growth."

jectured, for instance, that the Darwinian theory of natural selection was modeled after the competitive economic notions of Darwin's epoch and that Freudian theory reflects the patriarchal bourgeois attitudes of his society.[53] National and class differences in types of thoughts have also been considered, and Bertrand Russell once remarked, half seriously, that even animals employed in psychological research display the national characteristics of their observers.[54] Considering the amount of heterogeneous influence on each and every formulation of scientific truth it is curious that academic epistemology, with a few exceptions, has remained so detached from psychology and sociology.

(2) Divergences, East and West

We have seen earlier that Oriental religions and philosophies tend to stress the permanent and the passive. Passivity, which in its most advanced form is the immobility of the trance, is often paired with an organic rather than an analytical view of the world. In contrast, the Faustian man strives, directs his labor, and, in the Platonic tradition, prefers to immobilize not his body but the image of the world around him. Whereas Oriental philosophies are noncausal and organic, scientific world-views are those of the "frozen passage" of time: laws of nature and causality. This is the more interesting because there is a discrepancy between our thoughts and our experiences regarding the determination of events. While the method of exact science takes single causation for granted, in experience it is scarcely the case that an event is determined by one single cause. On the contrary, as Freud observed

> each event seems to be overdetermined and proves to be the effect of several convergent causes. Frightened by the immense complication of events, our investigations take the side of one correlation against another and set up contradictions which do not exist but only arise owing to a rupture of more comprehensive relations.[55]

Certain personality traits are consistently associated with what I have described above, seriously but not very rigorously, as Oriental and Western. I would like to illustrate the differences by two examples; one about acceptable truths in explanations of natural phenomena, the other about acceptable truths in explaining human behavior.

A recent report on science teaching in India revealed a deep-rooted pattern of multivalued truth among fifteen-year-old children in Chainput, India. What to Westerners appear as incompatible propositions, appeared as mutually acceptable verities to these children. They admitted the simultaneous validity of pairs of statements such as, "The deities break vessels of water in the sky, causing rain," and, "the sun evaporates water from the sea, producing water vapor which is cooled by the mountains to make clouds and rain."[56] Perhaps this wide latitude contributes to that state of the "Oriental mind" which was described by Lord Curzon as follows. "In character, a general indifference to truth and respect for the successful wile . . . a statuesque and inexhaustible patience, which attaches no value to time, and wages unappeased warfare against hurry." [57] Lord Curzon's point was, of course, that lying in the West was regarded as a moral fault. (This distilled impression of a statesman is dated. He could have hardly anticipated the rapprochement between the Oriental "respect for successful wile" and Western political, mercantile, and legal propaganda).

A sympathetic hearing is given by Joseph Needham to the question of why modern science originated in Europe and not in the Orient, especially in China, even though Chinese mathematical preparedness and mastery of nature chronologically preceded those of the West. The Chinese society, he writes, was constant, with a degree of homeostasis which resulted in a built-in stability while Europe had a built-in quality of instability.[58] In Chinese symbolism the emperor represented the Pole Star. He faced southward sitting on his throne, ruling successfully while theoretically doing nothing.[59] The European ideal of an emperor, had it ever been formulated, would no doubt have called for activity. Needham, following a neo-Marxist interpretation of history, prefers to assign the Faustian restlessness of the West to such socioeconomic factors as the struggle of independent city-states, the infighting of the aristocrats, and linguistic differences. He sees modern science as having evolved because of the demand of merchants for exact knowledge for the purposes of expanding trade [60] and expresses hope that the differential development of science in China and the West will eventually be understood through differences in social and economic patterns.[61] I do not believe that these elements alone could possibly be sufficient to account for the scientific revolution. The reasons must be much deeper. Curiously, as is often the case with seminal thinkers such as Needham, we can take our point of departure from some other work he has done, but one which, as far as I know, he has not yet carried sufficiently far.

In one of his superbly interesting papers he explores the relationship between human laws, as developed in the West and in China, and the laws of nature. He finds three main reasons why the concept of inviolable laws of nature did not evolve in China. They are the distaste for abstract, codified law; the identification of human law with good customs, mores, or instructions, hence their unintelligibility apart from humans; and the absence of creativity in the ideas of supreme beings.[62] These precepts reflect a state of mind quite different from the one in Basel in 1474, that permitted the burning alive of a cock for the "heinous and unnatural crime" of laying an egg. (We can add here many examples from our own epoch: attitudes against birth control, against homosexuality, and against certain sexual practices are still condemned on the basis of being unnatural.) Needham also shows that between the time of Galen, Ulpian, and the Theodosian Constitution on the one hand, and that of Kepler and Boyle on the other, "the conceptions of natural laws common to all man, and of a body of laws of nature common to all non-human things, had been completely differentiated." [63] Perhaps, concludes Needham, the state of mind that could prosecute a cock at law was necessary for the production of a Kepler.

(3) Psychological predispositions in the Christian West

This may indeed be the case, but the issue comes into better focus when it is restated differently, with emphasis on the confluence of certain psychological and cultural factors. As I see it, it was necessary (1) to produce a collective personality which preferred the use of repression as a mode of dealing with instincts and affects; (2) to reveal to this personality an acceptable link between himself and the timelessness of mathematics; and, finally, (3) to suggest to him the importance of the future in the history of man. I shall now consider these separate items together, for they amount to a single syndrome.

This syndrome might have its roots in the reevaluation of sexual mores following the teachings of Christ and the ministry of St. Paul. As subsequently developed, these attitudes emphasize the patristic and conjugal family, the unity of the father and the son, but with little room for the mother. Practices contemporary with Christ tended to venerate the power of sex in the male while condemning it as evil in the female. Christ seems to have been close to the Father but very distant from the Mother; he repudiated sex and made the sexless child the inheritor of the Kingdom of Heaven. Although he himself displayed much understanding for the (past) lust of penitent women, the main thrust of Roman Christianity degraded the feminine in life by upgrading the virginity of Mary. This is in clear contrast to the Chinese (and, generally, Oriental) organic view of sex, though not in contrast to the generally low social station of woman in the Orient.

We have seen that repression of instinctual drives favors a shift from genital knowledge, which explores people, to ocular knowledge, which explores objects and space. This whole, vast sweep of change in collectively approved personality reached its most repressed stage in the long-range effects of the Reformation. By the turn of the sixteenth century the daily routine of the Christian in general, and of the Protestant in particular, emphasized guilt and fostered anxiety by insisting on repression as a way of dealing with instinctual drives. Then, it encouraged sublimation through a world-view which interpreted the environment through the same type of controls it applied to the self: parsimony of expression. Since mathematics is par excellence, the representative of parsimonious thought, it suggested itself as the ideal domain in which the strict discipline of God's kingdom might be proved. That the application of mathematical laws turned out to be useful in the many activities which brought the merchant more power, money, and goods could only reinforce the faith of the believer. As for Plato, so for the Western mercantile economy, God was a mathematician, but for very different reasons. The systematic testing of mathematical

hypotheses against natural phenomena amounted to a continuous recon-
firmation of the timeless power and glory of the Christian God. It is not
surprising that the anti-Weberian social trend embodied in the youth and
romantic revolutions in America are, eo ipso, antiscientific. These impressions dovetail, perhaps even too neatly, with the three negations listed by
Needham and quoted above, as characteristic of the Chinese attitude to
law.[64]

(4) Theoretical predispositions

Beyond the psychological predispositions there are further elements in
Christianity which, I think, were instrumental in bringing about the scientific revolution: they may be described as theoretical predispositions. One is
the Christian view of man as separate from (rather than a part of) nature.
But for the rare phenomenon of St. Francis, humility towards nature is alien
to Western Christianity except in so far as nature is the work of God. Man
stands separate, as the image of God, subsumed under God's laws but not
under the natural order. Thus the laws of nature binding on all nonhumans
may be comfortably explored with profit but without threat to man. This
remained a tacit stance essentially unchallenged until the appearance of
positivism and behaviorism.

Another element of theoretical predisposition, which assisted the
emergence of the scientific revolution, is the central importance of the
dogma of resurrection in Christianity. Consider Kepler's profound conviction about the timeless mathematics of God's laws, "The Christian
knows," he wrote, "that the mathematical principles according to which the
corporeal world was to be created are coeternal with God." [65] How can one
reconcile the admiration for the timeless foundations of reality with the
stark emphasis on the importance of passing time? In many Oriental
philosophies timelessness is both a theoretical stance as well as a way of
life. In the Christian West the two are linked in the resurrection of Christ,
an event which connects the temporal with the timeless through the act of
suffering, even though the faithful did not necessarily put it this way.
Surely and admittedly, resurrection was a rare occurrence; it could happen
only to Christ, but it was sincerely promised for all. It was real enough
for the faithful who *knew* that there was a timeless world beyond the temporal one, and who certainly knew the reality of suffering. Everywhere in
Christendom, repeated day after day, alone and severally, morning and
evening, he heard and he said: "Credo in . . . remissionem peccatorum,
carnis resurrectionem, vitam aeternam." Guilt, time, suffering, and the timeless were thus neatly joined through the potentiality of resurrection for all.
To modern ears the dogma of resurrection is a curious story of wish-
fulfillment and, as such, it is easily discarded. But can we expect to make
sense of the emergence of early science without regard to the professed
beliefs of those who wrote that science? Certainly not. Rather, in those
beliefs we must seek the inspiration for, the approval of, and in due course

even the sources of disenchantment with science as the faith of truth. Although resurrection stories appear in many other religions, its power as a central dogma is unique in Christianity, as is the belief that man is not an integral part of the organic world.

Yet another source of theoretical predisposition which assisted the rise of mathematized science might have to do with the Judeo-Christian evaluation of history as Heilsgeschichte. I have already elaborated on the future-directedness of this understanding of man's fate. The promise of a better future for the oppressed Hebrews was at the origins of the belief that salvation history is a true report about the world. Likewise, the promise of a controllable future is at the source of the belief that the scientific description of the world is a true report. Thus the credentials of science derive from its capacity to accomplish certain things which mere belief in Yahweh could not: it can be the means of controlling the destiny of those whose collective perception of truth resides in the mathematized image of nature.[66] From an evolutionary perspective this is no more than an extension of behavior already known, namely, the power in all living to apply selection pressure upon the environment. Psychologically, emphasis on the future directs the energies of the libido toward distant goals and thereby assists in the sublimation of instinctual drives toward increased possession of things and increased control of time budgets.[67] In a homeostatic, self-contained and unchanging world, such as that of imperial China, salvation history was not called for, neither could there have been any deep-seated need for seeing in mathematical formulas the security of the future; that security was already vested in the emperor.

Those pioneers of the scientific revolution who employed religious truths as paradigms in the constructions of their science needed not to be aware of this connection. K. V. Thomas, using the methods of social anthropology, has inquired about the status of religion, magic, witchcraft, and astrology in England during the sixteenth and seventeenth centuries

••

Everywhere in Christendom, repeated day after day, alone and severally, morning and evening, the faithful heard and said: "Credo in . . . remissionem peccatorum, carnis resurrectionem, vitam aeternam." Guilt, time, suffering, and the timeless were thus neatly joined in the potentiality of resurrection for all. Can we expect to make sense of the emergence of mathematized science without regard to the professed beliefs of those who wrote that science? Certainly not. On the contrary, it is precisely in those beliefs that the origins of the paradigm of science should be sought.

After Pieter Brueghel (circa 1525–1569), engraving, probably by H. Cock. Wooded Village, P. I. Pagus Nemorosus, B.16. Courtesy, The Metropolitan Museum of Art, Harris Brisbane Dick Fund, 1926.

••

PAGVS   NEMOROSVS

and studied their relations to the rise of early modern science.[68] He documented the interaction between established religion, the tradition of magic, and the genesis of modern science and showed the strong influence of the then prevailing world views on the works of Gilbert, Boyle, Newton, and others. He leaves little doubt that while on the one hand, mechanical philosophy destroyed the popular credibility of magic, the religious practices and spiritual beliefs of the epoch served as paradigms in constructing that very philosophy.

(5) The scientific method

The Zeitgeist of an epoch, not unlike the personality of an individual or a group, can be delineated only with reference to other, different Zeitgeists—and this means the past. Thus, a prevailing philosophy of life, such as that of the Christianity of the Renaissance or that of the Reformation, is apt to be taken for granted in its own time. Knowing itself surely to be different from past world-views but believing itself as the final truth for all time to come, its tenets permeate the daily routine of its followers and guide their actions both consciously and unconsciously. This type of guidance or extrascientific suggestion is a necessary assumption beneath the paradigmatic model of the evolution of science as proposed by T. S. Kuhn. Science, he points out, grows by jumping from stable paradigm to stable paradigm, each providing model problems and solutions to a community of practitioners.[69] He identifies scientific revolutions with noncumulative developmental episodes in which older paradigms are replaced in whole or in part by incompatible new ones[70] and stresses that the choice of the new paradigms can never be settled by logic and experiment alone.[71]

In the history of science the series of paradigms, whose emergence Kuhn demonstrates and whose reality he defends, forms a creative process. We can easily recognize the hallmarks of this process; the stages are intrinsically unpredictable, they may be connected only by qualitative and not by quantitative judgements, yet they appear self-evident in retrospect. Some light may be thrown on the way such scientific paradigms come about if we recognize the parallelism as well as the differences between the ways that individual and communal creativities function. We recall the role of autogenic imagery created by the very complex as the repertory of conceivable conditions, both possibilities and impossibilities. In this chapter we are concentrating on one method which may be used to make practical selections from among these images, namely, that of veridicality. The method of checking truthfulness is known in various fields of learning as verification; it is generally done by comparing predictions with actual later events. It is here that an extended sense of time is necessary if the selection from among the images is to be done by the individual mind rather than by the species, under selection pressure working upon many generations. The scientific search for truth, known as "the scientific method," is an extension of individual learning in two ways: by promoting increasingly universal knowledge about the timeless (invariant) substratum of the world, and by drawing from the repertory of many sources of autogenic images (many minds) interconnected by communication.

The first of these has a curious, almost self-contradictory quality. It is the belief of scientists in general that their enterprise has been progressing toward the establishment of more and more universal laws. Using physics as an example, we can refer to our earlier deliberations about the successively increasing universality of kinematical laws: the Galilean transformation is a description of motion invariant for stationary observers; the Lorentz transformations are invariant for observers stationary or in translation; the general relativistic transformations are invariant for all types of motion: zero speed, translation, or acceleration.[72] Note, however, that whereas the Galilean transformations describe primarily the eotemporal portion of the physical world (for they apply mainly to macroscopic motion), both special and general relativity theory describe properties of the atemporal Umwelt (for they derive from the absolute motion of light). Similarly, quantum theory displays a universality unmatched in the classical physics of matter. But whereas classical physics dealt with the eotemporal world, quantum theory deals with the prototemporal world (for it is probabilistic). The foregoing expansions of regions of applicability may be considered improvements in depicting the timeless aspects of the physical world, until one realizes that their "timelessness" has a temporal structure. Thus, as the universality of physical laws increases, the rank of the Umwelts to which they are properly applicable, decreases. Following the general appeal of science, the life sciences and the sciences of the mind also have been seeking more universal laws. To the extent that they imitate physics in methodology and manner of thought, the increasing universality is bought by relinquishing their attachment to the biotemporal or nootemporal Umwelts. There is now a dire need for separate epistemologies for the life sciences and the sciences of the mind, dealing with the type of laws, degrees of indeterminacies and regions of undeterminacies that characterize the biotemporal and nootemporal Umwelts.

A second way in which science provides a means of extension of individual learning is through the linking of many minds. We remember that the complexity of the brain (which makes possible the creation of autogenic imagery) comprises a system of autonomous elements, the neurons, interconnected by biochemical and physical processes. The temporal program, which is the mind, is carried by appropriate coding of these processes. The complex of the intellectual community of man (which makes possible the creation of collective autogenic imagery) comprises a system of autonomous elements, the individual brains. The methods of interconnections among them are the channels of human communication. The temporal program, which is the communal intellect, is carried along these channels by appropriate codings in form of symbols and signs. The tightness and consistency of the rules and regulations applied to the flow of symbols and signs, expressed mainly in language and ritual, determine the texture of the communal intellect, as well as that of the quality of life of those who form the community. Thus we arrive at an integrative level above that of the mind. This level is commonly known as society.

(6) The integrative power and the limits of the scientific method

There are good reasons for introducing the idea of society as a new integrative level in connection with reflections on epistemology. Historically, the first universal language and ritual which has the potentiality of bringing about a coherent, new, and autonomous existent, i.e., a single society on earth, considering Homo sapiens as a single species, is that of science. All major religions have attempted to do so by word and by sword, as have most major civilizations. So far none has succeeded, and none is likely to succeed, for the very pluralism of ideas, folkways, and mores works against them. Science has some great advantages over religion and state in this enterprise. They stem from the fact that of all known systematic accomplishments of man, science has been the most successful in copying the functional description of the mind upon the external world. Thus, there is a continuity between the mode of operation of the mind and the scientific method. This claim needs some elaboration.

As does the individual mind, the collective intellect also forms within the body social models of those portions of the world which are external to the body social. These models enter into the substance of artistic, religious, philosophical, and scientific symbolism. Each mode of expression continuously verifies its content against reality. The methods of verifications and images of reality do, however, vary tremendously among the arts, the religions, and the philosophies as do their respective views of reality. But the methods of verification and assumptions about reality vary only slightly among the sciences, a feat that has been accomplished, so far, by focusing on lower Umwelts.

As does the individual mind, the collective perceptions of coherent groups also contain symbolic continuities of something not external to the group; to wit, their social identity. This is known to have been traditionally expressed in the mythic visions of the tribe or nation, and I would include here modern historiography as well as the speculations of witch doctors. Just as the individual must define his identity by reference to other individuals, so must nations, states, or ethnic groups. But the family of man has no siblings, or peers, and we know of no other comparable species against which we could delineate our collective identity. If an external focus is to be generated, it must be of a uniformitarian character, such as might conceivably derive from a scientific image of man. But for such an image to be sufficiently universal it would have to be dealt with mostly on the physical level, hence it would be likely to lose precisely those elements of manhood which distinguish man from other living creatures.

As does the perception of the individual mind, so does the collective perception of the world generate images which do, or do not, or cannot, have external reality. As in the individual mind, these come about through the wild combination of the many symbols and signs flowing in the channels that link the component elements, in this case, the individuals. I am thinking of the spectrum of tribal, regional, or national ambitions which can be tested against reality by such activities as the building of cities or the conquering of the world. Individuals have learned to select from among the possibilities their mind suggests by using rules of behavior and pursuing goals which they recognize as true or untrue, beautiful or ugly, good or bad, but, in any case, by qualitative judgement. These judgements are socially conditioned but must also abide by the restraints of life and matter. The individual mind developed its capacities for qualitative judgements during a period of millions of years and under tribal controls. But there does not yet exist a society with a sufficiently large number of individuals so tightly interconnected in a multiplicity of ways as to become an instance of the "very complex," though we may be headed that way. There is but insufficient evolutionary experience to assist in distinguishing the possible, the probable, and the impossible from among communal ambitions, as demonstrated by the frighteningly large swings in the quality of life that accompany the complexification of modern society, and by the degree of useless suffering, as social ideologies are tested by the most primitive means of trial and error.

During the evolution of man the individual could test his mind against reality, limited as he was in the pursuit of his fantasies by his environment and very largely by his peers, and had good chance of surviving such tests; if he did not, others would take over. There is no similar latitude available to national and ethnic groups, and none whatsoever for Homo sapiens as a single communal personality. Since society as such does not possess those qualitative traits which can guide the individual, hence can display no fear, intuition, empathy, sympathy or joy, these nonscientific types of knowledge must be supplied by the individual. As the mind has its freedom but is limited by the laws of life and matter, so society has its freedom but is limited by the laws of the mind. Against the deadly fantasies of a Stalin, man can assure his survival only through the individual courage of a Solzhenitsyn.

Thus, while seeking to understand the ways in which social perceptions of truth function, we find ourself referred back to the individual as the limiting and controlling condition of society. Employing the argument that we already used with respect to lower levels of integration, we may speculate that the (natural) laws of society as an integrative unit must arise from regions left undetermined by the mind. It is very difficult to discern the features of the next higher level which might so arise or even to outline these undetermined regions. It is much easier to see what has not been left undetermined, such as the mind's revolt against attacks upon its integrity. "If man could convince himself by natural science and mathematics that he is a predictable machine," wrote Dostoyevsky in his *Notes from the Under-ground,*

> he would once more, out of sheer ingratitude, attempt the perpetration of something perverse which would enable him to insist upon himself. And if he could not find means, he would contrive destruction and chaos, will contrive sufferings of all sorts, for the sole purpose, as before, of asserting his personality. He will launch a curse upon the world . . . to convince himself that he is not a machine. If you say that all this, too, can be calculated and tabulated—chaos and darkness and curses, so that the mere possibility of calculating it all beforehand would stop it all and reason would reassert itself—man would purposely become a lunatic, in order to become devoid of reason, and therefore, able to insist upon himself.[73]

In rather pedestrian terms, the mind, then, determines and protects its nootemporal Umwelt; one manifestation of this is its resistance to limiting truth to those phenomena appropriate only to the lower Umwelts. This suggests that in spite of our native tendency to the contrary, we must attempt to lift epistemology out of the context which would limit it to the study of the methods and boundaries of exact sciences, back to a theory of knowledge, as it has been conceived through the history of Western philosophy. Increasing interest in the biological basis of knowledge suggests that such a change has already begun. But we ought to go further and concern ourselves with knowledge applicable to all levels of the open-ended hierarchy of creative conflicts.

4. A Psychological Aside
Pertaining to the
Structure of Knowledge

THE SPECIFIC WAYS in which temporality is sliced into beinglike and becominglike components varies not only from epoch to epoch, or civilization to civilization, but within the structure of knowledge, from science to science.

Plato's God, the Geometrician, constructed the changeless order and became Leibniz' God-Architect of the changeless structure of the world. In our epoch the changeless law of nature itself has become the closest thing to the divine. Lawfulness, as we have often said, resides in predictability, and predictability, since the early science of the Renaissance, has been identified with that which "works." In the history of technology there is a clear continuity from pulleys, inclined planes, current-carrying conductors and vacuum pumps that work as expected, to machines that make beer cans as expected and other machines that light up the day with a brilliance "lighter than a thousand suns," as expected. Thus the great intellectual aim of Platonism to provide a rest for man's mind in timeless eternity has come to be transmuted into a vast effort to control man's fate through timeless laws of nature. There is little doubt that what has made this enterprise possible is the exactness of physical science. Since the discovery in Newton's epoch of how to separate the necessary from the contingent, physics has been striving to do with the necessary and has ignored the contingent in the tacit hope that all or, at least, most contingencies will eventually turn out to be necessities. I have already criticized this march from the nootemporal to the eo- , proto- and atemporal Umwelts and need only add that, as I see it, it is this drift toward the inanimate and the escape from the organic which has left the scientist's image, in Schrödinger's words, without "blue, yellow, bitter, sweet, beauty, delight, sorrow."[74] A break in the character of scientific knowledge appeared only with the Darwinian formulation of the principles of natural selection. Implicit in that principle is that such a fundamental feature of life as its very evolution may be explained in scientifically acceptable manner, without the simultaneous demand for strong predictive laws. Against the type of lawfulness employed in physics, biology admits contingencies, that is, a share of the intrinsically unpredictable.

How does this situation look in terms of temporal Umwelts? We will recall how successive integrative levels, each determined by its characteristic laws, arose from regions left undetermined by the lower integrative level. As we contemplate the biotemporal world, we see that it arose from what was left undetermined in the eotemporal and left, in its turn, many undetermined potentialities, out of which arose the nootemporal laws. What the evolutionist sees as the feature of becoming in life is the unpredictable content of all potentialities left undetermined by biological law. When we come to psychology, the issue of being versus becoming is sufficiently sharp to have precipitated two distinct general views. The reductionists insist that, if finally understood, the laws of the mind would reduce to those of matter. We have already discussed this issue in many different ways, and disposed of it. The rest of the brotherhood of psychologists inquire with great circumspection into the type of lawfulness which the mind uses to describe itself; again, we have already touched upon these matters.

Thus, the most rigorous forecasting is practiced by physics, the least rigorous by psychology. But this spectrum of the degrees to which being and becoming are given relative weights in the unitary experience of time, and along which the scientific modes of knowledge may conveniently be arranged, need not be limited to the sciences. It should, in fact, be extended to all modes of knowledge. This may be done by attending to the evolution of those signs and signals which we have identified with the information flowing in the channels that connect the individual elements (minds) of a society. John G. Gunnell, applying Cassirer's idea of symbolic forms to political philosophy, summed up the situation:

> The historical order of the development of symbolic forms in the West may
> be tentatively expressed as a progression from ritual to myth to language to

art to philosophy to science, and this corresponds to the differentiation of consciousness since "for each new problem that it encounters it constructs a new form of understanding." [75]

These historical steps survive in identifiable forms and coexist in the total body of knowledge at each epoch, including our own. They represent a spectrum of attitudes to time. A caricature of this spectrum would depict one of its terminals as the most precise knowledge of absolutely nothing, the other terminal as the most universal knowledge about everything which, however, cannot be communicated.

Let me contrast the imagined ends of this spectrum as those of pure analytical versus pure direct knowledge. The former strives to comprehend the world as totally predictable, one of being; the latter as totally unpredictable, one of becoming. Although all practical modes of knowledge are somewhere in between these two extremes, and the poles can never be found in their pure form, they are, nevertheless, significant, for they represent two recognizable trends in knowledge. Analytical thought tends to explore "outer space": astronomy, chemistry, life and mind of others, or the mathematical structure of the world. In contrast, direct knowledge tends to explore the "inner space": feelings, moods, intuitions, the life and mind of the self. In search of the biological and psychological sources of this polarity, I wish to associate the analytical and spatial modes of knowledge with the masculine principle of heterosexual life, the direct modes of knowledge with the feminine principle. I wish to draw some justification of these associations from the work of Erik Erikson on spontaneous play patterns of boys and girls between the ages of eight and twelve, for it is through such play patterns that the later play patterns known as modes of knowledge are first manifest.

The play material in his extensive tests consisted of a large number and variety of building blocks; four small figures of ordinary family members; uniformed figures (policemen, monks, etc.); wild and domestic animals; furniture, automobiles, etc.[76] The subjects were asked, one at a time, to construct on the table an "exciting scene." The results were some 450 scenes prepared with enthusiasm, once the task took over. Though Erikson's original interest was in the possible symbolism of the scenes as they might relate to stresses of puberty, soon it became apparent that the spatial modalities (that is, the nonverbal approach) bore deep affinity to genital modes. The configurations were then defined in terms of structural elements, the scenes photographed and analyzed by independent observers.

With hardly any exceptions, boys' scenes were houses or facades with protrusions and high towers. They were exterior, with the people and animals normally outside the enclosures. There were more automotive objects than in girls' scenes and they were infused with motion, accidents, contained high and collapsing structures, and ruins. Girls' scenes of houses were almost entirely of an interior nature: furniture, or enclosures built with the blocks, with people and animals within the enclosures. The enclosures consisted of low walls, sometimes with elaborate doorways. The scenes were generally peaceful and static except for occasional intrusion by dangerous men, or animals owned or controlled by a man.

> It may come as a surprise to some and seem a matter of course to others that here sexual differences in the organization of a play space seem to parallel the morphology of genital differentiation itself: in the male, an external organ is erectable and intrusive in character, serving the channelization of mobile sperm cells; in the female, internal organs with vestibular access, leading to statically expectant ova.[77]

In one of his earlier writings Erikson noted that cultures, starting with the differentiation produced by organic evolution, elaborate upon what is biologically given; they strive to divide social functions between the sexes in ways which are workable for the body, important for the group, and manageable by the individual ego.[78] This is an example of morphological projection upward, the discovery of new use for some already existing forms.

While the boys in Erikson's tests erected, constructed, and elaborated, the girls included, enclosed, and held safely. Such preferences go far beyond mere unconscious representations of sex organs by nonverbal communication; they project themselves into life-styles, world-views, and modes of knowledge. In this scheme the dominant male identity is based on fondness for what works, whether it builds or destroys. The male essence, we may say, is that of the architect, building the future in space. Dominant female identity is based on fondness of the inner, productive spaces; in them the art of creation is veiled. This is equivalent to favoring the organic approaches to things, to letting things happen according to a free order rather than forcing them to come about. Whereas the paradigm of male knowledge is carnal knowledge of another person, the paradigm of female knowledge is carnal knowledge of the self and nonself, both within. Whereas "outer space" can contain visible and tangible objects, "inner space" is not explorable by the senses. Hence, "inner space" must be identified with the idea of the "self" which, in its turn, is primarily temporal and only vestigially spatial. Thus, emphasis on "inner space" has an affinity to feeling, rhythm, changes, contingencies: to time. Concentration on "outer space" correlates with emphasis on surveying, walking around: on extension.

The modern intellectual history of the West has been a change from the symbolism of direct knowledge (the arts, religions, mythology, etc.) toward the symbolism of analytical knowledge; thus it has been a flight from time and a "flight from woman." This appropriate phrase is borrowed from a work by that title in which Karl Stern examines seven influential figures of modern man: Descartes, Goethe, Schopenhauer, Kierkegaard, Tolstoy, Ibsen, and Sartre.[79] To the extent that these men are taken as representatives of Western intellectual history, Stern discerns a historical trend to escape from what is wholesome, organic, intuitive—identified with the female—toward what is one-sided, inorganic and artificial—identified with the male. It is only a step from here to describe the exact sciences as masculine and the arts as feminine forms of knowledge.

Before the reader points, outraged, to the fact that a majority of the world's poets have been males, and that an increasing number of women are fine scientists, he ought to remember the following points. (1) We are talking about identifiable limiting principles useful in the interpretation of knowledge as a product of organic evolution. Knowledge itself is an integrated process for each individual, even though its components are identifiable—analytically. (2) The realization of the male and female principles is under very strong social control; it must be, for it pertains to the survival of the species. (3) The creativity of man is fired by certain unresolvable conflicts in his nature; one way to describe this conflict is to perceive it as between the male and female ways of knowledge. From this viewpoint every person is both androgynous and heterosexual: the female principle must be distinguished from woman, the male principle from man. (4) In our egalitarian era it is politically unpopular, and even distasteful, to stress *la différence*. This trend is itself an unfortunate victory for the male principle of static externality. Finally (5), the whole argument, stemming from Erikson's stress on the morphology of the genital differentiation, will sound less disturbing to the uninitiated if he remembers that the sexual act is the most important biological function of the individual, because it is through that act that he is able to overcome death. Specifically, males do so by injection, females by reception. It is not at all strange, therefore, that the morphologies of the organs involved in that act, one that has been practiced in the mammalian lineage of man for the past 200 million years, do show up as fundamental epistemological paradigms. An epistemology based on this type of insight ought really be called genetic epistemology; Piaget's approach, as noted above, though it is called genetic, is more accurately developmental.

Perception of the male and female principles in the structure of knowledge has a long prescientific history; it brings to mind the protochemistry of the alchemists, the yin and yang principles of the Tao, and also the androgynous God of the Book of Genesis: "So God created man in his own image, in the image of God created he him; male and female created he them" (Gen. 1:26). In Plato's *Symposium* Aristophanes sees in the male-female duality an original fullness torn asunder.[80] The creative duality of the male and female principles in nature is implied in all world religions, infused in all myths and embedded in the arts and letters. This rather vast demonstration is itself an example of direct knowledge. Its thesis may be placed in a scientifically respectable framework by pointing to the evolutionary coemergence of heterosexuality and death by aging. It would be rather strange if this profound and fateful evolutionary cleavage were *not* reflected in our individual and communal relationships to our environment as embodied in our modes of knowledge. In terms of Umwelts, we see here exemplified the absence of sharp limits between the nootemporal and the biotemporal; our noetic world is infused with "biological wisdom." Creativity on the noetic level of integration must also abide by the laws of the physical world: a sculptor rolling down the side of the hill will conserve his angular momentum for his body is so informed. A satisfactory epistemology must take all the level-specific restraints and undeterminacies into account. The writing of such an epistemology may not be easy, but it appears feasible.

In these reflections on the theory of knowledge we have invoked a good number of dualities. Some were heavily symbolic (such as the Dionysian-Apollonian duality), some emphatically biological (feminine-masculine), some political or social (Oriental-Occidental), some heavily psychoanalytic (oral-genital versus ocular) and at least a score of other pairs. Although these various qualities could be tabulated into two large groups, no such tabulation is intended. We are talking about intricately interwoven trends which, when considered in their details, often do indeed suggest the polarizations described, but they "go somewhere" only when considered together. Only the respectful overview of the whole may be judged realistic.

Such an overview suggests a need for a shift in the preferred modes of knowledge, for a change of emphasis. We are witnessing such a change in the emergence of a new natural philosophy based on a hierarchically organized complexity, wherein many of the working concepts of science, such as causation, time, or process, previously regarded as homogeneous, appear to have a hierarchical structure. New categories of scientific lawfulness are likely to come about to accommodate biology, psychology, and sociology and generally complement the static and physical by the organismic and dynamic. I regard this as a swing from the male principle back toward the female and an acknowledgement of their unresolvable conflict as expressed in their creative coexistence. Underneath this rediscovery I detect a reverberation in our methods and modes of knowledge of the coeval discoveries of time, sexuality, and death.

For Coleridge, no analytical thinker but an intuitive poet, man's existence centers on such a memory and that memory does not let him go.

> Since, then, at an uncertain hour,
> That agony returns:
> And till my ghastly tale is told,
> This heart within me burns.
>
> I pass like night, from land to land
> I have strange power of speech;
> The moment that his face I see,
> I know the man must hear me:
> To him my tale I teach.
>
> *The Rhyme of the Ancient Mariner*

5. A Mathematical Aside

Pertaining to the

Structure of Knowledge

EARLIER we considered some of the reasons for the "unreasonable effectiveness" of mathematics in natural science. We came to suspect that the universal power of mathematics derives from its essential primitiveness which, however, also limits the validity of mathematical description of nature to the Umwelts of lower temporalities. We found, for instance, that the density of the set of real numbers cannot be used to support the infinite divisibility of time, for the set of real numbers is too weak a scaffolding for the hierarchical structure of temporality. Quantum theory and its formalism describe a probabilistic, prototemporal world. These and similar thoughts bring to mind a remark made by Einstein in 1921: "As far as the mathematical theorems refer to reality, they are not sure and as far as they are sure, they do not refer to reality." [81]

I shall try to argue that mathematics, nevertheless, suggests something very important about the nature of time. But this suggestion does not come from any particular application of mathematics to one, or to a family of physical problems. It comes, instead, from those properties of pure mathematics which reveal its hierarchical limitations and intrinsic indeterminacies. To substantiate this claim we need to sketch the background from which valid statements about the properties of pure mathematics, as a system, are seen to arise.

Since the time of Archimedes, Euclid, and even Newton, the variety of notions and objects of interest to mathematics has increased tremendously; but the mathematical method has remained the same. This method comprises the observation of mathematical objects, and the abstraction from these observations of such properties as appear to be unchanging, as judged by the sense of time of the observer. The method of abstraction itself does not need to be accounted for as far as mathematicians are concerned, nor are they interested in the workings of the mind which produce, from the contemplation of abstractions, a small number of axioms. Then, well defined sets of rules are applied to the axioms so as to arrive at theorems and, by combining theorems, to obtain theories. The sets of rules must be such that the theories agree with the original axioms. This total thought process makes mathematics a deductive discipline, subject only to the test of logical consistency. Accordingly, deduction stands, or rather had stood, as the solid logical foundation of the science of mathematics.

That logic is a universal science which embraces the principles underlying all others is an idea due to Leibniz. In the nineteenth century, Boole, Dedekind, Gotlob Frege, and others attempted to derive arithmetics from logic. In our own time, A. N. Whitehead and Bertrand Russell have held that all of mathematics is reducible to logic and made this the central theme of their monumental *Principia Mathematica*. The doctrine that all the laws of mathematics (which, eventually, would involve geometry as well as all abstract forms of algebra) are derivable from and reducible to logic became known as the logistic thesis. As imagined, a final form of mathematics ought to be a consistent and completable edifice of the intellect, out of which all other knowledge would follow.

There are two crucial terms in this last statement. By completeness is meant that nothing that ought to be a theorem of a mathematical system would fail to be provable therein. For an interpreted system (wherein mathematical symbols stand for perceived properties of an Umwelt) completeness means that all true statements about the system, when expressed in the primitive terms of the system, are in fact theories of the system. If they are not, the system is incomplete. By consistency is meant that within the structure of any subdivision of mathematics, thus also in mathematics at large, there are no axioms from which logically contradictory statements may be drawn. For an interpreted system of mathematics a consistent theorem is one which does not lead to any statement about an Umwelt such that both it, and its denial, would be theorems of that Umwelt. If it does, the system is inconsistent because, in this pattern of thought, mutually contradictory statements cannot be true under any conditions.

That mathematics is consistent as well as complete, or at least completable, had been an article of faith underlying the foundations of mathematics all through the history of that discipline until 1931. That year Kurt Gödel showed that certain inherent limitations of the axiomatic method rule out the full axiomatization even of the ordinary arithmetic of integers, let alone the axiomatization of more complex systems. The mathematical principles of this work are known as Gödel's incompleteness theorem. With its surprising direction and intellectual strength, this theorem ruined the hopes of security exuded by the logistic thesis and offered, in its stead, an open-ended scheme which showed that mathematics is a hierachy of inductively coupled deductive structures.

In accordance with the logistic theory of mathematics, the *Principia Mathematica* of Whitehead and Russell created an uninterpreted calculus, that is, a system of instructions on how symbols may be combined in accordance with rules of operation, totally without regard to the significance or meaning of the symbols. Such a tool promised to make possible the derivation of all arithmetical truths from a set of axioms either already complete, or completable by adding a finite number of new axioms. Thus, by implication, it ascertained that no propositions could ever be found in mathematics that are undecidable, in principle. This comfortable view was challenged by Gödel's epic paper which bore the descriptive and critical title, "On Formally Undecidable Propositions of Principia Mathematica and Related Systems." [82]

Before attending to Gödel's conclusions, we should mention that his method of proof is in itself epistemologically significant. We learned that Galileo pioneered the mathematization of physics by switching back and forth between strictly verbal arguments concerning motion and mathematical representation of the same arguments. Thus, by the nature of his method, he abstracted from the wealth of our sense experience those relationships which fit the proto- and eotemporal Umwelts of mathematics. Descartes also used, in fact invented, another type of transformation when he assigned pairs of numbers to points in a plane and proceeded to prove geometrical theorems about points by proving algebraic theorems about numbers. Gödel proceeded by arithmetizing logico-mathematical statements, thereby making the most primitive truths of number the final arbiter of his very sophisticated theory.[83] Thus, the very substance of his theorem makes it an example of analytical, rather than direct knowledge.

The category of logic into which Gödel's work fits is known as proof theory, or metamathematics. The origins of metamathematics are associated with the work of the distinguished mathematician David Hilbert, who meant by it a theory of formalized proofs, the manipulations of logical formulas without reference to their meanings. Thus, from its very inception, metamathematics was an uninterpreted system. A fundamental demand on any theory of proofs, and one which is important for our subject, is that the methods employed in the proof must be in some sense less complex (more primitive) hence, presumably, less dubious than the methods of the proof whose validity is being tested. Thus it tacitly assumes that a series of deductive simplifications are possible by using some unchanging rules of inference.

With his ingenious method of reducing logical relationships to arithmetic ones, Gödel proceeded to work out what we are to mean (formally) by "unprovability," by "consistency," and by "unprovability of consistency." He concluded that if a formal arithmetic system is consistent, its consistency is unprovable in the system. In its simplest form this states that it is impossible to give a metamathematical proof of the consistency of an arithmetical system (sufficiently comprehensive to permit addition and multiplication—the simplest of operations) unless the rules of the proof differ in some essential ways from the rules employed by the arithmetical system. But inconsistency of a system means that it contains axioms from which logically contradictory statements may be drawn. Such an axiom, it will be remembered, is called a paradox. Gödel's proof offered evidence that this state of inconsistency is intrinsic in mathematics.

The existence of paradoxes in number theory was of concern to mathematical logic at the turn of the twentieth century. Foremost among them, and a representative one, is known as Russell's paradox. It states that there is a set which both is and is not a member of itself: an outright contradiction obtained from the basic premises of set theory, by legitimate logical steps.[84] To cope with this paradox and with other, related problems, the *Principia* introduced a "theory of types." The basic idea is that sets, sets of sets, sets of sets of sets, and so forth, form an array of categories or hierarchy of types such that members of any type are entities that comprise the lower, preceding types. To the type $n+1$ ("library") belong entities of the type n ("books"). The theory of types denies the meaningfulness of a sentence whose manifest content pertains to a level higher than that of its words. Thus, we cannot write propositions using a language appropriate to describe "books" and apply it to make statements about "libraries." This device anticipated, as I see it, the rigorous, formal proof offered by Gödel regarding the structure of mathematical knowledge.

In defiance of the completeness of mathematics implied in *Principia*, Gödel was able to construct a valid sentence about natural numbers, such that mathematicians would recognize it as true under its intended interpretations, but which could not be proved from the axioms of the system by rules of inference which were parts of the same system. Thus, quite clearly, there existed at least one theorem that should have been part of the system of the arithmetic of natural numbers, yet it was not; hence even the simplest of all arithmetic systems (that of addition and multiplication) was essentially incomplete. Perhaps, we could reason, adding the newly discovered axiom to the system on which arithmetic is based would remedy the situation. But, one of the consequences of Gödel's proof is that such an enlargement would not help, for there would always be further arithmetical truths not derivable from the enlarged set of axioms. Thus Gödel's proof offered evidence that the state of incompleteness is intrinsic in mathematics.

The intrinsic incompleteness and inconsistency of mathematics (in the sense here understood) thus permit the formulation of valid arithmetical propositions which are undecidable in the system, that is, they are neither provable nor disprovable within it. But, they, of course, may be provable or disprovable in a system of greater complexity than the one under consideration. Since the rules of such a higher system cannot be obtained by methods of deductive inference, they must be regarded as inductively coupled to the level of mathematics. This is what I meant by saying that mathematics after Gödel is understood as a hierarchy of inductively coupled deductive systems. As though these radical changes in our understanding of mathematical truth were not enough, Gödel's proof also amounts to evidence that makes the fulfillment of Hilbert's program in proof theory impossible. Namely, as we go from proofs to proofs of proofs, instead of being able to appeal to successively more primitive methods, the methods used to prove consistency must in some sense be more complex than the methods of the proof they are designed to authorize. Hence we must progress to potentially more rather than less suspicious arguments.

Whatever were Gödel's philosophical interests or convictions, his theory is one of arithmetic propositions whose explicit concern is limited to relations between whole numbers. Yet there is a remarkable isomorphism between Gödel's indeterminacy theorem (as it is sometimes called) and the theory of time as conflict.

On each level of the open-ended, hierarchical structure of mathematical knowledge we find paradoxical statements. In mathematics this is called inconsistency. In the theory of time as conflict these are the self-contradictory aspects of being and becoming, again on each integrative level. On each level of mathematical knowledge we find laws inexplicable on that level in terms of the sentences which can be written on that level. They may become explicable only in terms of sentences which correspond to the content of the next level. In temporality, the resolution of certain conflicts on each integrative level is possible only in terms of the content of the next higher level. In the proof theory of mathematics, subsequent proofs become progressively more complex, rather than less complex and potentially more, rather than less dubious. In the structure of temporality each level is more complex than the one beneath it, and on each level we find larger rather than smaller regions undetermined.

It looks as though the Whitehead-Russell logistic thesis is untenable, for it leaves out two essential features of mathematical knowledge: the inductive couplings between levels (or types), and the increasingly larger regions of incompleteness on each level. Logic, the science of implication, cannot cope either with inductive thought (that corresponds to creative emergence) or with undeterminacies (within which creative emergence may operate). I cannot assume that the isomorphism between the Gödelian principles and the theory of time as conflict is chance coincidence. I would rather see in it a demonstration of common origins in our psychobiological organization. Logic, then, would be one name for a set of unchanging rules learned by snails, horses, babies, and kings, individually and, in organic evolution, where it applies, as a species. Because of its evolutionary origins logic is in harmony with the strategy of existence (even if not in a pre-established way), otherwise it could not have selective advantages. Certain portions of logic have been translated into symbols and gathered in the science of mathematics. While the rules of mathematics must remain incomplete because they cannot accommodate becoming, they can and do provide regions for becoming, when considered as a total system of knowledge. And, as far as mathematical rules go, they do represent knowledge which individual and communal perception judges as timeless, hence in our earlier definition, true. But truth is not the only class of knowledge.

X

RELIGION, POLITICS, AND THE GOOD

IN THIS CHAPTER I shall take the ideas of good and evil as designating a quality of attitudes and emotions pertaining to such commitments as obligation, duty, morality, or judgement of values. I shall suggest that, when so understood, the good forms that class of knowledge or action which individual or communal perception judges as favoring conflict resolution by emergence; evil is that class of knowledge or action which individual or communal perception judges as favoring conflict resolution by catastrophe or regression.

1. Simple Thoughts about a Difficult Subject

THE POSSIBILITY that the moral sense of man which underlies the coherence of society is a product of natural selection was treated extensively by Darwin in *The Descent of Man*. He agrees with "our great philosopher, Herbert Spencer" that moral sense evolved by "experience of utility organized and consolidated through all past generations." [1] I am prepared to take the universal validity of evolution by natural selection for granted, but the question, then, remains: utility for what purpose? The immediate answer may be: The survival of the species. But the usefulness of the good for the survival of anyone—species or individual—is not at all clear cut; hence its role in natural selection remains obscure.

Consider, for instance, such behavior which on the human level is identified as friendship, care, love, jealousy, affection, hate, etc. Behavior which can be associated with these ideas is definitely manifest among animals. [2] I wish to illustrate, however, by an example, the subtle dangers in the uncritical assumption of continuity in kind. I have in mind the care and concern for the young. Parental affection among humans has been a perennial example of devotion and love. Especially in the form of maternal love it is something shared by all, or almost all forms of life. Although parent-offspring relationships vary from species to species and, among humans, from person to person, it is generally true that the young are profoundly appealing to mature and healthy adults of their species, and often also to adult members of other species. They must be appealing. Otherwise, among man for example, what adult would take the trouble of feeding and cleaning a baby, chasing the infant, teaching the child, or tolerating the adolescent? The charm of the young is a clear case of natural selection because, on the average, only those young things can survive that appeal to the adult—albeit, some only for hours, some for decades. Parental care is an example of evolutionary techniques where the future is favored over the present. Among humans, attitudes to children explicitly correlate with attitudes toward the future. R. J. Quinones makes this a point in his beautiful *The Renaissance Discovery of Time.* He stresses the link between the attention paid to children when the optimistic, Renaissance evaluation of the future prevailed, in contrast, for example, to the neglect of children that often went with pre-Renaissance fatalism.[3] Wild animals fiercely defend their offspring and, perhaps as an instinctive defense, seldom reproduce in captivity.

To die while protecting an offspring amounts to a biological defense of a biological future; it is life protecting itself and the argument of utility is clear, if not in detail, at least in broad lines. Thus, the lion protecting a lion cub, or the father protecting his cub scout, are taking good and useful measures. But, and this is an emphatic "but," ought one suffer and perhaps die for ideas, those "airy nothings" whose details are often not even understood and whose effects one is usually unable to foresee? Ought one suffer for causes which might, perhaps, benefit a stranger a century hence or a thousand miles away? If utility is unprovable, is the desirability of relevant conduct also indefensible? Can I depend on my instincts for the "when" and "how" in the defense of my "brain children" as I can for the defense of my biological children? Am I to imagine that the mechanism of natural selection may be transposed from the organic to the mental world without some substantial change in the form it must take? In an attempt to answer this question we must take, once again, a historical and hierarchical perspective.

Any attempt to trace the evolutionary origins of ethics so as to understand its significance in the strategy of existence must begin with human experience and only then revert to animal behavior. This is not an anthropomorphic bias but a methodological trick. We know quite well (even though not in an articulated fashion) the significance of ethics for man; all we need to assume is continuity along the evolutionary ladder. How far back along the chain of beings one cares to descend is a matter of taste. Perhaps even brooks run along gravitational gradients toward the open sea because they wish to return to their own kind, as Aristotle would have had it. And that, we might say, would be the praiseworthy (ethical) conduct for water, if it also had a choice of not flowing along the gradient. However, leaving matter alone, we may consider one of the simplest animal societies, that of the cellular slime mold. Its life cycle includes free-living and colonial stages. In the free-living or swarm mode the organisms act as independent amoebas; as the population density increases, however, they aggregate into slugs that move as a unit. The individuals interconnect by chemical communication; they produce a chemotactic substance which diffuses beyond the corporate limits of the slugs and induces free amoebas to follow the concentration gradient toward its source, and thus join the group. Temples and cities played analogous roles in archaic societies of man; they were navels, axes, and centers of the world toward which the individual tended to gravitate.[4] In myth and religion they often appear as earthly embodiments of heavenly archetypes, to be imitated. And they are usually spoken of as being "up." Hence, "Joseph also went up from Galilee" to his own city, Bethlehem, following the gradient of his mission.

It seems that individuals, in slime molds or in human society, more or less free, respond to certain enticements in their own interest which, when examined from a communal point of view, may appear to have been collectively inspired. An example of such enticements in man is the dispensation of pleasure and pain. The balance between these two is controlled by the complex struggle between individual and collective interests. For societies of simpler organisms the situation is not so prohibitively complex: they seem to be run by the tyranny of the statistical majority (starfish, for instance, sometimes tear off their tube feet if the feet oppose the motion of the organism). In return, as it were, members of a community so condition their environment as to promote the survival of other members of the group—on the average. The resulting communal capacity to exert adaptive pressure upon the environment is more substantial than that of an individual alone, not simply because the individuals are multiplied in number, but because coordinated collective actions directed outward can display a great strength in specialization, hence, in efficiency and in the capacity to survive. Thus, far below the level where free will and moral choice enter we can already find social control, coordinated aggression, and measures internal to society which support "socially praiseworthy" behavior. What we do *not* find is that intense and protracted conflict between the individual and the group which is imbued in human society. That sharp struggle is either absent in animal societies or, if not absent, at least insignificant in the evolution of those societies (if one can talk about the evolution of a particular animal society at all). Changes in groups of animals come about mostly in response to external pressure.

The situation for man is quite different. The polarization of interests among individuals, and between each individual and the group, is not only continuously present but, in many ways, it is at the source of social change. This is not the projection of a presumed American creed upon humanity at large, but only an emphasis on the importance and antiquity of intraspecific aggression. It is the intensity of the continuous balancing act between that aggression and the need for cooperation which sets human society apart from other societies. The underlying tension derives not only from conflicting present interests however, but, more importantly, from conflicting plans for the future. The separation of ethics into deontology (things I ought to do because they are intrinsically good) and teleological ethics (things I ought to do because they promote distant, future goals which are good) is not possible for groups whose members do not possess a sense of time. The two patterns of praiseworthy behavior are yet undifferentiated even in the sophisticated intention movements of the bees.[5] The ritualization of their dances, and the method of "voting with their feet," are signs as well as guarantees of complete unanimity to be reached sooner or later. In man deontological and teleological ethics are not only differentiated but are often in conflict.

But if experience teaches that in man the ideas of praiseworthy behavior, duty, responsibility, love, etc., are intimately tied to a conflict which does not exist in animal societies, then the utility of ethics envisaged by Darwin, as far as it concerns man, must have as its major referent something which itself is unique to man. Just as the utility of certain "good" behavior among animals pertains to its role in bringing about and protecting the biological offspring, so the ethical among man pertains to bringing about and protecting man's "brain children." We must remember that on each level all prior restraints must still be met; hence even the good will have to include that of life, but not only, and not predominantly that of life. This gets us back to our earlier theory that ethics is one of the three major methods we use to guide our selection of future courses from among the possibilities and impossibilities generated by the very complex. It is a learning process which we find expressed in the history of the normative teachings of religion and in the regulative functions of political bodies. It is to these two enterprises that we shall now turn.

2. The Need for Guidance in Conduct

SELECTION RULES from among the images of possible conduct were first taught by religions as values, associated with ideas of God and the Devil.

God and Devil in the Religious Vision

Inspection of primitive visual art leaves one with the gut impression that the concerns of paleolithic man in Europe, pre-historic Asia, Africa, or Oceania were not different from most of our own basic concerns. Suffering and joy seemed to have appeared to them no less fundamental and real than they do to the man of this century. The groping for beauty, order, and harmony expressed in primitive art imply, by contrast, an awareness of senseless cruelty in the world. The earliest written records, which follow in time the earliest paintings and artifacts, speak of struggle, suffering, and death and of passion directed to alleviate them. The early epics, which are likely to have arisen from prior verbal traditions, are of the same tenor. If archaic man had not felt the overbearing power of those things and conditions which he feared and hated, both within and without himself, we could hardly make sense of the emergence of magic and religion. Both of these institutions, throughout their long history, tried to give acceptable accounts for death, suffering, and injustic and tried to offer some means of controlling them.

Theories about the origins of religions are numerous, beginning with the self-serving report each religion gives about its own origins. Rudolf Otto introduced the idea of the holy as a true universal and postulated an instinctive sense of the numinous; this would make religion an autonomous function of the mind independent of any social or cultural usefulness.[6] Sir James George Frazer takes a more naturalistic approach in *The Golden Bough* when he observes that man has sought general rules so as to control nature, and in his search has come upon a great horde of maxims, some of them golden, some of them mere scum. The golden rules constitute, in his view, that body of applied science which we call art; the false rules constitute magic. Then he distinguishes religion from magic as "a propitiation or conciliation of powers superior to man which are believed to direct and control the course of nature and human life. Thus defined, religion consists of two elements, a theoretical and a practical. . . ."[7]

While Frazer saw clearly that this was one possible definition and demanded from himself only clear statements and consistency of views (which demands he fulfilled), he nevertheless had difficulties disentangling magic from religion, because magical and religious myths and rituals have so much in common. Differences between the two are treated by B. Malinowsky in an essay on "Magic, Science and Religion." He distinguishes between magic and religion by identifying the former as a practical series of acts directed to bringing about certain future events, and the latter as self-contained acts being themselves fulfillments of their purpose.[8] This is somewhat misleading. Magic can also be its own fulfillment and bring peace of mind through faith; religion may also contain practical series of acts prescribed to bring about some result. I would like to draw attention to a deeper-lying difference. This is in the way religion and magic separate necessity and contingency in the nature of time. Though they both claim mastery of contingencies, there is a gradation in the items which they regard as contingent. Magic tends to consider practically everything under its direct control;[9] religions generally posit intermediaries and claim power over the future only on abstract levels.

Recent historical scholarship pioneered by S. G. F. Brandon sees man's knowledge of time (understood as his awareness of coming into being and passing away) as the most important single source of religions.[10] Brandon points to the double-edged nature of the sense of time. On the one hand, it allows superior control over the environment; on the other hand, it functions as a continuous, disturbing reminder of passing. "I would define my conclusion as follows: religion is the expression of man's instinctive quest for security, which results from the sense of insecurity caused by his consciousness of time." [11] The reasoning, if not the specific conclusions, ought to sound familiar to the reader.

In our own epoch the good and the true are dogmatically separated, because science demands that truth be determined without value judgement. This lopsided arrangement is a dividend of the scientific revolution, hardly known before it, and quite unimaginable from the point of view of the practices of archaic man. To the contrary, rituals and myths have traditionally subsumed instructions for preferred behavior together with teachings about truth. The true and the good as a single mode of knowledge signaled the emerging loneliness of man; they mediated between his newly found identity and all that was not himself. It is this subjective feeling of loneliness which has given religion and magic their profound emotional importance and which has demanded (in addition to other advantages already discussed) societal existence as a way of human life.

As the individually and socially desirable goals served by rituals came to extend to longer futures, the symbolic transformation of experience embodied in the ritual had to become, by necessity, more abstract. Whereas the pre-acting of a hunt could have been considered potentially effective for the success of the hunt on the morrow, the liberation from bondage and deliverance from the Waters of Babylon demanded the whole symbolic structure of a Heilsgeschichte. I presume that there is a condition of mutual generation here: more abstract symbolism permits a longer lead-time; longer lead-times demand more abstraction. With the increasing autonomy of autogenic imagery, sometimes the goals themselves could change from physical things and conditions to intangible things and conditions. Thus, the good and evil of the here and now might have become transformed into the ethical ideas of distant futures.

We can arrive at similar conclusions via a different path. Magical and religious rituals, whether danced, painted, chanted, or recited are usually for the promotion of some purpose, hence the prevention of its opposite. They are thereby witness to the capacity of the celebrants to distinguish the desirable from the undesirable and try to sway fate appropriately. Magical practices are self-limiting in this respect, mainly because they propose to dominate nature by direct intervention. (This is one reason why science is so close to magic.) Religions were able to remove themselves, by and large, from the responsibility for direct and immediate results by enlarging the distance between the good, here and now, and the good on the distant horizon. They seem to have favored the distant reaches of time and space, and to conditions and things at those distances they attached the ideas of absolute good and evil. Since fate can display itself only in the course of time, it is not surprising that ideas of time, in some of the great religions, reveal themselves as two-faced deities. This dual significance of time was demonstrated by Brandon in a comparative study of ancient religions, appropriately titled "Time as God and Devil." [12] He relates how in Indian literature time is both creative and destructive (with an edge for destructivity) and is so deified in many and frightening ways; and how the Iranians deified time as Zurvan, progenitor of the two opposing cosmic principles of good and evil, light and darkness, creation and destruction. The Greeks, however, refused to elevate contradiction to Olympus and, in their oldest forms of religion, gave time a rather secondary role with only an occasional display of Iranian influence in Orphism.

As we just remarked, the good and the evil which rituals were designed to sway had to play themselves out in the course of time. But where long time periods are, history cannot be far; and where time and God and the Devil are, history and good and evil cannot be far. Perhaps it is in the evaluation of history as good or evil that we ought to search for the origins of teleological ethics.

In archaic humanity we find no such elements. Eliade gives many examples of how early man defended himself against all novelty and irreversibility which (our idea of) history entails, and how he did so through the use of myth and ritual celebrating the eternal return. [13] In our terms this is an expression of conservatism of what is now the unconscious; or a demonstration of the cyclicity of the eotemporal Umwelt. It is only with the slow coemergence of identity, language, advanced symbolism, etc., that nootemporal features could appear, including time and value judgements. Brandon gives us some profoundly new insight into the simultaneous growth of ideas of history, God, time, and value in his comparative studies of the concept of time in religious thought. [14]

He identifies five complex views. (1) Salvation from the burden of time by ritual perpetuation of the past. This is Eliade's myth of the eternal return—in many variations. Good and bad, while real enough, seem, on the average, to even out. (2) Time venerated as a deity, incorporating God and Devil—with examples given above. (3) Time as an illusion, a "sorrowful, weary wheel" is a collective escape from time. We saw examples of this earlier: the cyclic chronologies of Buddhism and Jainism, where good and bad wash out in the long run. We also saw it in a contemporary form as the pulsating universe of scientific cosmology, whose authority relieves man from the burden of asking at all whether good and evil exist in the world. (4) The view of Heilsgeschichte wherein final good awaits the group of the select at the fulfillment of history. And (5), history as a two-phased plan, which gave rise to the Christian view of time and ethics, and with which we will deal separately. To Brandon's five categories we may add a sixth:

those evaluations of time and evil which see history as a devolution toward a final, decadent stage which then might, or might not, be followed by an age of plenty.[15]

As one scans this schema of theories created to alleviate the feelings of loneliness and insecurity which were engendered by the knowledge of time, the impression is unavoidable that religions, no less than science, show certain epistemological properties, such as the personality traits of man projected on the heavenly screens. For instance, according to religious interpretations, the struggle of the forces of darkness against light, or good against evil is the frustration of the divinity. We know that the frustration tolerance of an individual is a measure of his maturity. God, or the gods, are then conceived of as the most mature beings; they are able to suffer greater and longer frustrations for larger and more distant goals than any mortal could. The divinity witnesses the struggles and setbacks of individual perfection (as in Buddhism) or the apocalyptic conflict of good and evil (as in Christianity) or the cosmic fight of light and darkness (as in Zoroastrianism) for it knows that when time from time shall liberate the world, the good (or the light, or the perfect) will win. Most religious evaluations of history may therefore offer a summum bonum, that particular good which will be seen as victorious and on whose bandwagon the faithful should jump. Conduct which promotes the anticipated final victory (or maintains the desired status quo) is usually seen as praiseworthy. Since such formulations imply foreknowledge, religious ethics has had an easier path than rational ethics for tying the good directly to the true.

Let us recall our earlier conclusion that the functioning of higher levels must always appear as mysterious from lower levels, for the lower strata cannot accommodate the lawfulness of the higher ones. If this is admitted, it would then follow that the regularities of the history of man (whatever they may be), and also all knowledge which has its origins in the collective manifestations of the mind, would tend to appear to the individual as mysterious. Things happen as though guided by a supernatural power. In this we might recognize Otto's idea that religion involves a sense of the numinous. From the "beyond" seem to originate our ethical instructions on how to select behavioral patterns from the vast repertory of autogenic imagery. The simplest form which such instructions have taken has been a binary one: that of good and evil. Out of the many sets of instructions we shall now reflect upon the nature of one, that of Christian ethics. For Christian ethics is the source of the idea of progress as the summum bonum of the good, and the idea of progress has come to command the attention and guide the aspirations of most peoples on earth.

Christianity, Progress, and the Good

The standard of conduct which the Sons of Israel were expected to uphold against the onslaught of evil was the terms of a covenant concluded between Yahweh and Israel, and summed up in the ethical decalogue of Exodus.[16] It was not an easy path to follow, as Job, passionate questioner of the purpose of life and protester against evil, was destined to find out. "Man that is born of woman is of few days, and full of trouble. . . . Behold I cry out of wrong, but I am not heard; I cry out aloud but there is no judgment." Eliphez the Temanite had a simple answer, almost Darwinian: life's purpose is the good; the good is its own reward, the bad its own punishment. This was too ethereal for Job, for he was deathly sick and had just lost his family. "I have often heard such things as these, you are all troublesome comforters." Job's misery, nevertheless, had a happy ending. Having persevered in faith, he lived "an hundred and forty years, and saw his sons, and his son's sons, even four generations" (Job 14:1; 19:7; 16:2; 42:16).

But earthly rewards for following the covenant were not always forthcoming, especially when such conduct put the individual in opposition to political power. There are reasons to believe that Jesus as a young child knew of the bloody conflict between the religious covenant and the power of the state and was aware that the Romans crucified some two thousand rebels to suppress an insurgence in Jerusalem during the first decade A.D.[17] There is even a whispered tradition that Jesus, as a child, helped his father, a carpenter, make the crosses upon which the Jews were crucified.[18] Whatever the truth of these details, the Christ story from which Christian ethics unfolds is a remarkably stable psychological constellation of archetypes. There are the carpenter and his pregnant wife; the flight from a tyrant; the struggle to abide by a covenant; the savagery of the law-and-order forces; the bearing of the cross; and the catharsis of crucifixion where man comes face to face with the impossibility of his ethical task and the emptiness of communication. This is a lasting drama of guilt and punishment (where who is guilty and who is being punished remain unspecified); sin and salvation, and the diametrical opposite of Job's long life on earth. Reward for suffering comes only after death.

Yet this summary segment of human fate was only a prologue to the single claim upon which Christian ethics originally was built: the dogma of resurrection. Rising from the dead may be something we never observe, but as an idea it is easy to comprehend; it is the negation of the fear of death, of mortal terror, and of the passing of time. In the resurrection of Christ the temporal is transformed into the timeless through suffering and self-sacrifice, setting thereby the standard of praiseworthy conduct for confessed Christians who wish to escape the tensions of their nootemporal Umwelts. The Hebrew Heilsgeschichte which was to have led to eventual completeness of man on earth was thus transformed into a salvation process for the individual and the community of Christians. The progress of the individual himself from the temporal to the timeless became identified with his struggle for Christian identity. The conflict came to be understood in terms of the contrasts between sin, evil, and hardship on earth on the one hand and, on the other hand, the purity and reward of the Christian heaven. Thus, the earthly salvation history of the Hebrew covenant and its earthly rituals were replaced by the heavenly beatitudes and, in due course, through the intellectual edifice of the Roman church, the beatitudes became incorporated in the ethical paradigms of the Christian West.

Under the authority of the Church, patristic and medieval Christian ethics remained continuous and coherent. St. Paul's cry that he was alive without the law once but when the commandment came he died (Rom. 7), was echoed in the introspective psychology of St. Augustine, preoccupied as he was with morality. His devaluation of the here and now of the body and concentration on the there and then of the soul is adumbrated in the famous hymn of his teacher, St. Ambrose. "My shameful thoughts must be controlled," he wrote, "and painful abstinence must tame, of wanton flesh the pride." * With many but minor variations moral philosophy through the middle ages maintained the image of man as existing apart from, and not as part of, the natural order. Justification of praiseworthy conduct was thereby limited to the substance of revealed religion and inspired writing. The situation remained unaltered until the emergence of natural science as a preferred method of seeking truth, when standards of conduct derived from the revealed truth came to look less and less convincing.

Within Christianity the break in heaven-orientedness came with the Reformation or, more precisely, with the post-Reformation reformulation of the principles that bind morality to faith. The Reformation itself may be symbolized by Luther's concern with the ubiquity of the Devil in our midst.[19] This strong emphasis on the undesirable features inherent in man eventually resulted in a refocusing of Christian attention from the heavens to man. It amounted to a new delineation of the boundaries between the timeless will of God on the one hand, and, on the other, the unpredictable contingency implicit in the free will of man. By its emphasis on good works here and now, the reevaluation of morals initiated by the Reformation underlies the later reevaluation of nature as the setting of that here-now, and, eventually, of nature as the sole depository of valid truth— even though Luther would never have intended this development.

The peculiarly strong concern of Luther with the Devil has been interpreted as a religious projection of what both Jung and Freud would recognize as the emergence of the repressed.[20] The preferred way of coping with the threat of the return of the repressed, in the practice of the Roman Church, has been the sublimation of drives and affects for the benefit of the eventual City of God, a preference expressed in the adoration of the Virgin Mary. In contrast, the Lutheran fear of the Devil, invariably a male character, seems to favor sublimation for the benefit of a city on

---

* Those familiar with the melodious quality of the Ambrosian chant together with the Latin severity of the message will see in their combination a demonstration of the patristic ambivalence for beauty and duty.

earth. Thus, the patristic and medieval ideal of conduct directed to ethereal, mystical, and distant goals nominally associated with the soft qualities of the beatitudes, came to be replaced by the Protestant ideal of conduct, directed to the building of the towns of wealthy merchants and associated with the harsh and realistic qualities of good deeds offered here and now, in barter with the devil within.

With the increasing store of goods on earth, it became worthwhile to attend to the future on earth, and the importance of heavenly reward became an appendage to the earthly goals, rather than a goal unto itself. The Christian evaluation of history, as we have seen, was characterized from its very beginning by an emphasis on time as having essential significance in the economy of God's purpose. With the Reformation, however, the relative weights of the City on Earth and City of God were changed. Though expressible in these simple terms, the change did not imply simplicity in the historic condition; as in other upheavals of life or thought, the Reformation comprised a large variety of mutually reinforcing changes. I have already pointed to the possible sources of the scientific revolution in the daily practices of Protestantism and discussed some of the correlations among mercantile economy, attitudes to sex, preferred modes of collective perception of truth, and the like. The Lutheran parsimony of time and emotion which figured so importantly in our interpretation of mathematized science as the acting out of personalities, is basically an ethical stance. In retrospect, it appears that the religious refocusing on earth, together with other mutually reinforcing elements of Renaissance life, was an essential step from the abstract future-directedness of early Christianity to its secularized derivative, the idea of progress. The Enlightenment rejected the supernatural but retained and assimilated the idea of progress, which remained the surviving heritage of the Judeo-Christian philosophy of history.

Progress and Heilsgeschichte do differ, however, in one important respect. Whereas the Christian interpretation of linear time justified Heilsgeschichte by teleological arguments in terms of a final, religious good, thereby easily tying moral philosophy and ethics to an understanding of history, it is very difficult to find an obvious, nonartificial entry for ethics in a world-view based on secular progress. History, in the latter world-view, is either said to be progressing toward an ethical goal somehow inherent in the laws of organic evolution, or else such progress is taken as a self-evident ethical truth without the need of further justification. For guidance in daily conduct, the secularized idea of progress must draw entirely, or mainly, on truth-as-science. Because of the parsimony of emotive depth in scientific truth, it then falls short of being able to deliver a code of conduct with sufficiently deep conviction. In spite of these difficulties the idea of progress remained the unquestioned metaphysical basis of the ethics of participatory democracies and of communist states until the last few decades of Western reaction.

The relationship between the idea of progress and views of time and history are intriguing. J. B. Bury, in a comparative study of the origins and growth of the idea of progress, properly ties the Platonic preference for immutability to the practical absence of the idea of progress in the Greek world-view. It was certainly not the absence of brilliance, but "the axioms of their thought, their suspiciousness of change, their theories of Moira [that human nature does not alter and their theories] of degeneration and cycles [that] suggested a view of the world which was the antithesis of progressive development." [21]

Bury is taken to task for these conclusions by Ludwig Edelstein in *The Idea of Progress in Classical Antiquity*.[22] Professor Edelstein died before his work was completed; thus no clear-cut conclusions are reached. With great erudition he marshaled armies of statements which, together and severally, suggest a Greek belief in the fundamentality of change. His remarks, while impressive and true, overlook the fact that the modern idea of progress (whose ancestry both he and Bury set out to trace) presupposes a teleology beyond the devices of the individual, together with the belief in the plausibility, or likelihood, of complying with the teleological goals. One of these elements alone is not sufficient. The social concept of progress came to fruition only after the political vision of humanity (originated by the progressless Greeks) became fused with two other elements (teleology, and confidence in man) in modern philosophy. The Judeo-Christian idea of Heilsgeschichte supplied the paradigm of teleological goals, but it is the Renaissance rediscovery of the potentialities of man that supplied the element of earthly plausibility. It is tempting to think that the idea of progress was suggested by the success of the cumulative effects of organized science. But the case is the opposite. The feasibility of cumulative knowledge was derived from the belief that historical process is teleological, and not vice versa.

Returning now to the three components of the twentieth century idea of progress (political vision, teleology, and confidence in man) we note that in the increasingly secularized faith in progress the self-evaluation of man became increasingly important. Humanistic optimism survived the attacks of Rousseau and remained viable through the infantile futuristic utopianisms of our days. Thus, the true forefather of "future research" was the Marquis de Condorcet (1743–1794) who held that insofar as errors in politics follow from errors in philosophy and, in turn, these follow from scientific errors, the process of history should be—and is—from headhunters to scientifically organized, Cartesian society, open to the future because of the infinite perfectibility of man. With Condorcet, Saint-Simon, and August Comte the search began for laws of progress, that is, unchanging principles that specify the way things and conditions favorable to man emerge.

During this search the idea of natural selection reverberated loudly and for a long time among thinkers concerned with conduct, society, and history. Herbet Spencer, in his *Progress: its Law and Cause* (1857, two years before *The Origin of Species*), held out for a linear, lawful progression of history from stars, to earth, to society, industry, art and science and earned Darwin's approval, quoted earlier. He also insisted on the infinite perfectibility of man and saw evil only as a transient characteristic and not a permanent necessity. Progress for him was a necessity that follows directly from the nature of man and is directed toward producing the greatest amount of happiness for the largest number of people. Praiseworthy conduct is that which allows each man to experience the highest enjoyment of life without preventing others from doing likewise. Always toward perfection is the mighty movement—toward a complete development and a more unmixed good.[23]

Friedrich Nietzsche, inspired by the possible social applications of Darwinism, thought otherwise and said so in *Beyond Good and Evil* (1886). Morality is either that of the master or that of the slave.

> Slave morality is essentially the morality of utility. Here is the seat of the origin of the famous antithesis 'good' and 'evil'. . . . Power is perceived by the slave morality as evil; a safe man is a good man, and 'good' and 'stupid' approximate one another.[24]

But, unfortunately, under the increasing pressure of slave morality championed by Christianity, the social evolution is from the *Übermensch* (superior man) to the *Untermensch* (inferior man). In a very rough way, we may say that beginning with evolution by natural selection Spencer turned his thoughts and feelings to the lawful, being-type component of time, while Nietzsche stressed becoming and unpredictability "über alles." These are echoes of Parmenides and Heraclitus.

By the end of the nineteenth century, and even more so by the last quarter of the twentieth century, organized knowledge had opened up a world of such vast dimensions, immense age, and of such curious lawfulness that the Hebrew Heilsgeschichte and Christian soteriology, upon which the original ideas of progress and the good were built, became untenable. The individual's hope, which used to be closely tied to his evaluation of his place in the universe, lost its cosmic anchorage and acquired in its stead certain ill-defined relationships to the impersonal laws of the external landscape. With this transformation, the bases for formulating norms of praiseworthy behavior were shifted from transcendental beliefs to science as the sole authority on truth. The malaise of modern man, as regards ethics, has followed.

3. Good and Evil, and the Political Vision

IN THIS SECTION we shall observe the evolution of yet another relationship between time and the good, that in political philosophy.

The Good of the Polis

We have already seen that self-identity was born as a necessity in extending man's predictive powers; likewise, when in the course of communal growth the tribe reached for more distant and greater advantages, the identity of the group was born. We also considered the unique difficulties involved in establishing the identity of mankind. Religious teachings were always rich in suggestions as to what that identity comprises but they never shed, and are not about to shed, the hallmarks of the heterogeneity of their origins. Their philosophies are still useful for maintaining local identity and color, but there is no single religion that would do for mankind. Religious morals, though they share much in common, have remained in many ways incompatible. Whitehead observed that in the long run, "that religion will conquer which can render clear to popular understanding some eternal greatness incarnate in the passage of temporal fact." [25] For contemporary man, rushing toward some kind of world-community, such an explanation seems to be promised not by the religious visions of the theologians but by reflections on the political vision that was born as, and when, philosophizing was invented around the Aegean Sea.

Now, if the world is good, remarked Timaeus, Plato's spokesman for cosmology, and his maker is good, his maker had to look to the timeless as a model upon which he was to fashion the universe. The opposite "which cannot be spoken without blasphemy" is that he looked to the temporal and the generated. Thus, says Timaeus, it is evident that the creator looked to the eternal, "for the world is the best of things that have become and the best of causes." [26] In this best of all worlds man could not do better than to form his social cosmos so as to resemble, as much as possible, the timeless order of the world at large. Gunnell has put it with unparalleled clarity:

> Political philosophy was at its inception, that is, with Plato, a reconstruction, on the level of conscious thought, of the vision of political order as a mediating space, a mesocosm, which would assimilate man to or integrate him with that which is most eternal and overcome historical existence, which in terms of the Greek experience meant essentially the fall from order. It is this theme which binds together Plato's speculative venture as it emerges in his political dialogues. [27]

The intimate connection between Plato's distaste for coming-into-being and the ideal of the polis erected to protect man from time is unmysterious: the collective perception of good, whatever else it does for the individual, no less than that of truth, is a bid for timeless order. The realization that only symbols can outlast an individual is part of our existential knowledge of time, perhaps part of the instinctive search for permanence postulated earlier. It is not surprising then, that the most lasting ethical instructions are couched in terms of symbols, usually myths, and become vulnerable almost in inverse proportion as they become actualized in codified law. Thus, Plato's *Laws* is, in a sense, a fable whose power resides in illuminating the consequences of a timeless cosmos of eternal forms. Although I doubt whether he ever put it in these words, his political views imply that what is good for the polis is good for the individual. Good is the highest order of reality on his totem pole of the divided line;[28] the line itself symbolizes all there is in the cosmos. Since the polis is to mirror the cosmos on earth, the good is no less the highest attribute of the polis than it is of the world. Why did this Platonic vision endure so long? Perhaps because Plato perceived, and implied throughout his dialogues, a convergence of symbolic forms equally applicable to the physical cosmos, to the state, and to the individual's intimate knowledge of temporal stress. This triumvirate rode the waves of Western intellectual history, reaching its pinnacle in the mathematization of knowledge and in the Communist and National Socialist views of the state.

Plato's influence and inspiration in Aristotle's *Nichomachean Ethics* is clear, but so is the difference in personalities. Unlike Plato, Aristotle formulated a specific doctrine, that of the mean, to help man decide about particulars. He agreed that the highest possible human achievement is the contemplation of eternal truths but insisted that on the next level of good were moral principles which are to guide us in everyday matters. These may be discovered by examining the actions of men, particularly of good men. The task is not easy, for "evil is infinite in nature, to use a Pythagorean figure, while good is finite." [29] But, by moderation and by observance of how a prudent man does things, guidance in conduct may be found. The naturalistic stance is evident: diligence and intelligence (in twentieth century terms this would amount to enough money and time for research) will discover what is good, for that information is already given in the world.

To the Neoplatonists the idea of good was largely associated with a sense of unity; Plotinus was said to have experienced the states of utmost good through ecstasies. The achievement of such unity was regarded as a return of man to the absolute good. St. Augustine, arguing against (hence influenced by) the neoplatonists of his day, had difficulty in reconciling temporal good with the timeless good of the Christian God. This Augustinian tension in ethics did not lessen until after the Renaissance, when the West had learned to live with opposites. This, more than a millenium after Augustine, prepared the way for an eventual relativization of values—in the continued company of absolutist trends.

Augustine knew all too well the imperfections of the city on earth, so he turned to the heavenly city. Nicollò Machiavelli (1469–1527) surveyed the city on earth and found it wanting but not irreparable. In his *Discorsi* he defends Romulus for having killed Remus for, as he sees it, the act was performed for the common good; thus Romulus ought to be classed not with Cain but with Moses and other founders of the state who had to kill to maintain the good of the collective. In the opinion of some scholars, the first articulate challenge to the Christian view of history as divine Heilsge-

schichte was that by Machiavelli when he placed the ruler in the flow of an
open-ended, rather than a predetermined history.[30] The ethics of Machiavelli's prince was to be determined not by an overall plan in the universe,
but by attending to the challenges of his time; man and the rules of man
came to be seen as capable of controlling their destiny.

Whereas Machiavelli perceived the good as that judged beneficial by
the prince for his state, two centuries later a genius of very different
temperament held diametrically opposite views. There exists a deontological
imperative of morality, said Kant, and categorical in that it admits no
exception. The imperative is this: "Act only according to that maxim by
which you can at the same time will that it should become a universal
law." [31] The jump from Machiavelli to Kant is a change of perspective:
from tribe to mankind. Hegel, in his *Grundlinien del Philosophie des Rechts*
(1821),[32] carried Kant's specifications further and saw the moral standards
(that which a person should want to become universal law) in the harmony
such behavior might achieve in terms of the multiplicity of social relations
of the individual. If in Hegel's dialectic we change "spirit" to "matter" we
have dialectical materialism and its ethics, with the Party Secretary as the
Prince.

Perhaps twenty-five years younger than Kant but of a pragmatic
English, rather than metaphysical German, temperament, Jeremy Bentham
(1748–1832) saw good as the greatest happiness of the greatest number and
began his famous treatise on morals and legislation (1823), sounding almost
like Freud:

> Nature has placed mankind under the governance of two sovereign masters,
> pain and pleasure. It is for them alone to point out what we ought to do,
> as well as to determine what we shall do. On the one hand the standard
> of right and wrong, on the other the chain of causes and effects, are
> fastened to their throne.[33]

Bentham's intellectual offspring, John Stuart Mill (1806–1873), refined the
undifferentiated category of pleasures praised in the hedonistic ethics of his
master and added his own utilitarian emphasis. Both, though especially
Bentham, devoted their lives to the modernization of British political institutions, each following his own evaluation of the balance of social and individual good, and each trying to put into practice the Socratic way of
living one's philosophy.

Within a generation after John Stuart Mill, tenets of moral philosophies became meaningless statements in principle, at least in the opinion of
that portion of academic philosophy which most closely mimics the methods
of exact science. Logical positivists concluded that insofar as ethical concepts are not testable by the axiomatic-experimental methods of physical
science, ethical concepts are pseudoconcepts and therefore unanalyzable.
And, as A. J. Ayer put it, since "ethical judgments are mere expressions of
feelings, there can be no way of determining the validity of an ethical
system, and, indeed, no sense in asking whether any such system is true." [34]

The chief causes of moral behavior are, in his summary view, fears of many types. These fears are fostered by society's concern for its own happiness, hence society encourages or discourages a certain conduct by the use of sanctions according to whether the conduct promotes or detracts from the contentment of society as a whole.

Our brief survey suggests that in the history of ideas the concept of good was tied by philosophers, more often than not, to the interests of the polis. As the city varied, the polis became secularized and so did the idea of the good. Plato's city was tied to the cosmos, Augustine's to the Christian God, Machiavelli's to Florence, Hegel's to the spirit. With the changing evaluation of man suggested and authorized by science, the connection between myth and man was officially dissolved. Good and evil became relative only to a humanity whose identity could not be defined. To replace the lost cosmos and the repudiated spirit, political myths are generated before our eyes. Mussolini's March on Rome, or the long march of modern China are not heavenly images, but mythical just the same.[35] And Plato's political vision as that which ties good and evil to society and to the timeless is still alive, if not in its early embodiments, certainly as a paradigm of law and order.

The Good of the Technopolis

Beginning with the industrial revolution, the religious and political visions of the good, derived from their respective ideas of the true, were joined by a third ideal: the technical vision of the good derived from the scientific idea of the true. Whereas religions tended to anchor their values in the transcendental, political theories mostly in the collective cosmos, technological man came to perceive the good in the products of his brain and brawn. His set of values had a long incubation period going back to the early high regard for the clock as the most important machine. The opinions of Marx and the findings of Mumford,[36] that the clock made possible the development of production and regulation, and that it marked the perfection toward which other machines had to aspire, may be seen in retrospect as a symbol of the Christian, especially the Protestant, pragmatic ethics. From the sixteenth century on the clockwork became the bourgeois, later the communist ideal, a quasi-religious object for both, and the paradigm of praiseworthy conduct. The history of the machine civilization which culminated in technological man, and of which the clock is such a distinguished citizen, is the story of a powerful efferent adaptive pressure by man upon his environment. It is also a good example of the autogenic functions of the mind: cities, highways, the production of metals, and the lighting up of streets at night are embodiments of symbols that were generated entirely within the confines of the internal landscape, with no prior correspondence in the external world.

Lewis Mumford in one of his inquiries in the history of modern technology sees three successive, but overlapping and interpenetrating phases in the development of the machine and machine civilization. He calls them the eotechnic, paleotechnic, and neotechnic phases.[37] The eotechnic phase was characterized by the use of water power and wood for material; in Europe this is roughly the period 1000–1750. Development during the paleotechnic period was characterized by the continuously increasing use of power chiefly for the purpose of decreasing the time necessary to perform a specific task. Since the time so "saved" could still be lost by disorganization, mechanical or human, the increasing regimentation of the work and workers became part of the over-all time-saving scheme.

Whereas the Christian "now" was valuable as part of the cosmic-teleological scheme, the political "now" as an instant of building the community, the "now" of technocracy became a commodity. Medieval man could be deprived of eternity; technological man can be deprived only of his present, because the technocratic "now" may be saved or wasted, owned or stolen, bought and sold. Some time during the nineteenth century, in the industrialized nations of the West, the use of the worker's time on the one hand and, on the other hand, his individual destiny in time came to be seen as two separate issues. Considering that in the same nations almost everyone is a worker of some sort, this amounted to a cleavage between what people do and what they are. This has probably always been the case as far as practice goes but not in principle; the industrial revolution ritualized this cleavage in principle and thus made a frontal attack on the integrity of the ego. This threat prepared the way for the rise of single-minded political ideologists that propose to bridge the gap between the self and the community by offering ethical concepts based on their versions of God, city, and machine. National Socialism championed the cause of a specific race; Communism deified the cause of a special class; the utopianism of modern participatory democracies the cause of the industrial merchant. In this kaleidoscope of trends, the state of contemporary man is complex but hardly less intelligible than the state of man in any other epoch, provided the observer is at peace with his perspective. I wish to offer such a perspective from the point of view of time as conflict.

Temporalities appropriate to technologies are those of physics: machines operate by atemporal, proto- and eotemporal laws. In their turn, however, machines are under the control of life and the passionate faculties of man, thus they serve, or ought to serve, bio- and nootemporal processes. But the power of industrial production, hence the promise of food and shelter for all, is seen to issue from processes of primitive temporalities. This fact tends to guide our views so as to suggest that functioning in the primitive Umwelts might be the saving grace for mankind. This impression is then reinforced by our instinctive search for strong predictive capacities—and, as we have learned, the powers of prediction are more reliable in the lower Umwelts than in the higher ones. Our ideas about time—including "my fate in time"—when evaluated under such guidance, tend to be reduced to temporalities appropriate to things or to livestock at the very best. But, things and livestock may be owned and stolen, bought and sold. Any idea of the good derived under similar inspiration is apt to be appropriate only for things and livestock and remain pedestrian from the point of view of the mind. Ethics so guided will concern the good of the machine and make the world subject to efferent adaptive pressure exerted by technical specifications.

It will be recalled that knowledge may be classified by reference to the personalities of people who create it and who, in their turn, are formed by it. Analogous arguments hold for ethics: individuals and groups of individuals have certain predispositions regarding praiseworthy conduct which, when evoked, produce a degree of consistency in response.* Thus, for instance, the good for the inhabitants of technocracies is defined by the views of people whose personalities correspond to the technological-scientific image of man. Perhaps axiology ought to be understood as the science of preferential behavior. This is not a new idea. It has been suggested and pursued in some detail by Charles Morris in several consecutive studies [38] and, no doubt, by others. Morris formulated thirteen "ways of life," which I would describe as moral preferences that an intelligent and interested twentieth century observer of the human scene might identify. He reaches no universal conclusions beyond modestly noting that there is an "orderliness and structure in the domain of human value," [39] a conclusion not in itself new. The motto of the book is taken from the sixth century B.C. Chinese philosopher Lao-tse: "In the affairs of man there is a system."

In Morris's work the system comprises the biological, psychological, and social determinants of value. Our interest would be to find out whether an identifiable relationship exists between views of time and judgements of the good in the particular context of the modern technopolis. Whether or not a sufficient amount of controlled information exists to deal scientifically with this matter has been considered briefly by Doob.[40] He leaves little room for doubt that the philosophy of values is developed side by side with temporal perspectives and that they interact continuously. But, because of the complexity of the questions and the methodological restraints of scientific reasoning he excuses himself from having to reach conclusions of sufficient universality. We must, therefore, leave his work (the best available scientific summary) alone in the hope that critical work will mature, in due course, and turn our attention to speculative opinions.

---

* Putting aside national and ethnic typologies reflected in daily wisdom, occupations have certainly been associated with typical constellations of Gestalt, behavior, and ethics:

> Soldiers brave, sailors true,
> Skilled physician, Oxford blue,
> Crooked lawyer, squire so hale,
> Dashing lover, curate pale.
> *Annotated Mother Goose*

We may begin by considering an example of well-documented journalism, such as Alvin Toffler's *Future Shock*.[41] This is an exploration of the confusion of universal values in America, produced by the flight from the conflicts of time into states of almost pure becoming. On some five hundred pages Toffler marshaled an overwhelming amount of evidence pertaining to the fragmentation of identity, life styles, and values. This breakup is the result of the frightening efficiency of the externalized functions of man: speech, touch, muscle, and deductive thought, all relegated to technology—though these are not his words. The book takes its title from the view that because of the incessant emergence of novelty, the individual finds himself in a different Umwelt day after day, hence lives in a continuous culture shock. The burden of the data comes from contemporary America with its peculiar protestant-pragmatic heritage, leaving open the question whether civilizations with different psychosocial bases would or would not generate the Western type of technocracy.

Let us for a moment imagine that Roger Bacon, who in 1278 was condemned to prison by his fellow Franciscans for certain suspected novelties of his teachings, has written a book called *Past Shock* on the dangers and debilitating effects of permanence. He could easily have done so. The ideas of praiseworthy conduct which guided the actions and informed the lives of his fellows, from their cradles to their graves, were moldy schemes of tradition. He might even have predicted revolts against this state of things because the nature of man, he would have argued, finds pure constancy an unbearable burden. Such a writer need not remain imaginary, however. There were many of them, and they produced that vast sweep of European change that became known as the Renaissance. Thus, though some of the features of "future shock" as described by Toffler are certainly characteristic of our epoch, and of our epoch only, the most important element of that condition is one of revolt against the ritualized prevalence of this or that view of time. But this is neither unique nor in any way new; on the contrary, it has been at the basis of many upheavals of social change. The Renaissance is only one example. The history of the idea of time which we have discussed on the preceding pages, illustrates many others.

An examination of the malaise meant by future shock suggests that the scientific and industrial revolutions have turned the oppressive medieval emphasis on tradition to an equally oppressive modern emphasis on no tradition; we have gone from time-as-being to time-as-becoming. The chaos of American values is the hubris detested by the Greeks. The continuous lip service paid to the future corresponds to the instinctive concern of all creatures with the future and resembles that of the adolescent who builds castles in the air. In this respect, it is important to note that only after his identity crisis has been weathered does the young man learn that aspect of continuity which carries him beyond the nest building of birds to the intentionality, care, and concern of man.

John Cohen, an able observer of the contemporary scene, described the situation of contemporary man by noting that we are emerging "from a period of neo-Zenoism which has lasted for two hundred years, a period during which things were made to last . . . into a neo-Heraclitian age . . . when only the ephemeral will count." [42] Continuity, as I see it, is supplied by the demands of the biological order. They pave the way for a swing back to the status of time-as-being as well as becoming. Meanwhile, the values of the society that concentrates on becoming, remain tied, as to a paradigm, to the demand of the child for immediate satisfaction, the pleasure principle sine qua non. From this tie follow the veneration of youth and of rapid achievement, and other related attitudes of the wealthy proletariat.

From the point of view of natural selection, a society in the frantic presentness of continuous becoming is detached from the tempo of the biological evolution of its environment. The origins of this detachment may even be traced back to the priestly world view of Genesis.[43] "So God created man. . . . male and female created he them. . . . and God said unto them, be fruitful, and multiply, and replenish the earth and subdue it: and have dominion over the fish of the sea, and over the fowl of the air, and over every living thing that moveth upon the earth." (Gen. 1:27–8). Certainly, the drive to control and dominate was not created but only expressed by the words of Genesis, but through them it acquired divine sanction. The crisis was precipitated when the human biomass and the forces of efferent adaptive pressures increased beyond any limits the Yahwehist could ever have imagined, and when the accountability of man for his destiny to Nature's God was lost. In this new world, life, that of man and of all other creatures, could not adapt fast enough to the demands of the internally generated imagery of the mind. This inability has led to an inversion of evolutionary roles, a condition which may best be described as the revolution of the environment. Nature at large, but especially the living world, is incapable of abiding by the standards of the good appropriate to the primitive Umwelt of the machine, a condition which the mind has been trying to foist upon the living. In ethical terms, this is a demand for the rewriting of some of the codes about what is to be judged as praise-worthy conduct.

Perhaps the Platonic vision of political good as an unchanging order had reached the end of its useful life. The good consistent with such a vision deteriorated to a shock of becoming, a caricature of progress, against which man must revolt because of his peculiar psychobiological organization, and nonhuman life revolts by retaining such rates of evolution as are possible to matter in living form.

The Good of Life

In the voices of an unorganized minority of the technocratic society, sometimes described as a "counterculture," [44] certain new preferences for conduct have become articulated during the middle decades of the twentieth century. These preferences may be associated with the mythological figures of Orpheus, Narcissus, and Dionysius—sensuousness, beauty, and celebration—rather than with the figure of Prometheus "who first transmuted atoms culled from human clay." The emergence of the counterculture represents the confluence of reactions to several traits associated with Protestant pragmatic ethics. They include such elements as cleanliness, punctuality, penuriousness, monogamy, and diligence; all of them are useful virtues of the merchant class because they assist in the pursuit of the devil within by chasing after goods without. Moral worth and praiseworthy conduct appropriate to this ethical system may be summed up by insisting on the command: subdue and rule over all things on earth. Work, which is regarded as the best means to achieve dominion over things, has been considered sometimes a curse, sometimes a blessing. "In the sweat of thy brow thou shalt eat bread" (Gen. 3:19), versus, "For thou shalt eat the labor of thy hands; happy shalt thou be." (Psalm 128). In most primitive societies work, in the modern sense, was undertaken only by women, as witnessed among other sources by the language: "labor" and "travail" signify "work," "pain," and "childbirth." St. Augustine, Luther, Calvin, Marx, and Marx's many children preached against idleness and praised work on moral grounds; they were the hardliners. Rousseau, Darwin, and Spencer saw work as biological necessity; they were the softliners.[45] We might add Aquinas and Descartes praising man the thinker rather than the worker; they were the brainliners.

By mid twentieth century, the association of creativity with what is organic, retained in our linguistic imagery of "labor," was lost from the concept of work. Productivity replaced creativity as the socially most desirable feature of daily activity. By productivity I mean the making of things or bringing about conditions which are more of the same, such as cheaper cars or a better glue; by creativity I mean the identification of patterns in the unpredictable. Emphasis on productivity was a victory for the hardliners who came to be the ones who determine what should constitute the collectively approved good for society. One eventual result was a powerful adaptive pressure exerted by man on his environment and, because of the reciprocity of natural selection, a correspondingly strong, though peculiar, adaptive pressure upon man. What had transpired may best be illustrated by a symbolic sketch of the American experience.

In the late eighteenth century William Blake saw the traditional values of continuity embrace a dedication to the new, in the vision of America:

> The Guardian Prince of Albion burns in his nightly tent:
> Sullen fires across the Atlantic glow to America's shore,
> Piercing the souls of warlike men who rise in silent night. . . .
> "America, A Prophecy" (1773)

A century later Walt Whitman could praise the multiplicity of choice and rugged individualism

> I hear America singing, the varied carols I hear. . . .
> Each singing to what belongs to him or her and none else
> > "I Hear America Singing" (1867)

Again a century, and the voices of two sensitive, latter day Hebrew prophets, Simon and Garfunkel:

> And in the naked light I saw
> Ten thousand people, maybe more
> People talking without speaking
> People hearing without listening
> People writing songs that voices never share
> No one dares disturb the sound of silence
> > Paul Simon, "The Sounds of Silence" (1964)

Reflected in the sequence of these three poetic reports is the gradual loss of the organic, and an eventual cry for identity, in a world of Toffler's modular man, hurry-up-welcome, rent-a-person, engineered message, and all the ugly manifestations of fractured and confused temporalities. This progress of increasing detachment from the scheme of things may be described as getting into a progressive rut (as fatal as the conservative rut) and is also known as the ecological crisis. Which one of these names we select for the same syndrome of detachment in social history depends on our temperament and training. Lynn White has traced the origins of the crisis to the emergence of an entirely novel democratic culture, itself the confluence in the nineteenth century of an aristocratic, speculative intellectual science with the empiricism of a lower-class, action-oriented technology. The issue, as he sees it, is nothing less than whether the democratized world can survive its own implications. He believes that it cannot unless it rethinks its axioms.[46]

In our technological societies the medieval view of man's superiority to nature lives on unharmed, even though not acknowledged; it is as powerful in the Christian West as it is in the Marxist portions of the world. The dogma of man's dominion over nature is built into all actions that are science based, even into those proposed to alleviate the situation. Thus we have the curious calls for more technology to curb technology, and for more science to control science. This is, as we have just reasoned, quite in the mainstream of Christian philosophy which has no tradition of preaching humility toward fellow creatures. The pantheism of St. Francis is rather an exception,[47] yet, if religion is needed, pantheism may be the only modality which can afford to reinsert man into nature and yet retain a claim to transcendence.

Would countries with different cultural traditions develop the same technocratic illnesses which are now displayed in the United States, if they were to industrialize to the same degree? The answer seems to be in the affirmative, except insofar as they may learn from the American experience. At least this was the conclusion of a number of well-qualified speakers at a recent UNESCO symposium on culture and science.[48] The sense of that symposium was voiced by Professor Shuichi Kato who described the dangers, in the technocracies, of a self-perpetuating "infernal machine which is crushing us and leading to the stifling of culture." He held that the pressure of the Third World countries could in principle be beneficial but is unlikely to be effective; hence, he concluded, changes must originate in the advanced countries themselves. This opinion, from a distinguished scholar of a country which displays the most remarkable contrasts of the strongly traditional and the wildly new, namely Japan, is sufficiently convincing—and is reinforced by the personal impressions of this writer. Thus, considering the possible fate of the "good of life" as a revolt against the "good of the technopolis," we might as well revert to the contemporary American experience.

I reasoned that the young of all species must appeal to the mature adult in order that its own survival be assured. Probably for the first time in the history of organic evolution, certainly in the evolution of advanced organisms, this demand has been partially eliminated in some segments of the American technocracy. In prior civilizations, from the urchins of ancient Rome to the nameless waifs who roamed the countryside after the Russian revolution, the young have been more often homeless than not; but the children of past havoc were still subject, in no uncertain terms, to the mores of their elders—at least if caught. Not so the well-fed and able minority of Roszak's "counterculture," the children of plenty, in command of their parents' wealth and of their society's tools but with no overt obligation to the past. Whereas displaced persons of war, flood, or pestilence desire nothing more than the reestablishment of prior conditions, displaced persons of the affluent society, those, that is, who do not vanish as the flotsam and jetsom of technocracy, tend to carry on as harbingers of a new, even though ill-defined and ill-understood, era. Such conditions should be ideal for the voice of a new drummer to be heard. The question concerns his identity.

The drummer does not need to be a metaphor, if we agree with Harvey Cox's complaint that what the technopolis had killed in man are the capacities for festivity and fantasy.[49] Festivity, he says, is the uniquely human capacity for revelry and joyous celebration; fantasy is the faculty "for envisioning radically alternative life situations." Festivity, he remarks, is predominantly past-related: we usually celebrate an event that has already occurred, seldom something that is yet to occur. Fantasy is predominantly future-related: we often fantasy the past, but the "radically new situation" aspect of imagination is usually connected with the future. He notes that following Luther and Marx and Aquinas and Descartes (the hardliners and the brainliners) man became regarded as a creature of work and thought, not as one of celebration and fantasy. The result is a "deformed man whose sense of a mysterious origin and cosmic destiny has nearly disappeared."[50]

Removing the faculties for festivity and fantasy amounts, therefore, to the removal of man from history, to a plunging of man into an eternal, timeless present. The ersatz festivity and ersatz beauty of organized media, spread only the "engineered message" and add insult to injury. "Imagineers" and "futurists" as suggested by the morphology of the words, attempt to set right by reductive rationality the wrong which reductive rationality has wrought. The alternative is to proclaim such visionary imagination that the claims of technical expertise must, of necessity, withdraw. "We must be prepared to entertain the astonishing claims men like Blake lay before us: that there are eyes which see the world not as commonplace sight or scientific scrutiny sees it, but see it transformed, made lustrous beyond measure, and in seeing the world so, see it as it really is." [51]

Those who have been fortunate enough to be close to a pilgrim of the counterculture cannot be left unmoved by the spectrum of its red, white, and black blues. Like Huxley's remark quoter earlier, on the impossibility of not reacting to the tom-toms, and Justice William O. Douglas's experience of reacting to the drums in the Himalayas, only the archstaid and the dull will not sense the throbbing emergence of . . . something.

> The answer my friend, is blowin' in the wind
> The answer is blowin' in the wind.
>
> Bob Dylan, "Blowin' in the Wind" (1962)

Still, the long range effects of the counterculture would have been negligible had its spirit not communicated itself to a larger and more widely based group. It did so by giving rise to the women's liberation movement. The genesis and dynamics of the women's movement in America, from the point of view of natural selection, are identical to those of the "youth quake." In the long and awe-inspiring history of the female of our species— even with many details missing—woman is seen to have been subdued by the cunning of the male, his physical strength as provider and as the "maker" of women, and woman has been retained as the reproductive organ of the family.

The early social status of the female is not well known and guesswork has been rampant from John Stuart Mill to Marx and Freud. The Biblical evaluation of woman, "and thy desire shall be to thy husband, and he shall rule over thee" (Gen. 3:16) has proved to be correct for a long period and formed, doubtless, the basis of a complex selective pressure exerted upon woman. In contemporary America the nature of this selective pressure has become substantially modified. Though women in the past, individually, have fought their wars, suffered their states, murdered their lovers and rivals and, occasionally wielded considerable power through their males, they were still subject in no uncertain terms to social mores promulgated by males. Not so the well-to-do, well-educated minority of American females, women in command of men's wealth and their society's tools but with no overt obligation to the male. Just as the youth of the counterculture among the American young, so has a vocal portion of American womanhood become the champion of the ascendency of woman or, more importantly, of the reinsertion of the feminine among the patterns of praiseworthy behavior.

The highly advertised causes of unrest among the educated, middle-class women of America, such as the tediousness of motherhood and house-wifery, or discrimination in the market place, while true enough, amount to no more than a caricature of the deep, underlying causes of the malaise. This cause is the oppressive rule of the masculine as the normative pattern of technocratic ethics. The dominant male identity, as Erik Erikson remarked, "is based on fondness for 'what works' and for what man can make, whether it helps to build or destroy." [52] He continues, "the play-constructing boys in Berkeley" (whom we met in chapter 9, section 4) "may give us pause: on the world scene, do we not see supremely gifted yet somewhat boyish mankind playing excitedly with history and technology, following a male pattern as embarrassingly simple (if technologically complex) as the play constructions of the preadolescent?" [53] As he sees it, the nuclear age has brought male leadership close to the limits of its adaptive imagination.

The unrest of the middle-class American female would have remained almost as inconsequential as that of the counterculture, had it not communicated itself into a more universal sphere, namely, that of a revolutionary reevaluation of what may be regarded as praiseworthy conduct in matters of sexual relations. Not that the post-technocratic ethics of sexual behavior is a result of any single cause: the scientific advances in birth control, the desirability of separating sexual pleasure from reproduction as the result of overpopulation, the breaking up of the traditional family structure, and many other elements figure in it in ways that are very intricate. But the slow yet apparently inevitable spread of new sexual ethics touches upon everyone on a biological level, simpler and more basic than that of the mind, and certainly universal to mankind. C. D. Darlington, in his massive study of *The Evolution of Man and Society,* considered the reason, instinct, and morals of paleolithic man. He concluded that "far and away the most important part of their behavior and morals [were] concerned with their breeding systems and it is here that we can see their profoundly instinctive basis," [54] a conclusion which could also have been reached from Freudian precepts. The importance of the biological contribution to behavior probably has not changed since paleolithic times. If we were to select a domain in which to look for a change from the good of the technopolis to the good of life, for signs of the reinsertion of the organic into the concept of labor, this should not be in politics, or economics, but in the very stuff of life expressed in sexual mores.

We have seen how the powers of the mind created a condition which imperils life. In the new, emergent ethics, even if its details are hazy, I perceive a reassertion of the powers of life over those of the mind, a corrective redress. Invoking the American counterculture, the women's liberation movement, and changing sexual ethics does not intend to imply a simple-minded cause-and-effect process. But I do believe that these transient social phenomena represent stages in the reaction to technocracy. Now, as long as the irritation with the overemphasis on becoming exists, the correcting forces will also be operative, pushing ethics from the technocratic-analytic to the organic.

4. The Good, Emergence, and War

WE WILL REVERT now to the problem of ethics in its evolutionary significance, mainly in the context of what, next to death and sex, has been mankind's most faithful companion: cruel conflict.

Problems of Stagnation

Although the deep involvement of good and evil in the affairs of man is all too evident, any attempt to tie moral standards to evolution by natural selection is difficult. Many of the qualities held most noble in man, such as charity, compassion, self-sacrifice, have no evident survival value for the individual. Social practices suggested by these qualities, such as the protection of the lives of the feeble, the ill, and the aged have, if anything, negative survival value for the species, as judged from the increase in the genetic load which they produce.[55] Other associated practices such as burying the dead, worshiping one's ancestors, or building comfortable homes for invisible beings have no clear-cut usefulness in adapting to the environment. In the face of these and similar difficulties, C. H. Waddington has sought to identify the function of ethicizing in man with the role it plays in promoting the general thrust of human evolution.[56] He reasons that human evolution, being almost entirely in the mental sphere, derives from man's capacity to transmit information from generation to generation by sociogenetic means. In its turn, this is possible only because the infant is an authority-bearing system of ethical beliefs. Biological wisdom, then, counsels ethical stances that encourage progress, "both of the mechanism of the sociogenetic evolutionary system, and of the changes in the grade of human organization which that system brings about." [57]

Waddington's reasoning certainly rings true, but it does come up against the same criticism of being static, or stagnant, as does the ordinary, afferent version of evolution by natural selection. Consider, for instance, that the working morals of the caveman, whatever the morals were, served him for one or two million years as stably as the "morals" (preferred behavior) of cockroaches served them for perhaps a thousand million years. Why, then, were the stable tenets of caveman morality abandoned for will-o-the wisps that counsel systematic bloodletting and cruelty, and produce mainly anxiety with only occasional states of satisfaction? The following up of dreams, and the adjusting of the norms of praiseworthy conduct to fit such dreams are difficult to justify as adaptive steps, because, instead of producing equilibria by way of redress, they are more likely to produce disequilibria in need of redress. Perhaps the usefulness of the good in the context of evolution is that of favoring disequilibria, but not just any kind—only specific kinds; namely, those which, on the average, favor conflict-resolution by emergence (leading to new conflicts) rather than by catastrophe or collapse (elimination of conflict). In this frame of thought, the question of ethics is shifted from what moral behavior is, to what it does.

Just as temporality and truth are relative to species-specific Umwelts, so, I believe, is the question of good. Consider the fact that perceiving the truth of an Umwelt, which all organisms do, remains useless unless it can be incorporated in the struggle for survival. This condition implies the possibility of misjudgement of the truth—useful truth—hence an attendant existential tension. The instances of tensions which we discussed earlier, those between the expected and the encountered, should thus be seen as a spectrum of tensions stretching, qualitatively, from the ruinous to the fulfilling. Applying a Heideggerian expression in an un-Heideggerian way, this is a spectrum of Sorge (concern, care), or rather, not even one spectrum but many spectra, corresponding to many Umwelts. That the burden of ethical problems concerns human actions, and that even evolutionary ethics must start with man and then grope backwards, does not signify that the concept of good may not be given any meaning for other life forms, but only that there is such an explosively large step in the continuity of ethics between animals and man that it looks like a discontinuity.

In the biological world good and bad somehow correspond to the extreme polarities of pleasure and pain, at the ends of a gradual scale of feelings. Organisms seem to have little difficulty in distinguishing various degrees between the two extremes, although in higher animals the issue is sometimes muddled by becoming mixed with behavior which might sometimes be described as guilt and pride. It is in the mind of man, in the internal landscape of symbolic forms, where the continuous spectrum of feelings, attitudes, and arguments tends to relinquish its continuity and polarize into good and evil. Certainly more people have died because of ideas, whether for or against, than for the possession of bread or company of woman.

Whereas the tendency to polarize the ethical spectrum is universal, the relativity of good and evil with respect to epoch and place is well known; this relativity per se has been perhaps the most intriguing, yet neglected, aspect of moral philosophy. As here envisaged, only ethical aspirations as such can be regarded as a universal need of man, amounting to a necessity to find guidance in the labyrinth of his autogenic imagery, but not the specific ways in which moral conduct is judged or implemented. Unavoidably, in addition to "horizontal" variations in ethical judgement among con-

temporary individuals and groups, we find a "vertical" hierarchical relativity. Thus, specifically, what is praiseworthy for society is not necessarily
so for the individual; what is morally right for the individual mind may be
wrong for the individual's life; what is good for the individual's life, may
be ruinous for the integrity of matter. If our speculative assumption is
correct about the usefulness of the good in favoring disequilibria that lead
to conflict-resolution, then we should find that favored behavior in animals
should subordinate matter to life, and that moral law in man should also
subordinate matter to life, life to mind, and mind to society. The problems
that underlie the identification of such a schema may be described as
epistemological—even if it addresses itself to ethics. (After all, we are
seeking truths about the good).

We have seen how the identity of self is a prerequisite for the separation of the knower from the known. The same can be said for the lover and
the beloved, and the moral actor versus the object of action. This demand
is implicit in the Freudian concept of the establishment of the superego, for
if there were no ego, then there would be nothing into which social standards could be incorporated. Piaget's findings on the origins of morals in
children do not contradict the emphatic prior need for identity.[58] The
ambiguities of moral choices with which the adult finally learns to live
remind us that the region left undetermined on each level (this time that
of man in society) is in some ways always broader than the regions left
undetermined on the lower levels.

The hierarchical structure of Umwelts suggests that when ethical arguments lead to cruel encounters among individuals or groups, such encounters be understood as many-leveled interactions. On each level we find
level-specific encounters (group with group, mind with mind, life with life,
matter with matter), each with its level-specific laws and various degrees
of undeterminacies. It is also easy to identify vertical conflicts which, however, in all but their simplest forms are prohibitively complex for practical
analysis. Thus, groups as supra-individual integrative units exert adaptive
pressure on other groups, on the minds of their own members, on all of life
within their reach, and on all of matter; in their turn, by the mutuality of
adaptive pressure, each of the subjects we designated as having been acted
upon, reacts. With a bit of fantasy, even chemicals may be described as
struggling with each other: common salt loses its crystalline identity when
dissolved in water.

Conflict, in the general sense of encounter here implied, seems to be
the ordinary mode of existence for animals and man alike. Among men,
cruel encounters may originate from any one, but usually from several of
these levels. This view differs from the purely Freudian stance which sees
the main causes of war in neurosis; from the Marxian theory of dialectical
materialism which sees it in economics, that is, in the control of material
goods; and from the Hegelian interpretation of strife as the working out of
the spirit alone.

Problems of War

Whatever the actual and/or stated causes of an armed conflict may be, during its course, the situation of unprivileged individuals will almost certainly deteriorate, and, if it lasts long enough, it will reach the dog-eat-dog situation: life regresses to defend life, even if in the beginning it looked as though people were fighting for ideas. The strong emphasis which all legal systems since the Code of Hammurabi (early second millennia B.C.) have placed upon controlling the taking of another man's life demonstrate the ubiquitous and perennial nature of cruel conflict, whether man against man or tribe against tribe. The essential question asked by philosophers has been whether or not war can be eliminated or replaced by some other social activity. Contemporary social science has not contributed substantially to the debate because, following the path of positivism, it has been generally adverse to speculative theories in favor of minutiae and "facts." Accordingly, broad enough views on war in the tradition of Montesquieu do not exist. There has been an increasing interest into inquiries about aggression in animals, following the inspired pioneer work of Konrad Lorenz [59] but, though the resultant findings might help remove the halo from around the head of Mars, they cannot lead to peace on earth. For scientific truth becomes action only through an active moral philosophy.

Stanislav Andreski, a sociologist, has credited war with producing or contributing to many of civilization's most cherished fruits: civil government, democracy, industrialization, culture in general, and the arts. He sees war as a product of civilization, but, unlike Freud, he remains on the surface and invokes only the basic desire for survival. He claims to see the ancestry of his thought in the work of the third century B.C. Chinese legalist scholar, Han Fei Tzu, who held that people increase fivefold in each generation but goods do not. "So the people fall to quarreling and though rewards may be doubled and punishment heaped up, one does not get away from disorder." [60] Han Fei Tzu and Andreski are pre- and neo-Malthusians who hold self-evident the claim that consumption increases faster than production, hence overpopulation with respect to resources is inevitable, as are all its cruel consequences.

An anonymous author has recently published a book called *Report from Iron Mountain on the Possibility and Desirability of Peace*.[61] The work has no scholarly pretension and is clearly a parody, but it does display a better understanding of man than do the many utopian reports of scientific peace research organizations. The author suggests substitutes for the functions of war, and constructs economic, political, sociological, and cultural models of activities which might fulfill this task. They include such measures as acceptable threats from outer space, slavery in a technologically modern and conceptually euphemized form, and perhaps purposeful blood-games. The proposals sound uncomfortably close to the matter-of-fact, journalistic reports of Alvin Toffler.

Reverting now to Malthus, we note that he connected time to moral philosophy by pointing to the change in population/goods ratio as the major determinant of ethical conduct; it was also Malthus who set Darwin and Wallace on the train of reasoning which led to the principle of natural selection. To Malthus himself, a mild-mannered minister, nature's solutions for the misery produced by the different rates of increase of man and food were war, epidemics, starvation, and vice. To him, the obvious means of control was through moral restraint, mostly in matters sexual. It is unprofitable to argue Malthusianism in its original form pro or con. He did not, for he could not, anticipate the revolutionary changes in production methods, in birth control, in the general attitude to sex that distinguishes reproduction from experience; and he could hardly have foreseen the subtle and profound biological measures toward population control taken by the "biological wisdom" of biomass in society: homosexuality, lesbianism, drugs, self-elimination of neurotics, and a trend toward impotence among certain large population-groups of technocracies. Neither did he command the vantage point for seeing his ideas as no more than examples of a wide range of rate variations intrinsic in the nature of evolution.

In chapter 6 I pointed to the different evolutionary rates of matter, life, and mind and stressed that the higher levels tend to outrun the lower levels by their rates of change. Since matter does not evolve rapidly enough to keep up with life, life to keep up with mind, mind to keep up with the collective needs of society, there ensues a series of vertical selective pressures. Though these pressures always do and did exist, they became significant only when the density of life became appreciable with respect to the density of nonliving matter, and when the density of Homo sapiens became appreciable with respect to the density of living and nonliving matter. Then and there substantial evolutionary inversions followed, wherein the lower levels were incapable either of changing or being changed sufficiently rapidly to suit the rates of evolution of the higher levels. The idea of such a broad spectrum of inversions may be called a generalized Malthusian principle. In some form or other it is likely to stay with mankind for all the foreseeable future—unless we begin to colonize some distant planets. And, just as the Malthusian conditions influence, even if they do not determine, the structure of human ethics on a broader basis, the nature of the good has been and will be, doubtlessly, influenced by the unavoidable stresses that follow from the differences among evolutionary rates.

I do not think that the identification of the lawfulness of wars is possible at this time, and certainly not through those methods of science (in whatever branch of science) which are properly capable of describing only the proto- and eotemporal Umwelts. This writer witnessed from a privileged position the coming into being and the resolution of World War II. First, the emotional power stored in the (mostly) unconscious content of language was released through primordial rituals of crowd behavior. Then followed the onslaught of civilized and uncivilized barbarians upon one another. The

intensity and magnitude of cruelty and suffering which members of our
species are ready to mete out as well as willing to endure is nothing short
of incredible. It became painfully clear to this witness (though perhaps
completely nonsensical to others) that one must evoke not the absence of
ethical judgement in man, but rather its presence, if one wishes to account
for the frenzy of the ethical animal. Our epoch is especially one of unparalleled moral indignation, where beneath the noise of armies and
merchants determined to save everyone, man might well perish.

> And here we are as on a darkling plain
> Swept with confused alarms of struggle and flight
> Where ignorant armies clash by night.
>   Matthew Arnold, "Dover Beach"

Since it is life which is threatened by the destructive advance of the
mind, what is called for by way of ethical norms is a redress of the balance
between the frightening capacity of the very complex and the limits and

----

*The cruelty which members of our species are ready to mete out and the
suffering which they are ready to endure in the name of ideas, are
nothing short of the incredible. It is painfully evident that we must
evoke not the absence of moral judgement in man but rather its presence, if we wish to account for the frenzy of the ethical animal. But the
utility of ethical judgement for the survival of the species, that is, its
role in natural selection, must remain obscure as long as moral choice
is regarded as essentially determined or determinable. It may, however,
be accommodated within a system of natural philosophy if one realizes
that unlike scientific truth, which suggests vistas of unchanging eternity,
claims invoked in support of moral choice pertain to man's capacity to
determine the future, hence they pertain to acts of creation.*

*Kaethe Kollwitz, The Carmagnole, circa 1900. Courtesy, Galerie St. Etienne,
New York. The Carmagnole is a social folk dance, sometimes introduced by a caller such as the drummer boy on the right, and is danced
around a real or symbolic victim. In this drawing the dance is around
the guillotine. The drawing is believed to have been inspired by the
words of Dickens in A Tale of Two Cities: "There was no other
music than their own singing. They danced to the popular revolutionary song, keeping a ferocious time that was like a gnashing of teeth in
unison." The song to which Dickens refers was probably the Jacobin
hymn that ends in the refrain:*

> *Vive le son, vive le son,*
> *Dansons la carmagnole,*
> *Vive le son du canon!*

capacities of matter in the living mode. All in all, a deemphasis on the noetic appears necessary.[62] But if this be so, and if principles of ethics do, in general, tend to be useful because they assist the emergence of new integrative levels, then the new level here indicated is not a further extension of the very complex, but something else.

5. Duty, Responsibility, and Temporality

UNLIKE the scientific search for truth which explores the regularities of the past and seeks from them permanent statements applicable to future contingencies, ethics must be directed primarily to the future, as it seeks to identify not what must be but what, among possible alternatives, ought to be. If we remain consistent with our prior reasoning, then what ought to come about are conditions and events which have emergence value. This demand is not stranger than the concept of survival value, an idea solidly embedded in the theory of evolution by natural selection. Emergence value is more difficult to identify, however, because whereas successful provisions for survival may be subject to immediate or almost immediate tests, the emergence of new forms and, especially, integrative levels can be tested only through longer periods. But longer periods are precisely what we usually cannot afford in acting upon moral principles. Can the evolutionary role of ethics be of any use in the generation of behavioral rules in daily life? Perhaps it might, but first we must understand the position of moral law in the theory of knowledge.

Let us recall that the laws of each integrative level had the following two general properties: (1) they arose out of conditions left undetermined by the laws of lower order (demanding, therefore, that the new regularities do not violate the earlier ones); and (2) each new level displayed an increasing region of undeterminateness of its own. How would these apply to moral principles? (1) They would have to abide by lower-order lawfulness. Man cannot be expected to fall upward, reproduce asexually, or dispense with his conscious/unconscious dichotomy. Ethical principles may, however, demand that a disciple climb a mountain, refrain from reproducing, or repress some content of his conscious experience. (2) Ethical principles qua laws of nature would have to leave a sufficiently broad region undetermined for yet another level of integration to arise. This cannot be illustrated at this time because we cannot distinguish between the limits of our ignorance and the necessary outer boundaries of ethical principles. We do not have enough members in our set of mankinds for a fair sampling. This poverty of sampling has been expressed by thinkers, such as Eric Voegelin, by saying that society is not an event in the external world nor an object of that world.[63]

Granting that epistemological rules such as these do apply to moral laws, how can we judge something for usefulness in emergence? Consider first an isomorphic problem pertaining to life. Can we test specific biologi-

cal structures or behavioral patterns in animals for their usefulness in organic evolution by natural selection? We may indeed do so, provided the
new conditions have been predetermined by man the experimenter. The
same goes for moral principles. We may be able to judge their usefulness
if somehow the new conditions are already known. This leads to circularity
which can be avoided only if we allow for the operation of creative thought;
this, in fact, is the way people have been getting to this problem in moral
philosophy from Plato, to Kant, to Lenin. It amounts to a juggling of ideas
of the true and the good until some combinations are found to suggest an
orderly process of principles. It is a process of reasoning which starts from
within, one that has a peculiar and well-established position in organic
evolution. If a trout could describe why it starves itself to death when
traveling upstream it might say that it has no choice ("it's my stomach, you
know"); also, it is "a far, far better thing" than he had ever done. Somehow,
action taken by the individual either because he had no choice or because
he judged it to be in his enlightened self-interest (such as to mate with the
prettiest nightingale in the watch), when considered from afar or in retrospect is likely to be seen as having promoted rather than hindered the
survival of the species in particular, and the progress of organic evolution
in general.

In the life of man the report of the trout finds no easy parallel. The
forthcoming death of an aging person might be inevitable but seldom a "far,
far better thing," because the mind believes itself to be in command of
eternal life. By this test necessities may be distinguished from ethical propositions. Whereas the former are ordinarily associated with cerebration, the
latter are in the form of feelings. According to the witness of many sensitive observers of the human theater, what people label as right decisions
taken along morally satisfying avenues, leaves them in a transient state of
feeling describable as being without conflicts. It is difficult to substantiate
such views experimentally, because laboratory tests must remain too feeble
to be significant. How many psychologists test a statistically significant number of subjects for their sentiments just before being beheaded to save a
friend? How many starve their subjects to death and monitor their personality changes? Or evaluate the self-images of ten thousand college
students (males) after they have slain another ten thousand college
students (males) subsequent to the second samples having raped the lovers
of the first group? While experimental work in this matter is difficult,
catechizing what specific actions may be classified as the sources of transient feelings of conflictless states, that is, the writing of moral laws, is
equally difficult. For some, at some time and some place, such principles
comprise the decision to live, or that of refraining from living; the decision
to embrace or that of not embracing. Yet perhaps our inquiry is not
altogether hopeless.

The subjective feeling of conflictlessness, or elation which often accompanies the discovery of truth (scientific or otherwise) may sometimes

be ascribed to the awe felt upon the contemplation of vistas of an unchanging, timeless eternity. But the elation by which we often recognize morally satisfactory acts is very seldom the result of glimpses of an unchanging world. Quite to the contrary, it is usually associated with some conditions which were created by that act. A moral action freely performed amounts to a determination of a future, up to then undetermined; it is not a view of, but a demonstration of sharing in, the act of creation.

These arguments do not remove reason from being instrumental in directing behavior, they only point to the limit of its usefulness. Stephen Toulmin has argued for admitting reason as the main, though not the only source of ethical judgement.[64] He sees the main advantage of rationalism in ethics in its superiority, by way of definiteness, as compared to pre-scientific ethics which "had nothing to offer to Everyman in his everyday problems but confusion." While this may well be true, a degree of confusion in moral judgement, unlike in scientific judgement, is unavoidable. Ethics based entirely or mostly on reason would be ill-prepared to meet the challenge of future conditions with no past history. Whereas some guidance by past experience is both necessary and desirable, the burden of decisions regarding the good must come, by the very condition of choice among undetermined states, not from knowledge understood but from knowledge felt. Only that way can it retain the necessary element of creativity.

These opinions resemble those of the intuitionists in ethics, but they differ from most of their suggestions. As here understood, moral principles can not be deontological, because they have a naturalistic basis in evolution, and because they are also utilitarian in the long run via their emergence value. But they are not teleological either, because they are not directed to a predetermined goal, and not even to a necessarily knowable one. Moral principles thus do not assume the existence of a class of values which once so identified would remain permanent. Only the ethical stress underlying moral decisions (the struggle, as it were, between good and evil) is given ontic status. These stresses are resolved by different people at different times into different rubrics of good and evil. Yet, the stance held here does not suggest that good is determined by some benefit to the greatest number, because quantitative conditions are largely irrelevant for the emergence of creative novelty in evolution.

It ought not come as a surprise that the idea of the good, with its nomothetic/generative aspects, also has a hierarchical structure. Consider, for instance, that in spite of the many opposing instructions in ethical principles, there exists a great deal of agreement among men about the referents of praiseworthy conduct. Even cross-culturally, sexual and family relations are usually strictly controlled, children and women protected more or less, the taking of life regulated and property safeguarded. These and similar concerns I would class as life protecting itself. Obligations which derive from them I would class as duty and regard as appropriate to the biotemporal Umwelt. Higher order obligations designed to protect man's "brain children," such as obligations that derive from teachings of political philosophies and advanced religions, and which implicitly assume man's awareness of his inevitable death and thus his sense of time, I would class as responsibilities.

Duty is binding on man somewhat as biotonic laws bind life: we must eat, yet not in any definite way, and one may even give up eating, though not without difficulty. Protection of the life of others on the biological level belongs to this category. Birds defend their young against predators but once the young is killed, there is no mourning; there is no symbolic world in which the dead baby keeps on living, their temporality is that of the creature present. The ethical and temporal states of primitive man, and of the primitive man surviving in each of us, are not much different.

Those moral obligations, which I called responsibility, concern mainly symbolic existents; since society's most important institutions are symbolic, responsibility is the obligation appropriate for man's noetic and social Umwelts. It is in this domain that the most profound emotions arise when questions concern moral conduct, and it is here where reason and our success in identifying truth as science can be of the least assistance. Whereas it is in the body social that a new and sufficiently advanced integrative level is likely to arise, one with such perspectives as to give guidance in moral conduct, the source of changes yet to be so manifested is in the individual mind, as I have argued earlier. Perhaps it is a case for society that parallels what for the mind is a regression in the service of the ego; society regresses to the individual mind to assure its survival and seek new paths. The parallelism between the social and individual retrenchment in service of social (and individual) integrity may be expressed in two declarations. The ancient Romans saw peace secured only if society was always ready for war: si vis pacem, para bellum; Freud saw the fullness of life secured only if the individual is always aware of his death: si vis vitam, para mortem.[65] Society cannot survive unless human life survives, and cannot further evolve unless the individual with his hierarchy of Umwelts and conflicts is permitted to explore freely those organizing principles of his autogenic imagery which we have described as pertaining to the good. But if that is the case, then there is no more important task to civilized society than to secure, first of all, the bodily well-being of its citizens, and second, to protect the free and creative play of the individual mind against the oppressive malefactor that the corporate interests of commissar, lawyer, merchant, banker, and soldier have come to represent.

XI

ARTS, LETTERS,
AND THE BEAUTIFUL

WE HAVE SPOKEN of timeless statements (truths) and of instructions for conduct (ethics) as guidelines for man in selecting future courses of actions from the vast store of his autogenic imagery. There exists yet another set of selection rules usually described as artistic and recognized by the universal experience of the beautiful. The selection rules of truth were precise though changing; the rules of praiseworthy conduct were much less precise and also changing. Although the generation and the contemplation of beautiful events, utterances, and objects have been recognized through the known history of man, the rules of artistic expression are so broad and so changing that the history of their variations is more striking than the scope of the rules themselves. This is not surprising, for artistic expressions and feelings for the beautiful pertain to the hazy, transitional regions between, on the one hand, the chaos of imagery of the very complex and, on the other hand, the conflicts classed as ethical and scientific knowledge. Popular parlance identifies artistic expressions with creative acts more often than it does scientific discoveries or ethical pronouncements. Perhaps this implies an awareness that the transformation of chaos into conflict is the most universally available mimesis of the cosmic process of creation. Ontogenetically and philogenetically, the three modes of knowledge—the true, the good, and the beautiful—are probably coemergent even if they may be discussed separately. In this chapter I shall examine artistic expression and the experience of the beautiful as the skill or capacity to roam freely among the many temporal Umwelts of man.

1. From Imitation
to Independence

PLATO was convinced that it is not by accident that certain things are called beautiful; he believed that we can recognize in them certain common characteristics which, therefore, ought to be definable. In the *Symposium* Socrates quotes Diotima of Mantineia on the essential nature of beauty as follows. He who wants to behold Beauty must use the beauties of this world as stepping stones in an unceasing journey toward the absolute, going from beautiful creatures to beautiful lives, from beautiful lives to beautiful truths, from beautiful truths "attaining finally to nothing less than the true knowledge of Beauty itself, and so knows at last what Beauty is." Beauty does not wane, nor wax, nor suffer change, hence, "this, my dear Socrates . . . is man's true home, with its vision of absolute beauty, if we have in this life any home at all."[1]

Eros, that leads the beholder in this journey, expresses its creative powers in fair words and deeds, such as in the poet's chants conceived in sacred madness. But, whether painter or poet, the artist participates in the beauty of the absolute by mimesis, that is both imitation and reenactment. Yet, the position of the beautiful in the scheme of things is not what untutored realism might suggest. A mirror, if it swings about, will also produce a sun and the stars, and the earth too, or at least their appearances.[2] A painter is also a producer of this sort. Consider, for instance, a bed made by a carpenter from a particular material: this is only an appearance of the reality of the bed as made by God. (The reality is the "bedness" of the bed.) The task of the artist is to create an image of this appearance. Thus, "the painter, the carpenter, and God are the three masters of the three kinds of bed."[3] Since the artist imitates not the real nature of things but only the manufactured thing, art is an imitation of appearances and not of facts. Thus it is twice removed from reality. This general scheme of art as mimesis remained an important element in the interpretation of artistic expression up to our own days, even if the question as to what is being imitated became an increasingly difficult one to answer. We will now try to trace some continuity from art as imitation to art as independence.

Recently the idea of mimesis as the source of art appeared in protoaesthetics, which is the study of prehuman manifestations of artistic expression.[4] The very notion of protoaesthetics demands that we replace Idea with the evolutionary Umwelt of an organism, a step which Plato would have abhorred. The vocalization of song birds is an interesting example for it suggests a linkage between, on the one hand, what appears to us as beautiful and, on the other hand, the evolutionary functions of the same—including, perhaps, self-expression by the bird. It is known that each avian species has a store of sound signals, sometimes as many as fifteen to eighteen types, each appearing in distinct functional context, and some with considerable variability. Interest in the study of these species-specific signals has been focused on the degree to which they are produced with or without examples from others. Among nonsinging birds a species-specific repertoire develops independently of examples; the same holds for true song birds as far as the simplest call notes go. More complex forms, functional in pair formation or in territorial defense, sometimes also develop spontaneously, though in most cases the development depends on imitation. But, as is the case with the European chaffinch, birds do not imitate just any song patterns, not even recorded and mixed-up chaffinch songs, but only such as resemble the "real chaffinch."[5] As Chomsky's grammatical deep structure with the child, so preferences for singing in certain ways are innate in the bird. R. H. Hinde and others[6] described the song learning as a gradual process approaching a Sollwert * but modifiable by external influences and subject to a maturing process. This Sollwert also appears to be the signature by which other individuals of the same species are recognized. Beyond these, however, birds improvise both by rearrangement of phrases and by the invention of new ones. Improvisation plays a role in the song development of many species with the song structure varying prior to crystallization. In some birds the result is a surprisingly large repertoire of songs, displaying sometimes as many as sixty variations.[7]

Among birds there are also examples of protoartistry which cater to visual satisfaction. The Australian bower bird constructs bowers to which it entices the females for the purpose of mating. The bowers are colorfully decorated with pieces of stones, shining pieces of any origin, bleached bones, colored fruit, and flowers which are not eaten but replaced when they wither. There are species-specific preferences for color; and some even paint their bowers with charcoal or fruit pulp.[8] Among larger animals, ape paintings have been studied in their relation to human art.[9] The absorption of chimpanzees in their painting activities is impressive and, in a way, moving; the resulting pictures are interesting. But whereas the sophistication of bird songs vis-a-vis music is substantial, the sophistication of chimpanzee paintings vis-a-vis human painting is nil. This striking contrast might be explained as follows. Bird songs relate primarily to sex and territory and only secondarily to self-expression. Sex and territory are essential for bird survival. Chimpanzee paintings have no obvious relation either to sex or to territory but only to self-expression. Hence there is no tension beneath them that would pertain to survival, nor can we seek much of a psychological charge for self-expression because the chimpanzee has only the whiff of an ego. So let us return to the birds.

Thorpe has seen enough evidence to make him suggest that improvisation in bird songs is similar to real musical invention. Following H. Szöke and others he noted that the audio spectrum acceptable to the human ear coincides with that of bird songs except for some cutoff at higher frequencies. Perhaps, then, bird songs might have formed the auditory background out of which human premusic gradually developed.[10] In an interesting etiological-anthropological argument, he points to the well-known prevalence of headdresses, other ornaments, and dances in primitive (and not so

* From the German, literally "a desired value." In our terms, a genetically determined model of the inner landscape; a paradigm.

primitive) cultures, which resemble the ceremonial displays and actions of birds. Both in birds and in man, the most striking songs, colors, and dances are associated with sexual pursuits. It may not be accidental at all that songs, colors, and dances of love strike us as beautiful. Thorpe wonders, "Does it all imply some fundamental unity between the mind and perceptual systems of groups as far apart as Insecta, the Aves and mankind?" [11] I believe it does.

In the cyclic order of life and in the functioning of the mind, the improvised songs are autogenic; some degree of improvisation must have existed *ab ovo*, in a very literal sense, otherwise how could even the simplest song have come, in the course of evolution, to have a Sollwert? Thus, already in the song of birds we encounter rudiments of artistic expression and of the beautiful: imitation and improvisation in the service of the cyclic order of life, of biological reproduction, and of the group. In man we will find a similar but much more sophisticated constellation of capacities made possible by his sense of time. The scope of imitation is enlarged by his perceptive faculties; improvisations are made more versatile by the broad spectrum of his cyclic biological order and by the wealth of his inner landscape. Moreover, these elements are mutually reinforced and intensified by his individual and collective knowledge of, and concern with, birth, sex, and death. The repertoire of thinkable and feelable conditions from which man may select the substance of his artistic expressions is immensely larger than that of even his closest animal cousins. In artistic expressions, he may choose any one or combinations of Umwelts. The imagery of common metaphors is not accidental: the artist may be as passive as a rolling stone, as happy as a bird, as vicious as a viper, or as innocent as a newborn babe, for he has all of these in him. Hence, he may freely share the experience and imitate or improvise in the Umwelts of his material and biological selves. He may also draw on the open store of his autogenic imagery and become virtually independent of his lower Umwelts, provided he dares enter the countryside which is truly and uniquely the home of his soul: that of poesy.

2. Aesthetic Adventures

ARTISTIC EXPRESSION and the experience of the beautiful may be seen to comprise unsystematic adventuring among the many Umwelts of man.

The Fine Arts

The wealth of art objects makes a precise definition of what constitutes artistic expression impractical; I would rather describe the various forms of the arts by means of their hallmarks. Thus, the hallmark of fine arts, if by that we mean painting, drawing, architecture, and sculpture, is the potentiality of instant encounter and appeal.

It is an undeniable fact that, except for wear and tear, an art object remains unchanging in time, in contrast to music performed or drama enacted. A well-manufactured book is also permanent but its message cannot be instantaneously encountered. This important immediacy of fine art has been hidden by questions pertaining to the static representation of dynamic expressions of life and mind (such as motion, feeling, or opinion). Since these are processes and not things, their stationary representations do pose problems. In his treatise entitled *Laokoon* (1766),* the German philosopher and dramatist Gotthold Ephraim Lessing distinguishes on this basis between the functions of sculptural and poetic art. He stresses that painting or sculpture can only represent a single moment of action, whereas poetry and drama extend through time. It follows that an objet d'art is confined to represent a telling instant of time, one whose present is pregnant with future and/or past.

The difficulties of this seemingly straightforward observation were posed in contemporary terms by E. H. Gombrich when he asked this: If a coherent drama were filmed, which frame would be best-suited to represent the story as a still?[12] The answer is that none might do. As a first reason for the likely failure, Gombrich points to the different sensitivities of camera and eye; secondly, he reminds us of the extreme complexity of man's temporal integration, which has no parallel in the mechanism of the camera. (In our concepts, Gombrich is invoking the complexity of the mental present.) As a partial conclusion he puts forth a principle of the primacy of meaning. Namely, we cannot judge the distance of an object in space before we estimate its size; likewise, we cannot estimate the passage of time in a picture without interpreting the event represented.[13] Earlier in his paper, Gombrich has noted in passing, "the idea that there is a 'moment' which has no movement and can be seized and fixed in this static form by the artist, or for that matter, by the camera, certainly leads to Zeno's paradox."[14] This remark deserves special attention.

It will be recalled that Zeno's paradox of the flying arrow ingeniously invokes and mixes the various temporal Umwelts. But, whereas Zeno's paradox implies an arrow which flies in our experience but is stopped in our imagination, in a painting, in sculpture, or in architecture the arrow is stationary in our experience but flies in our imagination. The connecting link between the stationary world of fine art and the dynamic image it invokes is Gombrich's "principle of meaning." The artist's gift is the capacity to retain a focus, in terms of meaning, while he freely travels among his (hence our) Umwelts. He functions as our time-roving ambassador. The way in which meaning may be introduced varies from artist to artist and object to object; but in all cases artistic expression is taken from the store

---

* After the famous Laocoön group of Pentelic marble in the Vatican.

of autogenic images of the mind. It then follows that artistic creation can be discussed only in retrospect and that aesthetic analysis tends to be explanatory and not generative; in different words, observations about completed objects of art do not amount to rules of creativity. As Henri Matisse put it: "Rules have no existence outside individuals: otherwise Racine would be no greater genius than a good professor." [15]

Gombrich has raised an interesting question which is relevant to our inquiry. "Why is symmetry experienced as static, asymmetry as unstable; why is any lucid order felt to express repose, any confusion movement?" [16] Our prior deliberations can answer this question. Symmetry implies predictability, which is necessity and is static. Asymmetry implies unpredictability and contingency, hence the unstable and the dynamic. Lucid order is to confusion as necessity to contingency, as being to becoming. The artist combines representations of being and becoming so as to imply the ecstasy of the forest, that of the dance, and sometimes that of the mushroom, sometimes with a great deal of the noetic mixed in, all brought into focus by meaning—yet never removing that ambiguity which Picasso once expressed in the remark that a green parrot is both a green salad *and* a green parrot.[17] Apparently the techniques of the artist offer him the method and means to touch upon several levels of temporality.[18] The need for ambiguity in art points to the individualistic character of artistic expression; it permits the spectator to insert details into niches of perception left undetermined by the artist. The Platonic idea of preexisting, eternal images already implies that much. But what does the viewer bring to the art object?

In a psychological study of artistic vision, Anton Ehrenzweig argued that the appeal of fine art is in bringing forth inarticulate "thing-less" forms in the mind of the viewer. His reasoning is interesting, for it is based on a hierarchical model of artistic perception, and something which is "thing-less" must be purely temporal. Every act of visual perception, according to Ehrenzweig, recapitulates the ontogenetic development in the visual motor pattern of the child, hence it has to run through infantile, undifferentiated stages of dreamlike structure before it is articulated into the final images which emerge into consciousness.[19] Therefore, the unitary picture which emerges is a composite of many levels of form differentiation. In terms of our philosophy, then, aesthetic experience draws upon all the perceptual Umwelts of man, that is, on the hierarchy of the many temporalities surviving in man. Back again in Ehrenzweig's perspectives, "we have come more and more to assume that consciousness continually oscillates between different levels of form differentiation." [20] These "oscillations" correspond to our vertical conflicts, hence the suggestion emerges that the artist creates or annihilates, but in any case manipulates, such conflicts. This sets him apart from most other professions; also the demands implied are so broad that he may properly be regarded as free.

At the one end of the spectrum are the many expressions of timelessness as being. Greek sculpture and painting, for example, whether archaic, classical, or Hellenistic leave one with the impression of order, that is of conflicts resolved, even if the manifest imagery is that of conflict. Greek art praises life and harmony even if it depicts or disapproves the behavior or state of God and man. Similar arguments may be applied to landscape and still-life painting. The latter is especially interesting for it is often filled with dead fowl, fish, game, and carcasses of all kinds. The carcasses imply conflicts solved rather than unresolvable: a feast is forthcoming, the atmosphere is silent. Picasso's "Death's Head with Leeks" (1945), suggests eternity and not death. The Greek resolve of seeing in eternity a praise of life is transmuted into eternity condemning life in such modern still-life paintings as Ivan Albright's "Poor room—there is no time, no end, no today, no yesterday, no tomorrow, only the forever, and forever, and forever, without end" (1959).

At the other end of the spectrum are the expressions of timelessness as becoming, a trait most notable in modern art since the analytical phase

---

Through the many means at his command, the artist conveys the experience of structured conflict rather than mere orderliness of the world. Thus, he imitates the cosmic emergence of creative conflicts from the primordial chaos, discussed in the chapter on cosmologies. By appealing to several levels of the mind along the hierarchy of conflicts, he roams freely among the temporal Umwelts of man.

Hieronymus Bosch (circa 1450–1516), The Owl's Nest. Courtesy, Museum Boymans-van Beuningen, Rotterdam. The immediate impression one gains of this late and beautiful work of Bosch is one of order in nature and an idyllic sense of peace. Yet in the view of many art historians the drawing represents the Dutch saying, "Het is een recht uilennest," or, freely translated, "things are a veritable mess." Be that as it may, three owls, some magpies, spiders, and some vegetation occupy an ecological niche on an old tree. On a broad landscape of sloping hills, on the left, there is a distant town with its church rising above the other buildings. Windmills left and right identify the Dutch countryside. On the left a cavalcade is led by a man on foot and some mounted on horseback. The man leading the group points to what appear to be other men working in the fields. The procession is passing by a wheel on a tall pole, a well-known instrument of execution. They seem to be following a road which will lead them by a lonely cross on the right hand side. The animals are totally involved in their own worlds, as the interconnected yet differing universes of owls, spiders, birds, and man go on. Some art historians believe that this is but the upper portion of a larger drawing. We may take their opinion symbolically: artistic beauty reports on a world which is usually beyond reach.

of cubism was invented by Picasso. The hallmark of this phase is the idea of "simultaneity," by which is meant the simultaneous revelation of more than one aspect of an object, both in painting and sculpture. In ordinary interpretation this is conceived of as an effort directed to a more complete presentation of the total image. Thus Metzinger's *Tea Time* or Picasso's *Female Nude* (1910—he did so many of them), or *Girl Reading* (1911), and any number of his paintings in the 1907–1919 period [21] combine frontal views and profiles, all ill-defined and interpenetrating surfaces which belong to the same structures—but not otherwise perceivable all at once. The mixing of views suggests a desire to stretch the mental present. I have argued earlier that the phylogenetic narrowing of the psychological present in man is necessary to secure his identity. Cubism works toward a breaking out from this limited present and incurs thereby the wrath of our perceptive system which does not know how to combine the wide creature present of the bee with the precise identity of man. Duchamp's *Nude Descending a Staircase* reminds one of George Gamow's *Mr. Tompkins in Wonderland.*[22] Mr. Tompkins observes a spreading array of *one* lion, for the constants of physics in his imaginary universe permit the direct perception of quantum mechanical uncertainties in spatial and temporal definitions. This does not usually bother earthlings, unless they are cubist painters whose Umwelt is often that of probabilistic prototemporality.

G. Schaltenbrand, writing from the point of view of neurology, finds that the gestalten of perceptive forms communicate among themselves in what he describes as a stream of resonance, from macroorganisms down to matter and light.[23] Perceptual distortions in modern art are ubiquitous, and many of them display this "stream of resonance" among forms that correspond to many Umwelts. A visit to any leading exhibit will convey a feeling of exploratory restlessness, in search of a world of broad sensual experience. Quite consistently with the view proposed, we find that primitive art exudes a similar uncertain fascination with the structure of perception. Its stylized heads, animals, and men speak of a world that was new and strange to the senses. Among contemporary artists, Chagall is one who succeeds remarkably in projecting a vibrating unity of integrative levels from countable but unorganizable prototemporality to the noetic organization of future, past, and present.

The profound significance of temporality for the fine arts, as sketched above, must be distinguished from the personification of time as a theme. The latter is usually an artistic expression of a philosophical, religious, social, or historical principle. We find a superb summary of the representation of time in Renaissance art in Erwin Panofsky's *Studies in Iconology.*[24] He shows how certain Renaissance figures became invested with meanings which had not been present in their classical prototypes, though often foreshadowed in literature. He traces scores of beliefs about, and representations of, time that finally combined in the popular image of Father Time. Time as Kairos (a decisive moment) in Greek myths, for instance, is represented by a winged man, the figure of opportunity; this image became mixed with the Mithraic *Aion*, also a winged figure, which stood for creativity. The sickle of Kronos (God of agriculture) was reinterpreted as that used by Chronos (father of all things) to castrate his own father, Uranus.[25] These and other mergers brought forth the Renaissance symbols of time as decrepitude and decay (the hourglass, the scythe or sickle, the crutches, the man eating his child). As people's evaluations of their own passing changed, so did the symbols: they were resurrected, modified, redrawn, somewhat as a species would change in its phylogeny. Thus in its origins

> half classical and half medieval, half western and half oriental, this figure [Father Time] illustrates both the abstract grandeur of a philosophical principle and the malignant voracity of a destructive demon, and just this rich complexity of the new image accounts for [its] frequent appearance and varied significance in Renaissance and Baroque art.[26]

Symbolization of the functions of time in fine art varies widely. Time appears as triumph, love, death, fame, chastity, virtue, truth, deceit, desperation, wealth, luxury, poverty; as the seasons or the rivers; as a revealer, a redeemer, a judge, and practically anything else the reader may think of. As a set of symbols, these figures speak not so much about art and artists as about the temporal concerns of man. Panofsky defined his own inquiry into iconographical synthesis as the interpretation of the symbolic values of art objects through familiarity (1) with the Weltanschauung of the artists, (2) with the general trend of the human mind, and (3) with the cultural milieu of the epoch. When so understood, the iconology of time appears as an important tool in seeking to understand the intellectual history of time. Iconographical synthesis and, by implication, iconographical inquiry into time and the fine arts were criticized by George Kubler who preferred to deemphasize, if not eliminate, the stress on the symbolic content of art and replace it with a study of morphological duration of series and sequences in art, regardless of the meanings of images. He sees the history of all man-made things as subsumable under the history of art and notes that figures and shapes are so distinctive that they suggest that artifacts "possess a specific sort of duration, occupying time differently from the animal beings of biology and the natural materials of physics."[27] He sees the formal sequence of things as consisting of gradually altered repetitions of certain spatial elements which he describes as bundles of fibers with changing cross-sections.[28] These bundles are the "shapes of time" in the domain of artifact forms. One is reminded of the bundles of fibers that fill treatises on organic evolution, an isomorphism which suggests that the fibers of forms are spatial models of changes in thing-archetypes—if things could be said to have archetypes. Temporality enters in a roundabout way, perhaps as in studies of morphological changes in organic evolution.

The discipline of painters, sculptors, and architects which traditionally forced them to represent the dynamics of experience in stationary images, has recently come to be challenged, as we have seen, beginning perhaps with cubism. Beyond cubism, the fine arts pressed in too many directions to be considered here, so I mention only two. One is the enthusiastic revival of fantastic and visionary painting. This has an impressive history going back at least to Bosch and Brueghel the Elder. Its mode of artistic expression, in the hands of a master, provides sufficient indeterminacy to permit the creation of conflict from chaos. Many of the drawings of M. C. Escher illustrate this point. The other postcubist development is kinetic or dynamic art, an outright revolt against the traditional temporal confines of fine art. A box made of untreated steel so that it may rust away [29] is the work of the Socratic carpenter who makes the bed-thing rather than the Socratic artist who copies the material image of the eternal idea. One need not be a believer in the Platonic theory of art, however, to realize that what happens here is a short-circuiting of artistic labor. The result is identical to the call of the microbiologist who asks us to look through his microscope and behold a beautiful world. Perhaps this is a new art form, but it is difficult to distinguish it from a put on. If kinetic art will do no more than reproduce the experience of watching a falling tree by felling a tree, its contribution to art will remain small; but more imaginative forms may be forthcoming.[30] Interestingly, kinetic art feeds (as does the mobile construct of the novel, as we shall see) into the film. It is to be hoped, however, that some artists, at least, will continue to respect the traditional temporal confines of the fine arts as they try to record in static matter the elusive feelings and impressions of their temporal journeys.

Music

Unlike paintings, statues, or buildings, musical compositions do not have the potentiality of instant appeal. This difference might have been the basis of Lessing's opinion that music is the art of time, a view that has since been often repeated.

For authoritative information on music and time we may turn to the repository of all things human.

> Music do I hear?
> Ha, ha! Keep time. How sour sweet music is
> When time is broke and no proportion kept!
> So it is in the music of men's lives.
> *Richard II* (V, iii, 41)

The duality of time-in-music versus the time-of-life and the harmony of dissonance between the two has served well as a philosophical-poetic idea, since Plato and Aristotle perceived the beauty of music as residing in the proportion and concord of true reality reflected in the soul. Plotinus wondered, "What man with music in his soul, beholding the harmony in the intelligible world, but must be moved by the harmony in sounds that are heard with the ear?" [31] To Augustine, harmony made music, *ars bene modulandi.*[32] To Schopenhauer, it was not like other arts, "the copy of the Ideas but the copy of the will itself, whose objectivity the Ideas are. This is why the effect of music is so much more powerful and penetrating than that of the other arts, for they speak only of shadows but it speaks of the thing itself." [33] From among our contemporaries, Susanne Langer perceives in the time of music a virtual time, an image of the passage of life, as future becomes present, present becomes past. She sees the task of the composer in the creation of virtual time, that is, an audible projection of the tensions and relaxations of existential stress.[34] To the psychologist Paul Fraisse,

> En définitive, le rhythme est né de la périodicité des mouvements et d'une organisation qui en dérive. Mais la création musicale tend, tout en gardant une nécessaire structure temporelle, à s'en affranchir dans un dynamisme créateur.* [35]

A current polarization of views and practices among composers and theoreticians of music brought out the paradox of time underneath the idea of music as an "art of time," namely, the question of predictability versus unpredictability. Through their dialogue, musicologists came to reenact the Eleatic debate concerning the nature of time stated as the difference between serial and chance music.[36] Serial composition leaves nothing to chance; it provides a total organization with predetermined mathematical certainty. It excludes subjectivity and promotes totally rational organization with unchanging rules, against which the composer's work may be checked. The organization embraces pitch, interval, meter, timbre, and dynamics as well. Formally opposite to serial composition is chance music, which, in the words of George Rochberg, is "the epitome of the unpredictable," where each sound defines only itself and nothing else. Composers of chance music "see music as the occurrence of unpredictable events, each moment of sound or silence freed of formal connection with the moment of before and after, audible only as a present sensation, an ensemble of musical happenings of undetermined form or length." In its extreme form serial music is Parmenidean, chance music is Heraclitean. From the point of view of time as conflict, only compositions that include both can be expected to convey the sense of complete temporality. The many and interesting remarks that musicologists have made recently in contrasting the flow and spatiality of music, the degree of individual self-awareness of the composer, its relation to society and personality, etc., follow from the character and wealth of the paradox of temporality, familiar by now to the reader, but, as far as I can ascertain, not so recognized in musicological treatises.

I speculated earlier that the origins of language involved the separation of the emotive from the intelligible in utterances, and that the former is the likely ancestor of music, the latter of language. If this be so, then music need not embody claims of articulate truths but only emotive truths.

---

* Most certainly, rhythm was born of the periodicity of motion and of the organization which may be derived from it. But musical creation, while retaining a necessary temporal structure, tends to escape from same through creative dynamism.

Articulate truths would demand an interpersonal (hence impersonal) judgement of beauty which, however, is not possible. For, although the sense of
beauty seems to be common to all man, its specific examples bear no such
collective authority even when enjoyed by many. Similar ideas have been
expressed, though not on an evolutionary basis, by British literary and art
critics, notably by I. A. Richards, who pioneered the emotive theory of art.[37]
His stance was challenged by Morris Weitz, who warned against taking the
separation of the noetic and the emotive too seriously, though he did stop
short of holding that music can make "truth claims." [38]

As we have seen, language is an essential tool for the definition of the
noetic self. Music plays an identical role in the definition of the emotive
self. But, although language and music are both in the "audio loop" of
man, their Umwelts are distinguishable. In language, those components
of speech which appropriately belong to the eotemporal Umwelt are functionally subordinate to the noetic role of speech; in music, it is the eotemporal and prototemporal features of the sound which are of primary
interest. That these lower Umwelts possess an evolutionary priority to the
nootemporal world may be exemplified with reference to the power of the
dance.[39]

The earliest archeological records, such as the famous "Sorcerer" wall
paintings in the Trois Frères cavern of France, suggest that rhythmic
motion must have been regarded by paleolithic man as a privileged activity.
Early ritual dancing is sometimes interpreted as evidence that the artist
believed that the representation of the dance would assist him in securing
continuous control over the future such as in the success of hunts.[40] A
likely source of this magic belief may be that dancing left the dancer, and
possibly the watcher as well, in an induced state of timeless ecstasy which
overcame the uncertainty of the present and filled the participants with a
sense of omnipotence about the future. The profound force of sympathetic
induction that leads to the loss of individuality and establishes timeless
belonging has been amply demonstrated in the twentieth century art of dic-
tatorial manipulation of the masses. Earlier, the sound of music and the
motion of the dance kept time for rhythmic group activity. For galley slaves
it was the drum, for the nineteenth century worker the workshop song, for
sailors the sea chanty, for soldiers the march.[41] But whereas some type of
dancing is common to all organisms from gyrating fish and marching pen-
guins to dancing bees, only in man may dance be directed to something
other than the needs of immediate survival, to wit, survival in the future.

Dance and music have impelled man to action rather than served as
depositories of findings about the world. This is what Tolstoy might have
had in mind with respect to music when he warned that certain compositions, such as Beethoven's Kreutzer Sonata,

> should be played only in certain grave, significant conditions, and only
> when certain deeds corresponding to such music are to be accomplished,

[because] to call forth energy which is not consonant with the place or time, and an impulse which does not manifest itself in anything cannot fail to have a baneful effect. On me, at least, it had a horrible effect. It seemed to me that entire new impulses, new possibilities were revealed to me in myself, such as I had never dreamed of before.[42]

Whereas a painter, sculptor, or a viewer creates, modifies, or beholds an external reality that one may feel as independent of the self, a composer or a listener creates or integrates music into that audio loop of sound generation and reception which ontogenetically and phylogenetically was functional in helping him establish his identity. It is probably this feature which makes musical experience so intimate, immediate, and individualistic. The close coupling between identity and music might also answer the question posed, but not answered, by Aristotle: namely, "Why do rhythms and melodies, which are mere sounds, resemble dispositions, while tastes do not, nor yet colours or smells?"[43]

If it is true that every act of visual perception brings into play several levels of the mind (as we assume it does), we should expect to find analogous features in auditory perception. In this respect let us note that the elements of music: tone, rhythm, melody, and the composition of a musical piece are conveyed by physically identical sense impressions that differ only in frequency. A tone consists of audio oscillations, perhaps, from 4 Hz to 15 KHz.; rhythms are in the domain of events per second; melody is describable only as events per minute; the total composition is a repeatable structure whose duration as a unit may be several hours. Thus, as one progresses from tone to rhythm to melody to composition, supersonic frequencies are left for the audible frequencies, then for the domain of the various short-term memories, and finally for long-term memory. In all of these domains we can encounter "something old and something new, something borrowed and something blue," or in philosophical terms, the coexistence of the expected and the encountered in the nature of temporality.

From bird songs to Brahms, the cyclic range of the *ars bene modulandi* embraces a substantial portion of the cyclic order of the biotemporal world. It certainly has features of the prototemporal world (events countable but unorderable) and of the eotemporal world (of pure succession). Across its broad spectrum, the experience of music extends into those regions of temporality where purpose and the mental present enter. Thus, within the boundaries of its technical range music may appeal simultaneously to processes both conscious and unconscious, both biological and noetic, and draw on elements predictable as well as unpredictable. If artistic expression truly derives from the capacity of the artist, via his specific skills, to roam freely among the temporal Umwelts of man, then music, with its intimate auditory entry into man's self-definition, is the art of arts, par excellence; not simply because it is not spatial, but primarily because it embraces all levels of temporality.

This music mads me: let it sound no more;
For though it have help madmen to their wits
In me it seems it will make wise men mad.

                                    Richard II (V, v, 62)

Tragedy

Achilles, urging the bereaved Priam to cease his grieving for Hector, God, and man, remarked that, "such is the thread the Gods for mortals spin, to live in woe, while they from cares are free." [44] Why it must be so is not asked, for the ranks of the gods personify cosmic forces not to be debated; they themselves are without unresolvable conflicts—and without creativity. In this frame of mind man accepts suffering as part of the nature of things, as animals accept their lot. A dog or a horse, if grossly abused, may display corresponding signs of subjective misery; but it will never give any indication which could be interpreted as an expression of suspicion that some universal injustice has been done. What in Homer is only adumbrated (the act of struggle by man against injustice) and is almost totally absent in animal life, is a type of conflict known since the Golden Age of Greece as tragedy.

For Aristotle, tragedy as a form of art originated with the authors of the dithyramb. [45] Others have conjectured that Greek tragedy is a mellowed extension of acts of ritual sacrifice. [46] This latter opinion is debatable, but even if it is correct, the practice of human ritual sacrifice is not a sufficient and probably not a necessary cause for the coming about of the tragic arts. The subject of the tragic play arises not from human abuse or struggle alone, but from certain conflicts among the many integrative levels that comprise the man's body, his mind, and his collective life. Thus, before tragedy may be conceived of as a form of art, it is necessary that writer, actor, and audience alike be cognizant of the potentiality of tragic conflict. Practically nothing is known of the tragedies written before Aeschylus (d. 525 B.C.). He is said to have discovered "tragic time," that is, the idea of continuous time and sequence of events in a play. According to de Romilly, time in Sophocles resembles Homer's interest in the immediate, and so does time in Euripides. However, in the latter, the day (which is the "immediate") becomes tragic because it comes to be felt as an isolated fragment of a broken *chronos* which, thereby, turns out to be irrational. [47] Such opinions suggest to me that only with the deeply personal concerns of tragedy, and not with the philosophies, did the classical Greeks approach the insight available already to the author of Job, namely, that human life incorporates a steady feud between God and man.

Aristotle's celebrated views on tragedy defined as well as directed its literary history.

A tragedy, then, is the imitation of an action that is serious and also, as having magnitude, complete in itself; in language with pleasurable accessories . . . in a dramatic and not in a narrative form; with incidents arousing pity and fear, wherewith to accomplish its catharsis of such emotions.[48]

Though theories of tragedy formulated in various epochs emphasized different aspects of these views, certain elements of the Aristotelian perspectives survived political, philosophical, and literary change. This viability vouches not only for the profound significance of tragedy among the literary arts but, more importantly, for the rooting of tragedy in something that is fundamental in the psychobiological organization of man and society. Tragedy, according to Aristotle, imitates action and life, happiness and misery, but not persons;[49] it does include characters but not for the purpose of portraying them, but because of their actions. Thus, the life and soul of tragedy is the plot, and characters are only secondary; its essential function is catharsis, which is the purification and purging of emotions. These and other Aristotelian arguments taken together imply that tragedy is built on a continuous comparison and contrasting of happenings, and on a steady reflection on future, past, and present in terms of memory and expectation. Thus, tragedy, as de Romilly emphasizes, always presents us with a more or less conscious philosophy of time.[50] It is not by chance, then, that tragedy and history were born at the same time.

The Renaissance displayed a new and energetic interest in tragedy, probably due to the confluence of several historical trends in the evaluation of man and the world. There was the explosively increasing belief in man's powers to control his fate, together with the realization of the darker side of man's nature, well represented by Machiavellianism. Tragedy reemerged with new force in the feelings of Rabelais, Montaigne, Spenser, Shakespeare, and even Milton. For these writers, says Quinones, allowing for some exceptions, victory over fate was not possible.[51] The anguished heroes of their plays, poems, or essays could either learn to live among the pieces of humanistic ideals wrecked by the natural insufficiencies of man, or they would perish. Quinones's sensitive report is a sober reminder to all that an image of the Renaissance which concerns the scientific revolution alone is a rather misleading one. In one of many beautiful passages, he illustrates his point with reference to the broken temporal continuity in Shakespeare's tragedies.[52] In the broken flow of time the predicament of the heroes and heroines is stirring, for although continuity as a value is maintained, under the impact of uncontrollable and unexpected events upon events, the continuity of collective and individual life is replaced by a narrowing concern. Quinones calls this "the tragic contraction" and remarks that often in the tragedies the most significant moment of life, that of death, is identified with the first moment; the final and the first hour become one.[53] The effect is an atmosphere of timeless determinism. In terms of our concepts, it is a regression to conditions which resemble those of the proto- and atemporal Umwelts. Let us consider the tragic contraction as formulated by Quinones, for it is a delicate and important feature of tragic time.

For Aristotle, epic poetry has no fixed limits in time but "tragedy endeavours to keep as far as possible within a single circuit of the sun." [54] This working rule, which derived from Aristotle's feelings about the nature of the tragic and his knowledge of the theater, was consummated two millennia later by the genius of Shakespeare in the "tragic contraction." [55] Only a few pages earlier I argued, following de Romilly, that tragedy at its very inception comprised a continuous comparison and contrasting of events and reflections on future, past, and present. How can these two trends, the narrowing and the broadening of time in tragedy, be reconciled? The answer follows from some of our prior thoughts. The tragic contraction is, on the surface, a narrowing concern expressible by going from years to hours to minutes. But more importantly, it corresponds to a regression along a hierarchy of Umwelts. Concentrating on a present task alone is not tragic; the lover, the poet, and the creative scientist all do that. It is the regression to the psychological and even physiological present that accomplishes the catharsis; it is a matter of removing the self from the heights of individuality to states next door to the worm-and-dust state of death. Tragedy can happen only to a creature who is capable of starting with the broad horizons of a mental present: memories of childhood, experiences of adulthood, and expectation of the future. From the point of view of the spectator, the process of tragic contraction is a journey through his temporal Umwelts, backward, and the tragic poet is the free, time-roaming ambassador.

In the fine arts, memory and expectation are called for as well as produced by visual stimulation and skillfull construction of a present; music inserts itself in the audio loop of the listener; tragedy can draw upon both of these methods. The spoken voice of the actor enters the auditory loop of self-definition; the imagery appeals to the visual faculties of the spectator, and memories and expectations are supplied ready-to-be-used. Thus, the playwright, not unlike the composer, sculptor, and painter, journeys freely among various levels of temporality, appealing simultaneously to the child, the adolescent, the young and the old, the biological and the mental, the mortal and the immortal. Let us call the temporal experience produced by the tragic contraction the "tragic present." Within the tragic present many temporalities conflict. All along there is likely to have been a conflict between, on the one hand, the eternal, timeless order of god, nature, the universe, or society and, on the other hand, the hero who knows time, for he anticipates his passing. The hero's creative imagination offers him rich worlds of possibilities which, however, become restricted externally by the laws of physics and biology, and internally by his sense of timeless destiny. When these two sets of laws, or instructions, are contradictory, the tragic conflict obtains. Thus, the tragic present comprises the conflicts of time recorded for examination. Nietzsche's idea of tragedy as the interplay of two form principles, the Dionysian or revolutionary with the Apollonian or conservative, is a colorful way of describing these conflicts.

Hegel had held that the conflict of tragedy is an ethical one. Insofar as tragedy often centers on questions of praiseworthy conduct, this is so—

but the ethical conflict is only part of the larger picture. Explicitly or implicitly, tragedies tend to imply conflicts larger than the individual and even society; it is the external order of things or the cosmic plan that is being challenged, even if the cosmos is no larger than the polis.

> Blow, winds, and crack your cheeks! rage! blow! . . .
> Crack nature's moulds, all germens spill at once
> That make ingrateful man!
> *King Lear* (III, ii, 1)

While the hero's problems may be unresolvable—hence the tragedy—his attempts to communicate his plight to the world at large are not so difficult. The "fatal flaw" in the character of the hero which A. C. Bradley saw as the source of tragedy is, as I see it, the fatal flaw of the world. Not the hero alone, but time itself is out of joint. Hence, the solution is affected by the reduction of integrative levels as a joint venture of man and nature. When the struggle in Ophelia's mind becomes unbearable, her mind is eliminated, reducing her to the level of life.

> . . . poor Ophelia
> Divided from herself and her fair judgement,
> Without the which we are pictures, or mere beasts.
> *Hamlet* (IV, v, 84)

Hotspur's conflicts are solved by reducing his body to matter.

> O! I could prophesy,
> But that the earthy and cold hand of death
> Lies on my tongue. No, Percy, thou are dust,
> And food for—
> *Henry IV* Part 1 (V, iv, 77)

Yet, those watching the play keep on living and, as in the motion picture *Never on Sunday*, "they all go to the seashore." Thus, temporal continuity is introduced by the viewer who is privy to the fate of Ophelia and Hotspur and experiences the feeling of the uncanny that characterizes regressive sharing. Here, somewhere, are the sources of the Aristotelian catharsis. The climb back from the feelings of lost minds and lives is a reemergence to higher orders. This reidentification is an essential element of tragedy. Without it tragedy changes into comedy or into some other dramatic form still without a name, a point which may be well illustrated in a brief discussion of the forms of modern drama.

As during the Renaissance, the modern self again encounters fundamental changes in its relation to the world, which are then reflected in changes of the dramatic form. I shall try to sketch the background of these changes in three different and slightly overlapping ways, with different emphases.

• With the scientific revolution, especially as it is embodied in the practices of mid-twentieth century America, arose an evaluation of temporality which limits its important features to such elements as are authorized by the discoveries of natural science. Forces stronger than man must, of course, be admitted; but they are dissociated from principles sympathetic or antagonistic to man and are associated, instead, with the impersonal operation of natural law. These, in their turn, are found to be probabilistic and—in an undefined and popular, hence disturbing way—relative. As a consequence of these shifts of authority, the idea of a reality higher than man (individually or collectively) but relevant to his aspirations has become unpopular and eventually outright taboo in the leading intellectual currents of the century. With the disappearance of a higher, but still humanly meaningful, order, the potentiality for tragic conflict vanishes.

• For the modern self time has lost its aspect of becoming. The fear of the unknown and the unknowable has been repressed (hence taboo), while arrogant certainty about man's powers to control his fate has changed time from an arena of conflict to one of commodity. But commodities are bought and sold and not heroically given or withheld; consequently, heroism has lost its value because it cannot be traded. Thus, for example, although instances of courage in the civil rights movement and in space exploration are plentiful in contemporary America, the power of imagination that would change bravery to heroism, and the generosity of acknowledgement that would change individual actions into collective feats, have all but vanished. Being a hero—other than the overnight variety—has become comparable to the thankless task of being a soldier in a forgotten war.

• The tragic present of rich temporality has been replaced by the "business present" that focuses entirely on the now, as the businessman sets his watch to the second. Tamás Ungvári claimed that no major poem, novel, or drama has been written since the nineteenth century that would glorify, praise, or celebrate time.[56] Except for the youth drama, such as the rock musicals and what is sometimes called underground poetry, this does indeed seem to be the case.[57]

The question then emerges: Why is time not celebrated in our time-conscious societies? The answer is that time has become an enemy to be killed, not a friend to be praised. The "business present" pays only lip-service to past and future; its essence is the removal of tensions associated with future and past, in sharp contrast to the tragic present with its wealth of temporal conflicts. Whereas the tragic present is informed of continuities and hidden necessities, the "business present" is informed only of discontinuities, that is, of chance. But chance is the hallmark of comedy, lawfulness the hallmark of tragedy; hence there has been a corresponding change in modern dramatic form, as may be illustrated by examples from the plays of two outstanding contemporary playwrights.

Dürrenmatt's *The Physicists* is an Oedipal story.[58] The characters, by free decisions, bring about precisely those conditions which their decisions were intended to avoid. Three physicists, acting independently, are frightened by the potential evils to which their knowledge may be put; hence, each decides to act so as to be placed in an asylum. Only in the madhouse can we be free, they say; outside the madhouse their thoughts would be dynamite. They are "prisoners but free," "physicists but innocent." As their nurses come to suspect the truth, each kills his nurse. But through the murders their secrets fall into the hands of an "insane female psychiatrist," who goes forth to rule the world by the very powers which the characters, called Newton, Einstein, and Möbius, had hoped to contain. Two of the physicists revert to their aliases and to their feigned madness. Only Möbius has the belated integrity to say: "I am poor King Solomon, once I was immeasurably rich. . . . Now the cities over which I rule are deserted . . . and somewhere round a yellow, nameless star there circles, pointlessly, everlastingly, the radioactive earth." [59] But, whereas Sophocles' *Oedipus Rex* is a tragedy of free will, guilt, despair, and responsibility, *The Physicists* is a comedy.[60] In the latter, things happen without anyone being responsible, simply because "we could not help it," and guilt is only the collective, diluted type. Oedipus experienced despair and grandeur; Möbius is an overeducated misfit who may be pitied but not admired. Using *King Lear* as an illustration, I pointed earlier to the cosmic aspects of tragedy: wind, rain, and all the spilled "germens" joined him as he raised his voice against the heavens; Möbius only squeaks and, except for the last few words, makes only scientific reports. In *The Physicists* the world is a piece of unfortunate happening, or as Brecht might have put it, an accident; as Ungvári put it, in modern drama the human condition is "a destiny without destination." [61]

What for Dürrenmatt is tragedy turned comedy, for Beckett is comedy turned into a type of drama without a name; let us call it antitragedy. His *Endgame* is a segment from the existence of four characters who vegetate in ashbins and in rocking chairs. Talk and motion cease to sustain a continuity beyond a few gestures or phrases:

Hamm:	What time is it?
Clov:	The same as usual.
Hamm:	Have you looked?
Clov:	Yes.
Hamm:	Well?
Clov:	Zero.
Hamm:	It'd need to rain.
Clov:	It won't rain.[62]

Because of the disconnectedness a tragic present cannot evolve, only chance may be manifest. The Umwelt of these humans is prototemporal: events are countable but unorderable. The play is not a comedy, for comedies are series of conflicts usually resolved in unexpectedly simple ways. Here the resolution cannot come either from heaven or from earth because there are no conflicts. The characters conduct a vita minima with only occasional echoes of long-forgotten truths, such as when they find a louse and want to kill it because from it humanity may start again. They talk about building a raft and floating down the river to join other mammals, for they themselves have regressed into hardly more than mammals and they desire company. The characters cannot live together, neither can they separate for "outside of here it's death," even though inside they can only "cry in darkness."

Endgame concludes with a mime for one player on an empty stage. "Desert. Dazzling light." The man tries to leave the stage on the left, then on the right, but he is thrown back by unseen hands. From above, below, left, and right objects of life descend or ascend (a tree, a carafe with water) while the voice of a whistle from the appropriate direction urges him to reach out. But as he does, the objects are withdrawn. In *The Tempest* and in *A Midsummer Night's Dream* things and people also appear, tantalize the lonely hero, and vanish, but these props are parts of a directed, purposeful design in the comic resolution of certain conflicts. The characters of *Endgame* live in the "unbounded yet finite" prison of four-dimensional space. We have seen the importance of this gnostic metaphor in physics; Beckett translated it into its human significance.[63] In *The Tempest*, the actors, "the cloud-capp'd towers," "gorgeous palaces," and the globe itself can turn into spirits, for they are such stuff as dreams are made of. The characters of *Endgame* are such stuff as nightmares are made of, and they do not go away because the play itself is a cosmology. In *Our Town* the props only represent and stand in for a world which is otherwise rich in things and events. But the ashcans of *Endgame*, the urns of *Play*, the mound of earth heaped up to the heroine's neck in *Happy Days*, and the gnarled tree and the road in *Waiting for Godot* are everything there is to the world. Operating from ashcans could be funny for Huckleberry Finn, or exciting on board ship in *Kidnapped*, but it can only be hell if the universe is nothing but the ashcan.

Waiting for Godot, also by Beckett, is again a segment from the existence of two characters; again, the dialogue is incoherent beyond a few exchanges. Thus, no harm would be done, for example, if the reader were to start in the middle of the play, go to its end and continue from the beginning until he reaches the point where he began. The characters are without memory. Their mental presents are limited to thirty seconds at the most, even though occasionally they do recall what appears to have been their early education. The conversation goes in overlapping circles. The second act is different from the first act only by number, as though they took place in Leibniz's world of indistinguishables. There are no causal connections, no chains that last for more than a few seconds, but only events whose sequence is determined by the spectator's sense of time. The Umwelt of *Godot*, as that of *Endgame*, is prototemporal.

Earlier I called Beckett's plays antitragedies. This expression needs some elucidation. Frederick Turner in his book on *Shakespeare and the Nature of Time* pointed to the various ways fate and time work in the tragedies.[64] In *Romeo and Juliet* fate works by bad luck; there is nothing in the makeup of the characters to suggest that things could not have turned out well. In *Troilus and Cressida* the end of the play unfolds from the psychological givens of the protagonists displayed quite early, but there are many people with less deterministic makeups, such as the self-searching Achilles. In *Othello* evil moral choices bring about the tragic end, but certainly not everyone is evil. In contrast to the worlds of these plays, the world of Estragon and Vladimir in *Godot* is, as I have reasoned, a complete cosmology; there are no alternate solutions to the no-solution which is appropriate to the prototemporality of their cosmos. Thus, the play is neither a tragedy nor a comedy but a drama, as it were, of negative tragedy.

Gunther Adams called *Waiting for Godot* a play of being without time. Inverting the Descartian cogito he recognized in the play a feeling of "I remain, therefore I am waiting for something." [65] This remark can guide us to some important points. Consider a cat, spending most of its life perched on a mantel, surveying the world. The things that happen to him in a year would hardly fill one hour of Charles Darwin's life on board the *Beagle*, pacing restlessly as he was known to do. Yet the cat is satisfied. It will be recalled that an animal of the cat's complexity lives in an ill-defined continuous now with a narrow physiological present: truly speaking, the cat "remains." But man, the chronosopher, must, for an authentic existence await as well as remain. Godot, says Anders, is not a person or even a thing but the name for waiting. Differently put, Godot is no more than the invented subject to the predicate "waiting for." [66] Estragon and Vladimir live in the temporal Umwelt of a cat and suffer because they are not cats. Whereas the power of tragedy is in the catharsis attendant to upward resolution of conflicts, the power of the antitragedy is in the blatant absence of such possibilities. As a rest in music, antitragedy reminds the spectator of what he is missing.

In tragedy, chance events reveal an underlying design; Job's ongoing feud with God is the creative trauma of man. In antitragedy the design reveals the chancelike nature of the play's universe: the antitragic drama is without feud, without creativity, without time. All this is frightfully consistent. While tragedy attests to conflicts which rise out of chaos, antitragedy attests to chaos itself. But, as Möbius says in *The Physicists*, "What was once thought cannot be unthought." Man cannot unthink his unresolvable, creative conflicts. Hence, when all is said and done, the antitragedy serves the same purpose as the tragedy. This latter view could be demonstrated by reference to Ionescu and to the theater of the absurd, and to the youth drama of rock musicals.

Poetry and the Novel

For the purposes of this subsection I will define poetry as dance put into words. The use of metaphor in this definition demonstrates that type of imagery whose frequent use sets poetry apart from prose; thus the definition itself is closer to poetry than to prose. The definition also suggests a link with the natural order, namely, that dance precedes speech both onto-genetically and phylogenetically; hence, the person informed of what poetry is can somehow build on his prior capacities to understand the claim. It also implies that poetry embodies some archaic elements which prose does not.

If mating dragonflies, dancing grebes, and rocking infants could verbalize they would probably sound like poets. Such claims must sound strange or useless to homocentric aesthetes, but not to some recent inter-preters of poetry. Stanley Burnshaw, for instance, sees poetry as the ex-pression of the poet's total organism. In his study of artistic expression entitled *The Seamless Web*, he perceives the body of man as an integral portion of the world at large, and mind as made by the body. Thus follows the monism of a seamless web. Burnshaw speaks of the many relationships between primal and civilized forces (following a neurological preference, we could speak of paleo- and cerebral cortices) and sees the origins of poetry—and by extension, of all the arts—in the conflict between the two. He regards this conflict a normative part of existence for the mind, a con-dition where neither force can wholly subdue the other; conflict leads to con-flict resolved, "resolved through innumerable modes of erstwhile truce, to enable the host to survive." [67] In terms of our own process of thought, poetry is seen to be tied both to dance and music on the one hand and on the other hand, to articulate speech. The binding forces between the two are the contradictions of Burnshaw's truce understood as contradictions between behaviors dominated by the paleo- and the neocortex, respectively. In the philosophy of time as conflict, the truce is between knowledge felt and knowledge understood. Their joint presence makes a poem, using Burnshaw's term, a "thinking song." [68]

Poetry vibrates between these two modes of knowledge; its metrical appeal is that of music, the linguistic portion communicates with the cog-nitive faculties. Herbert Spencer held that poetry is a form of speech for better expression of emotional ideas.[69] The complementary statement would be that poetry is a form of music for better expression of gnostic feelings. Both these conditions hold simultaneously. While poetry-as-music inserts itself in the audio loop of self-definition, poetry-as-language delivers the articulate message. But the two together are more than their linear sum. As in the cyclic order or in the functioning of the very complex, where separate modalities interact to produce "beat notes" which are originally parts of neither, so the musical and prosaic portions of a poem generate feelings and thoughts which are not within the original domains of language or music if taken separately. Thus, the spectrum of poetical form extends, on the one hand, even to operatic music and, on the other hand, to the prose chants of the Greek chorus. Poetry provides our time-roving ambassador with freedom not usually available to a composer of music or to a novelist. It is impossible to do justice here to the theme of time and poetry,[70] but I can certainly mention that three involvements may be distinguished in this respect. (1) Poetry as a mode of artistic expression communicates with a wide spectrum of our psychobiological organization. (2) Temporal organization of poems (including that of epic poetry) is akin to that of plays, the film and sometimes even the novel, hence, what we can say about the latter also applies to poetry—with some reservation. (3) Poems often incorporate philosophies of time; hence the frequent use of poetry on the previous pages.

As music and language gave birth to poetry, so epic poetry in its turn gave birth to the literary form known as the novel. Because of the continuity of the history of epic poetry up to modern times a precise identification of the birth of the novel is difficult. Some would consider *The Tale of Genji* by Lady Murasaki (tenth century Japan) the earliest example of the novel; some see its origins much later, such as with Cervantes' *Don Quixote* (1605) or even with Defoe's *Robinson Crusoe* (1719). Ordinary critical appraisal would describe both epic poems and novels as long narratives of considerable complexity and distinguish them by their different attitudes to reality. As I see it, the hallmark of the novel is its prosaic form and subject, in more than one meaning of "prosaic." The characters of the epic are judged to be lasting because they are men on the scale of god and devil; they last because of what they are. Consequently, their actions are appropriate to their vast superiority or inferiority to the average human. In contrast, the characters of the novel last primarily because of their actions and their thoughts, while they themselves grow or deteriorate in the process. In the prose narrative of the novel, unlike the often poetic narrative of the epic, the emotions of the reader must be reached through explicit appeal to reason and emotion, though a poetic style, often commanded by great novelists, does help. Putting it differently and more cautiously, in the novel the musical quality of poetry is replaced by temporal modalities which bring into play long-term memory and expectations. Since these are heavily conditioned by society, it is not strange to find that the traditional novel, a Western piece of art, insists on accommodating free will, fate, and causality as understood in the modern West. Consequently, the novel as a literary device displays the orientation of its characters to time and to the hierarchy of conflicts in temporality.[71]

It is probably these privileged features which made Georges Poulet give such prominent display to the views of novelists in his work on time in French intellectual life. In his *Studies in Human Time*,[72] he lets Flaubert, Baudelaire, and Valèry speak for the history of the rich French tradition of philosophical and literary self-search. Although their thought had led directly to French existentialism, Poulet sees the pinnacle of this intellectual motion not in the work of French existentialists, but in the writings of Proust. The work of Proust, he writes, "appears as a retrospective view of all French thought on time, unfolding in time, like the church of Combray, its nave."[73]

While the French intellectuals were trying to rebuild the nave of Combray and, finding the task impossible, turned eventually to the lostness of Sartre's existentialism, the British were expanding their empire and laboring to make their life more bearable. Dickens, Trollope, Tennyson, and the good Victorians were committed to a view of rationalism and secularism in which ideas of time, progress, and history were tied together; and, no less than their Renaissance forerunners, they felt not only the challenge but also the fear of change. Of special interest is a critical essay by George H. Ford on "Dickens and The Voices of Time." [74] His analysis derives from the keen observation of the balance between the musical and the articulate contents of prose in the novels of Dickens. Ford perceives in this balance a projection of the novelist's relation to the conflicts of time. He picks up an observation made by Graham Greene who described the delicate and exact cadences of Dickens as the novelist's "secret prose." Ford examines some features of this prose, such as the switching back and forth in tenses to halt or to move the moment, or the calling upon "the music of memory" to re-present what has been past, or shifting to the present so as to change the observer-reporter into a participant. The reading of Dickens with special regard to this "secret prose" suggested to Ford that the novelist, reflecting the spirit of the Victorian age, shared with that age a double awareness of the joy and anguish of becoming. Ford suspects that the source of the Dickensian manipulation of time and tense as means of escape from the progress of the age, might have been the trauma of the death of Mary Hogarth in Dickens's arms when she was seventeen and he was twenty-five. This event will be recognized by generations of readers in the death of Little Nell in *The Old Curiosity Shop*, when "time itself seemed to have grown dull and old, as if no day were ever to displace the melancholy night." Death taught man about temporality, as we have amply witnessed: tempora mutantur et nos mutamur in illis (times change and we change with them) for Victorian novelists and all others. As far as Dickens goes, while his "secret prose" serves him as a means of expressing timelessness, his articulation praises progress for the benefit of the polis.

In spite of Dickens's ambivalence about past and future, his stories are linear and usually project into a happy, though imprecise future. With Thomas Hardy and Henry James, this straight-line idea of time, progress, and history began to break down or at least become suspect; happy endings ceased to be the rule. The characters, it seems to me, were less interested in running the empire than in repairing themselves. In an essay on history and time in the English novel, J. H. Raleigh observed that as the novel turned inward, characters came to endure rather than become, nature insisted on being cyclic; the individual's time became "existentialized," and the time of the universe became circularized. [75] Both the French and English novels, after having struggled with change and permanence, drifted toward modalities of time rather uncharacteristic of the early novel. It is this shift which eventually came to challenge the novel as an outdated literary form, as we shall see. First, however, a few words are in order about a man for whom the novel was a stable and majestic literary image of reality.

Thomas Mann's *Joseph and his Brethren* begins with a prelude entitled, "Descent into Hell," which can carry all but the very dull into worlds not to be found elsewhere.[76] In it, metaphorically, Mann is looking back from the apex of a parabola along the curve to see whether it meets the asymptote. In psychological terms, he is describing the difficulty of regressive sharing. The time which we see is through the eyes of evolving humanity. As we pass our sight backward to Joseph, to the Town of Ur, to paleolithic man, and even to his apelike ancestors, we cannot find a boundary where time stops because time molds into timelessness. This phylogenetic situation is analogous to the ontogenetic situation of the individual. As we plumb the depth of our memory we cannot reach the day of our conception or birth because the organism that was conceived and born did not know time. On our backward journey with the historian, or with the psychologist, we pass various levels of temporality or, as we learned to call them, the temporal Umwelts of man. Mann was a magistral time-roving emissary among these many levels of temporality.

The tetralogy *Joseph and his Brethren* is a metaphysical novel which treats time in a biblical setting. Even the tempo of reading is determined by the style of the novelist: the book cannot be read in a hurry. There is also a dreamlike quality to the narrative, even when it speaks of the immediacies of passion and knowledge. Future events appear to have been foreseen, past events are reenacted, yet there is still enough reality in them to make one suffer the uncertainties and loneliness of Joseph in his real and symbolic wells, and of the heat of the desert. Although all is set *in illo tempore,* Joseph's life is in the mental present; forever he relives his resurrection and knows that it has happened "already," and also awaits his resurrection and knows that it is "not yet." He is man making history and getting lost in it.[77] The tetralogy is a narrative illustration of Picasso's remark already quoted: "a green parrot is also a green salad *and* a green parrot."

The Magic Mountain is similarly time-roving, set within the confines of a sanatorium. As Dickens, so Mann also speaks through both the musical and the discursive modes of prose. He informs us that to him the novel is like a symphony, a work in counterpoint, a thematic fabric. Mann says he followed Wagner in the use of the leitmotiv, which he carried over into the work of language.[78] The narrator in the story, speaking "by the ocean of time," holds that time is the medium of narration and the medium of music, "thus music and narration are alike." [79] The rhythm and structure of music is imaged in the rhythm and structure of Mann's long narrative, calling into use long-term memory. Recurrent characteristics, objects, and events are employed to introduce a degree of ambiguity among future, past, and present. But Mann also appeals directly to the intellect when he raises the question "What is time?" He considers some philosophical queries: If there were no motion would there be time? Do time and space depend on one another? What is to be meant by limits of time, whether zero or infinity? Toward the end of the book Naphta, son of Jacob by Rachel's handmaiden, Bilah, makes a frontal attack on Zeno and insists that dimensionless points cannot add up to a line; eternity or zero time have nothing to do with duration. In short, time is not in the domain of knowledge understood but in the domain of knowledge felt.[80] We shall leave the other works of Mann alone. *The Beloved Returns* and *Buddenbrooks* are rich sources of poetic insight, but for us one important point has been sufficiently illustrated. The novel as a literary form is well suited for the exposition of problems of time and man because it can speak, through the genius of some authors, to the many temporal Umwelts of man.

But if the novel is so successful as artistic expression, can it also be used to put forth ideas diametrically opposite to Mann's? It certainly can, as may be illustrated by Kafka's *The Trial* and *The Castle*. A comparison between Joseph and the central figure in these two novels is interesting. Joseph was the man of Heilsgeschichte, a man of destiny, an individual wending his way through the labyrinth of time toward a final resurrection. Kafka's K is a victim rather than a hero; he is only the remnant of an individual, an almost indistinguishable member of an impersonal society; his position vis-a-vis reality and time is similar to that of the characters of antitragedies. Joseph was separated from Jacob and pursued by his brothers; K is alienated from society and removed as a nonperson from the flow of time. Joseph could be and was, in fact, redeemed; K cannot be. To Joseph fate was mysterious as it implied the incomprehensibility of an existing Divine plan; the fate of K comprises pure incomprehensibility for its own sake. Joseph came to be liked and loved because he could say "good morning" in a hundred different ways and because he was handsome of appearance; his reality was often harsh but earthbound. K's reality is that of the dream wherein events do not form causal chains and the changes in the flow of time are comparable to those a lizard might experience: when the lizard is in the sun, external things appear to him to happen slowly; when it is in the shade things happen rapidly. The world which became known as Kafkaesque is one of proto- and eotemporality. As in Beckett's antitragedies the suffering of K, and of the reader, derive from foisting of these Umwelts upon the noetic Umwelt of man.

The modern novel is as true an image of "future shock" as its literary ancestor, the classical novel, was an image of the care and concern of precomputerized man. It is as though the primitive temporality which, for the victims of the antitragedy, was a matter to be borne in abject suffering, had now become a worthy way of life. France, which produced Proust and with him the glory of her intellectual history, also produced the most articulate spokesman of the antinovel school of novelists, Alain Robbe-Grillet. In clear and purposeful words he argues against the traditional form of the novel because, as he sees it, that form is based on notions which have become obsolete.[81] (1) Character should be renounced because social progress minimizes the importance of the individual. The novel has lost its hero and new discoveries are needed to replace the role of the hero. (Poetry has not been so challenged probably because its appeal to the cyclic order, an evolutionary level more primitive than that of articulate reasoning, is a more significant portion of verse than it is of prose.) (2) Story should be regarded as unimportant and even irrelevant, and writer and reader must learn to concentrate on the present only. (3) Commitment to ideas or conditions external to the antinovel should be renounced in favor of complete commitment to the writing itself. (4) Form and content are irrelevant. The notion that a writer has something to say and knows how to say it is obsolete; a good writer "has nothing to say" for he has only a way of speaking.

These instructions counsel a regression from nootemporality to the temporality of simpler organisms through the loss of continuity and identity. The inevitable consequences, though not in chronosophical terms, have been properly drawn by Robbe-Grillet in an essay on "Time and Description in Fiction Today"[82] and may be summarized as follows. In the new novel the place of description is opposite to its traditional place. Whereas description used to provide the framework for action and a background setting, the purpose of description in the new novel is to confuse, hence, to deemphasize the background. Yet, it does have a role, namely, the production and suggestion of the double movement of creation and destruction. Since long-term temporal structures should be absent, any attempt to reconstruct a story should lead to contradiction. The antinovel exists only in the mind of the writer and the reader; no world is offered, for the reader must invent it. Unlike in conventional narratives, "in the modern narrative, time seems to be cut off from its temporality. It no longer passes. It no longer completes anything." [83]

When Robbe-Grillet observes that the real, the false, and the illusion are subjects of all modern works he only rediscovers the concern of Socrates with the three types of beds. This should sound familiar to the reader, as should the philosophy of timelessness which emerges as the hallmark of the antinovel. In its frontal attack on objective truth, the antinovel might be one of the many reactions to the rational madness of science. But, whatever its historical origins, we can readily recall our extensive deliberations about temporality, causality, and identity as evolutionary coemergent qualities. It is to be expected therefore, that, when the new novel regresses from nootemporality, it also dissolves the qualities which make up temporality. Thus, although coincidences which appear meaningful to the reader do happen, they are not meaningful to the characters who live in a world of magic causation. Connective principles are primitive, coherence extending beyond a few phrases is absent. Identity vanishes and people become indistinguishable, hence mutually replacable, as are sulfur atoms. Goals and human values also vanish, for they are inexpressible in a prototemporal

Umwelt (of literature) which, while robbing man of his dignity, misses the precision which gives the quantum mechanical image of the world (appropriate for prototemporality in science) both usefulness and beauty. The world of the new novel is a modern gothic horror. It is unlikely that many proponents of the new novel recognize in it a display of the same "terror of history" which Eliade identified as already manifest in archaic man. Perhaps this latter day regression is, in the development of art forms, a *reculer pour mieux sauter*, a drawing back of the searching spirit for a better leap, even if the direction and the usefulness of the leap are at this time uncertain.

Though we have been discussing it as a literary phenomenon, the new novel is of course an image of social attitudes. The demand to regress into lower temporalities had brought with it the revolt against responsibilities including, but not limited to, such matters as schooling in writing and reading. The metamorphosis of the novel signifies an opposition not so much to the novel as such, but more broadly to the civilized skills and practices of reading, writing, arithmetic, remembering, and individual planning. Because of its potential complexity, the novel is the most likely target of the revolt as literature itself undergoes a "future shock." Alvin Toffler has documented the substantial decrease in the length of time through which a book appears popularly valuable, expressed in the radical decrease in the number of hardcover books in favor of paperbacks and in the decrease in the length of time the general public remains interested in any given subject.[84] Reading and writing themselves may become obsolete skills, at least in the more advanced countries. We witness the unholy alliance of three trends. The shrinking span of interest combined with that "engineering of the message" which makes the material digestible with minimal effort and with no effect other, or deeper, than the one desired by the financial sponsor.[85] In addition, because of the vast amount of printed material, each single work becomes a smaller portion of things knowable—whether the knowable is relevant, or not. At its limits, a message of ephemeral interest amounts to a pebble in a chaos, even if it is a blizzard of pebbles. In a curious way this brings to mind the remark I made in chapter 4, that in physics relativity theory replaced the absoluteness of rest with the absoluteness of motion.

Anthony Burgess hopes that

> the huge sales of the late Ian Fleming will, in a few years' time, be no more than a wistful lip-licking publisher's memory. Meanwhile, the more quietly meritorious, the moderate seller with something permanent to give, will go steadily but unspectacularly on, finding a lasting place in the history of the novel.[86]

Perhaps so, but only for a limited audience. We seem to be headed for a cleavage in taste and education. Only the learned and the privileged will know how to read and write at any level of sophistication, or perhaps at all. The literary prophets of our epoch: Aldous Huxley, George Orwell, Ortega y Gasset, and others seem to believe so, perceiving, as they do, the flight of the masses from the terror and responsibility of knowing time. But how will the priests of Olympus be able to communicate with the *idiōtēs* which, in ancient Greece, designated illiterate persons? The trend in attitudes to time, the changes just reviewed, and the fact that many disciples of the new novel (and some disciples of the old, such as Burgess) have been imaginative writers of movie scripts, suggests the answer: through the cinema.

The Film

I suggested earlier that the archaic separation of the intelligible from the emotional in language be identified with the origins of music and language; then I suggested that poetry combines both, and noted the appearance of the novel, in very recent times, as an offshoot of epic poetry. Unlike these artistic expressions, the cinema, sometimes called the eighth art,* arose not as a simple derivative of any preexisting form of art but rather as a confluence of many arts and trades. Its birth expresses a desire to behold a moving but focused image of a world characterized by motion and change. From its ancestors the motion picture inherited a variety of means of artistic appeal and, by combining them, generated a broad spectrum of new means of its own. The film provides its creator, the filmwright, with a freedom of time-roving unparalleled in any other art. By "filmwright" I mean the person or persons who are most instrumental and essential in the creation of a specific movie. The parallel with playwright is not complete. Because of the nature of the motion picture, the filmwright of a specific movie may be its director, or its cameraman, the person who wrote the original story, or the one who wrote the script. Perhaps in all movies "filmwright" is a collective noun: though sometimes one man's genius may predominate. Thus in Laurence Olivier's *Hamlet* the filmwrights are Shakespeare and Olivier; in documentary pictures it is often the cameraman or even the narrator.

As a form of art the film is no more nor less mystifying than a cave painting or the plaintive song of a flute; but from the point of its technique—and the way this technique influences the film's character as an art—it is much more complex than music, literature or the fine arts. A look at its origins will help us understand the sources of its power as a traveler among temporal Umwelts. The history of film has many and peculiarly disparate branches. The "motion from motion" branch of this history harks back to ancient techniques of the projected image: Chinese shadow plays produced with perforated balls, Japanese mirror tricks, Javanese shadow puppet plays, and the technically more sophisticated, though artistically less rewarding, shadow shows of nineteenth century Europe. These and related techniques produced images of motion from figures which themselves moved. The Zoetrope or the "wheel of life" of 1832, a parlor toy, starts off

---

* The origin of the phrase, the seven arts, is in its medieval use. It meant: grammar, logic, rhetoric, arithmetic, geometry, music, and astronomy. This meaning survives in our "liberal arts" and the various academic degrees in "art." In contemporary use it came to mean: painting, architecture, sculpture, dance, music, theater, and literature.

the "motion from rest" branch of film history. It comprised a series of pictures showing consecutive phases of motion, placed inside a revolving drum and viewed through a combination of one stationary and several moving slots. Clowns, jugglers, and animals came to life when viewed through consecutively coinciding slots. Thus in the Zoetrope and in its improved version with mirrors, called the Praxinoscope, the impression of motion was created from stationary images. The ancestry of the projection equipment itself goes back to Father Athanasius Kircher (the same man who constructed magnetic clocks) who in 1646 invented the magic lantern.[87] The principles of his magic lantern were combined with slides which actually moved in the Lampascope, a mid-nineteenth century invention. In 1853 an Austrian artillery officer named von Uchatius lined up several magic lanterns in a semicircle, all focused on one spot, each with a phase picture, and moved a torch rapidly behind them, producing thereby the impression of motion from pictures at rest.

Starting in the early decades of the eighteenth century, the film's history came to combine photography with earlier optical tricks which employed lenses and mirrors. The peepshows of late eighteenth century survive in the Mutoscopes of amusement parks but, more importantly, in an essential feature of the psychology of the motion picture: namely, that it gives a brightly lit and sharply delimited image against a dark background. Edison's Kinetoscope of 1893 combined several of the various techniques just discussed, with other features that reflect the skill and ingenuity of the inventor. He used transparent photographic film of phase pictures running continuously between a magnifying lens and a light source. A revolving shutter made it possible to view each moving frame momentarily. The crucial step in the technique of the motion picture was the invention in 1893 of a mechanism that looks like (and is called) the Maltese Cross. This makes it possible to project a stationary image for a brief period (about 1/100 of a second), cover up the optics for another comparable period while the film is moved, then display the next stationary phase picture.

Impression of motion may be produced through the sequential presentation of stationary images if they follow each other sufficiently rapidly. That this is the case was first reported to the Royal Society in 1824 by the English physician and philologist Roget (remembered for *Roget's Thesaurus*) in a paper entitled "The Persistence of Vision with Regard to Moving Objects." The phenomenon with which we are dealing here is species specific. For man, flicker fusion is achieved at rates higher than 18 stationary frames per second; for the fighting fish (Betta) the rate is 30 frames per second, minimum.[88] We may recall here other data on the creature present, discussed earlier, such as the snail judging a surface continuous if it receives sense impressions at rates higher than four times a second. Von Uexküll distinguished between rapid motion picture and slow motion picture animals.[89] Thus, at the very basis of the motion picture technique we come upon Zeno's paradox, for motion appears to have been constructed from elements of no motion.* Here we may happily refer back to chapter 8 and point out that it is the no-motion, the phase picture which is an abstraction, and not the motion. Looking back at the history of motion picture technique, we note that the motion-from-motion branch of the skill is quite ancient; a great deal of ingenuity and technical preparedness was necessary before the flying arrow could be stopped on the screen, to be recreated as a moving image of reality.

We shall turn now to some features of cinematography which make it very effective in reaching the lower levels of temporality.

1. Consider camera movement that corresponds to achievable translation, rotation or scan of the head. This provides the viewer with moving imagery of the world which he might also have seen without the camera—except that the viewer's head remains stationary. There is no associated kinesthetic experience. Infants raised on motion pictures, like immobilized kittens raised in an otherwise normal environment, would not be able to perform coordinated motion; the space of the cinema is unreal.

2. Although proportions of objects on film are ordinarily the same as those observed in real space, the absolute size of the projected image is usually either larger or smaller than that of the actual object. Hence the relation of the image to the observer is unnatural. Furthermore, the image is starkly framed. These details, together with (1) make for a dreamlike world; they address the mind through images which cannot shed a degree of unreality.

3. Zooming techniques, even with camera motion restricted as in (1), endow the viewer with the fast wings of birds—or even faster wings of imagination—as the camera attacks or escapes from a distant object.

4. The tempo of action is determined by the filmwright more completely than by the playwright or novelist. This stems from the possibility of attention to details unavailable on the stage, and from the hypnotic power of the screen, unavailable to a writer. It follows that the amount of time the viewer is permitted to spend on reflection can be closely controlled.

5. Imposing an unnatural tempo on change, such as in slow or fast motion, tends to be uncanny or awe-inspiring and may also be intellectually revealing. By such means appeal may be made directly to knowledge felt and to knowledge understood, separately or together.

Before completing our chronosophical analysis of the film, I wish to consider a few great moving pictures to illustrate certain points.

Slaughterhouse Five (1972) consists of episodes from the life of Billy Pilgrim, quite out of temporal sequence. The viewer, if familiar with recent world history, may assemble these episodes into a temporal order and interpret the mixing up as caused by the mental imbalance of a sensitive hero. Yet, each episode is in the present, in two ways. One follows from

---

* Television-type scanning adds a further refinement here by adding yet another memory element—another present, if we wish—that of the retentivity of the phosphor.

what Robbe-Grillet calls the grammatical limitations of film: it always talks in the present tense. The other way is more subtle. The viewer is unable to anchor himself in the flow of time, for it is impossible to tell whether a particular episode is an unproblematic "now," or memory, or prophetic foresight. The total effect is a timeless, existential statement sub specia aeternitatis about the impotence of man.

The chronology of *Last Year at Marienbad* (1961) is the nightmare of a trial lawyer and the delight of an artist. As the film progresses any hypothesis formed by the viewer as to whether he is watching a present, past, or future episode is sooner or later challenged. There is not even a continuity as regards the identity of a narrator; the sentiments expressed are sometimes his, sometimes hers. Since in its total gestalt the film is one of decadent beauty and power, many interpretations may be and have been advanced. It is a dream, some say, or a recollection of dreams and events, or a narrative with personal asides. Did the man and woman really meet last year in Marienbad, as it is sometimes implied, sometimes doubted, by the film? Is the claim of this prior meeting only a civilized way to proposition a woman or catch a man? Is it a game, or is it the confusion of someone or someones? Robbe-Grillet, who wrote the story and worked with Alain Resnais, the director, insists that

> the universe in which the entire film occurs is, characteristically, that of a perpetual present which makes all recourse to memory impossible. This is a world without a past, a world which is self-sufficient at every moment and which obliterates itself as it proceeds. [This man and woman exist] only as long as the film lasts. There can be no reality outside the images we see, the words we hear.[90]

Thus, he concludes, the duration is not a condensed version of a more extended real duration. What Robbe-Grillet is saying is that the Aristotelian unity of time must be strictly applied. I think he faces the difficulty of evaluating his own creative processes. Just as his arguments on the new novel echo the terror of history of archaic man, so *Last Year at Marienbad* is a rendering of the Cinderella story: archetypal, hence timeless. The feelings and characters are those of man and woman; they have already met in the embrace and ecstasy of love, if not last year in Marienbad then last millennium in the Roman seaport of Ostia.

Hiroshima, Mon Amour (1959) has a complex chronology but one that can be reconstructed ex post facto. The unique power of the film in manipulating many levels of temporality is mainly in the camera work. For example, the film opens with a scene which suggests the horrors of burnt flesh, a natural association invoked by the title. As the camera withdraws, the texture changes to what looks like molten lava pouring forth from the womb of the earth and suggests the throes of geological creation. Thus, from the immediate present of atomic war our vistas open to the primordial power of material creation, soon to realize that the motion we are watching is a different paradigm of becoming: a man and a woman in the act of creating new life. Whereas on the legitimate stage a copulating pair is unlikely to amount to more than a blue show, for the filmwright the same pair can offer an expression of the tragic present.[91]

Persona (1966) deals with time and identity. Two young women of similar appearance but with different world-views are drawn together by a combination of sympathy, envy, and need. As their intimacy progresses they approach complete identity; in one scene they become interchangeable. Since this process amounts to a regression to a prototemporal Umwelt, it is resisted by the ego. An episode which involves the sucking of accidentally spilled blood introduces the force of biological wisdom into the confused minds of the women, and they return to their lives of unresolvable conflicts.

The Seventh Seal (1956) has the hallmarks of a medieval piece of art, as though it were a book of hours recited as a ballad. Against the darkness of suffering, ignorance, and evil which the Church both created and was caught in, stands the holy family of itinerant actors: Jof, Mia, and the child. A Knight, back from the Crusades, a man who has known evil and sought but not found God, tricks Death into taking him and his entourage in exchange for the lives of Jof, Mia, and Mikael. As does the visible light in the works of Rembrandt, so the visible and poetic light falls on the beauty of Mia, on the innocence of Jof, and on the promise of Mikael.

Mia: I want Mikael to have a better life than ours.
Jof: Mikael will grow up to a great acrobat—or a juggler who can do the one impossible trick.
Mia: What's that?
Jof: To make one of the balls stand absolutely still in the air.
Mia: But that's impossible.
Jof: Impossible for us—but not for him.[92]

This kind of juggling, visually and in time-roving, is the art of the filmwright. As the story ends the silhouettes of the characters appear on a distant hilltop, set against the dark, stormy sky. "Death, the severe master, invites them to dance. He tells them to hold each other's hands. . . . First goes the master with his scythe and hourglass, but Skat [a traveling actor] dangles at the end with his lyre." The viewer has the impression that he is witnessing not the eighth art, but all the seven other arts rolled into one.

The first two of the films just considered (*Slaughterhouse Five* and *Last Year at Marienbad*) demonstrate the universality of statements achievable by making the film convey a timeless present. *Hiroshima, Mon Amour* illustrates the power of camera work in transforming a pedestrian "now" into a tragic present of vast temporal horizons. *Persona* travels among the many layers of temporality as personal identities devolve and evolve. *The Seventh Seal* expresses metaphysical convictions about life, death, time, and man. It is also possible to use the freedom of the film to deal with Weltanschauungs and time profoundly yet unobviously. I do not have in mind the barrage of science-fiction movies about time which tend to be pedestrian and

sophomoric. I am thinking, instead, of such masterpieces as Hiroshi Teshi-
gara's *The Woman in the Dunes* (1964).

On the journalistic level it concerns a young entomologist who misses
a return bus from a lonely seashore. He is put up by villagers in the house
of a woman where he becomes a prisoner. The house is in a sand pit. His
protests are of no avail, he is unable to climb out, and is forced to remain.
Three years after his disappearance the police list him as legally lost. On
the philosophical level the house in the dunes is the Taoist version of the pit
of Joseph. But whereas Joseph, as an Ur-Faust, resurrects by taking the
right action, the entomologist resurrects by learning inaction. As he slowly
faces and comprehends the inevitable, he becomes one with the body of the
woman and with the feminine soul of reality. He relinquishes his goals in
time in favor of harmony with all creation, learns to shun the complexity
of his desires directed toward the control of the future, and withdraws into
the present. He accepts the unknown and allows the element of becoming
to enter his life. His temporality changes from a business present to a
conflict that is harmonious, even though it appears to embody logical contradictions. Into this peace of mind then storms the call of life: the woman
is with child. With a gentle touch of humor, as the man learns to live with
the present and remains voluntarily in his pit, the woman becomes future-
directed.

An 1842 trade card of the first daguerreotype studio claimed that their
pictures reflected "such exactness & truth as to establish an identity of
character; the minutest details may be perfectly represented. As well for
momentary expression of countenance. . . ." [93] The artistic growth of
machine-made images from the daguerreotype to the film reminds one of
the much longer evolutionary history of language. In the beginning, language also was used for no more than identification of friend or foe, food
or poison, or call of the mate. But as language for the playwright, so also
the motion picture for the filmwright opened up possibilities of expression
whose sentient and noetic content could not have existed without the motion
picture. It ceased to be the sum total of its artistic and technical features
and came to generate "beatnotes" among its constituent principles. The film
itself is Zeno's arrow: it both moves and does not move, hence it is capable
of addressing man's Umwelts from the atemporal to the nootemporal. One
of the many ways in which the sources of power of the film may be described is by stressing that the motion picture can make the self-evidence
of temporality incomprehensible, hence disturbing. The self-search and self-
recognition which follow are the substance of the experience of the beautiful.

3. The Freedom of
the Beautiful

KANT has clearly sensed something paradoxical about aesthetic judgement.
In critical philosophy, pure reason constitutes phenomenal necessity as de-

termined by a priori laws of creative understanding (Verstand). Practical reason amounts to a freedom of action where common sense (Vernunft) acts as the moral guide. Aesthetic judgement is certainly not a necessity (it cannot be reasoned); neither is it something of common sense (its validity cannot be demonstrated), yet such judgements still carry, or at least imply, universal truths. This curious "now you see it, now you don't" feature of aesthetic judgement might have appeared to Kant less paradoxical had he been a post-Darwinian scholar. He correctly stressed that aesthetic judgement represents a harmonious interplay of cognitive faculties but could not have seen it in the perspective of organic evolution whose stages, or integrative levels, remain represented in man.

We spoke earlier of truths and ethics as guidelines which assist man in selecting future courses of actions from his store of autogenic imagery. To these we may now add the sense of beauty and regard it not as a set of instructions (which, as I have reasoned, aesthetic judgements very seldom are), but rather as a play with those images, a practice of arranging them, perhaps even testing them for meaning and significance. The sense of beauty is a way of learning to live with what may be called the metaphorical imperative. By this I mean the necessity of communicating in metaphors, themselves being symbolic transformations of experience. In *Romeo and Juliet*, Mercutio, mortally wounded, cries out, "I am hurt. A plague o'both of your houses! I am sped" (III, i, 96). He did not envisage that cultures of the bacteria Pasteurella pestis would be delivered to the residences of the Montagues and the Capulets, even if he had known about such creatures. Just the opposite; it is the name of the bacteria that was derived from the more general idea of *pestis*.

Poetic language (spoken, sculptured, or painted) experiments with the most general impressions of the world and man and learns to arrange them in ways that "ring true" or "ring right," but alas, without invoking either truth or ethics. In due course, out of such general outlines which, as Kant would have said it (sopra), represent a harmonious interplay of cognitive faculties, articulate instructions for preferred behavior may arise and only much later, utterances of truth. Thus the experience, experiment, play, or skill which we recognize as the beautiful must remain by its very nature an inquiry free of artificial restrictions. Physics can search for and find universally valid patterns in a mole of mercury because it concerns itself only with the lower Umwelts; art must be permitted to embrace all of man's Umwelts, and then some. A. I. Solzhenitsyn, in his Nobel lecture on "Art—for Man's Sake," diagnosed the sources of those fears which lead to organized suppression of the freedom of artistic expression. Such fears stem from the awareness that the beautiful can communicate the condensed essence of experience on a more universal basis than any other known modes of communication, in spite of the virtual nonexistence of anything that would resemble general rules of art.

Arnold Hauser in his famous *The Social History of Art* warns against art's advancing too close to chaos, even if the purpose is that of rescuing artistic expression from chaos.[94] As I see it, he warns against careless regression into primitive Umwelts. Yet it is precisely this regression that is at the basis of creative advance for man in general, and artistic experimentation, in particular. But of course, regression alone is not enough: it is the artist's fight back up, as it were, the experience of creative trauma which makes possible that time-roving appeal which is the strength and very material of the beautiful. Although there is beauty in order, an evaluation of artistic beauty as the identification of order in chaos is not sufficient; equations of planetary motion also identify order in chaos, and they are also beautiful—but not artistic. The work of the artist (dancer, writer, or potter) whether intended to put the world afire or bespeak feelings and thoughts tender and intimate, creates conflicts from chaos, and not order. In this function the artist imitates the cosmic process of the emergence of unresolvable conflicts. Because of this quality of cosmic imitation it is the beautiful which offers the most direct knowledge of the open-ended hierarchy of the unresolvable conflicts of time.

XII

TIME AS CONFLICT

IN THIS CHAPTER I shall attempt to sketch some of the salient features of the theory of time as conflict. Thus, this is not a summary or review chapter but only a refocusing of our attention upon a few of the many new concepts which were developed and found useful in the book, and upon their interdependence.

What we unproblematically call "time," though a single and coherent feature of the world, is, in fact, a hierarchy of distinct temporalities corresponding to certain semiautonomous integrative levels of nature. The well-known difficulties of giving a discursive report about the nature of time stem mostly from the hopelessly confused image that obtains when the hierarchical character of time is unrecognized. That this dynamic stratification has continued to be neglected is, nevertheless, understandable. Since the difficulties of regressive sharing make a direct appeal to experience regarding the nature of lower-level temporalities unattractive, even though there is evidence that they are fully represented in man, the existence of these temporalities below the noetic is not intuitively obvious.

In spite of the conceptual barriers, we did succeed in exploring the major features of temporal levels below the noetic by drawing upon the large store of humanistic insight and scientific understanding regarding the nature of time. We would not have been able to do so, however, and could not have hoped for a coherent theory based on many different ways of knowing, had it not been for our serious appeal to the *Umweltlehre*, or the study of species-specific universes, first put forth by the biologist Jakob von Uexküll over half a century ago. We have examined this principle and extended its validity, originally limited to biology, to several levels of reality and modes of knowledge, by taking advantage of the potentialities inherent in the idea. Through the informed use of the Umwelt principle it was found possible to interpret statements which appear epistemological in character as amounting, in fact, to ontological claims. The integration of many disparate views could then proceed on the basis of the assumption that the many aspects of reality are coherent and that universal schemes underlying these different aspects are discoverable.

435

1. Temporalities

THE semiautonomous levels of nature which we found useful to distinguish comprise those of electromagnetic radiation, of indistinguishable
particles, ponderable mass, the cyclic and the aging orders of life, the
mental functions of the individual, and the communal functions of society.
The temporal Umwelts determined by these integrative levels must be
understood as coexisting and interpenetrating; nevertheless they are sufficiently distinct that they may be meaningfully identified and analyzed.

We have named atemporal all those Umwelts, be they physical, physiological, or psychological, in which it is impossible to identify succession or
the passage of time when it is imagined that the determination is performed
from within the Umwelt. In atemporal worlds, as exemplified by the electromagnetic substratum of the universe, everything happens at once. We have
given the name prototemporal to all Umwelts which comprise indistinguishable entities. The members which make up and determine a prototemporal
world are countable but not orderable, a condition exemplified by the
particulate features of the physical world, by some more advanced forms
of matter, and even by some states of mind. This world is an imperfect
speaker which can talk only in probabilities; its lawfulness may be described
as that of controlled randomness. In a prototemporal Umwelt time and
space are not sufficiently differentiated; hence histories (as seen by a nootemporal observer) are interchangeable with simultaneous group actions.

We gave the name eotemporal to worlds of pure succession or, what
is equivalent, of pure asymmetry. Eotemporal Umwelts display the diadic
relationship of before/after but not a preferred direction of time. They are
characteristic of much of the physical world and also of the cyclic order
of life. We can also find eotemporal conditions projected, or mapped, into
certain world-views and attitudes toward history. In this substratum of pure
succession, simple living organisms determine a present, for it is necessary
for them to coordinate their cyclic functions in a "now" if they are to maintain their inner coherence. Biotemporality of the simple, cyclic order thus
is a triadic time, since it provides a present to which the eotemporal asym-

••

*Only a wayfarer born under unruly stars would attempt to put into practice
in our epoch of proliferating knowledge the Heraclitean dictum that
"men who love wisdom must be inquirers into very many things indeed." Indeed, the classic function of all philosophies is to offer a
comprehensive view of the many things that make up our one world,
and thereby assist and guide man in his search for meaning and order
in his life.*

Hokusai, Katsushika (1760–1849), The Scribe. *From "Album of Sketches."
Courtesy, The Metropolitan Museum of Art—Gift in Memory of
Charles Stewart Smith, 1914.*

••

metry of time may be anchored. The evolution of living forms extends the significance of the biotemporal "now" as the spectrum of the cyclic complement of life widens, leading eventually to a differentiation between the linear (aging) and the cyclical ("immortal" or eotemporal) orders of life.

Complexity as a variable is the determinant of the nootemporal Umwelt. This is the temporality which common usage sometimes describes as human time. The very complex, exemplified by the brain, extends the if-then relationship of conditional probability, already operational in biogenesis, into the long-term memory of man. The capacity of the very complex to form autogenic imagery, and its command of long-term memory, lead to the possibility of designing, as it were, long-term expectations.

As already implied, we can ordinarily talk about presents only for or above the biotemporal level. Among living creatures we may distinguish presents of increasing sharpness as we ascend along the evolutionary ladder. They often overlap and are nested but are nevertheless recognizable. We have noted the physiological present, the creature present, the psychological present, and the mental present.

Whereas for life-forms of increasing complexity the "presents" sharpen through the evolving capacity for attention, there is also an evolutionary broadening of its domain. Thus in man we can distinguish between the measurable span of attention and an ill-defined region of memories and expectations which, added to attention, make up the mental present. We may extend the idea of the present below the biotemporal but only formally, not phenomenologically. Thus, we might decide to mean by "present" any logically or instrumentally definable simultaneity. In the cosmological domain such a present is exemplified by the electromagnetic universe; in the microscopic domain, by the chronon.

The hierarchical organization of time suggests, and can accommodate, several levels of causation. In an atemporal world causality cannot be given any meaning. A prototemporal world can offer no more than hazy connectedness; its laws, as already mentioned, are probabilistic, with space and time not yet distinguishable. An eotemporal Umwelt displays the classical action-reaction principle of deterministic causation. Because of its purely asymmetrical nature with no preference for a direction of time, cause and effect within an eotemporal Umwelt are interchangeable and indistinguishable. With biogenesis we witness the separation of two aspects of causation: connectedness and intentionality. The emergence of the present in the triadic form of time is a corollary of the differentiation of final causation from deterministic causation, for the biotemporal present is defined in terms of the needs (hence, of the goals) of the organism. In its turn, final causation grows into multiple causation with the complexification of living matter and leads eventually to the potentiality of free will among the functions of the very complex.

The gradually sharpening separation of the intentionality of living things from the increasingly sophisticated connectivities (lawful necessities) illustrates nomogenesis, that is, the coming about of laws of nature. We may think of natural laws as determinants of various Umwelts. As we rise from atemporality to nootemporality the increasing store of principles of nature are at the roots of the multiple controls under which higher Umwelts have to function. For instance, systems properly classified as biotemporal are, nevertheless, restrained by the regularities of the atemporal, prototemporal, and eotemporal worlds, while also exhibiting some level-specific principles. We may add here the example of an interesting, cosmic recapitulation process. The stages in the cognitive growth of the child repeat in a rather loose manner the evolutionary development of the different methods of connectedness. In the mind of the small child and in primitive societies, as in the lower Umwelts, connectedness independent of intentionality does not exist; they are not yet differentiated.

The hierarchical structuring of time also suggests that beginnings and endings are themselves so organized. In an atemporal world they are meaningless. Prototemporal beginnings and endings are conditions which one might be able to infer theoretically but could not actually locate in time, even in principle, at least not in the experiential time of the clock-watcher. Eotemporal beginnings and endings involve the coming into being and going out of being of ordering per se, and would again have to be inferred theoretically. But, since eotemporality is pure succession, an eotemporal beginning could not be distinguished from an eotemporal ending. In the aging order of living things beginnings and endings may be identified, though not without ambiguity, with the births and deaths of individuals or species. Yet we would not call the appearance of the one-toed horse the beginning of time, or the death of a rat its ending. In nootemporality beginnings and endings are tied to my birth and my death, but again we would hesitate to identify the beginning and end of time with these events. In summary, there is nothing in nature that would correspond to the sharp contrast which we feel we ought to mean by a time/timeless interface.

The hierarchical structuring of time, that is, its progression along temporalities, may be found recapitulated in the development of the child's concept of number and of his number skill. Starting from number "one" as the archetype which symbolizes the coming into being of the self versus the nonself, the child first masters cardinality. His earliest capacity is the understanding of sets whose members are countable but indistinguishable, hence unorderable. In these specifications we recognize the determinants of a prototemporal Umwelt. Ordinal numbers are mastered next. They form series of pure succession, that is, pure asymmetry, thus they symbolize the eotemporality of an Umwelt. The introduction of zero provides an absolute separation between two semiinfinite rays, both purely asymmetrical, though symmetrical about zero. This new mathematical quality corresponds to the interface between the eotemporal and the biotemporal, with its emergent "now," but still without preferred direction. These and related features are probably at the roots of the awesome power of number in natural science. For it is reasonable to assume that members of mathematical sets, which are symbolic entities, can be put into one to one correspondence with physical entities of the Umwelts whose essential properties they image. But the same "lowly origins" of number also suggest that no matter what sleight of hand one applies to numbers, or how sophisticated an exercise one performs with them, they cannot be made to lead to valid statements about biotemporality and above, for they do not image the essential properties of these Umwelts.

2. Transcendences

THERE ARE at least two well-established theorems which pierce the interface between the proto- and eotemporal worlds. The variables of the H-theorem are prototemporal, whereas the theorem in its integrated form speaks about pure succession (though not about the unidirectionality of time) hence about eotemporal conditions. Heisenberg's uncertainty principle is usually interpreted as pertaining to single particles, hence it is pertinent to the prototemporal world. But the variables of the equations are macroscopic, eotemporal quantities. The usefulness of these two theorems and the difficulties associated with their proper interpretation stem from their double citizenship. If one considers the substance of the H-theorem together with an "inverse H-theorem" which may be postulated for open systems, the two together show the continuity of the principles which minimize the rate of entropy production. These principles, already manifest in the physical world, are extended in the biotemporal world as entropy-producing systems come into being within closed thermodynamic systems.

Nature seems to disfavor transitional states between integrative levels; Umwelts between the well-established levels of temporality appear to be unstable. Forms of matter between nonliving and the living are not known, probably because they are short-lived, metastable forms. Regions between mind and life are known but they occupy rather unfavored positions, disclaimed, as it were, both by mind and by life; they form the content of the unconscious, a buffer zone between living and mental functions. Yet it is precisely within such metastable interfaces that the transcending processes must be going on. It is here that particulate matter emerges from radiation, ponderable mass forms from particles, life rises from matter, and mind emerges from life. The instability of forms in the regions of transcendence guarantees the identities of the regions so separated, the autonomy of their laws, and the distinctness of their temporalities. This hierarchical stratification of nature again has a certain correspondence in mathematics. Namely, as it is now understood, the world of symbols known as mathematics does not comprise a continuous set of rules but consists, instead, of inductively coupled regions of deductive structures.

Some of the ways in which one level may give rise to the next may, nevertheless, be explored or at least convincingly guessed. There are reasons to believe that early life comprised no more than the combination of physiological clocks which copied environmental periodicities. The necessity to coordinate several such periodicities amounted to the definition of a "now" in terms of the needs of the organism. We must assume that the cyclicities operating within the intimate confines of the organism had to produce, sooner or later, some new internal periodicities, by linear superposition and nonlinear generation. The new periodicities did not need to correspond to any cyclic phenomena in the external world. They were life's first autonomous rhythms arising from established functions of nonliving matter. This impressive combination of continuity and creativity finds a formal expression in the appearance of open systems which reduce the rate of entropy increase in the closed system of which they are part.

The emergence of the mind is a process analogous to biogenesis. The central nervous system, but especially the brain, came to form models of the external world obtained by sense impressions. The coordination of these models amounts to the definition of the mental present. We must assume that the coexistence of many models within the intimate confines of the brain had to generate, sooner or later, new models of events, things, and of conditions. These new images did not need to correspond to anything in the external world and, sometimes, could not have had any correspondence, even in principle, because of the restraints of lower-level lawfulness. These new models were man's first creative thoughts arising from the functions of living matter.

Going one level higher, individuals in communities, connected not by physiological processes (as are neurons) but by the flow of symbols through communication channels, also tend to form images of things and conditions external to the group, images of the group itself, as well as autogenic images with no prior correspondence in reality. The aggregate of these symbols comprises the substance of, and determines the next higher integrative level above the noetic: that of man in societies. In all of these examples of transcendency the technique involves the revolutionary new use of already existing structures and functions: a common method which does, however, take different forms.

3. Future and Past

PARALLEL with the hierarchy of transcendences, we find an increasing polarization between future and past in terms of a growing store of intractable uncertainties associated with the future but not with the past. The polarization begins in the atemporal world as an asymmetry due to uncertainties of the future, consequent upon the finite propagation velocity of causal connections (as perceived by a nootemporal observer). The repertory of uncertainties is enlarged with the proto- and eotemporal contributions, for, in addition to modulated radiation, messages may now also be carried by configurations of moving particles and by macroscopic matter. Final causation sharpens or, so to say, legitimizes the future/past polarization by providing the biotemporal "now" to which local past and future may be referred. Multiple causation operative in advanced organisms and, finally, the free will of man further increase the sharpness of separation between past and future.

As though opposing this trend of increasing uncertainties in the future, we find, along the hierarchy of integrative levels, a growing intensity of intentionality that is a capacity to control the future and thereby decrease the uncertainties which inform the present. With imaginative liberty, perhaps even the bunching together of prototemporal particles into eotemporal masses may be taken as demonstrating this process. Certainly, the replacement of microscopic by macroscopic laws in the physical world decreases the uncertainties in some physical parameters. Above the eotemporal world, we note that life has been characterized from its very beginnings by a striving to increase its control over lengthening periods of future time and thus decrease the uncertainties and attendant tensions of the present. We may point to the formation of colonies and multicellular organisms which perform such a task, and to the steadily complexifying evolutionary forms with their growing specialization of functions and division of labor. The evolutionary emergence of heterosexuality and death are biological ways of overcoming the limitations of the soma, by providing living offspring as the means of controlling the future of the species. A more efficient way to do the same is through the functions of the mind, which is thus seen as extending the power of intentionality beyond the death and spatial limitations of the individual. Whereas the body records its learning in the biological forms of the genes, the mind records its learning through the formation of inanimate symbols. We should point here to the technique of the mind in the emergence of a syndrome of mutually reinforcing skills, capacities, and modes of knowledge. Articulate language, writing, toolmaking, long-term memory, creative imagination, identity of the self—all cooperate in the extension of the regions of intentionality, and all contribute to the sense of time. Further along the hierarchy, societies try to extend their capacities for control by further specialization, increased memory storage, and the use of symbolic transformation of social experience.

Throughout the course of organic evolution we may observe the growing refinement of the eotemporal action-reaction principle, leading in the domain of life to the dynamics of the mutuality of adaptation. Among other features, this dynamics is characterized by increasing time-economy or increasing rate of change. This aspect of evolution is so pronounced that, had we not known of time, we might have identified it as the variable in some sophisticated calculus of variations. There are reasons to believe that this increasing rate of evolutionary change guides judgements of usefulness manifest in animal behavior and value judgements manifest in man.

The increasing capacity of organisms, along the evolutionary hierarchy, to act with intentionality is probably at the basis of the asymmetrical evaluation which the behavior of living things indicates, when seen in their relationship to other living things below and above them in sophistication and complexity. Using terms which are most appropriate only to man, we might nevertheless note that organisms behave vis-a-vis more complex organisms as though the behavior of the more advanced organisms were surprisingly unintelligible or mysterious. Looking toward lower levels, the behavior of less advanced organisms might be described as suggesting intelligibility, expressed, for instance, in the facility of many organisms to manipulate simpler organisms. In the case of man, intelligibility can be taken non-metaphorically: the behavior of lower organisms may be understandable, but not generally sharable. Although our lives and thoughts include proto-, eo-, and biotemporal components, we have no easy way of imagining the Umwelt of a DNA molecule. We may feel sympathy for a gene, or for a ladybug, but we cannot feel empathy. Looking now upward, above the noetic, toward the manifestations of the communal lives of man, the forces that move history are much more likely to be felt to be mysterious than the desires that move the guinea pig. From this asymmetry in the nature of knowledge, it follows that if we wish to make assertions about nootemporality with the same authority that we command when making assertions about lower temporalities, we find that we cannot do so. We miss the platform from which we might behold the complete phenomenon of man.

4. The Strategy of Existence

THE PRINCIPLES which control the integrative levels of nature that we have considered seem to leave certain regions of possibilities in each level unrecognized, hence undetermined. It is from these undetermined regions that the next higher integrative level may, in each case, arise. For instance, the laws of gravitating matter are located in areas left undetermined by the regularities of flat space-time; principles of nature unique to life can operate only in regions unrecognized by eotemporal regularities. Whatever principles may be unique to the mind, they can be expected to control only those regions of reality which were left uncontrolled by the laws of physics and of life. Analogous arguments hold for any regularities history may conceivably possess. In all cases, nomogenesis must take place within the previously undetermined regions. We can easily identify such regions ex post facto. We can see, for instance, that life may arise from regions of the inorganic which are uncontrolled by physics; there are no physical laws against the evolution of the ostrich, and we know of ostriches. But it is doubtful whether the ostrich can divine the emergence of identity and language; likewise it is difficult for the mind to discern the boundaries of its own limitations. (We might still be able to do so by seeking the main features of the next higher integrative level above the noetic; that is, by inquiring about the three great cultural continuities of man: his search for the true, his struggle with the good, and his desire for the beautiful.)

A corollary of these epistemological features is that every integrative level, except the atemporal, must be under multiple controls: those unique to itself and those operating on the lower levels. (Control imposed from above, such as life determining the fate of matter, cannot be classed as necessity or lawfulness, but only as contingency). Consistently with this scheme of things, each temporality must subsume or, so to say, accommodate lower-order temporalities. Our nootemporal Umwelt cannot show any features which would somehow negate, rather than extend, the salient qualities of the atemporal, proto-, eo-, and biotemporal Umwelts. For instance, the world of our experience must permit the existence of reversible processes, or probabilistic laws, or of atemporal phenomena.

On each integrative level we may discern certain conflicts. The concept of "conflict" here contemplated is taken to subsume a wide spectrum of oppositions, from the logical to the phenomenological. In some cases this conflict is best described as that between the nomothetic and the undefined (or generative) aspects of an integrative level; sometimes as that between level-specific necessities and contingencies; sometimes between Umwelt-specific aspects of being and becoming; sometimes (as often is the case when discussing organic growth) as that between the expected and the encountered. These various conflicts often manifest themselves through a degree of stress, though the stress takes many different forms. For instance, for living systems we have a tension between the forces which tend to remove matter from the inorganic and those which tend to return matter to it; this is the continuous struggle between growth and decay. This conflict must be regarded as unresolvable, because life lasts only while, and only as long as, the conflict lasts. In the case of the mind, we find a continuous stress between the forces which tend to promote and maintain the integrity of the self and the forces which tend to destroy it. For the rich nootemporal world of man, this struggle may sometimes be described as that between knowledge felt and knowledge understood or, in ordinary terms, between passion and knowledge. In the conscious experience of man this struggle has been ceaseless, for through individual and communal experience we seem to have learned that knowledge untamed by passion is dangerous, while passion uninformed by knowledge is useless. Here again, identity and conscious experience last only as long as the struggle between passion and knowledge lasts. With a purposefully broad phrase, we have called the family of the many stresses associated with unresolvable conflicts those of existential tension. In the philosophy put forth in this work we have insisted that only existential tension be given ontic status, but not the many ways it may be broken down into beinglike and becominglike projections, for that breakdown is heavily influenced by cultural traditions and linguistic habits.

It seems that each emerging integrative level, with its new form of temporality, offers a solution, by containment rather than by regression to the unresolvable conflicts of the prior level. Thus, the conflict of life, which is resolved upon death by the return of the body to the inorganic, that is, by collapse, is resolved by containment through the evolutionary emergence of the mind. The unresolvable conflicts of the mind which find a solution by regression, for instance when senility sets in, are also resolved by containment through the communal organization of individuals. I would assume that unresolvable conflicts characteristic of society may also be identified, perhaps as those between the forces of social coherence and those of the identity of the self. At the lower end of the spectrum I would regard biogenesis as the resolution of some unresolvable conflict of the physical world, perhaps that between those processes which tend to decrease the rates of entropy production, and the universal trend represented by the second law of thermodynamics. The emergence of self-organizing systems may be regarded as having resolved this conflict by containment.

While new integrative levels do seem to offer solutions by containment of the conflicts of the levels immediately beneath them, they also represent increases in existential tension. The reason for this is that increased complexity, greater freedom in the modes of causal connectedness, and, in organic evolution, better adaptation itself generate an increase in the quantity and in the kinds of situations wherein the expected and the encountered might differ. (Among the most sophisticated life-forms this can sometimes be described as an increase in the possibilities of misjudgements.) The opposite path, that of decreasing existential tension, accompanies conflict resolution by collapse or catastrophe: society may relapse to the anarchy of uncoordinated individuals; the mind may return to the level of life; life may return to matter and even ponderable matter may return to energy in its radiative form.

Based on the material at hand we conclude that the strategy of existence seems to favor the resolution of certain conflicts by containment, through the creation of new integrative levels which themselves display certain unresolvable conflicts, rather than by the elimination of the conflicting opposites. This strategy is corollary to a preference for increased rather than decreased existential tension. Conflict resolution by mutual collapse of certain opposites would soon lead back to the equilibrium of preconflict chaos; on the average, only increasing existential tension can guarantee a self-maintaining and open-ended process. These thoughts are not intended to answer such ancient queries ("unto the Jews a stumbling block, and unto the Greeks foolishness") as why there is a world. They propose only a theory of natural philosophy which sees the world as one of semiautonomous integrative levels, created through the dynamics of emerging, unresolvable conflicts. In this dynamics, time itself is hierarchically ordered. It subsumes several levels of temporality, each contributing some new and unique qualities to the noetic temporality of man.

The Crossroads
Pleasantville, New York
September 10, 1974

Abbreviations for Works
Frequently Quoted

Biological Clocks	*Biological Clocks*. Cold Spring Harbor Symposia on Quantitative Biology, Cold Spring Harbor, N. Y. vol. 25, 1960.
Brain	J. C. Eccles, ed., *Brain and Conscious Experience* (New York: Springer Verlag, 1966).
Nat. Phil. Time	G. J. Whitrow, *The Natural Philosophy of Time* (London and Edinburgh: Nelson, 1961).
N. Y. Acad. Sci.	Roland Fischer, ed., "Interdisciplinary Perspectives of Time," *New York Academy of Sciences, Annals*, vol. 138, art. 2, pp. 367–915 (1967).
Patterning of Time	L. W. Doob, *Patterning of Time* (New Haven: Yale University Press, 1971).
Political Phil.	J. G. Gunnell, *Political Philosophy and Time* (Middletown, Conn: Wesleyan University Press, 1968).
Relativity	A. Einstein, H. A. Lorentz, H. Weyl, and H. Minkowski, *The Principle of Relativity*. A Collection of Original Memoirs, tr. W. Perrett and G. B. Jeffrey (New York: Dover Reprint, n.d.).
Study v. 1	J. T. Fraser, F. C. Haber, and G. H. Müller, eds., *The Study of Time, Vol. 1* (New York: Springer Verlag, 1972).
Study v. 2	J. T. Fraser, N. Lawrence, eds., *The Study of Time, Vol. 2* (New York: Springer Verlag, 1975).
The Voices	J. T. Fraser, ed., *The Voices of Time*. (New York: Braziller, 1966).

All Greek classics, unless otherwise specified, are those in *The Loeb Classical Library*, Cambridge, Mass: Harvard University Press.

Notes and References

I / THE INTELLECTUAL QUEST

1. Hans Joachim von Schumann believes that the preponderance of auditory imagery in Homeric dreams suggests that Homer was blinded in early childhood. "Phenomenologische und psychoanalytische Untersuchung der Homerischen Träume," *Acta Psychotherapeutica, Psychosomatica et Ortopaedagogica*, III, No. 3 (1935), pp. 205–19.

2. H. Fränkel, "Die Zeitauffassung in archaischen griecheschen Literatur," *Zeit. f. Ästhetik u. Allg. Kunstwissenschaft*, Bd. 25, *Beilageheft* (1931), pp. 97–117.

3. G. J. Whitrow, "The Concept of Time from Pythagoras to Aristotle," *Ithaca*, 26 VIII–2 IX (1962), p. 499.

4. C. Whitman, *Homer and the Heroic Tradition* (Cambridge, Mass: Harvard University Press, 1967), p. 247.

5. E. Auerbach, *Mimesis*, trans. W. R. Trask (Princeton: Princeton University Press, 1953), p. 23.

6. *Political Phil.*, p. 75.

7. J. de Romilly notes this in pointing to other recent studies. See her, *Time in Greek Tragedy* (Ithaca: Cornell University Press, 1968), p. 4.

8. *Hesiod, Works and Days*, p. 41.

9. Hesiod, *Theogony*, p. 87.

10. I am indebted to Prof. J. G. Gunnell for helping me clarify these details of a complex and colorful myth.

11. K. Freeman, *Ancilla to the Pre-Socratic Philosophers* (Cambridge, Mass: Harvard University Press, 1957), "Heraclitus of Ephesus," fr. 85.

12. *Ibid.*, fr. 67.

13. *Ibid.*, fr. 49A.

14. On this see F. M. Cornford, *Plato and Parmenides* (London: Routledge and Paul, 1964), p. 199.

15. Freeman, *op. cit.*, "Empedocles of Acragas," fr. 17 and *passim*.

16. See the excellent summary on Empedocles in Freeman, *The Pre-Socratic Philosophers* (Oxford: Blackwell, 1953), p. 172.

17. Freeman, *op. cit.*, "Philolaus of Tarantum," fr. 4.

18. Aristotle, *Metaphysics* 986a.

19. To what extent the Pythagorean concepts derive from Oriental ideas is an interesting subject for speculation. Whitrow sees in the Pythagorean doctrine remnants of the Babylonian belief that all names are associated with number. Whitrow, *op. cit.*, p. 500.

20. Freeman, *op. cit.*, "Zeno of Elea," fr. 4 is the original. See also the discussion of Zeno in Freeman, *The Pre-Socratic Philosophers*, (Cambridge, Mass: Harvard University Press, 1959), p. 153.

21. Aristotle, *Physica* 239b.

22. *Nat. Phil. Time*, p. 135 ff.

23. Plato, *Timaeus* 37d-37e.

24. Plato, *The Republic* vii. 526.

25. Plato, *Meno* 85d.

26. Plato, *Phaedrus* 245c.

27. Plato, *Timaeus* 51a.

28. Aristotle, *Physics* iv. 218a–24a.

29. *Ibid.*

30. Aristotle, *Problems* 916a.

31. Aristotle, *Physics* 233a.

32. *Ibid.*, 222b. Cf. Thalen, when asked, "What is the wisest?" answered "Time, for some of the things it has found already and some it will discover later." Plutarch's *Convivium* 9. 153d.

33. Aristotle, *On Coming to be and Passing Away.* II, 11.

34. P. Ariotti, "The Concept of Time in Antiquity," *Study v. 2.*

35. See the masterful writings of S. G. F. Brandon, especially *History, Time and Deity* (Manchester: University Press, 1965), p. 106. Also M. Burrows, "Ancient Israel," in R. C. Denta, ed., *The Idea of History in the Ancient Near East* (New Haven: Yale University Press, 1955), p. 99.

36. Sigmund Freud, *The Standard Edition of the Complete Psychological Works* (London: Hogarth Press, 1964), vol. 23, "Moses and Monotheism," p. 105.

37. Brandon, *op. cit.*, p. 133.

38. Burrows, *op. cit.*, p. 127.

39. Brandon, *Man and His Destiny in the Great Religions* (Manchester: University Press, 1962), p. 196.

40. Brandon, *History, Time and Deity, op. cit.*, pp. 137–40.

41. K. Jaspers, *The Origin and Goal of History* (New Haven: Yale University Press, 1953), p. 1.

42. Brandon, *op. cit.*, p. 203.

43. From Plotinus, *The Third Ennead.* trans. McKenna. ii. 7. 7–11. 320–38.

44. Marcus Aurelius, *Meditations* xi. 1., trans. G. Long, (New York: Burt, n.d.), p. 273.

45. St. Augustine, *The City of God* xii. 13.

46. *Ibid.*, xi. 6.

47. *Ibid.*, xii. 18.

48. St. Augustine, *Confessions.* trans. Pusey, Everyman's Library, xi. 14.

49. *Ibid.*, iv. 11.

50. *Ibid.*, i. 1.

51. *Ibid.*, xi. 36. He invited the reader to think of the rhythm and melody of a hymn: *Deus Creator omnium.* . . . This remark of St. Augustine and his reference to rhythm are often quoted in the intellectual history of time, and deservedly so. However, no one seems to have noticed the kinship of this understanding of time to the interest of St. Augustine in music. See reference 32, chapter 11.

52. "Quid est tempus? Se nemo a me quaeret, scio, si quaerenti explicare velim, nescio!"

53. J. F. Callahan, *Four Views of Time in Ancient Philosophy* (Cambridge, Mass: Harvard University Press, 1948), p. 204.

54. These introductory thoughts are based on Brandon, *Man and His Destiny in the Great Religions, op. cit.*, pp. 241–2.

55. L. Massignon, "Time in Islamic Thought," *Man and Time*, ed. J. Campbell (New York: Pantheon Books, 1957), p. 110.

56. Averroës Cordubensis, *De Physico Auditu*, Libri Octo. Lib. iv, Summa 3, "De Tempore," cap. 3, comm 98. Quoted in: P. Ariotti, "Celestial Reductionism of Time," *Studi Internazionali de Filosofia*, vol. IV, 1972, pp. 93–100.

57. Cf. *The Voices*, p. 625, Note 200.

58. J. Needham, "Embryology in Antiquity," *History of Embryology* (New York: Abelard-Schuman, 1959), chap. 1.

59. Those interested may follow up P. Ariotti, *op. cit.* Also J. M. Quinn, *The Doctrine of Time in St. Thomas* (Washington, D.C.: The Catholic University Press, 1960).

60. For a complete translation of this famous edict see R. Lerner and M. Mahdi, eds., *Medieval Political Philosophy: A Sourcebook* (Glencoe, Ill.: The Free Press of Glencoe, 1963), pp. 335–54. The original edict is a helter-skelter listing of propositions, consecutively numbered. The translation in this source book is from P. Mandonnet's work on Siger de Brabant and shows the propositions grouped and renumbered by Mandonnet. My reference to propositions 83–92 is to the Mandonnet renumbering.

61. For background, see James A. Weiskeipl, *The Development of Physical Theory in the Middle Ages* (New York: Sheed and Ward, 1959), pp. 58–63 & *passim;* also E. J. Dijksterhuis, *The Mechanization of the World Picture* (Oxford: Oxford University Press, 1961), pp. 162–3, 172–5; also Annaliese Maier, "Die Subjectivisierung der Zeit," *Philosophia Naturalis,* 1 (1950), pp. 364–5. I am indebted to Prof. F. C. Haber for his assistance in locating the background material.

62. See the translation of Ockham's *Philosophia Naturalis,* IV and its interpretation in H. Shapiro, *Motion, Time and Space according to William of Ockham* (St. Bonaventura, N.Y.: The Franciscan Institute, 1957), p. 96 ff.

63. *Nicole Oresme and the Medieval Geometry of Qualities and Motions,* ed. and trans. M. Clagett (Madison: University of Wisconsin Press, 1968), p. 273.

64. The précis on Duns Scotus, Marbres, and Nicholas Bonet are from P. Ariotti, "Celestial Reductionism of Time," *op. cit.*

65. San Isidore de Sevilla, *Etimologias,* trans. L. C. y Gongora (Madrid: Biblioteca de Autores Cristianes, 1951), p. 320. In other editions or translations, see Bk. XIII, "Parts of the World," chap. II, par. 3. On Bede, see *Bedae Opera de temporibus,* C. W. Jones, ed. (Cambridge, Mass.: The Mediaeval Academy of America, 1943).

66. M. Maimonides, *Guide for the Perplexed* (New York: Dover, 1956), p. 121.

67. On the impressive continuity of this idea, traced by Alexander Koyré back to the 12th Century, see G. J. Whitrow, *Structure and Evolution of the Universe* (New York: Harper & Brothers, 1959), p. 60.

68. Nicolaus Cusanus, *Of Learned Ignorance,* trans. G. Heron (New Haven: Yale University Press, 1954). See chap. 1, part 1, "How 'knowledge' is 'ignorance.' "

69. I am following W. Pagel, "Paracelsus and the Neoplatonic and Gnostic Tradition," *Ambix,* vol. VIII (Oct., 1960), p. 151 ff.

70. W. Pagel, *Paracelsus* (Basel: Karger, 1958), p. 76.

71. W. Pagel, "J. B. Van Helmont, De Tempore, and Biological Time," *Osiris,* vol. 8 (1949), pp. 346–417.

72. D. W. Singer, *Giordano Bruno* (New York: Greenwood Press, 1968), p. 3. For a brief but fresh and interesting evaluation of Bruno, see Stillman Drake, "Copernicus, Philosophy and Science," (Norwalk, Conn: The Burndy Library, 1973).

73. *Ibid.,* p. 80.

74. *Ibid.,* p. 96.

75. G. Bruno, *Cause, Principle and Unity,* trans. J. Lindsay (London: Background Books, 1962), p. 149.

76. Singer, *op. cit.,* p. 179. The unknown eyewitness and author of the Vatican manuscript had judiciously forgotten that Christ before Pilate, "held His peace and answered nothing" (Mark 14:61).

77. *Sir Isaac Newton's The Mathematical Principles of Natural Philosophy*

and His System of the World, trans. A. Motte, ed. F. Cajori (Berkeley: University of California Press, 1960), p. 6.

79. Fill a pail with water and mount it so that it may be spun along its vertical axis. As long as the water does not rotate, its surface remains flat; when it begins to rotate with the pail, it will exhibit a concave surface. In both cases, the behavior of the water depends on its rotation with respect to the environment at large and not with respect to the vessel. Hence, as Newton perceived it, the water is informed of an absolute framework external to its immediate framework, the pail. "The effects which distinguish absolute from relative motion are the forces of receding from the axis of circular motion." *Ibid.*, p. 10.

80. *Nat. Phil. Time*, p. 35.

81. See the series of fine papers by P. E. Ariotti in which he traces the rise and fall of celestial reductionism of time, that is, the long and impressive intellectual effort in Western thought to see time as somehow derived from the motion of the stars of the heavens or the bobs of pendulums. "Celestial Reductionism of Time," *Studi Internazionali di Filosofia* (Torino, Italy; Autumn, 1972), pp. 91–120; "Aspects of the Conception and Development of the Pendulum in the Seventeenth Century," *Archive for History of Exact Sciences*, vol. 8 (1972), pp. 330–407; "The Conception of Time in Late Antiquity," *Int. Philosophical Quart.*, vol. 12 (Dec., 1972), pp. 526–53; "Toward Absolute Time," *Annals of Science*, vol. 30 (1973), pp. 31–50; "The Concept of Time in Western Antiquity," *Study v. 2*.

82. E. Wigner, *Symmetries and Reflections* (Bloomington: Indiana University Press, 1967), p. 40.

83. "Mr. Leibnitz's Fourth Paper," sec. 6, ed. H. G. Alexander, *The Leibniz-Clarke Correspondence* (Manchester: Manchester University Press, 1956).

84. *Ibid.*, "Mr. Leibnitz's Third Paper," sec. 6.

85. A. A. Luce and T. E. Jessop, eds., *The Works of George Berkeley* (London: Nelson, 1948), vol. 4, "DeMotu," secs. 55, 62–5.

86. He had no difficulties, he wrote, in the operational use of the idea of time, but when he wanted to take time concretely exclusive of particulars, he found himself "lost and embrangled in inextricable difficulties." This is the quotation used at the beginning of this chapter.

87. E. Mach, *Principles of Human Knowledge*, vol. 2, sec. 97, 98.

88. *The New Science of Giambattista Vico*, trans. T. G. Bergin and M. H. Fisch, (Ithaca: Cornell University Press, 1968), para. 2, p. 342.

89. The reader is likely to recognize the ideas of Bacon, Adam Smith, Hegel, Kant, and Marx. For entropy see, e.g., F. N. Arumi, "Entropy and Demography," *Nature*, vol. 243, (1973), pp. 497–9. "The behavior must conform to some natural laws in the same way as nonsocial systems do. This article explores the feasibility of using equilibrium statistical theories in the description of large population aggregates." Compare this with "It is divine providence that institutes commonwealths and at the same time the law of the gentes." Vico, *op. cit.*, par. 629.

90. I. Kant, *Kant's Inaugural Dissertation and Early Writings on Space*, trans. J. Handyside (La Salle, Ill: The Open Court, 1929), p. 58, par. 14, sec. 5.

91. *Prolegomena to Any Future Metaphysics* (New York: Bobbs-Merrill, 1950), par. 21.

92. *Ibid.*, par. 31.

93. Kant, *On History* (New York: Bobbs-Merrill, 1963), p. 11.

94. J. G. von Herder, *Reflections on the Philosophy of History* (Chicago: University of Chicago Press, 1968), pp. 82–104.

95. G. W. F. Hegel, *Philosophy of History*, trans. Sibree, (New York: Dover, 1956), p. 457.

96. *Ibid.*, p. 72.

97. *Ibid.*, p. 77.

98. *Ibid.*, p. 33.

99. Hegel, *Phenomenology of the Mind*, trans. J. B. Baillie, (New York: Humanities Press, 1966), p. 96.

100. *Ibid.*, p. 80. The negativity as well as masculinity of time are also expressed by Hegel's contemporary, William Blake, who regarded "time and space as real beings, a male and a female. Time is a man, space is a woman, and her masculine portion is death." In: "A Vision of the Last Judgement," *The Portable Blake* (New York: The Viking Press, 1963), p. 667.

101. Hegel, *op. cit.*, p. 104. Cf. J. N. Findley, *Hegel: A Re-examination* (London, 1958), p. 146. "Time, so far from being unreal, is the very form of that creative unrest which represents Spirit as it becomes conscious of itself." See also the fine paper of Wolfe Mays in *Study v. 2*, "Temporality and Time in Hegel and Marx."

102. J. Needham, "Science and China's Influence on the World," *The Legacy of China*, ed. R. Dawson (Oxford: Oxford University Press, 1964), p. 307.

103. A. M. Granet, *Das Chinesische Denken*, trans. M. Porkert (München: Piper, 1971), p. 108.

104. *Ibid.*, p. 78.

105. *Ibid.*, p. 83.

106. Nathan Sivin, "The Chinese Conception of Time," *Earlham Review* (1966), vol. 1, pp. 82–91.

107. *The Voices*, pp. 92–135.

108. *Ibid.*, p. 104.

109. *Ibid.*, p. 97.

110. It might be useful for the uninitiated to have a look at the following work and then perhaps give up as totally confused: K. K. Mandal, *A Comparative Study of the Concepts of Space and Time in Indian Thought*, (Varanasi, India: Chowkhambe Sanskrit Series Office, 1968).

111. M. Eliade, "Time and Eternity in Indian Thought," in, J. Campbell, ed., *Man and Time, op. cit.*, p. 177.

112. Brandon, *History, Time and Deity, op. cit.*, p. 32.

113. Eliade, *op. cit.*, p. 181.

114. See as a rare example, Hajime Nakamura in, "Time in Indian and Japanese Thought," *The Voices*, p. 77.

115. Kitaro Nishida, *Intelligibility and the Philosophy of Nothingness*, trans. R. S. Hinziger (Honolulu: East-West Center Press, 1966), Frontispiece.

116. *Ibid.*, p. 163.

117. The quote is from the translator's comments, *Ibid.*, p. 26.

118. J. M. E. McTaggart, *The Nature of Existence* (Cambridge: Cambridge University Press, 1927), vol. 2, p. 10.

119. I have dealt with this problem in a preliminary fashion in terms of the question of "unexpected truths." See J. T. Fraser, "Time and the Paradox of Unexpected Truths," in *Akten, XIV, Internationalen Kongress für Philosophie*, (Wien, Herder), vol. IV, p. 395. Also *The Voices*, 524 ff.

120. That these rules may be stated in the concise form of propositional calculus does not alter the validity of the following discussion.

121. "Mr. Leibnitz's Second Paper," sec. 1, ed. H. G. Alexander, *op. cit.*, p. 16.

122. S. Kierkegaard, *Sickness Unto Death*, trans. W. Lowrie (New York: Doubleday, 1954), p. 147.

II / THE EMPIRICAL SEARCH

1. This is from Stanza 44 of Voluspá, "The Prophecy of the Seeress," from *Codex Regius,* a collection of Eddic poems written in Iceland. *The Poetic Edda,* trans. L. M. Hollander (Austin: University of Texas Press, 1962), p. 9.

2. A. Marshack, *Roots of Civilization* (New York: McGraw Hill, 1972), has many rather stunning illustrations; among them the text tends to get lost. For a concise earlier statement see his "Lunar Notation on Upper Paleolithic Remains," *Science* (1964), pp. 743–5.

3. For a critical review of Marshack, *Ibid.*, see *Scientific American*, vol. 227 (1971), p. 117.

4. G. C. Hawkins, *Stonehenge Decoded* (New York: Doubleday, 1965). For speculations on an even earlier calendrical structure in Ireland, dating possibly to the third millennium B.C., see J. Patrick, "Midwinter Sunrise at New-grange," *Nature*, vol. 249 (1974), pp. 517–19.

5. On this and the early history of clocks see F. A. B. Ward, *Time Measurement*, Part I (London: Her Majesty's Stationery Office, 1961); and *List 36* of the Time Measurement Collection, London, Science Museum, 1964.

6. For some delightful examples, see B. Chandler and C. Vincent, "A Sure Reckoning," *Bulletin* (The Metropolitan Museum of Art, Dec., 1967).

7. M. P. Nilsson, *Primitive Time Reckoning* (Lund: Gleerup, 1930), p. 1.

8. *Ibid.*, pp. 256, 358 & *passim.*

9. H. A. Lloyd, "Timekeepers: An Historical Sketch," in *The Voices*, p. 399.

10. J. Needham *et al.*, *The Heavenly Clockwork* (Cambridge, Mass.: Cambridge University Press, 1960), p. 199.

11. S. A. Bedini, "The Scent of Time," *Transactions American Philosophical Society*, NS, vol. 53, Part 5 (1963), p. 40.

12. As late as the sixteenth century some Northern Italian public clocks counted hours in the "Jewish manner," with the twenty-fourth hour at dusk. Other clocks were on "Bohemian time," which was the same as Jewish time, but north of the Alps; also "German time," which was twelve unequal hours from sunrise to sunset and sunset to sunrise.

13. Needham *et al.*, *op. cit.*, p. 200.

14. If the reader has difficulty identifying "11 o'clock" as a duration as well as an instant, "the 11th hour," he should think of a shorter period, such as "the 11th second," which is both an "instant" and a "duration."

15. Ward, *op. cit.*, p. 14.

16. Anyone who examined one of these masterpieces will appreciate the epithet. Cf. A. C. Crombie, *Medieval and Early Modern Science* (New York: Doubleday, 1959), vol. 1, p. 91.

17. See, e.g., M. Eliade, *Cosmos and History* (New York: Harper, 1954), p. 86.

18. P. Bohannan, "Concepts of Time among the Tiv in Nigeria," in John Middleton, ed., *Myth and Symbolism* (New York: Natural History Press, 1967), p. 318.

19. L. Massignon, "Time in Islamic Thought," in J. Campbell, ed., *Man and Time* (New York: Pantheon, 1957), p. 109.

20. This division is that of *Vishnu Purana*, trans. H. H. Wilson (Calcutta, 1964), p. 494. There are several others. See, e.g., K. K. Mandal, *A Comparative Study of the Concepts of Space and Time in Indian Thought* (Varanasi, India: The Chowkhanba Sanskrit Series Office, 1968), pp. 32–5.

21. Needham, "Time and Knowledge in China and the West," in *The Voices*, p. 100.

22. Recent work based on faunal and archeological evidence sees the origin of the 260 day cycle in the interval, very close to 260 days, between transits of the zenithal sun measured at the Mesoamerican cultural center located at about 15° N. The ritual importance of 260 days, however, is not limited to this astronomical privilege, but connects with other astronomical coincidences, partly fortuitous, partly necessary. V. H. Malmstrom, "Origin of Mesoamerican 260 Day Calendar," *Science*, vol. 181 (1973), pp. 939–41.

23. J. E. S. Thompson, *The Rise and Fall of Maya Civilization* (Norman: University of Oklahoma Press, 1966), p. 167. Periodicity was so deeply embedded in the imagery that the Mayan priests ceased to differentiate between past and future events. History was consulted for signs preceding important events, for it was believed that if initial conditions were restored everything would start anew.

24. In a recent book of scholarship and charm, *Time and Reality in the Thought of the Maya*, trans. C. L. Boilés and F. Horcasitas, (Boston: Beacon Press, 1973), Miguel León-Portilla described the central feature of Mayan culture as *chronovision*.

25. For an excellent survey of the calendar in the context of the history of astronomy see G. Sarton, *A History of Science*, vol. 2 (New York: Norton, 1959), p. 320 ff.

26. O. Cullman, *Christ and Time*, trans. F. V. Filson, (London: SCM Press, 1967), p. 17 ff.

27. Chandler and Vincent, *op. cit.*, p. 154.

28. *List 36*, Science Museum, *op. cit.*

29. Sarton, *op. cit.*, vol. 1, p. 75.

30. *Vitruvius*, ed. and trans. F. Granger (New York: G. P. Putnam's Sons, 1934), ix. 8.

31. See Chapters 21, 22, and 24, Vol 3, by D. J. Price in Charles Singer, ed., *History of Technology* (Oxford: Oxford University Press, 1954). Cf. note 45 below.

32. D. J. de Solla Price, "The Water Clock in the Tower of the Winds," *Am. J. Archeology*, vol. 72, No. 4. (Oct, 1968).

33. Needham, *The Heavenly Clockwork, op. cit.*, p. 85.

34. Magnetic timepieces, though never very popular, belong in the same category. See S. A. Bedini, "Seventeenth Century Magnetic Timepieces," *Physis*, vol. 11 (1969), p. 37.

35. Bedini, "The Compartmentalized Cylindrical Clepsydrae," *Technology and Culture*, vol. 3 (1962), pp. 115–41.

36. Ward, *op. cit.*, p. 14.

37. Quoted by Needham, *op. cit.*, p. 155.

38. The impression on the front binding is the image of an incense clock of a type that was popular in medieval China. (Reproduction courtesy of Silvio A. Bedini, from his "The Scent of Time." *Transactions of the American Philosophical Society*, new.ser. v.53, pt. 5, 1966, p.11.) It is called "The Greatly Elaborated Incense Seal Clock" and was included in *Hsin Tsuan Hsiang-P'u*, or "The Newly Compiled Handbook of Aromatic Incense," from which this illustration is taken. It shows the design of a continuous groove carved into hard wood; the length of the entire groove is believed to have been about twenty feet. Incense made from a variety of aromatic powders according to prescribed recipes was placed into the groove and lit at one end of the path, probably in the center of the seal. The incense then burned for perhaps twelve hours, telling the time of day by the changing scent. According to writings dated 1329 A.D. and signed by a "Retired Gentleman of the Central Studio" who cannot otherwise be identified, the seal design was presented in the form of a diagram to a

certain Tsou Hsiang-hu, an official of the Prefecture of Yü-li, lover of literature, and particularly proficient in *I Ching*, or "The Book of Changes."

39. Bedini, "The Scent of Time," *loc. cit.*, p. 22.

40. *Ibid.*, p. 28.

41. Bedini, "Time and Light," *La Suisse Horlogere*, April, 1964 and January, 1965.

42. A. S. Eddington, *The Nature of the Physical World* (Cambridge, Mass: Cambridge University Press, 1928), p. 99.

43. This translation from Cicero's *Tusculanae Disputationes I, 63* is quoted from D. J. de Solla Price, "On the Origin of Clockwork, Perpetual Motion Devices and Compass," in *Bulletin 218*, U. S. National Museum, Washington, D. C. (1959), p. 89.

44. Price, "Clockwork before the Clock and Timekeepers before the Timekeeping," in *Study v. 2*.

45. Early clocks did not have dials. That familiar face was invented in 1344 by Jacopo Dondi of Choggia, Italy, who was awarded the title Del Orologio for his labors. His epitaph reads in part: "Gracious reader, advised from afar from the top of a high tower how you tell the time and the hours, though their number changes, recognize my invention." H. Alan Lloyd, "Timekeepers: An Historical Sketch," *The Voices*, p. 391.

46. Needham, *The Heavenly Clockwork*, *loc. cit.*, p. 49.

47. Needham, "Time and Knowledge in China and the West," *loc. cit.*, p. 107.

48. See the fine exposition of F. C. Haber, "The Cathedral Clock and the Cosmological Clock Metaphor," in *Study v. 2*.

49. An entrance to the literature on the history of these clocks may be made through *The Voices*, pp. 388–400; also J. T. Fraser, ed., *Timekeepers and Time*, (New York: Springer Verlag, 1975).

50. On this, see Bedini, "Galileo Galilei and the Measure of Time," in *Saggi su Galileo Galilei* (Firenze: Barbera, 1967). Galileo's readiness, according to his student and biographer Viviani, derived from his familiarity with the measure in music, having been taught music by his father. We may recall here that the original meaning of the musical "takt" was beat or pulse.

51. Galileo Galilei, *Dialogues Concerning Two New Sciences*, trans. Crew and de Salvio (New York: Dover, n.d.), p. 96.

52. Bedini, "Galileo Galilei and the Measure of Time," *loc. cit.*

53. Crystal and atomic clocks are often used to control other processes without the intermediary step of displaying time or the time of the day.

54. H. M. Smith, "Dissemination of Astronomical and Atomic Time," *Nature*, 221 (1969), p. 221.

55. Although gravitational forces in atomic resonances are usually electrical forces, such resonances are not independent of the gravitational field in which they take place. General relativistic effects come to mind first. Thus, for example, all transition frequencies between magnetic substates are dependent on the rate of rotation of the laboratory about the direction of the magnetic field. This feature may even be employed such as in the design of nuclear gyroscopes and related devices. (See USA Patents 3,103,621 "Optically Pumped Magnetic Resonance Gyroscope and Direction Sensor" and 3,103,620 "Direction Sensor," both of 1963, J. T. Fraser, inventor). An ideal atomic clock, if it employs magnetic resonances, must be mounted on a platform stable with respect to the fixed stars.

56. Smith, *op. cit.*, p. 222. "No agreement" means this. The separation of events E_1 from E_0 when measured in terms of astronomical seconds is a number different from the same separation as measured in atomic clock seconds.

57. G. M. Clemence, "Time and Its Measurement," *American Scientist*, 40 (1952), pp. 260–9.

58. J. A. Carroll, "An Absolute Scale of Time," *Nature*, 184 (1959), p. 260.

59. One might add here the many dating techniques as de facto clocks: Carbon 14 decay, dendrochronology, archeomagnetism, thermoluminescence, fission tracks, and others. See H. N. Michael and E. K. Ralph, eds., *Dating Techniques for Archeologists* (Cambridge, Mass: M.I.T. Press, 1971). The important point is that these processes are useful because they function according to known laws, even if they are stated, as in the case of Carroll's clock, in probabilistic forms.

60. For an example of the clock versus theory problem in modern physics see Shinko Aoki, "A Note on the Variability of Time Standards Due to Relativistic Effects," *Astronomical Journal*, vol. 69 (1964), p. 2213. See also p. 393 ff. of C. W. Misner, Kip S. Thorne, and J. A. Wheeler, *Gravitation* (San Francisco: Freeman, 1973).

61. G. Polya, in his *Patterns of Plausible Inference* (Princeton: Princeton University Press, 1954), vol. 2, p. 100, poses the following problem in chance and conjecture. In a watchmaker's window three out of four clocks are less than two minutes apart. It is assumed, though not stated, that the clocks are indistinguishably identical. Why do they not show the same time? One conjecture is that all four were set on time; one is out of order, the other three are inaccurate. Or, perhaps, the three out of the four agree only by chance. What is the probability of the latter condition? This amounts to asking for the probability of three out of four clocks being within two minutes out of the possible total of $12 \times 60 = 720$ minutes. The probability of such a condition is very small: about one out of 10,000. It would be more reasonable to assume, therefore, that an external agent, the clockwatcher, interfered with chance and, sometime in the past, he had set all the clocks in phase. The problem of Charles V was to produce such identical clocks.

62. On this, see F. C. Haber, "The Darwinian Revolution in the Concept of Time," *Study v. 1*, p. 391.

63. L. Whyte, *Medieval Technology and Social Change* (Oxford: Oxford University Press, 1962), p. 124.

64. J. D. North, "Opus quorundam rotarum mirabilium," *Physis 8* (1966), p. 337.

65. Quoted from a contemporary document in Bedini and Madison, "A Mechanical Universe," *Transactions American Philosophical Society*, NS, vol 56, part 5 (1966).

66. Needham, *The Heavenly Clockwork, op. cit.*, p. 171.

67. *Summa Theologica*. Part I of II. Question 13. Article 2. Reply to Objection 3.

68. He wrote that the heavens move continually like a mechanical clock, but without violence, with the ratios of celestial motions resembling the gears of the clock. Irrationality among the cycles, the "ratio of ratios" introduces a basic numerical indeterminacy, thus a degree of freedom. See M. Clagett, ed., *Nicole Oresme and the Medieval Geometry of Qualities and Motions* (Madison: University of Wisconsin Press, 1968), pp. 6–11; Also *Nicole Oresme, De Proportionibus Proportionum and Ad Pauca Respicientes*, ed. Edward Grant, (Madison: University of Wisconsin Press, 1966), note 75 on p. 53.

69. Johannes Kepler, *Gesammelte Werke*, V. von Dyck and M. Caper, eds. (Munich: Beck, 1938) vol. XV, p. 146. Letter to Herwart von Hohenburg, Feb. 10, 1605.

70. Alphons Borelli, *De Motu Animalium* (Rome, 1681), vol. 2, p. 185. This is his Chapter 14 "De Animalis Generatione," proposition 185 and 186. My

attention was drawn to this curious suggestion by Father S. L. Jaki's *The Relevance of Physics* (Chicago: The University of Chicago Press, 1966) p. 288.

71. This apt phrase is due to L. Mumford, *Technics and Civilization* (New York: Harcourt, Brace and World, 1963), p. 15.

72. This point is made by R. L. Gregory in his delightful book, *The Intelligent Eye* (New York: McGraw Hill, 1970). See his Chapter 9 on "Seeing How Things work," illustrated with a drawing by Charles Babbage of a mechanical clockwork.

73. Quoted by J. Ford in a literary paper on "Dickens and The Voices of Time," *Nineteenth Century Fiction*, vol. 24, (1970), p. 441.

74. "Its nature is such that it is based on a combination of half artistic handicraft and direct theory. . . . German authors of the sixteenth century called clockmaking 'learned handicraft' (i.e., not for the guilds) and it would be possible to show from the development of the clock how entirely different relations between technical learning and practice was the basis of the handicraft from what it is, for instance, in large scale industry." Letter of Marx to Engels, Jan. 28, 1863, in *Karl Marx and Frederick Engels, Selected Correspondence 1846–1895*, trans. D. Torr (New York: International Publishers, 1942), p. 142.

75. H. Reichenbach, *The Philosophy of Space and Time* (Dover, n.d.), p. 119 ff. Earlier (p. 23) Reichenbach notes that a system can only be closed to a certain degree of exactness. Then he follows a type of quantitative reasoning useful in physics. He says that closedness may be achieved to any desired degree by increasing the ratio of internal to external forces by technical manipulation, until the latter becomes negligible. If there are any forces that still penetrate the insulating walls, these he sets to zero by definition. My argument shows that approximations of this type are not valid for timekeepers.

76. This graceful allusion to the clepsydra is quoted, without source identification, in the equally graceful book by Alice Morse Earle, *Sundials and Roses of Yesterday* (Detroit: Singing Tree Press, 1969, [originally published in 1902]), p. 54.

77. Andrew Marvell specified a rule to time transformation between men and women:

> Had we but world enough and time,
> This coyness, Lady, were no crime. . . .
> An hundred years should go to praise
> Thine eyes and on thy forehead gaze. . . .

This is an epistomologically dangerous example, however, for one must distinguish biological clocks as observed, from time as experienced.

78. A. N. Whitehead, *The Concept of Nature* (Cambridge: Cambridge University Press, 1964), p. 52.

79. R. G. Collingwood, *The Idea of Nature* (Oxford: Oxford University Press, 1967), p. 19.

80. B. L. Whorf, *Language, Thought and Reality* (Cambridge: M.I.T. Press, 1956), p. 215.

81. *Ibid.*, p. 215.

82. Whitehead, *op. cit.*, p. 178.

III / THE SEEKER

1. For a scholarly summary and a sympathetic, but very cautious, evaluation of this confusion, see *Patterning of Time*, p. 407.

2. In *The Monist*, vol. 53 (1969), p. 478.

3. See and compare the many entries under "Perception" in English and English, *A Comprehensive Dictionary of Psychological and Psychoanalytical Terms* (New York: David McKay, 1964).

4. This is recognized by Chomsky in his formal and substantive universals of grammar, which are the elements brought to the language by the child. See his *Aspects of the Theory of Syntax* (Cambridge: M.I.T. Press, 1965), p. 27, and all of his later publications on linguistics.

5. For a detailed exposition of the communal role in the repression of meanings of words and the significance of this process in the formation of language, see the little known but excellent work of T. Thass-Thienemann, *The Subconscious Language* (New York: Washington Square Press, 1967).

6. M. Proust, *Remembrance of Things Past* (New York: Random House, 1934), vol. 1, p. 5.

7. Quoted with approval by L. von Bertalanffy in *Robots, Men and Minds* (New York: Braziller, 1967), p. 91. See also his note 13.

8. Jakob von Uexküll, *Theoretische Biologie* (Berlin: Springer, 1928), p. 100. See also his *Umwelt und Innenwelt der Tiere* (Berlin: Springer, 1921), p. 218–9.

9. As an entry to the literature, see the information about time perception suffused through Doob, *Patterning of Time*. See also J. Cohen, *Homo Psychologicus* (London: Allen & Unwin, 1970), ch. 5 and its references.

10. J. Cohen, "Subjective Time," *The Voices*, p. 260. Cf. A. J. Sanford, "A Periodic Basis for Perception and Action" in W. P. Colquhoun, ed., *Biological Rhythm and Human Performance* (New York: Academic Press, 1971), pp. 197–201.

11. J. Cohen, *et al.*, "Interdependence of Temporal and Auditory Judgements," *Nature*, 174, (1954), p. 642.

12. R. Efron, "The Measurement of Perceptual Durations," in *Study v. 1*, p. 213.

13. The lower limit of stimulus identity is sometimes taken as demonstrating the quantized nature of time perception. On this see J. A. Michon, *Timing and Temporal Tracking* (Soesterberg, Netherlands: Institute for Perception, 1967); also see his subsequent papers in *Study v. 1*, p. 242, and *Study v. 2*. For a sympathetic evaluation see Cohen, *Homo Psychologicus, op. cit.*, p. 112.

14. M. Konishi, "Time Resolution by Single Auditory Neurons in Birds," *Nature*, 222 (1969), p. 566.

15. G. A. Brecher, "Die Entstehung und biologische Bedeutung der subjektiven Zeitenheit—des Momentes," *Zeit. Vergl. Physiologie*, vol. 18, (1932), pp. 204–43. See also J. von Uexküll and G. Kriszat, *Streifzüge durch die Umwelten von Tieren und Menschen* (Frankfurt: S. Fischer, 1970).

16. J. B. S. Haldane, "Animal Ritual and Human Language," *Diogenes*, 4 (1953), p. 61. This argument presumes a unity of danced messages. If this unity exists, then the bee should be able to deliver the balance of a danced report, if interrupted. The work of Karl von Frisch, at least to the extent reported in his delightful *Bees* (Ithaca: Cornell University Press, 1950), leaves this question open.

17. *Nat. Phil. Time*, p. 78.

18. W. H. Thorpe, "Ethology and Consciousness," in J. Eccles, *Brain*, p. 480.

19. Cohen, *Psychological Time in Health and Disease* (Springfield, Ill: Thomas, 1967), p. 34.

20. *Patterning of Time*, p. 12 ff.

21. Ernst Mach, *The Analysis of Sensations*, trans. C. M. Williams, revised, S. Waterlaw, (New York: Dover, 1959), see especially chap. 12.

22. E. J. Gibson, "The Development of Perception as an Adaptive Process," *American Scientist,* 58, (1970), pp. 98–104.

23. J. Eccles, *The Brain and the Unity of Conscious Experience* (Cambridge: Cambridge University Press, 1965), p. 12.

24. J. Cohen and I. Christensen, "A Note on Intermodal Phenomena," *IKON,* suppl. 56, (Jan/Mar, 1966), p. 37–52.

25. Haldane, *op. cit.,* p. 68. The factual content of this paragraph is due to Haldane; the interpretation is that of the author unless otherwise specified.

26. Bees which have found possible sites for a new home, return and entice others to survey their findings; upon return, members of the surveying party again dance. Those bees so far uninvolved watch the return of the scouting groups and join in the dance of one or another. During a period which may last several days some of the sites become favored over others, and when unanimity is reached, the swarm sets off. Haldane quotes Lorenz, who found similar behavior in geese: a noisy conversation of polysyllabic sounds slowly reduces to monosyllabic sounds and when unanimity is reached, the flock takes off, honking. Haldane, *op. cit.,* p. 69.

27. F. Engels, *The Part Played by Labour in the Transition of Ape to Man* (Moscow: Foreign Language Publishing House, n.d.).

28. See *Philosophical Investigations* (Oxford: Blackwell, 1968), Part I. Propositions 258, 259, and 380 seem to imply that private language is inconceivable; propositions 269 and 275 permit components to words which carry privileged disclosures, hence they do constitute a private language.

29. Proust, *op. cit.,* p. 617.

30. *Ibid.,* p. 5.

31. "in fact, it may well be that children go through detectable stages in this respect . . . and in some societies adults allegedly do not draw a sharp line between the two levels of psychic reality." *Patterning of Time,* p. 289. This tentative guess may be changed to an affirmative statement by anyone familiar with children and primitive societies in our own epoch.

32. E. H. Lenneberg, ed., *New Directions in the Study of Language* (Cambridge: M.I.T. Press, 1966), p. 2.

33. T. E. Seboek, "Goals and Limitations of the Study of Animal Communication," in Seboek, ed., *Animal Communication* (Bloomington: Indiana University Press, 1966), p. 3 & *passim.*

34. See L. W. Gregg, "Similarities in the Cognitive Processes of Monkeys and Man" and its cross-references to other papers in L. E. Jarrard, ed., *Cognitive Processes in Nonhuman Primates* (New York: Academic Press, 1971), pp. 155–64.

35. L. Carmichael, "The Early Growth of Language," in Lenneberg, *op. cit.,* p. 9.

36. E. Lenneberg, "On Explaining Language," *Science,* 164 (1969), pp. 635–63. See also his, "A Biological Perspective of Language," in Lenneberg, *New Directions in the Study of Language* (Cambridge: M.I.T. Press, 1966), pp. 65–68.

37. C. Cherry and R. Wiley, "Speech Communication in Very Noisy Environment," *Nature,* 214 (1967), p. 1164.

38. It seems that our mind can fill "gaps in a conversation" from an external source of white noise. This fact points to possible sources of hallucination in sensory deprivation. The brain might be employing its internal Brownian motion as a source of noise out of which the mind can create articulate patterns of speech and inform the subject about details of a nonexistent reality. It is only a step from here to the ghost of King Claudius, created from the white noise in the brain of Hamlet, Prince of Denmark.

39. See his two companion papers: *Brain*, vol. 86, part 2 (1963), pp. 261–84 and 285–94.

40. Editorial on "Memories of Amnesiacs," *Nature*, 227 (1970), p. 17.

41. A. Ehrenzweig, *The Psychoanalysis of Artistic Vision and Hearing* (New York: Braziller, 1965), p. 96 ff.

42. *Ibid.*, p. 99.

43. J. J. Gibson, "The Problem of Temporal Order in Stimulation and Perception," *J. Psychology*, 62 (1966), pp. 141–9. The quotation is from p. 145.

44. J. J. Gibson, *The Senses Considered as Perceptual Systems* (Boston: Houghton Mifflin, 1966).

45. See his masterly, *Biological Foundations of Language* (New York: Wiley, 1967).

46. H. Kalmus, "Analogies of Language of Life," *Language and Speech*, vol. 5 (1962), pp. 15–25.

47. H. Nakamura, "Time In Indian and Japanese Thought," in *The Voices*, p. 78.

48. B. L. Whorf, *Language, Thought and Reality* (Cambridge: M.I.T. Press, 1956), p. 59 ff.

49. C. Kluckhohn, "The Scientific Study of Values," in *Three Lectures* (Toronto: University of Toronto Press, 1958), no pagination.

50. J. Piaget, *Six Psychological Studies*, trans. A. Tenzer and D. Elkind (New York: Random House, 1967). See scattered entries under "Conservation." See also his *Genetic Epistemology* (New York: Columbia University Press, 1970), p. 43.

51. Concepts such as "external reality," "continuity of the self," etc., arise through sight, touch, and the other senses. Sight normally predominates by establishing what is "outside" and by negation, what is not outside, though it might be superseded. Helen Keller, for instance, established her identity in a world of hearing, touch, and smell alone. Cf. the role of vision (and light) in the entrainment of biological clocks and the role of light in physics—to be discussed later.

52. R. L. Gregory, *The Intelligent Eye* (London: Weifenfeld, 1970), pp. 160–6.

53. R. L. Gregory, "Origins of Eyes and Brain," *Nature*, 213, (1967), p. 369.

54. G. Schaltenbrand, "Consciousness and Time," *N.Y. Acad. Sci.* p. 632.

55. Piaget, "Time Perception in Children," *The Voices*, pp. 202–16.

56. R. A. Hinde, *Animal Behavior* (New York: McGraw Hill, 1966), p. 76 ff.

57. D. M. Mackay, "Perceptual Stability of Stroboscopically Lit Visual Field Containing Self-Luminous Objects." *Nature*, 181 (1958), p. 507. For further elaboration see Gregory, "Eye Movements and the Stability of the Visual World," *Nature*, 182 (1958), pp. 1214–6.

58. Spinoza, *Ethics*. Proposal xliv. Bk. II. "On The Nature and Origin of Mind."

59. Spinoza, *Improvement of Understanding* (Washington, D.C.: M. Walter Dinn, 1901), p. 3.

60. Plato, *Protagoras*, 321C.

61. Cf. *The Voices*, pp. 590–2. Also, J. T. Fraser, "The Interdisciplinary Study of Time," *N. Y. Acad. Sci.*, pp. 822–47. See, especially, pp. 844–5; and *Study* v. 1, pp. 499–502.

IV / THE ROOTS OF TIME IN THE PHYSICAL WORLD

1. R. Tolman, *The Principles of Statistical Mechanics* (Oxford: Oxford University Press, 1938), p. 65. For a recent summary see J. L. Leibowitz and O. Penrose, "Modern Ergodic Theory," *Physics Today* (Feb. 1973), pp. 23–29.

2. W. Heisenberg, *The Physical Principles of Quantum Theory* (New York: Dover, n.d.), p. 153 ff.

3. See, e.g., McCrea on "Why Are All Electrons Alike?," *Nature*, 202 (1964), pp. 527–8.

4. A basketful of identical balls may be weighed and, if the weight of one ball is known, their total number calculated. But if they were to be placed in numerical order, each would have to be endowed with identity, such as with the picture of a different symbol. But then they would cease to be identical. Cf. the curious problem of indiscernible bishops at the Council of Nicea, in *Nat. Phil. Time*, p. 142.

5. In conventional notation $dS = dQ/T$. Here dS is the change in entropy, dQ the change in heat, T is absolute temperature. In practice the situation is more complicated, for between any initial and final conditions the temperature is likely to change; thus T itself is a function of some variable common to Q and T. Then S is obtained by integrating that variable. The construction of this quantity is ingenious. An energy or temperature *difference* could not in itself have carried information on a temporal trend. By incorporating the Kelvin scale, Clausius combined energy difference with an absolute measure. Consider, e.g., bodies A and B isolated from the rest of the world and from each other at temperatures $T_A > T_B$. When placed in contact, heat will flow from A to B. Each loss dQ by A is an entropy loss of $dS_A = -dQ/T_A$, and a gain of $dS_B = dQ/T_B$. Throughout the heat flow, except for the last instant, T_A remains larger than T_B. Hence the entropy loss dS_A is always smaller than the entropy gain dS_B. It follows that there is a residual net entropy gain.

6. See G. J. Whitrow's fine article "Entropy" in *Encyclopedia of Philosophy* (New York: Macmillan and the Free Press, 1967).

7. What we sense as heat and measure as temperature is the sum of the kinetic energies of particles. The experiment described in note 5 corresponds to the averaging of velocities among the particles of blocks A and B. It is conceivable, but very unlikely, that only the high energy particles of the cold block would collide with only the lower energy particles of the hot block. Then the hot block would become hotter and the cold block colder.

A simple model of entropy-increasing phenomena, and one which is easily quantifiable, is described by M. S. Watanabe, "Time and the Probabilistic View Of the World," in *The Voices*, p. 528 ff. Consider a tray separated into two equal compartments by a wall with an opening. Place a number of identical balls in one of the compartments, and shake the table, permitting each ball to run around randomly and, if conditions are correct, roll through the opening. How will the distribution of the balls vary as time passes? Assigning zero to the condition of all balls in one compartment and maximum entropy to equal distribution of balls, Watanabe calculated the history of entropy to the system using a Monte Carlo simulation of randomness. He showed that after a transient period, depending mainly on the numbers of balls, the system will almost always be close to maximum entropy. In nonabstract terms, the balls will tend to equalize between the two compartments and remain close to that equilibrium.

Boltzmann knew that probabilities combine by multiplication while entropy combines by addition; hence their relationship is logarithmic. He incorporated this in his famous equation $S = k \ln W$ where S is the entropy of the system, W

is the probability of a specific state and k came to be known as Boltzmann's constant. This equation is inscribed upon his gravestone in Vienna. It is also shown on a Nicaraguan stamp of forty centavos commemorating the ten mathematical formulas which changed the face of the earth.

8. Watanabe, *op. cit.*, p. 532.

9. This is the "branch system" interpretation proposed by H. Reichenbach in *The Direction of Time* (Berkeley: University of California Press, 1956), p. 118.

10. A. S. Eddington, *The Nature of the Physical World* (Ann Arbor: University of Michigan Press, 1958), p. 68.

11. *Ibid.*, p. 79.

12. *Ibid.*, p. 101. It is in this context that he devised the idea of an entropy clock, mentioned in (2.1.3.).

13. *Leibnizens Mathematische Schriften*, ed. C. I. Gerhardt, (Halle: Schmidt, 1863) vol. 7, p. 18.

14. I. Kant, *Critique of Pure Reason*, trans. N. K. Smith, (London, 1958), p. 218 ff. This is his "Second Analogy of Experience: Principle of Succession in Time in accordance with the Law of Causality."

15. *John Stuart Mill's Philosophy of the Scientific Method*, ed. E. Nagel (New York: Hafner, 1950), p. 198 f.

16. D. Park, "Are Space and Time Necessary?" *Scientia*, 105 (1970), pp. 1–13. Cf. E. J. Zimmerman, "Time and Quantum Theory," in *The Voices*, p. 490 ff.

17. J. Piaget, *The Child's Conception of Physical Causality* (New York: Harcourt Brace & Co., 1930), pp. 258–73. Should be studied together with a careful exegesis, such as J. Flavell, *The Developmental Psychology of Jean Piaget* (New York: Van Nostrand, 1963).

18. Cf. T. Thass-Thienemann, *The Subconscious Language* (New York: Washington Square Press, 1967), p. 140 ff. & *passim*.

19. See its brief discussion in S. Weinberg, *Gravitation and Cosmology* (New York: Wiley, 1972), p. 595 ff. Cf. the idea of gravitational fine structure and "inertial spectroscopy" in J. T. Fraser, "Some Consequences of a Linear Vector Theory of Inertial Fields," *J. Franklin Institute*, 272, (1961), p. 460–92. Geoerges Lemaitre built a cosmogony on the idea that the present universe is the result of the radioactive disintegration of an atom. See *The Primeval Atom: An Essay on Cosmogony*, trans. B. H. and S. A. Korff, (New York: Van Nostrand, 1950). Our interest is in the probabilistic nature of the fireball. The universe so understood is a contingency and not a necessity.

20. H. Salecker and E. P. Wigner, "Quantum Limitations of the Measurement of Space-Time Distance," *Physical Review*, vol. 109 (1958), p. 517 ff.

21. H. T. Flint, "The Quantification of Space and Time," *Physical Review*, 74 (1948), pp. 209–10.

22. See, e.g., A. Grünbaum's paper on "Modern Science and Zeno's Paradoxes of Motion," in R. M. Gale, ed., *The Philosophy of Time* (New York: Doubleday & Co., 1967), pp. 422–94, and Gale's fine introductory summary, pp. 387–96.

23. A. N. Whitehead, *Organization of Thought* (Cambridge: Cambridge University Press, 1917), p. 146.

24. Whitehead, *Concepts of Nature* (Cambridge: Cambridge University Press, 1964), p. 60.

25. Such a limit has been suggested for a fluid-type cosmological model. P. T. Landsberg in *Study v. 1*, p. 106. Its approximate value for certain reasonable assumptions is 10^{-44} sec.

26. H. Mehlberg, Review of H. Reichenbach's, *The Direction of Time* in *Philosophical Review*, 71 (1962), p. 104.

27. I. Prigogine, *Introduction to the Thermodynamics of Irreversible Processes* (New York: Interscience, 1967), p. 91.

28. Prigogine, "Time, Structure and Entropy," in J. Zeman, ed., *Time in Science and Philosophy* (New York: Elsevier, 1971).

29. Hendrik Van Loon once remarked that all men that ever lived could be piled in a box with a capacity of one cubic mile and dumped into the Grand Canyon without so much as an echo from the universe.

30. The Michelson-Morley experiment consisted of a comparison of light velocity measurements made along and perpendicular to the direction of the earth's motion. These conditions correspond to two observers with different velocities with respect to a hypothetical ether at absolute rest. It is instructive to speculate, as did Whitrow ("Time, Gravitation and the Universe: the Evolution of Relativistic Theories," Inaugural Lecture, May 22, 1973, Imperial College of Science and Technology, University of London), what might have happened if such results were to have been presented to those debating the rival merits of the Copernican and Ptolemaic systems. In all likelihood, the results would have been interpreted as evidence against the Copernican hypothesis that the earth is in motion.

31. On this see G. Holton, "Einstein and the 'Crucial' Experiment," *Am. J. Physics*, 37 (1969), p. 969.

32. A. Einstein, "On the Electrodynamics of Moving Bodies," in *Relativity*. p. 37.

33. We need two identical clocks at relative rest at positions A and B. Send a signal from A to B and reflect it back. Assume that the times of travel AB and BA are equal. The exact time at B is assumed to be obtained when one half of the total travel time is added to the time of A's signal. If A sends his signal at 2 P.M. and sees it arriving back at 3 P.M. then the time it reached B is assumed to have been 2:30 P.M. Whitrow pointed to the circularity of assigning time to a distant event so as to establish what we are to mean by time at a distance. Yet he found the prescription warranted in spite of its arbitrariness because it isolates one set of instructions among other alternatives and proves to be consistent and useful for the theory. *Nat. Phil. Time*, p. 200.

34. Velocity is ordinarily measured as distance/time. The reciprocal quantity is equally valid. Speedometers of cars calibrated in inverse velocity might read from "∞ min/mi" (for stationary vehicles) to "0.5 min/mi" as the speed limit. Selection between the two scales is a matter of convenience and not one of principle.

35. It is a three-dimensional model of planar motion with time plotted along the third Cartesian coordinate. It achieved a frightful popularity as a symbol of relativity theory, sometimes identified as a sandglass. It has been a source of annoying confusion for the educated layman who is seldom informed that the static light cone stands for a moving sphere.

36. A. S. Eddington, *The Mathematical Theory of Relativity* (Cambridge: Cambridge University Press, 1960), p. 13 ff.

37. "Proper time" is an unfortunate translation of Minkowski's original phrasing. The integral of the world line from a fixed origin P_0 "bis zu dem variablen Endpunkte P gerührt, nennen wir die *Eigenzeit* des substantiellen Punktes in P." [italics his]. *Fortschritte der Mathematischen Wissenschaften*, Hrsg. Otto Blumenthal, Heft. 2., (Leipzig, Teubner, 1922), p. 62. *Eigen* signifies belonging to, or deriving from, the self, such as an estate, a pair of shoes, or an opinion.

38. Whether the limiting nature of c is a consequence of the principle of the constancy of c assumed by Einstein, or whether the constancy must be regarded as a consequence of the limiting nature of velocity c is a subtle epistemological problem.

39. The precise details of events during the first $1/1000$ second A.B.E. are not known; a meticulous determination of happenings in that period, as is the metier and joy of physics, is not yet possible. (A.B.E. stand for "after the beginning of expansion," a sufficiently antiseptic phrase used by some physicists to circumvent the necessity of talking about the instant of creation). Whatever ideas do exist are too fluid and too uncriticized to be taken seriously in a study of time. We may note, however, following the clear understanding formulated by Misner, that "proper time near the singularity is not a direct counting of simple and actual physical phenomena, but an elaborate mathematical extrapolation" (C. W. Misner, Kip S. Thorn, and J. A. Wheeler, *Gravitation*, New York: Freeman, 1973, p. 814). This must, of course, be the case, for we are dealing with proto- and atemporal worlds. If a violently radiating source is assumed, there are a few nonradiative mechanisms available, in principle, to sop up thermal energy. However, for various and complex reasons, even if they are assumed to have come about within a few milliseconds A.B.E., within another very short period of (unanalyzed) time one is again left with radiations of zero rest mass (*ibid.*, p. 736). Dominance of matter over radiation is not reached until perhaps 10^{10} seconds A.B.E. when the universe had cooled to about $10^{12}°$K. The cogency of the arguments and the broadly based evidence regarding the structure of temporality put forth in this book counsels that we regard the primordial universe as radiative and atemporal.

40. In *The Nature of the Physical World, op. cit.*, p. 48.

41. That proper time replaced velocity is not to be construed as a confusion of proper time with four-velocity, a four-vector defined as $u_i = dx_i/ds_i$ in conventional notation. The replacement is a conceptual and operational one, as discussed.

42. *Relativity*, p. 94.

43. Einstein, "Reply to Criticism," in P. A. Schilpp, *Albert Einstein: Philosopher-Scientist* (New York: Tudor, 1951), p. 685.

44. The principle of equivalence states that it is impossible to distinguish by means of physical experiments between the effects upon a small mass due to its inertia or due to a gravitational field. Zero gravitational field and zero acceleration are, of course, included.

45. A. Einstein, *Sidelines on Relativity* (London: Methuen, 1922), p. 36.

46. Cf. Whitrow's discussion on cosmic time, in *Nat. Phil. Time*, pp. 256–61.

47. See the colorful discussion of G. Sarton in *A History of Science*, vol. 1, (New York: Wiley and Sons, 1964), p. 115.

48. The term *Astralgeometrie* seems to have been coined by a certain *Rechtsgelehrte* Schweikart who wrote a paper on the theory of parallels in 1807. He intended to imply by its use that the crucial tests for his geometry may be found only at interstellar distances. When it was brought to the attention of Gauss by a common friend named Gerling, Gauss wrote to Gerling that the phrase was very much after his own heart. Subsequently he made the term his own. (F. Gauss, *Gesammelte Werke*, Berlin: Springer Verlag, 1910. vol. 10, p. 31 and vol. 8, p. 178 ff.)

49. T. L. Heath, ed., *The Thirteen Books of Euclid's Elements*, vol. 1 (New York: Dover, 1956) p. 155.

50. "I am becoming more and more convinced that the [intrinsic] necessity of our [Euclidean] geometry cannot be proved, at least not by human reason

nor for human reason. Perhaps in another life we shall attain other insights into the nature of space, such as are not possible now. Until then we must class geometry not with arithmetic which is purely a priori, but with mechanics. . . ." Letter of Gauss to Olbers, April 28, 1817. F. Gauss, *Gesammelte Werke, op. cit.*, vol. 8, p. 177.

51. *Ibid.*, p. 873. The sides of the triangle were 69, 85, and 197 kms., the sum of the angles exceeded 180° by 14.85" which was within his experimental error.

52. See his article on "Space-Time" in *Encyclopedia Britannica*, © 1969.

53. A. Einstein, *Relativity: The Special and the General Theory*, trans. R. W. Lawson, (New York: Peter Smith, 1920).

54. A. Einstein, "The Foundations of the General Theory of Relativity," *Relativity*, p. 113.

55. *Ibid.*, p. 114.

56. This view but with a different emphasis is expanded by Carl Meng, "Theory of Relativity and Geometry," in Schilpp, *op. cit.*, p. 463 ff.

57. In the journal "Fortnightly Review" (1875), quoted in toto in Eddington, *Space, Time and Gravitation* (New York: Harper Torchbooks, 1959), p. 192. "The reality corresponding to our perception of the motion of matter is an element of the complex thing we call feeling. What we might perceive as a plexus of nerve-disturbances is really in itself a feeling; and the succession of feelings which constitutes a man's consciousness is the reality which produces in our minds the perception of the motions of his brain. These elements of feeling have relations of nextness or contiguity in space, which are exemplified by the sight-perceptions of contiguous points; and relations of succession in time which are exemplified by all perceptions. Out of these two relations the future theorist has to build up the world as best he may. Two things may perhaps help him. There are many lines of mathematical thought which indicate that distance of quantity may come to be expressed in terms of *position* in the wide sense of the *analysis situs*. And the theory of space-curvature hints at a possibility of describing matter and motion in terms of extension only." [Italics his]

58. The orbit is not reentrant, resulting in a slow advance of the perihelion.

59. Eddington, *op. cit.*, p. 187.

60. Einstein, Infeld, and Hoffman noted the asymmetry between the electromagnetic and the inertial world in their paper on "Gravitational Equations and the Problem of Motion," *Annals of Mathematics*, Series 2, vol. 39, (1938), p. 65. They showed that the motion of bodies, conceived in the idealized form of singularities, follows from Einstein's field equations. In contrast, the ponderomotive law in electrodynamics is an ad hoc postulate, logically independent of Maxwell's equations.

60a. Cf. M. S. Watanabe, "Advanced potential is a probabilistic inference based on a mathematically concocted future which may or may not occur in the real sequence of events in time—a retrodiction based on a fictitious state." "Conditional Probability in Physics," *Progress of Theoretical Physics, Supplement, Extra Number* (1965), p. 159.

61. On this see J. T. Fraser, "Some Consequences of a Linear Vector Theory of Inertial Fields," *op. cit.*, p. 480. See also, J. B. Barbour, "Relative-distance Machian Theories," *Nature*, vol. 249 (1974), pp. 328–9.

62. First of Newton's four letters to Richard Bentley, dated Dec. 10, 1692. Note that these words were written less than a century after Giordano Bruno was burned at the stake, "on account of his wicked words" about the infinity of the universe. Newton's letters from: W. H. Turnbull, ed., *The Correspondence of Isaac Newton* (Cambridge: Cambridge University Press, 1961), vol. 3, p. 234.

63. Empedocles had already described the world as a fluid. One may have two distinct attitudes to this coincidence of words. One is that it is due to the limited vocabulary of languages, therefore, it is without significance. The other is that the common metaphor masks some tacit common understanding, attendant to our methods of perception. Empedocles does not have to be credited with divine foresight, nor should the coincidence of metaphors be belittled. Both descriptions stem from utterances of a creature which had hardly changed between 500 B.C. and 1700 A.D.

64. See equation 9–14 and section 9.6 in H. Bondi, *Cosmology* (Cambridge: The University Press, 1960).

65. The first problem was considered in its simplest form in 1826 by the astronomer H. W. M. Olbers, who noted something inconsistent in the fact that the night sky is dark. Namely, if the distribution of stars in the universe is assumed uniform and, on the average, they are assumed to be of the same luminosity, then it can be shown that each added increase in the volume of a vast sphere adds just enough light to compensate for the loss of light caused by the increase in distance. Hence the radiation density everywhere in a static infinite universe should equal the radiation density of an average star and the night sky should be as bright as the day sky. "On the Transparency of Cosmic Space," in *Astronomische Jahrbuch*, (1826), reprinted as Appendix III in S. L. Jaki's provocative study *The Paradox of Olbers' Paradox* (New York: Herder & Herder, 1969). On the cosmological red shift, see the text below. The third problem is that any cosmology written after 1905 must accommodate the special theory of relativity.

66. Einstein, "Cosmological Considerations in the General Theory of Relativity," in *Relativity*, p. 177.

67. K. Gödel, "An Example of a New Type of Cosmological Solution of Einstein's Field Equations of Gravitation," *Rev. Modern Physics*, 21 (1949), pp. 447–50.

68. For a critical discussion see W. H. McCrea, "Doubts about Mach's Principle," *Nature*, 230 (1971), pp. 95–7.

69. The rotation rate of an inertial system defined by the motion of the planets agrees with the rotation rate of an inertial system anchored in the fixed stars within an uncertainty of 0.4 seconds of arc/century. L. I. Schiff, "Observational Basis of Mach's Principle," *Rev. Modern Physics*, 36 (1964), pp. 510–11. This is sufficient accuracy to regard Mach's principle as empirically confirmed at this level of understanding of inertia.

70. For an introductory discussion see G. J. Whitrow, *The Structure and Evolution of the Universe* (New York, Harper Torchbooks, 1959), p. 39. For a critical evaluation, J. D. North, *The Measure of the Universe* (New York: Oxford University Press, 1965), p. 386 ff.

71. The presently accepted value of the rate of increase of velocity, known as Hubble's constant, is 50 km/sec/Mps or, roughly, 100 km/sec for each 6.3 million light years.

72. Bondi, *op. cit.*, pp. 81–6.

73. *Ibid.*, p. 140.

74. *Ibid.*, p. 149.

75. See Karl Popper's spirited criticism in *Of Clouds and Clocks* (Washington, D.C.: Washington University Press, 1966).

76. R. H. Dicke *et al.*, "Cosmic Black Body Radiation," *Astrophysical Journal*, 142 (1965), pp. 414–29. The quotation is from p. 415; see also Dicke *et al.*, "Radiation from the Origin of Time," *Nature*, 215 (1967), pp. 1056–7.

77. Dicke, "Cosmic Black Body Radiation," p. 415.

78. See McCrea's survey and reflections in, "A Philosophy for Big Bang Cosmology," *Nature*, 228 (1970), pp. 21–4.

79. C. Misner, "Absolute Zero of Time," *Physical Review*, 186 (1969), pp. 1328–33.

80. Cf. Whitrow, *What Is Time* (London: Thames and Hudson, 1972), p. 133 ff.

81. *A History of Science*, vol. 1, (New York: Wiley, 1964), p. 13. Sarton echoed Francis Bacon who held that "There is no excellent beauty that hath not some strangeness in the proportion" (*Of Beauty*).

82. H. Minkowski, "Space and Time," address delivered in 1908, in *Relativity*, p. 88. The term "imaginary" designating $\sqrt{-1}$ carries no metaphysical bias about time. Its use probably dates back to Heron of Alexandria in the first century A.D. and to Cardano in the sixteenth century who regarded the square root of the negative numbers as meaningless. One could write relativity with imaginary space axes and a real time axis. It is the incommensurability between imaginary and real quantities which concerns us here.

83. C. Muses, "Time, Experience and Dimensionality," *N.Y. Acad. Sci.*, p. 649.

84. The literature on this is disturbingly large. For a scholarly overview see *Nat. Phil. Time*, p. 215 ff. & *passim*. See also the annotated bibliography through 1958, *The Clock Problem (Clock Paradox) in Relativity*, comp. M. Benton (Bethesda: U.S. Naval Research Laboratory, 1959). A well-informed summary in book length is that by L. Marder, *Time and the Space Traveler* (London: George Allen and Unwin, 1971).

85. Whitrow, *What is Time*, p. 139.

86. On the debate see the editorial in *Nature* entitled "Faster and Colder," vol. 214 (1967), p. 1069; past writings may be traced from there.

87. Paul Valéry, *History and Politics*, trans. D. Foliot and J. Mathews, (New York: Pantheon Books, 1962), *Collected Works*, vol. 10, p. 372. Compare the quotation from Valéry with an editorial in *Nature* (vol. 122 1969, p. 719) entitled "Beautiful Symmetry," reporting about a lecture by Tsung-Dao Lee, distinguished discoverer of the parity violation in particle physics. The editorial writer stresses that physics depends fundamentally on invariance principles, or symmetry, because of "the compelling simplicity of using symmetry to understand the physical world." Against this reign of symmetry are set the violations of certain symmetry principles in particle physics. The writer cites T. D. Lee as holding that the implications of certain asymmetries in particle physics, to the philosophy of physics are not yet clear. But "Professor Lee thought that the asymmetries that seemed part of modern physics had a natural attraction to the human mind." This opinion represents a gift of Valéry's sage to Western physics.

V / TIME CONTAINED: COSMOLOGIES

1. The world of an observer A and the observed ∼A makes sense only from the viewpoint of the observer B removed from both A and ∼A; but an observer B then has its own ∼B, etc.

2. Plato, *Timaeus*, 30B–30C.

3. *Ibid.*, 33B.

4. F. M. Cornford, *From Religion to Philosophy* (New York: Harper & Brothers, 1957), p. 53. Cf. the detailed expositions in *Political Phil.*, chapters 2 and 4.

5. If the reader wishes to pursue the idea of cosmology as order and auditory harmony, he should read the chapter entitled "Heaven and Earth" and follow up the references in T. Thass-Thienemann's *The Subconscious Language* (Washington Square Press, 1967).

6. Cornford, *Principium Sapientiae* (New York: Harper & Row, 1965), p. 194.

7. There are some striking parallels between the steps of creation in Genesis and those of Ionian cosmogony. In both we find these steps: the opening gap and void over which moves the spirit of God, giving birth to night and day; generation of the starry heavens; separating dry land from the sea; creation of the sun, moon, and planets; the bringing forth of creatures having life. Cf. *ibid.*

8. Ernst Cassirer, *The Philosophy of Symbolic Forms,* trans. R. Mannheim, (New Haven: Yale University Press, 1955), vol. 2, p. 96. See the whole, superb chap. 2 on "Foundations of a Theory of Mythical Forms, Space, Time and Number."

9. Thass-Thienemann, *op. cit.*, p. 62.

10. Eric Voegelin, "Immortality: Experience and Symbol," *Harvard Theological Review,* vol. 60 (1967), p. 277.

11. G. de Santillana and H. von Dechend, *Hamlet's Mill: An Essay on Myth and the Frame of Time* (London: MacMillan, 1970).

12. S. G. F. Brandon, *Creation Legends of the Ancient Near East* (London: Hodder & Stoughton, 1963), p. 5.

13. Thass-Thienemann, *op. cit.*, p. 133. For the detailed argument the reader must turn to the source. The results are employed here because they are convincing.

14. The term "internal landscape" is a free and extended use of the phrase suggested by Claude Bernard in 1865. He proposed to describe as *milieu interior,* the stabilized internal conditions of higher organisms. (*Introduction á l'étude de la Medicine Experimentale,* Paris, J. B. Bailliere et Fils.) In our present use it means an internal model of the external world which in its most primitive form is topologically equivalent to the outer landscape but physically very different from it.

15. Brandon, *op. cit.*, p. 17.

16. A. Heidel, *The Babylonian Genesis* (Chicago: Chicago University Press, 1967), p. 18. These are from Tablet I, lines 1–9. I omitted the parentheses which signify words obliterated on the tablets.

17. To the author of a "Hymn of Creation" in the Rig Veda, the law of contradiction was unknown:

> There was then neither non-existence nor existence;
> There was no air, nor sky that is beyond it.
> What was concealed? Wherein? In whose protection?
> And was there deep unfathomable water?
> . . .
> None knoweth whence creation has arisen;
> And whether he has or has not produced it;
> He who surveys it in the highest heaven,
> He only knows, or haply, he may not know.

18. See, e.g., C. H. Long, ed., *Alpha, The Myths of Creation* (New York: Braziller, 1963).

19. Cornford, *Principium Sapientiae, op. cit.*, p. 193.

20. *Ibid.*, p. 27.

21. *Herodotus,* II. 53.

22. Goethe's attack on Newton is a good example of the bitter struggle between what may be called cosmologies of letters and cosmologies of numbers. On this see S. L. Jaki, *The Relevance of Physics* (Chicago: University of Chicago Press, 1966), pp. 39–45.

23. "In principio creavit Deus Coelum & Terram (Gen.I.1.) quod Temporis principium (justa nostra Chronologiam) incidit in poctis illius initiatum, quae XXIII diem Octobris praecessit in anno periodi Julianae 710. . . . Anno ante eram Christianam 4004." James Ussher, *Annales Veteris et Novi Testamenti a Prima Mundi Origine Deducti. . . .* (Bremae, MDCLXXXVI). p. 1.

24. "Dr. Clark's Fourth Reply," in H. G. Alexander, ed., *The Leibniz-Clark Correspondence* (Manchester: Manchester University Press, 1956).

25. St. Thomas Aquinas, *Summa Theologica* (New York: McGraw Hill, 1967), part I, question 46, art. 2, reply 1.

26. Quoted in G. Sarton, *A History of Science* (New York: Wiley, 1952), vol. 1. p. 37.

27. J. Campbell, ed., *Man and Time* (New York: Pantheon Books, 1957), p. 325.

28. Mircea Eliade, *Cosmos and History* (New York: Harper & Row, 1959), p. 3.

29. *Ibid.,* p. 52.

30. *Ibid.,* p. 59.

31. R. H. Dicke, *et al.,* "Cosmic Black Body Radiation," *Astrophysical Journal,* 142 (1965), p. 415.

32. Eliade, *op. cit.,* p. 85.

33. *Ibid.,* p. 89.

34. Robert Graves, *The Greek Myths* (New York: Braziller, 1955), sec. 7a. Cf. *The Voices,* p. 274.

35. Trans. M. Savill, (New York: Ungar, 1949), p. 396.

36. Lucretius, *The Nature of the Universe,* trans. R. E. Latham, (Penguin Books, 1951), p. 27.

37. Plato, *Timaeus,* sec. 29A.

38. A. Heidel, *op. cit.,* Tablet IV, lines 129, 130, 138.

39. *Ibid.,* Tablet VI, lines 5–6.

40. *Ibid.,* Tablet V, lines 1–5.

41. Gunnell, *op. cit.,* p. 42.

42. Hesiod, *Theogony,* lines 116–13B.

43. *Ibid.,* lines 700 ff.

44. Cf. the discussion on Hesiod and history in Gunnell, *op. cit.,* p. 87 ff.

45. E. Voegelin, *Plato* (Baton Rouge: Louisiana University Press, 1966), p. 200.

46. *The World of Archimedes,* trans. T. L. Heath, (Dover reprint of 1897, Cambridge University edition), p. 221.

47. Sarton, *op. cit.,* p. 420.

48. Lucretius, *op. cit.,* p. 92.

49. Nicolaus Cusanus, *Of Learned Ignorance,* trans. G. Heron, (New Haven: Yale University Press, 1954), p. 107.

50. Letter from Andreas Osiander to Copernicus, April 20, 1541. *Apologia Tychonis Contra Ursum,* p. 219 in *Johannes Kepler Gesammelte Werke,* ed., Max Caspar (München: C. H. Beck'sche Verlagsbuchhandlung, 1940), vol. 10.

51. Galileo, Galilei, *The Starry Messenger,* trans. S. Drake (New York: Doubleday, 1957), p. 21 ff.

52. Read the Preface to the First Edition in *Sir Isaac Newton's Mathemati-*

cal Principles of Natural Philosophy and His System of the World, trans. A. Motte
and F. Cajori, (Berkeley: University of California Press, 1934).

53. I. Kant, *Kant's Cosmogony,* trans. W. Hastie (New York: Greenwood
Press, 1968), p. 123. This is the translation of Kant's "Allgemeine Naturgeschichte und
Theorie des Himmels. . . ." (1755). It makes delightful and interesting reading.
For its historical background, see G. J. Whitrow, "Kant and the Extra-galactic
Nebulae," *Quart. J. Royal Astronomical Soc.* vol. 8 (1967), pp. 48–56.

54. H. P. Robertson, "Geometry as a Branch of Physics," in P. A. Schilpp,
ed., *Albert Einstein: Philosopher-Scientist* (New York: Tudor, 1951), p. 317.
Then he puts forth one of the simplest illustrative demonstrations of the quality
of the curvature of physical space, based on some of E. P. Hubble's earlier work.
Assuming that the distribution of galaxies is homogeneous and isotropic, volume
counts of galaxies either do or do not fit volume segments of a Euclidean sphere.
Based on Hubble's count, they do not fit, but they do fit a hyperbolic space with
positive curvature. Recently, informed opinion regarding the type of space
curvature that best fits the universe became somewhat less certain, but this is
unimportant. Robertson's point is that the existence of space curvature is
determinable from within the system.

55. A. S. Eddington, *Space, Time and Gravitation* (New York: Harper
Torchbook, 1959), p. 190.

56. On this, see any good topology book.

57. On this see chap. X, part IV of R. C. Tolman, *Relativity, Thermodynamics and Cosmology* (Oxford: Clarendon Press, 1958).

58. On this see the keen analysis of J. D. North, *The Measure of the
Universe* (New York: Oxford University Press, 1965), p. 234, 386 & *passim.*

59. A. Sandage, "The Redshift-Distance Relation—II," *Astrophysical
Journal,* 178 (1972), pp. 1–24.

60. The difficulties of the finity/infinity disjunction have been well known
to philosophy. For the meaning and difficulties of the idea of infinity in scientific cosmologies see North, *op. cit.,* chap. 17 & *passim.* The point in the present
text is that finity and infinity are logical contraries.

61. Trans. D. Hardy, (Macmillan, 1941), p. 257.

62. Heidegger is a world unto himself not only regarding his own thoughts
but also concerning his private vocabulary. See either *Being and Time,* trans.
J. Macquarrie and E. Robinson (London: SCM Press, 1962); or the comparative
study of Charles M. Sherover, *Heidegger, Kant and Time* (Bloomington: Indiana
University Press, 1971).

63. L. Bull, "Ancient Egypt," in *The Idea of History in the Ancient Near
East,* ed., R. C. Denton (New Haven: Yale University Press, 1955), p. 33.

64. *Herodotus, The Histories,* trans. A. de Sélincourt (London: Penguin
Books, 1972), I. 1.

65. Cf. the remark of Joseph Needham in *The Voices,* p. 132. "The enlightenment secularized Judeo-Christian time in the interests of the belief of
progress which is still with us, so that today when 'humanists' or Marxists dispute
with theologians they wear coats of different colors, the coats (to an Indian
spectator, at least) are actually the same coats worn inside out."

66. C. D. Broad, *Scientific Thought* (London: Routledge and Kegan Paul,
1952). See his introduction.

67. I. Kant, *On History,* ed., L. W. Beck (New York: Bobbs Merrill, 1963),
p. 18.

68. On this, see D. W. Dauer, "Nietzsche and the Concept of Time," in
Study v. 2.

69. The basic work, though easily available, is out of reach for anyone

but possibly historians on sabbatical leave. An abridgement of high esteem, though naturally without much of the supporting detail, is D. C. Somerwell's *A Study of History* (New York: Dell, 1965), 2 vols. My comments are based mostly on the "Argument," p. 390 ff. in Somerwell's abridgement. See, however, vol. XII, *Reconsiderations of A Study of History* (Oxford: Oxford University Press, 1961). This quote vol. 2, p. 393.

70. *Civilization on Trial* (Oxford: Oxford University Press, 1948), p. 15.

71. For a sympathetic review, see P. A. Sorokin, "Toynbee's Philosophy of History," *J. Modern History*, 12 (1940), pp. 374–87. A bibliography of critical writings may be found in J. C. Rule and B. F. Crosby, "Bibliography of Works on Arnold J. Toynbee, 1946–1960," *History and Theory*, vol. 4 (1965), pp. 212–33.

72. J. S. Mill, *A System of Logic*, 8th edition (New York: Longmans, Green, & Co., 1956), bk. VI, chap. 10, par. 8.

73. See, e.g., W. Dilthey, *Pattern and Meaning in History*, ed., H. P. Rockman (New York: Harper and Row, 1961).

74. See his *Essays in the Philosophy of History* (Austin: University of Texas Press, 1965).

75. C. G. Hempel, "The Functions of General Laws in History," (1942); Reprinted in P. Gardiner, ed., *Theories of History* (New York: The Free Press, 1969), pp. 344–56.

76. Karl Popper, *The Logic of Scientific Discovery* (New York: Basic Books, 1961), p. 59.

77. A. J. Toynbee, *A Study of History*, vol. 12 (New York: Oxford University Press, 1964), p. 229.

78. *Ibid.*, p. 228.

79. *Hymn of the Atharvaveda*, tr. and comments by R. T. H. Griffith, Varnasi (India), Schokhamba Sanskrit Series Office, 1968, book X, chap. 8, verses 39–40. Cf. Brandon, *Man and His Destiny in the Great Religions* (Manchester: Manchester University Press, 1962), p. 314–5.

80. "Time as Good and Evil," *Bull. John Rylands Lib.* 47, no. 1, (Sept. 1964), p. 18.

81. *Vishnu Purana*, trans. and illus. H. H. Wilson, (Calcutta: Punthi Postak, 1964). The cosmologies are most explicitly expressed in book VI, chap. 3–5.

82. *Ibid.*, p. 493. We learn that

At the end of a thousand periods of four ages the earth is for the most part exhausted. . . . The eternal Vishnu than assumes the character of Rudra, the destroyer who enters into the seven rays of the sun, drinks up all the waters of the globe, and causes all moisture whatever, in living bodies or in the soil, to evaporate; thus drying up the whole earth. . . . Thus fed, through his intervention with abundant moisture, the seven solar rays dilate to seven suns, whose radiance glows above, below, and on every side, and sets the three worlds . . . on fire. . . . The destroyer of all things, Hari, in the form of Rudra, who is the flame of time . . . proceeds to the earth, and consumes it also. A vast whirlpool of eddying flame then spreads to the region of the atmosphere, and the sphere of the gods, and wraps them in ruin.

Rudra, having consumed the whole world, breathes forth heavy clouds; resembling vast elephants in bulk, overspread the sky, roaring, and darting lightnings. Some are as black as the blue lotus; some are white as the water-lily; and some are yellow; some are of a dun colour, like that of an ass; some like ashes sprinkled on the forehead; some are deep blue, as the lapis

lazuli; some azure, like the sapphire; some are of bright red, like the lady-
bird; some are of the fierceness of red arsenic; and some are like the wing
of the painted jay. Mighty in size, and loud in thunder, they fill all space.
Showering down torrents of water, these clouds quench the dreadful fires
and deluge the whole world. Pouring down in drops as large as dice,
these rains overspread the earth. The world is now enveloped in darkness, and all things, animate or inanimate, having perished, the clouds continue to pour down their waters.

The wind is then reabsorbed, and he of whom all things are made, the
lord by whom all things exist, he who is inconceivable, without beginning
of the universe, reposes, sleeping upon Sesha, in the midst of the deep.

When the universal spirit wakes, the world revives. . . . Awaking at the
end of his night, the unborn, Vishnu, in the character of Brahma, creates
the universe anew, in the manner formerly related to you.

83. I. Nicholson, *Firefly in the Night, A Study of Ancient Mexican Poetry
and Symbolism* (New York: Grove Press, 1959), p. 56. Note that in the witness
of our language the appearance of the sun, such as in the morning, is often
associated with violent change. The "thick darkness" is "cut" at "day-break."
In Late Greek the plural *charamata* means "dawn," while its singular *charagmos*
means "incision, cut." In Hungarian, sunrise is *hajnal hasadás*, "rending the
sky." Features of the Universe are depicted through the inner landscape of man.
See Thass-Thienemann, *op. cit.*, p. 155 & *passim.*
84. The similarity to the four elements of the Greeks is obvious, but does
not demand actual connection beyond the fact that both imageries arose from
the activities of Homo sapiens. The same argument goes for the universality of
dance, death, and creation; the dance of Shiva, versus the "Lord of the Dance"
currently popular in the American romantic youth movement: "I danced in
the morning when the world was begun. . . ."
85. Nicholson, *op. cit.*, p. 119.
86. *The Elder Edda and the Younger Edda*, of Saemund Sigfusson and
The Younger Edda of Snorre Stuleson, trans. I. A. Blackwell, (London: Norroena
Society, 1907), p. 327.
87. P. Lawrence, *Road Belong Cargo* (Manchester: Manchester University
Press, 1964), p. 75.
88. For one record of this modern quasi-religious cult the reader might
consult *The Flying Saucer Reader*, by J. David (New York, The New American
Library, 1967).
89. See the monumental collection of Stith Thompson, *Motif Index of
Folk Literature* (Bloomington: Indiana University Press, 1955).
90. The quotation is from *Udana*, verse 80, quoted in R. E. A. Johansson,
The Psychology of Nirvana (London: Allen & Unwin, 1969), p. 51.
91. Brandon, "Time and the Destiny of Man," *The Voices*, p. 152.
92. Text of *The Book of the Dead* by E. A. T. W. Budge, (New York:
Dutton, 1923), vol. 1, p. 178.
93. The day of the Last Judgement is expected/unexpected, "But of that
day or hour no man knoweth, neither the angels in heaven, nor the Son, but
the Father" (Mark 13:32). This gives rise to the curious condition I have called
"The Judgement Day Paradox." J. T. Fraser, "Time and the Paradox of Unexpected Truths," in *Akten, XIV Intern. Kongress f. Philosophie* (Wein, 1968),
vol. 5, pp. 395–402.
94. L. Baeck, *The Essence of Judaism* (New York: Schocken, 1970).

95. See the extensive treatment of this problem in Brandon, *Jesus and the Zealots* (Manchester: Manchester University Press, 1967).

96. The title of Frank Kermode's study of the role of the apocalyptic in Western literature, *The Sense of an Ending* (New York: Oxford University Press, 1967).

97. See any good translation of the *Critique of Pure Reason*.

98. *Nat. Phil. Time*, p. 31 ff.

99. *Ibid.*, p. 32. The psychologically minded reader may consider in this context Thass-Thienemann's remark that counting is a "dead activity."

> If the steady flow of time is, so to say, 'frozen,' it becomes a dead sequence of things which can be counted. It has been observed in clinical practice that, with the loss of futurity . . . the feeling of incapacity overcomes the depressed person. He prefers to leave every task unfinished, leave the door open, and so forth, because every finished act leads to a new task in the process of time. . . . Another frequent symptom . . . is the compulsion to count. For instance, they count the stairs . . . , their flow of time becomes fragmented. . . . The sequence of numbers offers a mechanized substitute for the living time experience. Time becomes arithmetic. . . . Time, in the plural *many times*, is just the negation of the thesis of Heraclitus that time is the irreversible unique moment of presence which never returns.

T. Thass-Thienemann, *Symbolic Behavior* (New York: Washington Square Press, 1968), p. 82.

100. North, *op. cit.*, p. 392 ff.

101. This was formulated by Russell in 1937 in connection with his studies of Zeno's paradox of Achilles and the Tortoise. In B. Russell, *The Principles of Mathematics* (New York: Norton, 1938).

102. Shandy would arrive at a day when he would have to begin including in his diary the complete records of the day when he first began his diary, and copy the complete records of his first day's diary entry. Karl Popper has shown that the series of all of Tristram Shandy's actions when numerically represented (including this increasingly repetitive feature) do not converge. "Indeterminism in Quantum Physics and in Classical Physics," *Brit. J. Phil. Science*, 1 (1950), p. 173. This side issue, however, is irrelevant here.

103. *Nat. Phil. Time*, p. 149.

104. For examples see, R. Schlegel, *Completeness in Science* (New York: Appleton-Century-Crofts, 1967), chap. 6, for some interesting problems in the application of transfinite numbers to the physical world.

VI / TIME EXTENDED: LIFE

1. J. Needham *et al*, *Heavenly Clockwork* (Cambridge: Cambridge University Press, 1960), p. 171.

2. E. T. Hatto, ed., *EOS An Inquiry Into the Theme of Lovers' Meetings and Partings at Dawn in Poetry* (The Hague: Mouton & Co., 1965), p. 709 ff.

3. H. Bretzl, "Botanische Forschungen des Alexanderzuges," (Leipzig, 1903) quoted in E. Bünning, *The Physiological Clock* (Heidelberg: Springer Verlag, 1954), p. 10.

4. For the reasons—or educated guesses—see *Biological Rhythms in Psychiatry and Medicine* (Chevy Chase, Md.: U.S. Dept of Health, Education

and Welfare, Nat. Inst. Mental Health, Public Health Publication 2088, 1970),
p. 150.

5. S. Bedini, "The Instruments of Galileo Galilei," in *Galileo, Man of Science*, ed. E. McMullin, (New York: Basic Books, 1967), p. 257 ff.

6. Quoted from unidentified source in J. Cohen, *Homo Psychologicus* (London: Allen & Unwin, 1970), p. 115.

7. *Biological Clocks*, p. 160 ff.

8. J. L. Cloudsley-Thompson, *Rhythmic Activity in Animal Physiology and Behavior* (New York: Academic Press, 1961), p. 39.

9. *Ibid.*, p. 43.

10. *Ibid.*, p. 47.

11. C. P. Richter, "Inherent Twenty-four Hour and Lunar Clocks in Primates," *Communications in Behavioral Biology*, part A, no. 5 (1968), pp. 305–31.

12. C. P. Richter, "Astronomical References in Biological Rhythm," *Study v. 2*.

13. Cloudsley-Thompson, *op. cit.*, p. 82.

14. The interested reader may begin by consulting the *Journal of Interdisciplinary Cycle Research* (Amsterdam); see also the extensive references in F. Halberg, "Physiologic Considerations Underlying Rhythmometry, with Special Reference to emotional Illness," in *Symposium Bell-Air III* (Geneve: Masson et Cie, 1968), pp. 73–126. Also, consult *Biological Rhythm and Human Performance*, a title accurately descriptive of its subject, ed. W. P. Colquhoun (New York: Academic Press, 1971).

15. J. Aschoff, "Circadian Rhythm in Man," *Science*, 148 (1965), p. 1427. Cf. Richter, *Biological Clocks in Medicine and Psychiatry* (Springfield, Ill.: Thomas, 1965), p. 52.

16. W. Blunt, *The Compleat Naturalist—A Life of Linneaus* (New York: Viking Press, 1971), p. 198. In 1633, that is, a century before Linnaeus, Father Athanasius Kircher, a Jesuit mathematician and inventor, was said to have exhibited to many persons a "mysterious root" (obtained from an Arab merchant in Marseilles) which was said to turn as the sun turns, and could serve "as a most perfect clock." S. Drake, "Galileo Gleanings XII: an Unpublished Letter of Galileo to Peiresc," *ISIS*, 53, part 2 (1962), p. 206.

17. Bünning, *op. cit.*

18. For a tabulation see Arne Sollberger, "Biological Rhythm," in *Documenta Geigy*, "Rhythm in Medicine," (Geigy Pharmaceutical Company, n.d.), p. 2.

19. E. T. Pengelley and S. J. Asmundson, "Annual Biological Clocks," *Scientific American*, (April, 1971), pp. 72–80.

20. E. Gwinner, "A Comparative Study of Circannual Rhythms in Warblers," *Symposium on Biochronometry*, ed. M. Menaker (Washington, D. C.: National Academy of Sciences, 1971), p. 405.

21. M. Menaker, ed., *Symposium on Biochronometry* (Washington, D. C.: National Academy of Sciences, 1971), p. 428.

22. J. R. Udry and N. M. Morris, "Distribution of Coitus in the Menstrual Cycle," *Nature*, 220 (1968), pp. 593–6.

23. "Effects of Sexual Activity on Beard Growth in Man," *Nature*, 226 (1970), pp. 869–70.

24. V. T. Wynn, "Absolute Pitch: A Bimensual Rhythm," *Nature*, 230 (1971), p. 337.

25. R. P. Kimberly, "Rhythmic Patterns in Human Interaction," *Nature*, 228 (1970), pp. 88–90.

26. M. Young and J. Ziman, "Cycles in Social Behavior," *Nature*, 229

(1971), pp. 91–5. The selection of the basic concepts in this study ought to sound familiar to the reader. What the authors call "episodic time" has no quantitative measure, while that which they recognize as a cyclic sequence, does. These are the Eleatic polarities of becoming and being.

27. B. Goodwin, "The Cell as a Resonating System," *Towards a Theoretical Biology,* ed. C. H. Waddington (Chicago: Aldine, 1969), vol. 2, p. 140.

28. Richter, *Biological Clocks in Medicine and Psychiatry, op. cit.,* p. 65. Normally repressed periodicities sometimes give rise to curious diseases such as the alternate-day squint in children. Here the eyes are alternately crossed on one day and normal the next, with transition phases occurring within minutes during the waking state, though they usually occur at night. Richter, "Clock Mechanism Esotropia in Children Alternate Day Squint," *The Johns Hopkins Medical Journal,* vol. 122 (1968), pp. 218–23.

29. Richter, *Biological Clocks in Medicine and Psychiatry,* p. 93.

30. V. G. Bruce, "Environmental Entrainment of Circadian Rhythms," *Biological Clocks,* pp. 29–48.

31. For the role of light in entrainment, see M. Menaker, ed., *Symposium on Biochronometry, op. cit.,* parts 2–4. For a popular summary consult Gay Gaer Luce, *Body Time* (New York: Pantheon, 1971), chap. 9, and the extensive bibliography on "Light: Photoperiodism, Reproduction and Circadian Rhythms," p. 351 ff.

32. M. Menaker and A. Eskin, "Entrainment of Circadian Rhythms by Sound in Passe Domesticus," *Science,* 154 (1966), pp. 1579–81.

33. H. B. Dowse and J. D. Palmer, "Entrainment of Circadian Activity Rhythms in Mice by Electrostatic Fields," *Nature,* 222 (1969), pp. 564–6.

34. M. K. McClintock, "Menstrual Synchrony and Suppression," *Nature,* 229 (1971), pp. 244–5.

35. C. Cherry and R. Wiley, "Speech Communication in a Very Noisy Environment," *Nature,* 214 (1967), p. 1164.

36. M. Menaker, "Synchronization with the Photic Environment Via Extraretinal Receptors in the Avian Brain," *Symposium on Biochronometry, op. cit.,* pp. 315–32.

37. "Adaptive Functions of Circadian Rhythms," *Biological Clocks,* pp. 345–56.

38. *International Journal of Biometeorology* (1966), pp. 119–25. See also J. L. Cloudsley-Thompson, *Animal Conflict and Adaptation* (London: Foulis, 1965), and the article by the same author on "Adaptive Functions of Biological Rhythm," in *Documenta Geigy,* "Rhythm in Medicine," p. 5.

39. M. Lindauer, "Time Compensated Sun Orientation in Bees," *Biological Clocks,* pp. 371–77.

40. K. Hoffmann, "Experimental Manipulation of the Orientational Clock in Birds," *Biological Clocks,* pp. 379–87. Also, K. Schmidt-König, "The Sun Azimuth Compass," *Science,* 131 (1960), pp. 826–28.

41. That this is the case is far from enough to solve the mystery of much of animal navigation. What the experiments indicate is the remarkable capacity of the animal to make correction for the time of day. However, while the clocks of the homing pigeon can be systematically rephased to an artificial sun, once the birds are let go, no clock-shifted pigeon was ever observed to fly to or investigate a false home area. See "Homing," in *Nature,* 232 (1971), pp. 86–7.

42. A. D. Hasler and H. O. Schwassmann, "Environmental Cues in the Orientation Rhythm of Fish," *Biological Clocks,* pp. 429–31.

43. G. Birukow, "Innate Types of Chronometry in Insect Orientation," *Biological Clocks,* p. 405.

44. B. C. Goodwin, "Temporal Order as the Origin of Spatial Order in Embryos," *Study v. 1*, pp. 190–99.

45. The approximateness of many rhythms to external rhythms is interesting. Perhaps the circa-diem nature of circadian clocks is caused by the intrinsic inaccuracies of the clocks. Yet, another reason might be conjectured, in analogy to communication and servo-systems. For two systems to be synchronous it is necessary that one of them should run faster or slower than the other one, so as to develop the difference to be employed as an error signal. Only through a nonzero signal can a practical, rather than ideal, system be synchronized with a Zeitgeber.

46. Goodwin, *op. cit.*, p. 198. Cf. Richter, "Astronomical References in Biological Rhythms," *Study v. 2*.

47. It is Beatrice M. Sweeney's, *Rhythmic Phenomena in Plants* (New York: Academic Press, 1969).

48. L. Edelstein, *Ancient Medicine,* ed. and trans. O. and C. L. Temkin (Baltimore: Johns Hopkins Press, 1967), pp. 4, 70 & *passim.*

49. *Hippocrates,* "Airs, Waters, Places" II. 20–26.

50. R. Dubos, *Man Adapting* (New Haven: Yale University Press, 1965), p. 33. Cf. Needham's belief that biochemistry will blend with morphogenesis. J. Needham, *Order and Life* (Cambridge: M.I.T. Press, 1968), p. xv.

51. For a critique of specific etiology, see David Bakan, *Disease, Pain, and Sacrifice* (Chicago: University of Chicago Press, 1968), p. 11.

52. Goodwin, "Biological Control Processes in Time," *N.Y. Acad. Sci.*, pp. 752–55.

53. H. J. Curtis, *Biological Mechanisms of Aging* (Springfield, Ill.: Thomas, 1966), pp. 9, 120.

54. B. L. Strehler, *Time, Cells and Aging* (New York: Academic Press, 1962), p. 36.

55. A. Comfort, *Aging: The Biology of Senescence* (New York: Holt, Rinehart & Winston, 1964), p. 158.

56. *Ibid.*, p. 117.

57. *Ibid.*, p. 160.

58. L. N. Edmunds, "Persistent Circadian Rhythms of Cell Division in Euglena," in M. Menaker, ed., *Symposium on Biochronometry, op. cit.*, pp. 594–611.

59. G. Schaltenbrand, "Die Krisis des Zeitbegriffes," *Zeit in Nervenärztlicher Sicht* (Stuttgart: Enke, 1963), p. 9.

60. Bünning, *Physiological Clocks, op. cit.*, p. 128 ff.

61. It may seem that subsequent theories change only our understanding of the world, but not the world "as it is." Such a view misses the point, however, that for man and beast the known Umwelt *is* his world; when man knows more about that world, his world changes.

62. See, e.g., Erik H. Erikson, "Growth and Crises of the Healthy Personality," *Personality,* ed. C. Kluckhohn and H. A. Murray (New York: Alfred A. Knopf, 1965), chap. 12.

63. With a very few exceptions (see notably J. Michon, "Processing of Temporal Information and the Cognitive Theory of Time," in *Study v. 1*, pp. 242–58) all psychological theories proposing to account for man's awareness of time invoke and base their arguments on the existence of physiological clocks. Unavoidably, they remain silent on such matters as questions of before/after, or future/past/present. Why they must remain silent is clear. If one begins with an atemporal scheme, all higher levels must be postulated; if one begins with an eotemporal scheme (such as a clock), one will ossify on a level of before/after.

64. See H. Kalmus, "Periodic Phenomena in Genetic Systems," *New York Academy of Sciences, Annals,* vol. 98 (1962), art. 4, pp. 1083–95.

65. The distinction between "immortal germ cells" and "mortal soma" drawn by Weisman, has somewhat blurred since the finding of somatic cells in invertebrates which are thought to give rise to germ cells. Comfort, *op. cit.,* p. 217.

66. H. J. Curtis, "Biological Mechanisms Underlying the Aging Process," *Science,* vol. 141 (1963), pp. 686–94. Also R. Kastenbaum, "Theories of Human Aging," *The Journal of Social Issues,* 21 (1965), no. 4, pp. 13–36. For a criticism of the somatic mutation theory see, A. Comfort, "Feasibility in Age Research," *Nature,* 217 (1966), pp. 320–22.

67. Kastenbaum, *loc. cit.,* p. 16; and Curtis, *loc. cit.*

68. For a discussion of Bidder's work see Comfort, *The Biology of Senescence,* p. 15 ff.

69. Kastenbaum, *loc. cit.,* p. 18.

70. S. Gelfant and J. G. Smith, Jr., "Aging: Noncycling Cells, an Explanation," *Science,* 178 (1972), p. 357–61.

71. Comfort, *The Biology of Senescence, op. cit.,* p. 51 ff.

72. Curtis, *Science,* 141, p. 688.

73. B. L. Strehler, *Time, Cells and Aging* (New York: Academic Press, 1962), p. 22.

74. H. B. Green, "Temporal Stages in the Development of the Self," *Study v. 2.* Cf. her stages IX–XI for American subjects.

75. For a summary view see M. E. Cumming, "New Thoughts on the Theory of Disengagement," in *New Thoughts on Old Age,* ed. R. Kastenbaum (New York: Springer, 1964), pp. 3–18.

76. Although gerontological theories are usually expressed in the language of scientific detachment, careful attention to details will inevitably reveal the obvious: unlike studies dealing with amoeba, the observer here is also the observed.

77. Kastenbaum, *Journal of Social Issues,* 21, p. 31. See also his "Is Old Age the End of Development?" in his, *New Thoughts on Old Age,* pp. 3–18.

78. The naive and good-willed Captain Gulliver is first asked by the Luggnaggians what he would do if he were a Strudlebug. The answer is all happiness: maintain order among nations, increase knowledge and wisdom, discover perpetual motion and the universal medicine, and promote "many other great inventions brought to the utmost perfection." J. Swift, *Gulliver's Travels* (Everyman's Library Edition, 1946), p. 224. But Strudlebugs, like Frankenstein's monster, did not turn out that way. They retained all the infirmities of old people: they became living relics of ages misunderstood, beggars, and wards of the state. They ceased to be able to communicate, and they lived in the constant horror of all things that arise from the "dreadful prospect of never dying. They were not only opinionative, peevish, covetous, morose, vain, talkative: but uncapable of friendship, and dead to all natural affection. . . . Envy and impotent desires are their prevailing passions." Strudlebug marriages are dissolved by the courtesy of the state "for the law thinks . . . that those who are condemned without any fault of their own, to a perpetual continuance in the world, should not have their misery doubled by the load of a wife." *Ibid.,* p. 226. This passage was brought to my attention by a remark in Kastenbaum, *op. cit.,* p. 13.

79. Strehler, *op. cit.,* p. 86. See also Strehler and A. S. Midvan, "General Theory of Mortality and Aging," *Science,* 132 (1960), pp. 14–21.

80. Strehler, *Time, Cells and Aging, op. cit.,* p. 92 ff.

81. L. J. and M. Milne, *The Ages of Life* (New York: Harcourt, Brace & World, 1968), p. 313.

82. A. C. Leopold, "Senescence in Plant Development," *Science,* 134 (1961), pp. 1727–32.

83. New World Saturniid moths have five species with similar activity patterns. Two of the species are bad-tasting and loudly colored: they live to 45 days on the average after laying their eggs. Three species have concealing coloration and are palatable to predators. They have an average of 13 days of postreproductive life. In the interpretation of A. D. Blest, the palatable moths die rapidly (by hyperexcitation), thus giving predators less chance on the average to learn to recognize the members of the species as desirable; the unpalatable moths stick around longer and are conspicuous so as to teach predators to keep away from their species. Blest, "Longevity, Palatability and Natural Selection in Five Species of New World Saturniid Moths," *Nature,* 197 (1963), pp. 1183–86. See also L. J. and M. Milne, *op. cit.,* chap. 11.

One might then ask, why does not natural selection favor lower birth rate, making the species scarce and unavailable for food? The answer might be statistical: the chances of survival of the species are probably enhanced by more individuals and more early deaths compared with fewer individuals living longer.

84. Comfort, *The Biology of Senescence, op. cit.,* p. 257.

85. D'Arcy Thompson, *On Growth and Form* (Cambridge: Cambridge University Press, 1968), p. 35.

86. N. Bohr, "Light of Life," *Nature* (April 1, 1933), p. 457.

87. T. Thass-Thienemann, *The Subconscious Language* (New York: Washington Square Press, 1967), p. 269 ff.

88. N. Kazantzakis, *The Last Temptation of Christ* (New York: Simon and Schuster, 1960), p. 496. "Temptation" itself is a time concept, probably related to "stretching" and "extending." The title, "The Last Temptation," carries with it the archaic imagery of achieving timelessness. Thass-Thienemann, *Symbolic Behavior* (New York: Washington Square Press, 1968), p. 372.

89. S. Freud, *The Standard Edition of the Complete Psychological Works,* ed. J. Strachey (London: Hogarth Press, 1955–64), vol. 14, p. 289.

90. A fairly recent bid for "The Vitality of Death" observes the profound effects of impending death upon the value system of the dying and stresses the phenomenological nothingness which characterizes "my death." Peter Kastenbaum, *Journal of Existentialism,* 5 (1964), pp. 139–66. This paper is an example of that curious American existential trend wherein the terror of death, while admitted, is repaired, as in the custom of prettying up the dead bodies before they are displayed in funeral parlors. See also *Vitality of Death,* by the same author, (Greenwood Press, 1971).

91. P. Tillich, *The Courage To Be* (New Haven: Yale University Press, 1952), p. 169.

92. Freud, *op. cit.,* p. 290. He continues: "It becomes as shallow and empty as, let us say, an American love affair, in which it is understood from the first that nothing is to happen."

93. See a recent debate in *Science* demonstrating the slow removal of death from the domain of social taboos: "Death, Process or Event" by R. S. Morison, vol. 173 (1971), pp. 694–98; and a commentary thereon by L. R. Kass in the same volume, pp. 698–702. Both articles are sincere, compassionate, groping—yet unaware or afraid of transcendental ethics.

94. See the essay by David Bakan on *Disease, Pain, and Sacrifice, op. cit.* This is a brave and well-informed work, a contribution to philosophical anthropology. Readers not afraid of Gestalt theory, psychoanalysis, or the Book of

Job will profit from following up Professor Bakan's inquiry into the biological, psychological, and existential aspects of suffering.

95. J. P. Richter, ed., *The Literary Works of Leonardo da Vinci* (London: Phaidon), vol. II, p. 242. Entry #1162 corresponding to "Morals" from his MS. Similar thoughts are expressed in modern cloak in the motion picture version of Arthur C. Clarke's *2001, A Space Odyssey*. As a frontal lobotomy is performed on Herald the Evil Computer, first it shows signs of mortal terror. Then as its components are picked off one by one its voice gets higher. It finishes by reciting in the voice of a kindergarten child: "I was born on January 9, 1913 in Yorba Linda, California. . . ."

96. *Time,* Jan. 24, 1969, p. 52. Cf. Susanne Langer's sensitive introduction to her *Mind,* vol 1 (Baltimore: The Johns Hopkins Press, 1967), p. 3.

97. Whittaker Chambers, *Witness* (New York: Random House, 1952), p. 14.

98. Recent tests revealed that Western subjects tend to associate the past with the spatial directions of "to the back," "to the bottom," and "to the left;" the future is imagined "to the front," "to the top," and "to the right." B. Aaronson and P. Mundschank, "Some Spatial Stereotypes of Time," paper presented at the meeting of the Eastern Psychological Association, 1968. (Private communication.) It is very doubtful whether this discloses anything profound about temporality, but it certainly demonstrates interesting social conditioning.

99. I was led to realize the ubiquity of this imagery through the paper by Aaronson and Mundschank (*ibid.*), and through a paper by R. Kastenbaum, "The Righteous Mind and Its Sinister Future," presented at a joint meeting of the American Association for Public Opinion Research and the World Association for Public Opinion Research, Santa Barbara, California, 1968; (unpublished manuscript).

100. J. Needham, *Time, The Refreshing River* (London: Allen & Unwin, 1943), pp. 207–32. Needham distinguishes between *order* and *organization* (p. 224). The thermodynamic principle of order is "separatedness" (building blocks separated according to color) versus thermodynamic disorder (all the blocks mixed up) to be distinguished from patterned mixed-up-ness (bricks made into buildings).

101. *Ibid.*, p. 230.

102. The interested reader might want to refer back to our references in the last subsection of chap. 4, sec. 1, and then proceed to a treatment on open systems in L. von Bertalanffy, *General System Theory* (New York: Braziller, 1968), chap. 6 & *passim.*

103. Goodwin has seriously questioned the usefulness of thermodynamical concepts, especially entropy, in their present formulations, to the analysis of living organisms. B. C. Goodwin, *Temporal Organization in Cells* (New York: Academic Press, 1963), p. 62 ff. Although concepts more sophisticated than our present thermodynamical, informational, or organizational entropy are certainly needed, and presumably will eventually be formulated, the arguments given here may need only to be refined but not invalidated.

104. M. S. Watanabe, "Learning Process and the Inverse H-Theorem," *IRE Transactions, PGIT,* IT-18 (1962), pp. 246–51.

105. C. E. Muses shows that beneath the many action laws of physics (such as Hamilton's principle of least action, LeChatelier's law of displacement against stress, Fermat's principle of minimal optical paths) lies the unifying principle of minimal entropy increase. "Aspects of Some Problems in Biological and Medical Cybernetics," *Progress in Biocybernetics,* ed. N. Wiener and J. B. S. Haldane (New York: Elsevier, 1965), pp. 243–48.

106. I. Prigogine, *Introduction to Thermodynamics of Irreversible Processes* (New York: Wiley, 1967), p. 92.

107. W. W. Forrest and D. J. Walker, "Change in Entropy During Bacterial Metabolism," *Nature*, 201 (1964), pp. 49–52.

108. Aristotle, *On the Soul*, 412a.

109. Aristotle, *Historia Animalium*, 588b.

110. J. Needham, "Mechanistic Biology and The Religious Consciousness," in *Science, Religion and Reality*, ed. J. Needham (New York: Braziller, 1955), pp. 223–61. Cf. the dogmatic review of L. Ya. Blyakher, "The Development of Notions of the Material Basis of Living Structures," in Academy of Sciences, USSR, Institute of the History of Natural Sciences and Technology, *Life Phenomena*, IPST trans., US Department of Commerce (Springfield, Va., 1966), pp. 3–44.

111. J. S. Huxley proposed the following general definition: "Evolution is a self-maintaining, self-transforming, and self-transcending process, directional in time and therefore irreversible, which in its course generates ever fresh novelty, greater variety, more complex organization, higher level of awareness, and increasingly conscious mental activity." "Evolution, Cultural and Biological," in *Current Anthropology* (Chicago: The University of Chicago Press, 1956), p. 3. I think this definition is in need of more ancillary explanations than the one given in the text.

112. This is the conclusion of G. Sarton, *A History of Science* (New York: Wiley, 1952), vol. 1, p. 535.

113. This may be clearly seen from the analysis of Hegelian cosmology by R. G. Collingwood in, *The Idea of Nature* (New York: Oxford University Press, 1967), p. 131.

114. Hegel discusses this in sec. 249 of the *Encyclopaedia*. See the recent excellent translation by A. V. Miller, *Hegel's Philosophy of Nature* (Oxford: The Clarendon Press, 1970); also B. Glass *et al.*, eds., *Forerunners of Darwin 1745–1859* (Baltimore: The Johns Hopkins Press, 1968), entries under Hegel. Also, M. J. Petry, *Hegel's Philosophy of Nature* (London: Allen & Unwin, 1970), pp. 25–6. I am indebted to Prof. Wolfe Mays for the lead to the Petry analysis.

115. For a masterful summary of the genotype/phenotype relationship see chap. 10 and *passim*, in E. Mayr, *Animal Species and Evolution* (Cambridge, Mass.: Harvard University Press, 1963). See also the instructive diagram reproduced on p. 10 in Gavin de Beer, *Embryos and Ancestors* (New York: Oxford University Press, 1962).

116. This is from a letter by Darwin to G. C. Wallich, quoted in Gavin de Beer, *Charles Darwin* (New York: Doubleday, 1967), p. 271, without source identification. The belief that life has been imported from other places is a belief in secular form. If taken seriously, it either begs the question or amounts to rigid vitalism for it implies that life always existed.

117. T. Dobzhansky, *The Biology of Ultimate Concern* (New York: New American Library, 1967), p. 47.

118. For a brief evaluation of this suggestion see Sir Gavin de Beer's discussion on the origin of life in his *Atlas of Evolution* (London: Nelson, 1964), p. 135 ff. This has also been suggested by I. A. Oparin, *Life: Its Nature, Origins and Development*, trans. Ann Synge (New York: Academic Press, 1966).

Details of such a scheme, as far as I know, have not been constructed. One reason for the difficulty may be illustrated as follows: Consider that electrons in stationary orbits remain in association with particular nuclei in interstellar space, on the average, for 10^3 sec; let us call these "favored" states. Less favored states, such as those of neutral pi mesons, have an average life time of only

10^{-16} sec. Similar arguments can be given for stars which remain stable for millions of years versus stars, less "favored," which undergo gravitational collapse and disappear in a four-dimensional wormhole. In organic evolution unfit organisms can survive for periods comparable to the period of survival of the fittest. Thus, life itself is only about 3×10^{16} sec old, and an unfit species is likely to exist for a number of multiples of 10^7 sec. (which is one year). The fit/unfit ratio of lifetimes in the physical world is perhaps 10^{19} sec., or even more; the same ratio for life is 10^9 or less. The Oparin and de Beer suggestions are not to be confused, however, with evolutionary continuity based on biochemical relations such as put forth in I. M. Lerner, *Heredity, Evolution and Society* (San Francisco: Freeman, 1968), p. 45.

119. Oparin, *op. cit.*, p. 202.

120. Dobzhansky, *op. cit.*, p. 43.

121. M. Polányi, "Life's Irreducible Structure," *Science*, 160 (1968), pp. 1308–12.

122. See A. D. Hershey, "Genes and Hereditary Characteristics," *Nature*, 226 (1970), pp. 697–700.

123. L. H. Hartwell, "Genetic Control of the Cell Division Cycle in Yeast," *Journal Molecular Biology*, 59 (1971), p. 183–94.

124. C. F. Ehret and E. Trucco, "Molecular Models for the Circadian Clock," *Journal Theoretical Biology*, 15 (1967), pp. 240–62.

125. See, for example, R. L. Gregory, "Origin of Eyes and Brains," *Nature*, vol. 215 (1967).

126. Joseph Needham speaks of "organizers" which direct the ectoderm to form, for instance, head ectoderm, then the lens of the eye, etc., thus gradually delimiting its potential roles. See Needham, *Order and Life* (Cambridge: M.I.T. Press, 1968), chap. 2.

127. On this see the literature of biosemiotics. Start with F. S. Rothschild, "Posture and Psyche," *Problems in Dynamic Neurology*, ed. L. Halpern (Israel Academy of Sciences: Jerusalem, 1963), pp. 475–509 and its references.

128. In C. H. Waddington, *The Ethical Animal* (New York: Atheneum, 1961), p. 52.

129. G. G. Simpson, *This View of Life* (New York: Harcourt, Brace & World, Inc., 1964), p. 189. See also W. M. Elsasser, "Acausal Phenomena in Physics and Biology," *American Scientist*, 57 (1969), pp. 502–16; and P. Weiss, "One Plus One Does Not Equal Two," *The Neurosciences*, ed. Querton, Melnechuk, and Schmitt (New York: Rockefeller University Press, 1967), pp. 801–21.

130. N. Bohr, "Light and Life," *Nature*, vol. 28 (1933), p. 458.

131. R. W. Gerard coined the word *org* to mean "those material systems of entities which are individuals at a given level but are composed of subordinate units, lower level 'orgs', and which serve as units in superordinate individuals, higher level 'orgs.'" "Units and Concepts of Biology," *Science*, 125 (1957), pp. 429–33.

132. W. M. Elsasser, *The Physical Foundations of Biology* (London: Pergamon, 1958).

133. On this see "Hierarchical Structures," ed. L. L. Whyte, A. G. Wilson, and D. Wilson (New York: Elsevier, 1969). In this, see the excellent summary by Donna Wilson, p. 287 ff., disguised as a bibliography. Also, P. A. Weiss *et al.*, *Hierarchically Organized Systems in Theory and Practice* (New York: Hafner, 1971).

134. C. Darwin, *Origin of the Species* (New York: Collier Books, 1967), p. 134 ff.

135. T. Dobzhansky, *Biology of Ultimate Concern*, p. 60. As an illustra-

tion of Dobzhansky's remark about poets, here is Pablo Neruda's "The Turtle,"
A *New Decade, Poems 1958–67* (New York: Grove Press, 1969), p. 99.

> Patriarch, long
> hardening
> into his time,
> he grew
> weary of waves
> and stiffened himself
> like a flatiron.
> Having dared
> so much
> ocean and sky, time and terrain,
> he let his eyes droop
> and then slept,
> a boulder
> among other boulders.

136. Simpson, *op. cit.*, p. 70.

137. *Ibid.*, p. 80.

138. Mayr, *op. cit.*, p. 181.

139. D. J. Merrell, *Evolution and Genetics* (New York: Holt, Rinehart & Winston, 1962), p. 399.

140. L. von Bertalanffy, *Robots, Men and Minds* (New York: Braziller, 1967), p. 84.

141. J. S. Huxley, for instance, lays particular stress on the increasing independence from the environment in the case of evolving animals, in his *Evolution, the Modern Synthesis* (New York: Harper, 1942). But still, he sees only one single line along which sustained evolutionary progress is truly a possibility and that is the lineage of man. (See his masterly discussion on "Evolutionary Progress" [chap. 10]). The basis of this hope is in the conceptual powers available to man for interacting with his environment—and that, as he observes it, is but a single thread (p. 572). My stress on the mutuality of adaptation emphasizes that whether or not a species is led into a blind alley, there is no single evolutionary change without involving some environmental change; and this relationship is independent of the relative position of the organism on the evolutionary ladder. Note also the title of Rene Dubos' *Man Adapting* (New Haven: Yale University, 1965) which is quite descriptive of its contents.

142. There are a number of ideas current, or at least known in evolutionary biology which must be distinguished from those suggested here. "The fitness of the environment" is a concept whose value is that if "taken in moderate doses [it illustrates] the existence of limitations to the range of properties of living systems, and [shows] that the direction of organic evolution must have been guided to a considerable extent accordingly." H. Blum's *Time's Arrow and Evolution* (New York: Harper, 1962), p. 86. Efferent adaptation cannot be taken in moderate doses but must be regarded, as explained in the text, as the dynamically equal but opposing counterpart of afferent adaptation. The "fitness of the environment," in the sense explained by Professor Blum, remains a limiting and guiding factor except that now the environment itself is also seen as evolving under the selective pressure of organisms.

My emphasis on the reciprocity between organisms and their surroundings must also be distinguished from L. L. Whyte's concern with *Internal Factors in Evolution* (New York: Braziller, 1965). His internal factors are those selective processes which insure ordering until ordinary Darwinian selection can take over (p. 58). Thus "internal" is spatial, for instance, intrauterine, rather than outward directed or efferent as conceived in the present scheme.

143. L. van Valen, "A New Evolutionary Law," *Evolutionary Theory*, vol. 1, pp. 1–30, (1973).

144. A concept whose origins are usually, but erroneously, attributed to Chardin. The term originated with the geologists of the turn of the nineteenth century, signifying all the life on this globe as a coherent unit.

145. H. Selye, "The Physiology and Pathology of Exposure to Stress," Annual Report on Stress, *Lancet*, 275 (1958), pp. 205–8. For a summary review see "The Evolution of the Stress Concept" in *American Scientist*, 61 (1973), pp. 692–99.

146. Selye, *The Stress of Life* (New York: McGraw Hill, 1956), p. 31 f. Cf. the Freudian idea of the ambiguity of tension reduction in P. E. Slater, "Prolegomena to a Psychoanalytic Theory of Aging and Death," in R. Kastenbaum, ed., *New Thoughts on Old Age, op. cit.*, p. 33.

147. A. E. Needham, "How Living Organisms Repair Themselves," *New Scientist*, 259 (1961), pp. 284–87.

148. H. Kalmus, "Axioms and Theorems in Biology," *Nature*, 198 (1963), pp. 240–43. Cf. the decline of massive, slowly changing woody trees in favor of annuals with rapid population turnover through which advantaged genetic changes can spread much faster (ref. 82, above). See also the work of Maynard Smith who showed that the advantage of sexual reproduction over asexual reproduction is the faster rate at which favorable mutations can spread in a given population. "What Use Is Sex?" *Nature*, 230 (1971), pp. 209–10.

149. T. Dobzhansky, *Mankind Evolving* (New Haven: Yale University Press, 1962), p. 319.

150. Oparin, *op. cit.*, p. 12 f.

151. For one brief discussion of time available versus time necessary see H. Kalmus in *The Voices*, p. 345 ff. For a type of calculation which often appears in the works of evolution when time available is discussed, see F. B. Salisbury, "Natural Selection and the Complexity of the Genes," *Nature*, 224 (1969), pp. 342–43. The fancy of folk imagination seems to run the other way: in folk literature animal characteristics are sometimes accounted for through the belief that the creator ran out of time and could not complete them. Stith Thompson, *Motif Index of Folk Literature* (Bloomington: Indiana University Press, 1955–58), entry A 2286.1.0.1.

152. *Ethics*, part 3, props. 4 and 6; part 4, prop. 20.

153. J. Needham, *Order and Life*, p. 70.

VII / THE ORGAN OF TIME SENSE

1. N. Kleitmann, *Sleep and Wakefulness* (Chicago: University of Chicago Press, 1963), pp. 363–70.

2. I am using *sensation* with the preferred meaning, i.e., "the process or activity of apprehending colors, sounds, tastes, etc." while keeping in mind that "nowhere has the intrusion of philosophic viewpoints worked more havoc with psychological terminology than here." English and English, eds., *A Comprehensive Dictionary of Psychological and Psychoanalytical Terms* (New York: David McKay, 1964).

3. K. Freeman, *The Pre-Socratic Philosophers* (Oxford: Blackwell, 1946), pp. 261–74.

4. Plato, *Phaedrus* 248d–e.

5. Descartes, *Meditations on First Philosophy* (1641), in *The Philosophical Works of Descartes*, tr. and ed. E. S. Haldane and G. R. T. Ross (New York: Dover, 1955), Meditation II.

6. *Ibid.*, Meditation VI.

7. *Ibid.*

8. Descartes, *Passions of the Soul* (1649), *loc. cit.*, art. XXXI.

9. T. Hobbes, *Hobbes' Leviathan*, reprinted from the edition of 1651, (Oxford: Clarendon Press, 1967), chap. 6.

10. K. Marx and F. Engels, *The German Ideology* (New York: International Publishers, 1947), p. 14.

11. M. Schlick, *Allgemeine Erkentnislehre* (Leipzig: Springer, 1925), sec. 32. Trans. Gillian Brown in *Body and Mind*, ed. G. N. A. Vesey (London: Allan and Unwin, 1964).

12. J. J. C. Smart, "Sensations and Brain Processes," *Philosophical Review*, 68 (1959), p. 145.

13. H. Feigl, *The "Mental" and the Physical* (Minneapolis: University of Minnesota Press, 1967).

14. G. Berkeley, *Of the Principles of Human Knowledge*, part 1, sec. 1. The quote is from *The Works of George Berkeley*, ed. A. C. Fraser (Oxford: Clarendon Press, 1901).

15. G. G. Simpson, *This View of Life* (New York: Harcourt, Brace & World, 1964), p. 105.

16. *Ibid.*, p. 107.

17. I. A. Oparin, *Life*, trans. A. Synge (New York: Academic Publishers, 1961).

18. W. James, "Does 'Consciousness' Exist?" *Journal of Philosophy, Psychology and Scientific Methods*, 1 (1904). From James, *Essays in Radical Empiricism* (London: Longmans Green & Co., 1912).

19. Spinoza, *Ethics*, part II, def. 1.

20. *Ibid.*, part II, prop. 7.

21. *Ibid.*, part III, prop. 2.

22. N. de Malebranche, *Dialogues on Metaphysics and Religion*, tr. M. Ginsberg (London: Allen and Unwin, 1923), part VII of the Seventh Dialogue.

23. See note 21, p. 609, *The Voices*.

24. J. Cohen, "Cyclopean Psychology," *Hibbert Journal*, 59 (1961), pp. 236–44.

25. See, for example, the "double-language" view of Herbert Feigl, sometimes called the psychoneural identity. "The Mind-Body Problem in the Development of Logical Empiricism," *Revue Internationale de la Philosophie*, 4 (1950).

26. J. C. Eccles, *The Neurophysiological Basis of Mind* (Oxford: Clarendon Press, 1965), pp. 265–66.

27. R. L. Gregory, "Origins of Eyes and Brains," *Nature*, 214, (Jan. 28, 1967), pp. 369–72.

28. On this, see W. H. Thorpe, "Ethology and Consciousness," in *Brain*, p. 476.

29. The experimental situation is complicated, but there seems to be no record of anyone ever having reported an experience streaming backward in time. See, e.g., W. Penfield, "Speech, Perception and the Uncommitted Cortex," in *Brain*, pp. 217–37.

30. *Patterning of Time*, pp. 288–89.

31. J. C. Eccles, "Conscious Experience and Memory," in *Brain*, pp. 314–44.

32. Neurons either fire under a sufficiently large input signal or they do not. Hence, weaker signals would be transmitted at longer intervals.

33. H. C. Longuet-Higgins, "Holographic Model of Temporal Recall," *Nature*, 217 (1968), p. 104. The principle of the hologram involves the recording, on photographic plates, of the interference and diffraction patterns produced by an object in monochromatic light. Although the individual images so obtained appear to be meaningless, they contain the necessary visual information in a distributed fashion. When the reference monochromatic light is reinserted, the original image is reproduced, as far as an observer is concerned.

34. See Longuet-Higgins and the subsequent exchange in *Nature*, 225 (1970), pp. 177–78. For a discussion of the limitations of holographic memory see H. M. Smith, *Principles of Holography* (New York: Wiley, 1969), pp. 217–26.

35. E. R. John, *Mechanisms of Memory* (New York: Academic Press, 1967), p. 418.

36. *Ibid.*, p. 342.

37. *Ibid.*, p. 420.

38. Brain, Lord, "Some Reflections on Brain and Mind," *Brain*, 86 (1963), p. 392.

39. K. Popper, "Indeterminism in Quantum Physics and in Classical Physics," Part II, *British Journal of Philosophy of Science*, 1 (1950), pp. 173–95.

40. Norbert Wiener took it as a self-evident fact of reality that no predictor can operate on the future of a time series, hence all prediction operators must be intrinsically lopsided. *Extrapolation, Interpolation, and Smoothing of Stationary Time Series* (Cambridge, Mass.: M.I.T. Press, 1966), p. 12 ff. Wiener's approach involves the separation of stationary time-series (akin to our idea of stationary processes) from what we have called "creative processes." The latter are unrecognized by Wiener, except by inference. In any case, his analysis of prediction, mixing rigorous predictability with statistics, should not be confused with Popper's analysis of self-prediction.

41. D. M. MacKay, "On the Logical Indeterminacy of a Free Choice," *Mind*, 10 (1960), p. 31; "Cerebral Organization and the Conscious Control of Action," in *Brain*, pp. 422–45; *Freedom of Action in a Mechanistic Universe* (Cambridge: The University Press, 1967).

42. This rather difficult conceptual exercise may be elucidated by the following thought experiment. Suppose that I give you on January 12th a complete description of your brain, including some predictions which I judge to be inevitable. One of them is this: "On May 7th, after having studied my 600 volumes, *you shall decide* to exclaim: 'All your descriptions and predictions are correct.'" On May 7th, having finished studying the volumes, you must still believe that the choice is up to you, even if you exclaim as predicted, and thereby confirm my calculations. For, if you would believe *before* your exclamation that my predictions are correct, this would be against my own precept that your decision will be taken *on* May 7th. Thus, while it is appropriate for me to believe in the truth of my proposition, it is inappropriate for you to accept my view as certain, for if you did, this would make my argument invalid.

43. MacKay, "Cerebral Organization and the Conscious Control of Action," *loc. cit.*, p. 434.

44. *Ibid.*, p. 425. For a critical analysis of MacKay's ideas, see P. T. Landsberg and D. A. Evans, "Free Will in a Mechanistic Universe?" *British Journal of Philosophy of Science*, 21 (1970), pp. 343–58 and the ensuing correspondence.

45. MacKay, "Cerebral Organization and the Conscious Control of Action," *loc. cit.*, p. 432.

46. Eccles, *The Neurophysiological Basis of Mind, op. cit.*, p. 276.

47. A very appropriate phrase due to Christian von Monakow. *Die Lokalisation im Grosshirn und der Abbau der Funktion durch Kortikale Herde*, (Wiesbaden: J. F. Bergmann, 1914). Instead of picturing the engram as a static trace, he urged us to consider it as a sequence of impressions (pp. 72, 223 ff.) and assumed the existence of a physiological basis for such chronogenic localization (p. 303).

48. K. S. Lashley, "In Search of the Engram," *Symposium of the Society for Experimental Biology*, (Cambridge, 1950), vol 4, p. 454–83.

49. E. H. Lenneberg, "A Biological Perspective of Language," *New Directions in The Study of Language*, ed. E. H. Lenneberg (Cambridge: M.I.T. Press, 1964), p. 81.

50. As an entry to the concern of general systems theory with complexity, see L. von Bertalanffy, *General System Theory* (New York: Braziller, 1968), chap. 2. Also, "The Place of the Brain in the Natural World" by W. R. Ashby in *Currents in Modern Biology*, 1 (1967), pp. 95–104.

51. William Ross Ashby, the distinguished cyberneticist, wrote an editorial shortly before his death in which he spoke of "richly interactive systems." *Behavioral Science*, 18 (1973). In it he points out that the doubling of the size of a complex system, such as 10^{10} neurons to 2×10^{10} neurons would increase its information capacity not by a factor of 2, but by perhaps 10^6. Ashby is talking here about total information capacity, which is a static quantity, and not about such dynamic (temporal) manifestations (and, possibly measures) of complexity as the number of connections which may be made during time spans important in the cerebrating activity of the subject.

52. von Bertalanffy, *op. cit.*, p. 19, 55 ff.

53. Thorpe, *op. cit.*, p. 493.

54. For one approach see J. C. Eccles, *The Brain and the Unity of Conscious Experience* (Cambridge: Cambridge University Press, 1965).

55. H. A. Simon, "The Architecture of Complexity," *The Sciences of the Artificial*, ed. H. A. Simon (Cambridge: M.I.T. Press, 1969), p. 117.

56. On this see Thorpe, *op. cit.*

57. H. Bergson, *Time and Free Will* (New York: Harper Torch Books, 1960), p. 183.

58. See, e.g., F. Mucho, *Berdyaev's Philosophy* (New York: Doubleday: Anchor Books, 1966).

59. If skilled surgeons were able to interchange the heads of two healthy adults, would the two heads brag about new bodies, or the bodies about new heads? The question may be called the Frankenstein identity syndrome. With a neurological bias in dealing with the mind-brain problem the answer would favor the heads. There have been unconfirmed reports about dog brains and those of a monkey having been kept functioning for several days after their removal from the cranium. A. Toffler, *Future Shock* (New York: Bantam Books, 1970), p. 214. Did the severed, but well-kept brain cry out for mercy from the insane search for knowledge by all other brains? The reports do not say. Concern with the ethical implications of improved techniques in neurosurgery has often been expressed. See, *e.g.*, G. Schaltenbrand, "Die Manipulation der Person," *Studium Generale*, 22 (1969), pp. 494–512.

60. "The Logic of Abduction," *Charles S. Pierce, Essays in the Philosophy of Science*, ed. V. Thomas (New York: Liberal Arts Press, 1957), pp. 245–46.

61. J. von Neumann, *The Computer and the Brain* (New Haven: Yale University Press, 1958), p. 52.

62. J. Eisenbud, "Why Psi?" *The Psychoanalytic Review*, (Winter, 1966–67), p. 151.

63. Wm. Gooddy, "Asymmetry of Cerebral Function," *Journal of Mental Science*, 107 (1961), p. 431–37.

64. J. Cohen, *Humanistic Psychology* (New York: Collier Books, 1962), p. 110.

65. E. H. Lenneberg, "Language in the Light of Evolution," *Animal Communication*, ed. T. Sebeok (Bloomington: Indiana University Press, 1968), p. 611.

66. See Lenneberg, "A Biological Perspective of Language," in *New Directions in the Study of Language, op. cit.*, pp. 65–6.

67. Lenneberg, "Language in the Light of Evolution," *loc. cit.*

68. On this consult Lenneberg, *Biological Foundations of Language* (New York: Wiley, 1967), p. 265. This is a superb, scholary work to which the interested reader is enthusiastically referred.

69. E. Cassirer, *An Essay on Man* (New Haven: Yale University Press, 1945), p. 132.

70. Stith Thomson, *Motif Index for Folk Literature* (Bloomington: Indiana University Press, 1955–58). See various entries under language, such as dumbness as punishment for breaking taboo (C 944), dumbness as curse (D 2021.1), loss of speech as punishment (Q 451.3), change of language for breaking taboo (C 966).

71. The importance of feedback on the control of skilled movements in general and in audio feedback in particular becomes painfully apparent when it is cut off or upset. It is known, for instance, that singers keep the constancy of their notes by continuous correction to a Sollwert. When it comes to more complex (modulated, and not strictly cyclical) audio modalities, such as speech, interference with the feedback can totally interrupt the speech. This is known as the Lee effect. Lenneberg, *Biological Foundations of Language*, p. 110.

72. J. Piaget, *The Language and Thought of the Child* (New York: Meridian, 1955), p. 146.

73. See, e.g., R. A. Hinde, ed., *Non-Verbal Communication* (Cambridge: Cambridge University Press, 1972), which contains some perceptive surveys of our present knowledge of communication in man and animal.

74. Some students of language have insisted that language is not denotative (transferring information about some things) but connotative (orients the recipient within his own cognitive domain without pointing to entities independent of the hearer). See, e.g., H. A. Maturana and F. Verga, *Autopoietic Systems* (Le Hague: Moutton, 1973). These are interesting points but irrelevant to our concern. In communicating with speech, the vibrations of the air, or some similar functions, must originate somewhere and arrive somewhere.

75. *Political Phil.*, see especially chapter 2 and its extensive references.

76. B. L. Whorf, *Language, Thought and Reality* (Cambridge: M.I.T. Press, 1956), p. 143.

77. *Ibid.*, p. 216.

78. *Ibid.*, p. 247 f.

79. *Brain*, vol. 78 (1955), p. 669.

80. N. Chomsky, *Language and Mind* (New York: Harcourt, Brace & World, 1968), p. 1.

81. N. Chomsky, *Aspects of the Theory of Syntax* (Cambridge: M.I.T. Press, 1965), p. 27 ff.

82. One of the greatest English writers, Joseph Conrad (Korzeniowski), was born in Poland and landed at Lowestoft, Suffolk, at the age of twenty-one, knowing no one and speaking no more than a few words of English. See his

fascinating biography by J. Bainer, *Joseph Conrad* (New York: McGraw Hill, 1960). The author of the work *Of Time, Passion, and Knowledge* (New York: Braziller, 1975), never studied grammar but learned his English by reading the speeches of Winston Churchill. The assumption of linguistic universals even underlies that peculiar device known as a dictionary of the English language. How else is one supposed to consult it for spelling or for meaning in context if one does not have a "feeling" as to where to look and does not already know or guess most of the words in it?

83. Chomsky, *Language and Mind*, p. 62. It has been calculated that the number of grammatically well formed twenty-word sentences in English is of the order of 10^{20}. R. A. Hinde, *Animal Behavior* (New York: McGraw Hill, 1970), p. 336.

84. K. Lorenz, "Kants Lehre vom apriorischen in Lichte gegenwärtiger Biologie," *Blätter f. Deutsche Philosophie*, vol. 15 (1941), pp. 94–125. Quoted approvingly in Chomsky, *Language and Mind*, p. 81.

85. S. G. F. Brandon, *Man and His Destiny in the Great Religions* (Manchester: Manchester University Press, 1963), p. 8.

86. See Eccles's superb reasoning in "Conscious Experience and Memory," *Brain*, p. 314 ff.

87. For instance, A. C. Hardy in *The Living Stream* (London: Collins, 1965), p. 260.

88. J. M. Goldrich, "Separation and the Sense of Time," *Omega*, vol. 3 (1970), p. 15.

89. *Patterning of Time*, p. 247 ff. American subjects.

90. *Ibid.*, p. 252 and *passim*. Doob deals with this question under the headings of "acquiring temporal potential" and "orientation to time." He discusses several suggestions regarding the development of the notions of the three temporal categories, stating his own inclination for the priority of the present, followed by the simultaneous development of future and past.

VIII / OUT OF THE DEPTHS

1. St. Augustine, *Confessions*, tr. E. B. Pusey (New York: Dutton, 1946, Everyman's Library), Bk. X, chap. 15.

2. See his essay, "The Diseases that Deprive Man of his Reason. . . ." (1512), tr. G. Zilborg in H. E. Sigerist, ed., *Four Treatises* (Baltimore: The Johns Hopkins Press, 1941), p. 135–212.

3. Eduard von Hartmann, *Philosophy of the Unconscious*, tr. W. C. Coupland (London: Kegan, Paul & Trench, 1931), p. 1.

4. For an enjoyable introduction, consult R. I. Watson, *The Great Psychologists* (New York: Lippincott, 1968), chap. 20.

5. R. S. Peters, ed., *Brett's History of Psychology* (Cambridge: M.I.T. Press, 1965), p. 578.

6. See L. L. Whyte's fine summary on the "Unconscious" in *Encyclopedia of Philosophy* (New York: Macmillan, 1967) P. Edwards, ed., and *The Unconscious Before Freud* (New York: Doubleday, 1962), a slender volume also by L. L. Whyte. For more detailed documentation see H. F. Ellenberger, *The Discovery of the Unconscious* (New York: Basic Books, 1970) pp. 1–181.

7. English and English, *A Comprehensive Dictionary of Psychological and Psychoanalytical Terms* (New York: David McKay, 1964), entry under Unconscious.

8. There is an interesting epistemological similarity between relativity theory and the Freudian theory of the unconscious. Specific predictions of the former, as for example, the "diseases" of clocks and rods, are extremely difficult to observe and their everyday significance is negligible. The most important effect of relativity theory is a profound change in our views of the nature of the physical world. Similarly, Freudian analysis, though often astonishing, is too clumsy to be useful in modern popular democracies. Its revolutionary effect is the profound change it brought about in our evaluation of man in the natural order.

9. S. Freud, *The Complete Introductory Lectures* (New York: Norton, 1966), p. 538.

10. C. S. Hall, "Empirical Evidence for the Timelessness of the Unconscious," (unpublished paper, 1958).

11. M. Bonaparte, "Time and the Unconscious," *International Journal of Psychoanalysis,* 11 (1940), p. 427.

12. W. C. Lewis, "Structural Aspects of the Psychoanalytic Theory of Instinctual Drives, Affects, and Time," *Psychoanalysis and Current Biological Thought,* N. S. Greenfield and W. C. Lewis, eds., (Madison: University of Wisconsin Press, 1965), pp. 151–79.

13. S. Freud, "Beyond the Pleasure Principle," in *The Standard Edition of the Complete Psychological Works of Sigmund Freud,* ed. J. Strachey (London: Hogarth Press, 1955), vol. 18, p. 28.

14. *Patterning of Time,* p. 287.

15. On this subject, see L. Dooley, "The Concept of Time in Defence of Ego Integrity," *Psychiatry,* 4 (1941), p. 15.

16. *Ibid.,* p. 22.

17. Norman O. Brown based a complete theory of history on the symbolic struggle between Eros and Thanatos. See his *Life Against Death* (Middletown: Wesleyan University Press, 1959).

18. Freud, "Beyond the Pleasure Principle," *loc. cit.,* p. 45.

19. *Political Phil.,* p. 27. See chapter 2 with its enlightening notes. In a different field of knowledge, Freud argues with reference to clinical evidence that the present attitude of the Ucs. toward death is the same as the Cs. attitude of early man. That is, the Ucs. functions as though it were immortal. Freud, *The Standard Edition of the Complete Psychological Works,* vol. 14, p. 289. It is questionable whether this early state of mind really deserves to be identified as conscious experience. I prefer the *fragestellung* in the text.

20. S. Kierkegaard, *Works of Love,* tr. H. and E. Hong (New York: Harper & Row, 1962), p. 253.

21. See chap. 7, "The War Years," E. Jones, *The Life and Work of Sigmund Freud* (New York: Basic Books, 1955), vol. 2.

22. C. G. Jung, *Memories, Dreams and Reflections* (New York: Pantheon, 1963), p. 176.

23. *Ibid.,* p. 390. The existence of a collective Ucs. is an unstated assumption in all psychoanalytic theories, for unless there were some things common in the function and form of the Ucs. of all men, it would not be possible to develop a science to deal with it. Some time ago Géza Roheim quoted approvingly the findings of dream analysis in anthropological fieldwork which demonstrated that no special knowledge of a culture is required to understand the dreams of members of different cultures. "Techniques of Dream Analysis and Field Work in Anthropology," *Psychoanalytic Quart.,* 18 (1949), p. 471. Yet in Freudian psychoanalysis there is a strong strain of disapproval against the Jungian views

of collective Ucs. in the interpretation of behavior. The disapproval is often linked with the alleged racist political views of Jung.

24. Jung, *op. cit.*, p. 381.

25. Cf. the call for a pre-geometry in C. W. Misner, K. S. Thorne, and J. A. Wheeler, *Gravitation* (San Francisco: Freeman, 1973), chap. 44.

26. Jung, *op. cit.*, p. 380.

27. K. Lorenz, "Kants Lehre vom apriorischen in Licht gegenwärtiger Biologie," *Blätter f. Deutsche Philosophie*, 15 (1941), pp. 94–125, quoted in N. Chomsky, *Language and Mind* (New York: Harcourt, Brace & World, 1968), p. 81.

28. On this issue see, *Number and Time*, by the Jungian scholar Marie Louise von Franz (Evanston: Northwestern University Press, 1974), and its references.

29. J. Flavell, *The Developmental Psychology of Jean Piaget* (New York: Van Nostrand, 1963), p. 309 ff.

30. *Ibid.*, p. 371.

31. *Ibid.*, p. 311.

32. *Ibid.*, p. 359.

33. E. Wigner, "The Unreasonable Effectiveness of Mathematics in the Natural Sciences," *Symmetries and Reflections* (Bloomington: Indiana University Press, 1967), chap. 17, p. 222.

34. Thornton Wilder's beautiful creation, *The Bridge of San Luis Rey*, is an intuitive literary examination of the concept of synchronicity, for it is directed to an inquiry about meaningful, noncausal connections. See *The Voices*, p. 233–34.

35. C. G. Jung and W. Pauli, *The Interpretation of Nature and the Psyche* (New York: Pantheon, 1955), pp. 1–144.

36. C. G. Jung, *Collected Works*, ed. M. Read, M. Fordham, and G. Adler (New York: Pantheon, 1953). For a criticism of synchronicity from the point of view of neurology, see G. Schaltenbrand, ed., *Zeit in Nervenärztlicher Sicht* (Stuttgart: Enke, 1963), pp. 5–7. For support, cf. A. Koestler's *The Roots of Coincidence* (New York: Random House, 1972).

37. E. Neumann, *The Origins and History of Consciousness* (New York: Pantheon, 1964), p. xvi.

38. Freud, *Civilization and Its Discontents*, in *The Standard Edition of the Complete Psychological Works*, vol. 21, p. 69.

39. *Ibid.*, p. 71.

40. A favorite tool of science fiction writers, time travel, deals with the imagined ability of a person to transport himself into a world yet to come, or into a world we recognize as already dead. Neglecting such problems as the ever-changing spatial position of the earth or the sudden creation of absolute vacuums, we can try to imagine our traveler in the future. So as to have a confirmation of his journey he must return to the present, because waiting until he is met "in the future" would make the annihilation of the traveler necessary for a period measured by his own sense of time. Upon his return, then, he might decide to interfere with his future-as-seen. If he can be successful, he must have visited only a potential future; if he cannot possibly succeed, then the world is completely deterministic, that is, timeless. Both cases would disqualify his venture as time travel. Neglecting again the embarrassing details of transfer, the traveler might decide to visit the past. Unlike his prior journey, he does not need to return to the present from the past, because he can close the communication loop by interfering with his own future. For instance, through judicious actions he can prevent his own coming about in his mother's womb,

and thus make himself nonexistent. If, however, one must postulate that he cannot interfere with his future (our present) because the nature of the world is such as to make this impossible, then the world is again an aggregate of static timeless processes. It follows that, whether or not he can interfere with history, the time machine he is imagined to have employed is an intellectually dishonest device. The difficulties revealed by this thought experiment suggest a fundamental connection between time and free will.

41. Freud, "The Uncanny," *The Standard Edition of the Complete Psychological Works,* vol. 4, p. 368. *Cf.* the English *homely:* "kind," "warm," "simple," "ugly."

42. *Ibid.,* p. 377.

43. E. Leach, "Anthropological Aspects of Language, Animal Categories and Verbal Abuse," *New Directions in the Study of Language,* ed. E. H. Lenneberg (Cambridge: M.I.T. Press, 1966), pp. 23–64. The argument is too complex to be reproduced here, but it is straightforward and, in a way, uncanny.

44. There is strong analogy here to the attitudes toward minority groups. If the minority group lacks qualities esteemed by the majority, the privileged group holds the other one totally responsible for its own fate. Ipso facto, members of the group also become disgusting and undesirable.

45. J. Cohen, *Human Robots in Myth and Science* (London: Allen & Unwin, 1966), p. 50 ff.

46. Leach, *loc. cit.,* p. 45.

47. Freud's examples are almost identical to those of Jung given in support of the concept of synchronicity. While Jung attempts to find a connection of such events through their meaning in the observer's mind, Freud tries to identify analytically the sources of the strange feelings.

48. Freud, "The Uncanny," *loc. cit.,* p. 371, stated early in the paper, but it anticipates the result.

49. A. Ehrenzeig, *The Psycho-Analysis of Artistic Vision and Hearing* (New York: Braziller, 1965), chap. 5, p. 103 ff.

50. For a pioneer work, see W. Stern, *Psychology of Early Childhood* (New York: Henry Holt, 1924), p. 112. For a recent paper which argues, de novo, the primacy of futurity in the sense of time, see J. R. Nuttin, "The Future Time Perspective in Human Motivation and Learning," *Proceedings of the 17th International Congress of Psychology* (Amsterdam: North Holland, 1974), pp. 60–82.

51. J. Cohen, *Behavior in Uncertainty* (London: Allen & Unwin, 1964), chap. 10.

52. In a more universal context see G. Rochberg, "The Structure of Time in Music: Traditional and Contemporary Ramifications and Consequences," in *Study v. 2.*

53. A. Huxley, *The Devils of Loudon* (New York: Harper & Row, 1952), p. 321.

54. R. Fischer, "A Cartography of the Ecstatic and Meditative States," *Science,* 174 (1971), p. 897–904. The concepts are somewhat unfamiliar. "Ergotropic arousal denotes behavioral pattern preparatory to positive action and is characterized by increasing activity of the sympathetic nervous system and an activated psychic state." "Trophotropic arousal results from an integration of parasympathetic with somatomotor activities to produce behavioral patterns that conserve and restore energy, a decrease in sensitivity to external stimuli, and sedation." For a quantitative earlier work see, T. Thompson and C. R. Shuster, *Behavioral Pharmacology* (Englewood: Prentice Hall, 1968).

55. A. Huxley, *The Doors of Perception* (New York: Harper, 1954).

56. *Patterning of Time,* pp. 282–85, 312 and *passim.*

57. Whether lovers leaping off lovers' leaps will forever be united and their prior misery ended, I do not know; but at least at one level of understanding, these are the usual reasons for the acts. Thus, I do not hesitate to go along with Doob's not so cautious comment that, "various drugs, therefore, may be ingested by persons who have a deliberate, conscious desire to renounce some kind of activity of the every day world; who in some respects are unable to defer gratification . . . , who anticipate satisfaction from the drug-induced experiences. . . ." *Patterning of Time*, p. 282.

58. Freud, *Civilization and Its Discontents, loc. cit.,* p. 64.

59. R. B. Blakney, *Meister Eckhart: A Modern Translation* (New York: Harper, 1941), sermon 12, p. 151.

60. *The Collected Works of St. John of the Cross*, tr. K. Kavanaugh *et al.* (New York: Doubleday, 1964), p. 718.

61. *Ibid.*, p. 719.

62. For one attempt, see C. T. Tart, "States of Consciousness and State-Specific Sciences," *Science*, 176 (1972), pp. 1203–10. See also, chapters 6–13 in H. Yaker *et al.*, eds., *The Future of Time* (New York: Doubleday, 1971). For an intelligent and honest journalistic account see, W. Braden, *The Private Sea—LSD and the Search for God* (Quadrangle, 1967). A forward-looking criticism of the lopsided legal situation may be found in a book review on "Official Views on Marijuana," by J. Kaplan. "It is a scandal that the shabbiest type of research, uncontrolled and primarily based upon clinical impressions, is published and given the widest publicity today, so long as it points to 'possible dangers' in the drug, while at the same time better designed studies which place the issue into perspective tend to be ignored." *Science*, 179 (1973), p. 169.

63. For Tillich's practical attitudes to agape and eros, see Hannah Tillich, *From Time to Time* (New York: Stein & Day, 1973).

64. It was recently suggested from biographical and medical evidence that the Gothic drama of *The Strange Case of Dr. Jekyll and Mr. Hyde*, depicting as it does the hero/villain's struggle for identity, sanity, and creativity, was itself written while Stevenson was under the influence of cocaine. M. G. Schultz, "The 'Strange Case' of Robert Louis Stevenson," *Journal of the American Medical Association*, 216 (1971), pp. 90–4.

IX / EPISTEMOLOGY AND THE TRUE

1. Along the line we find four objects and their corresponding cognitive faculties. The lower part has an upper portion of material objects recognized through perception; also a lower portion, known through conjecture, made up of images and reflections of the objects just above. The upper part of the line also has a lower portion which contains provisional knowledge (*dianoia*) of the forms attainable in mathematical science; the upper portion is that of complete knowledge (*episteme*) of the total hierarchical system of forms. Plato, *The Republic* 509d.

2. J. Locke, *An Essay Concerning Human Understanding*, collated and annotated by A. C. Fraser, 1894 (New York: Dover Reprint, 1959), book 2, chap. 1, sec. 3.

3. *Ibid.*, sec. 4.

4. *Ibid.*, footnote 4, p. 123.

5. *Ibid.*, book 2, chap. 14, sec. 2.

6. B. Russell, *Human Knowledge* (New York: Simon & Schuster, 1948), p. 158.

7. *Ibid.*, p. 504.

8. *Ibid.*, p. 507.

9. M. Polányi, *Personal Knowledge* (Chicago: Chicago University Press, 1964), pp. 63–5 & *passim.*

10. G. Sarton, *The History of Science and the New Humanism* (Bloomington: Indiana University Press, 1962), p. 162.

11. J. Piaget, *Genetic Epistemology* (New York: Columbia University Press, 1970).

12. J. Piaget, "Time Perception in Children," *The Voices*, p. 202.

13. J. Piaget, *The Child's Conception of Physical Causality* (New York: Humanities Press, 1951), see last chapter.

14. *Biologie et Connaissance.* Its final chapter is translated in H. G. Firth, *Piaget and Knowledge* (Englewood Cliffs: Prentice Hall, 1969), pp. 193–202.

15. *Ibid.*, p. 201. These assumptions are the fundamental tenets of *Biologie et Connaissance* and are the reasons for calling Piaget's epistemology vitalistic monism. The title of the section from which these paragraphs were prepared is called "Life and Truth."

16. *Ibid.*, p. 200.

17. Piaget is frequently criticized for his poor methodological policies and inadequate statistical tests (in addition to his obscure and careless writing). Such criticism is prompted by Piaget's claim that his research is scientific; hence, in the minds of the critics, it should be precise and clear. Freudian exposition is usually remarkably clear and its empirical content well identified. But, since Freud's approach is unabashedly organic and qualitative, it cannot be faulted for statistical inadequacies.

18. See, for instance, the substance of N. O. Brown's, *Life Against Death* (New York: Random House, 1959).

19. T. Thass-Thienemann, *The Subconscious Language* (New York: Washington Square Press, 1967).

20. The subject matter of Thass-Thienemann's work is firmly bound to the linguistic analysis of words in all their shades of meaning and to their changes through their known histories. It combines the understanding of a linguist with that of a psychologist. I find the basic tenets well established and the execution of the argument scholarly, hence the results credible. In the discussion which follows I shall reference only pages of conclusions or special points and refrain from giving examples of the material on which such results are based. The reason for this is that individual examples, if given, might easily sound like bad puns. Thass-Thienemann's arguments become convincing by reinforcement, as he demonstrates the existence of certain trends in hosts of related words and in a score of languages. Here is an example: that genital knowledge may "discover" the "naked truth," as compared to oral and ocular knowledge, sounds like a remark addressed to a lady philosopher before propositioning her. It becomes a statement in epistemology-as-psychology only after it is examined in its reflections through many languages and some three millennia of usage. The interested reader must consult the original.

21. Thass-Thienemann, *op. cit.*, p. 74 ff.

22. J. Cohen, "Ideas of Work and Play," *British Journal of Sociology*, 4 (1955), pp. 312–22.

23. K. Mannheim, *Ideology and Utopia*, tr. L. Worth and E. Shills (New York: Harcourt, Brace, n.d.), p. 240.

24. K. Mannheim, *Essays in the Sociology of Knowledge*, tr. and ed. P. P. Kecskemeti (London: Routledge, 1964), p. 137.

25. *Ibid.*, p. 141 ff. and p. 171 ff.

26. See, e.g., the volume of readings edited by J. E. Curtis and J. W. Petras, *The Sociology of Knowledge* (New York: Praeger, 1970), and start with the long introductory essay.

27. C. Lévi-Strauss, *The Savage Mind* (Chicago: University of Chicago Press, 1966), p. 248.

28. *Ibid.*, p. 269.

29. H. G. McCurdy, *The Personal World* (New York: Harcourt, Brace & World, 1961), p. 556.

30. *Patterning of Time*, p. 32.

31. J. E. Orme, *Time, Experience and Behavior* (New York: American Elsevier, 1969). See his chap. 4.

32. *Patterning of Time*, p. 222.

33. For an annotated bibliography as of 1965, see K. H. Craig, "Of Time and Personality," delivered at the *Symposium on Human Time Structure*, Meeting of the American Psychological Association, 1965. Unpublished manuscript.

34. R. H. Knapp, "A Study of the Metaphor," *Journal of Projective Techniques*, 24 (1960), pp. 389–95.

35. "Personality and the Psychology of Time," *Study v. 1*, pp. 312–19.

36. *Ibid.*, p. 316 ff.

37. *Patterning of Time*, p. 222 & passim; and Orme, *op. cit.*, p. 57.

38. Aristotle, *Politics* 1340–b–26.

39. R. S. Westfall, "Newton and the 'Fudge Factor,'" *Science*, 179 (1973), pp. 751–58.

40. M. Kline, *Mathematical Thought from Ancient to Modern Times* (New York: Oxford University Press, 1972), p. 873.

41. A. Roe, "The Psychology of the Scientist," *Science*, 134 (1961), pp. 458–59.

42. L. Dooley, "The Concept of Time in Defence of Ego Entegrity," *Psychiatry*, 4 (1941), p. 15.

43. *Science News*, 98 (April 17, 1971), p. 265.

44. Roe, *loc. cit.*, p. 459.

45. M. D. Austin, "Dream Recall and the Bias of Intellectual Ability," *Nature*, 231 (1971), p. 59. The terms in quotes were coined by Einstein who considered combinatory play a central feature of his productive thought.

46. L. Hudson, "Arts and Sciences," *Nature*, 214 (1967), pp. 968–69.

47. E. C. Ladd and S. M. Lispet, "Politics of Academic Natural Scientists and Engineers," *Science*, 177 (1972), pp. 1091–1100.

48. N. Pastore, "The Nature-Nurture Controversy: A Sociological Approach," *School and Society*, 57 (1943), pp. 373–77.

49. See, e.g., "Creationists and Evolutionists," *Science*, 178 (1972), pp. 724–29.

50. *Collected Scientific Papers of James Clerk Maxwell*, W. D. Niven, ed., (Cambridge: Cambridge University Press, 1890) vol. 2, p. 220.

51. For an overview, see R. A. Levine, *Culture, Behavior, and Personality* (Chicago: Aldine, 1973).

52. C. D. Darlington, *The Evolution of Man and Society* (New York: Simon & Schuster, 1971), p. 678.

53. The references to Darwin, Freud, and Russell are from R. K. Merton, "Paradigm for the Sociology of Knowledge," in Curtis and Petras, eds., *The Sociology of Knowledge*, p. 365. Currently Freud's male-centered philosophy has come under attack by women's liberation critics of psychoanalysis. For a recent book-length study of the relationship between truth as seen by one man,

and his social milieu, see A. Janik and S. Toulmin, *Wittgenstein's Vienna* (New York: Simon & Schuster, 1973).

54. "Animals studied by Americans rush about frantically and . . . at last achieve the desired result by chance. Animals observed by Germans sit still and think, and at last evolve the solution out of their inner consciousness." Quoted in Merton, *op. cit.*, p. 371.

55. S. Freud, "Moses and Monotheism," *The Standard Edition of The Complete Psychological Works of Sigmund Freud*, ed. J. Strachey (London: Hogarth Press, 1964), vol. XXIII, p. 107.

56. F. E. Dart and P. L. Pradhan, "Cross-Cultural Teaching of Science," *Science*, 155 (1967), pp. 439–56.

57. Quoted in *The New York Times*, Sept. 12, 1965.

58. J. Needham, "Science and China's Influence on the World," *The Legacy of China*, ed. R. Dawson (Oxford: Oxford University Press, 1964), pp. 234–308. "Europeans suffered from a schizophrenia of the soul, oscillating forever unhappily between the heavenly host on the one side and the 'atoms and void' on the other." *Ibid.*, p. 307. Cf. ref. 102, chap. 1.

59. *Ibid.*, p. 252.

60. See his *Clerks and Craftsmen in China and the West* (Cambridge: Cambridge University Press, 1970).

61. J. Needham, "Science and Society East and West," in *Society and Science*, M. Goldsmith, ed. (New York: Simon and Schuster, 1965), p. 149.

62. J. Needham, "Human Law and the Laws of Nature," *Technology, Science and Art: A Common Ground* (London: Hatfried College of Technology, 1961), pp. 3–26.

63. *Ibid.*, p. 11.

64. Needham points to the authoritarian assurance of Latin clergy, inherited from Rome and opposed by the explosion of the Reformation, that felt the need to formulate timeless, axiomatic propositions. "China, however, was algebraic and 'Babylonian,' not geometrical and 'Greek,' so opposition (to political propositions) tended to be practical and approximate rather than theoretical and absolute." "Time and Knowledge: In China and the West," *The Voices*, p. 623.

65. "rationes creandorum corporum mathematicas Deo coaeternas fuissee . . . id sciunt Christiani." *Harmonices Mundi* IV.1. Max Caspar, ed., *Johannes Kepler Gesammelte Werke* (München: C.H.Beck'sche Verlagsbuchhandlung, 1940), vol. VI, p. 219.

66. One need not proceed via the study of time to see this. Nicholas Rescher reaches similar conclusions in his *Scientific Explanation* (Glencoe: Free Press, 1970), appendix II. Warren Weaver directly equates scientific explanation with being in the position to predict and control the future in "Scientific Explanation," *Science*, 143 (1964), pp. 1297–300.

67. The authority of timeless mathematics seems to appeal more to those with less earthly authority. Mathematicians and natural scientists in the United States seem to differ in ethnic and class origins. Eminent scientists are predominately Protestant and native born from English, Scottish, and German stock. Among the mathematicians studied 38 per cent were Jewish, 16 percent Catholic; more than 50 percent were foreign born or first generation Americans. While natural scientists have been drawn from the upper and middle classes, more than 50 percent of the American born mathematicians were sons of blue-collar workers and farmers. *Science News*, 98 (July 4, 1970), p. 12.

68. K. V. Thomas, *Religion and the Decline of Magic* (London: Weidenfeld, 1971).

69. T. S. Kuhn, *The Structure of Scientific Revolutions* (Chicago: University of Chicago Press, 1962), p. X.

70. *Ibid.*, p. 91.

71. *Ibid.*, p. 116. Koyré had similar ideas when he held that "good physics is not made a priori. Theory precedes fact. Experience is useless because before any experience we are already in possession of the knowledge we are seeking." A. Koyré, *Metaphysics and Measurement* (Cambridge, Mass.: Harvard University Press, 1966), p. 13.

72. This sequence has been recognized by David Bohm who sees the history of physics as a history of new kinds of descriptions of the world, each an extension of our perceptual faculties, each increasingly abstract, and each a more broadly invariant formulation of changing (physical) truths. See the fine appendix to his, *The Special Theory of Relativity* (New York: Benjamin, 1965).

73. F. M. Dostoyevsky, *Notes From the Underground*, tr. B. G. Guerney, reprinted in C. Neider, ed., *Short Novels of the Masters* (New York: Rinehart & Co., 1948). Quote is from pp. 145–46.

74. Quoted without source identification in L. Mumford, *The Pentagon of Power* (New York: Harcourt, Brace & Jovanovich, 1971), p. 59.

75. *Polit. Phil.*, p. 8. (The quotation is from Cassirer.)

76. E. H. Erikson, *Childhood and Society* (New York: Norton, 1963), p. 97 & *passim*. Also *Identity, Youth and Crisis* (New York: Norton, 1968), p. 261 ff.

77. Erikson, *Identity, Youth and Crisis*, p. 271.

78. Erikson, *Childhood and Society*, p. 108 and its references.

79. K. Stern, *The Flight from Woman* (New York: Farrar, Strauss & Giraux, 1965).

80. This is an amusing derivation of the two sexes from a single, four-legged being. It is also an account of the unique position of sex organs in man which make face to face copulation possible. Furthermore, it claims to account for the subsequent division of temperament depending on whether a man or a woman derives from the male or female parts of the original living thing. "Each of us when separated, having one side only, like a flat fish, is but the tally-half of a man, and he is always looking for his other half." Plato, *Symposium*, 191D. Of the original creature only its name survives: the androgyne, the man-woman.

81. The quotation is from Einstein's 1921 lecture on "Geometrie und Erfahrung." It appeared as part 2 in A. Einstein, *Sidelines on Relativity*, tr. G. B. Jeffrey and W. Perrett (London: Methuen, 1922).

82. Gödel's Proof was first published in the *Monatschefte für Mathematik und Physik*, 38 (1931), pp. 173–98 under the title "Über formale unentscheidbare Sätze der Principia Mathematica und verwandte Systeme I." He intended to write a second part, but that was never published. In preparing those portions of the discussion which pertain to the ordinary interpretation of Gödel's proof, the following works were consulted. B. Meltzer's translation of the original paper (Edinburgh: Oliver and Boyd, 1962), which also contains a valuable introduction by R. B. Braithwaite; also E. Nagel and J. R. Newman, *Gödel's Proof* (New York: New York University Press, 1964); R. Schlegel, *Completeness In Science* (New York: Appleton-Century-Crofts, 1964), chap. 5; S. F. Baker, *Philosophy of Mathematics* (Englewood Cliffs: Prentice Hall, 1964), *passim;* and L. Menkin, "Are Logic and Mathematics Identical?" *Science*, 138 (1962), pp. 788–94.

83. It consists of assigning numbers to basic signs, to series of basic signs, to series of such series, and so forth, reminiscent of the levels of Russell's hierarchy of types. He assigned consecutive integers to logical symbols, consecutive prime numbers (beyond integers already chosen) to numerical variables, squares of consecutive prime numbers to sentential variables, and cubes of consecutive

prime numbers to predicate variables. This ingenious process, known as Gödel numbering, leads to a unique one-to-one representation of all formulas and sequences of formulas which occur in his proofs.

84. (1) Assume that regardless of what x stands for, a set can always be constructed such that x is a member thereof, if and only if certain conditions are fulfilled. (2) There must therefore exist a set, let us call it m, such that x is a member of m if and only if x is a set that is not a member of itself. (The set of all horses is not itself a horse; while the set of all printable pictures is itself a printable picture). Now, m itself is either a member of m, or it is not a member of m. Suppose (3a) that it is a member of m; then it does not fulfill the condition (2) that anything must fulfill to belong to m, therefore m *cannot be* a member of itself. Suppose (3b) that m is not a member of itself, hence m *is* a member of itself.

X / RELIGION, POLITICS, AND THE GOOD

1. C. Darwin, *The Origin of the Species and the Descent of Man* (New York: Modern Library, n.d.), p. 492 ff. M. T. Chiselin has traced the path of Darwin's thought about the evolution of morals as recorded in Darwin's unpublished notebooks. The essence of his finding is that Darwin envisioned his theory as a comprehensive system which should, therefore, account for the ontogeny and phylogeny of behavior. See "Darwin and Evolutionary Psychology," *Science*, 179 (1973), pp. 964–68.

2. See, e.g., the interesting summary by W. Thorpe, "Ethology and Consciousness," in *Brain*, pp. 470–505, and the extensive works of Konrad Lorenz.

3. R. J. Quinones, *The Renaissance Discovery of Time* (Cambridge, Mass.: Harvard University Press, 1972), p. 17 & *passim*. Cf. Plato, *Laws*, 721B

4. M. Eliade, *Cosmos and History* (New York: Harper & Row, 1959), p. 6 ff.

5. J. B. S. Haldane, "Animal Ritual and Human Language," *Diogenes*, 4 (1953), pp. 61–73.

6. R. Otto, *The Idea of the Holy*, tr. J. Harvey (Oxford: Oxford University Press, 1957).

7. J. G. Frazer, *The Golden Bough* (New York: Macmillan, 1925), p. 50.

8. B. Malinowski, "Magic, Science, and Religion," *Science, Religion and Reality*, ed. J. Needham (New York: Braziller, 1955), p. 85.

9. In chapter 4 of *The Golden Bough* (*op. cit.*), Frazer asks how it could happen that men of great intelligence and keen perception have not detected the fallacy of magic. One of the reasons he gives is that the ceremonies are intended to bring about a condition which could, in almost all cases, point to the coming about of the hoped-for condition, even if too late. A ceremony intended to make the rain fall is always followed by a rainfall, sooner or later (p. 59). There is an interesting temporal parameter hidden here. Before the idea of causation as an essential part of the validity of prediction could be appreciated, the observer had to have a certain degree of keenness about the importance of time. Only after he began to insist that the period of time connecting cause and proffered effect be connected by a continuous chain of actions filling that time, could scientific causation be employed as a test of actions.

10. For an introductory summary see S. G. F. Brandon, "Time and the Destiny of Man," in *The Voices*, pp. 140–57 and the references to Brandon's earlier writings.

11. S. G. F. Brandon, "A New Awareness of Time and History," lecture delivered at the 1966 Gallahue Conference at Princeton Theological Seminary on *Religious Pluralism and World Community,* unpublished ms.

12. S. G. F. Brandon, "Time As God and Devil," *Bulletin of the John Rylands Library,* vol. 47 (1964), pp. 12–31.

13. Eliade, *op. cit.,* p. 47 ff.

14. See, mainly, his *History, Time and Deity* (Manchester: Manchester University Press, 1965).

15. Cf. Needham, in *The Voices,* p. 111 ff.

16. Exodus 20:2–17. There exist another "Ten Commandments," the ritual decalogues of Exodus 34:14–26. That was also part of the bargain.

17. S. G. F. Brandon, *Jesus and the Zealots* (Manchester: Manchester University Press, 1967), p. 29 ff. This is a masterful study of the political factor in primitive Christianity.

18. Known to this writer in his childhood. Also some other curious hearsay, such as that Judas was a dedicated disciple of Christ, willing to sacrifice himself so that the master might meet his self-appointed fate; it forms the background to the life of Jesus in N. Kazantzakis' moving religious novel *The Last Temptation of Christ* (New York: Simon & Schuster, 1960). The same theme was recently revived in the modern passion of *Jesus Christ Superstar.*

19. E. Erikson, *Young Man Luther* (New York: Norton, 1958), p. 242 ff.

20. N. O. Brown, *Life Against Death* (New York: Random House, n.d.) p. 232 ff. Whereas a personified Devil has been slowly disappearing from among men, yielding to the many concepts of evil, it was recently resurrected on the pages of *Osservatore Romano* following the concern of Pope Paul VI with that very disappearance. The Devil, it is said, is a "perfidious and astute charmer who manages to insinuate himself into us by way of the senses, of fantasy, of concupiscence, of utopian logic, of disorderly social conduct." *The New York Times,* December 17, 1972.

21. J. B. Bury, *The Idea of Progress* (New York: Dover, 1955), p. 19.

22. L. Edelstein, *The Idea of Progress in Classical Antiquity* (Baltimore: The Johns Hopkins University Press, 1967).

23. H. Spencer, "Progress: Its Law and Cause," reprinted in *Herbert Spencer on Social Evolution,* ed. J. D. Y. Peel (Chicago: University of Chicago Press, 1972), pp. 38–52.

24. F. Nietzsche, *Beyond Good and Evil,* in *The Complete Works of Friedrich Nietzsche,* ed. O. Levy (Edinburgh: Foulis, 1911), vol. 12, sec. 260.

25. This often quoted and rather profound utterance is in A. N. Whitehead, *Adventures of Ideas* (New York: Macmillan, 1956), p. 41.

26. Plato, *Timaeus* 27C.

27. *Polit. Phil.,* p. 15.

28. Plato, *The Republic,* 508A–513E

29. Aristotle, *Nichomachean Ethics,* II. vi. 14

30. The references to Machiavelli's writings are from Quinones, *The Renaissance Discovery of Time* (Cambridge, Mass.: Harvard University Press, 1972), pp. 175–80.

31. I. Kant, *Foundations of the Metaphysics of Morals,* tr. L. W. Beck (New York: The Liberal Arts Press, 1959), p. 39.

32. In an epoch when the world of the law is often upheld and its spirit or intent is neglected, the interested reader will find inspiration in Hegel's *The Philosophy of Right,* tr. T. M. Knox (Chicago: Encyclopedia Britannica, 1953), *Great Books of the Western World,* vol 46.

33. Jeremy Bentham, *An Introduction to the Principles of Morals and Legislation* (New York: Doubleday, 1961), p. 17.

34. A. J. Ayer, *Language, Truth and Logic* (New York: Dover, 1946), p. 113.

35. It is impossible to elaborate here upon this fascinating theme. The interested reader may consult as a first source T. J. J. Altizer, *et al.*, eds., *Truth, Myth and Symbol* (Englewood Cliffs: Prentice Hall, 1962). See especially the article by Gregor Sebba on "Symbol and Myth in Modern Rationalistic Societies."

36. See reference note 74 of chapter 2. Also L. Mumford, *Technics and Civilization* (New York: Harcourt, Brace & World, 1963), chap. 1.

37. *Ibid.*, p. 110 ff.

38. See the quantitative and evaluative summary of his work in, C. Morris, *Varieties of Human Value* (Chicago: Chicago University Press, 1956).

39. Individual preferences were sought from some fifteen hundred college students from the USA, Canada, Japan, pre-revolutionary China, India, and several European countries. The 13 ways evolved pragmatically from seven combinations of three basic components of human personality postulated by Morris. They are the Dionysian (immediate satisfaction), the Promethean (manipulating the external world) and Buddhistic (self-controlling). Cf. the Knapp metaphor tests discussed earlier.

40. *Patterning of Time*, pp. 73–9 & *passim*.

41. A. Toffler, *Future Shock* (New York: Bantam Books, 1970). For a philosophical analysis, see "Future Shock, The Emerging Conflict Between Two Temporal Paradigms," unpublished paper by Ferrel M. Christensen.

42. J. Cohen, *Homo Psychologicus* (London: George Allen & Unwin, Ltd, 1970).

43. On this see the sensitive book by John Black, *The Dominion of Man* (Edinburgh: Edinburgh University Press, 1970).

44. The term is that of Theodore Roszak, *The Making of a Counter-Culture* (New York: Doubleday, 1969). "I have colleagues in the academy who have come within an ace of convincing me that no such things as 'The Romantic Movement' or 'The Renaissance' ever existed—not if one gets down to scrutinizing the microscopic phenomena of history. At that level, one tends only to see many different people doing many different things and thinking many different thoughts. How much more vulnerable such broad-gauged categorizations become when they are meant to corral elements of the stormy contemporary scene and hold them steady for comment! And yet that illusive concept called 'the spirit of the times' continues to nip at the mind and demand recognition." p. i. This observation made by Roszak illustrates one of the problems of positivistic reasoning: not seeing the obvious.

45. See chap. 7 in Cohen, *op. cit.*, and its references.

46. L. White, "The Historical Roots of Our Ecologic Crisis," *Science*, 155 (1967), pp. 1203–207.

47. The eighteenth century English poet Christopher Smart was never taken seriously, though that might have to do with the quality of his poetry rather than with its message.

> For I will consider my Cat Jeoffrey,
> For he is a servant of the Living God. . . .
> For he knows that God is his Savior. . . .
> For a mouse is a creature of great personal valor.
> "Jubilate Agno"

48. UNESCO, *Report on the Symposium on Culture and Science*, Document SHC. 71/CONF/1/15, Paris. "Distribution Limited." Prof. Abdelwahab

Bouhdiba of the University of Tunis held that the less-advanced countries cannot afford the luxury of disillusionment with science, "at present 'cultivated' by the developed countries, for science is, in their case, a necessary avenue to development." Science has been associated with the colonial powers, he said, so that many countries of the Third World even after achieving independence, regard science as an ornament, a means of acquiring prestige. The governments of many new countries "refer to the objectivity and rationality of science to justify undertakings and decisions totally devoid of such characteristics, while the country's own particular culture—laws, patterns of social behavior, ethics— tends to be relegated to the sphere of folklore." A subsequent discussion by other participants generally confirmed the view that, by whatever paths industrialization is achieved, the machine carries with it its particular ethos that tends to suppress historical and local differences.

49. H. Cox, *The Feast of Fools* (Cambridge, Mass.: Harvard University Press, 1969).

50. *Ibid.*, p. 15.

51. *Ibid.*, p. 240.

52. E. Erikson, *Identity, Youth and Crisis* (New York: Norton, 1968), p. 261.

53. *Ibid.*, p. 274.

54. C. D. Darlington, *The Evolution of Man and Society* (New York: Simon & Schuster, 1969), p. 62.

55. For a compassionate scientific discussion on the balance between social cost versus social contribution resulting from man's ethical interference in evolution see chap. 12 in T. Dobzhansky's, *Mankind Evolving* (New Haven: Yale University Press, 1962).

56. C. H. Waddington, *The Ethical Animal* (Chicago: Chicago University Press, 1960), p. 59.

57. *Ibid.*, p. 213.

58. His reasoning, to the extent that it can be extracted from his writing, seems to be this. The good is a summary idea of all moral sentiments; all moral sentiments are rooted in the respect felt by one individual for others. (This latter represents the major force in what Piaget calls the relation of cooperation, in opposition to the relations of constraint). Clearly, no respect can be extended from one individual to others without self-identity. J. Piaget, *The Moral Judgment of the Child* (London: Routledge & Kegan Paul, 1932), p. 102, 402 & *passim.*

59. The reader not familiar with the pertinent work may inform himself through the popular summary given by K. Lorenz, *On Aggression*, tr. M. K. Wilson (New York: Harcourt, Brace & World, 1966), and then consult the records of the ensuing debate. Lorenz and other ethologists of similar persuasion have held that aggression in man is likely to be instinctive. We have argued their point with reference to demonstration in studies in animal behavior. They have come under attack mainly by biologists who lean towards the interpretation of life in terms of precepts resembling those of physical science, that is, life interpreted in terms which cannot accommodate value or conflict. The curious blindness to the ubiquity of war in history can be explained only as a flight from the unresolvable conflicts of time.

60. S. Andreski, *Military Organization and Society* (Berkeley: University of California Press, 1968), p. 17. Using secondary sources, he erroneously identifies Han Fei Tzu as having lived in the fifth Century B.C. In fact, he died in 233 B.C.

61. Special Study Group (pseud.), *Report from Iron Mountain on the Possibility and Desirability of Peace* (New York: The Dial Press, 1967). It is a

curious commentary on the naiveté of the prevailing moralistic utopianism, hiding behind the security of exactness.

62. The reader might profit from reading an exchange of letters between Albert Einstein and Sigmund Freud on the subject, "Why War?" S. Freud, *The Standard Edition of the Complete Psychological Works*, J. Strachey, ed. (London: Hogarth Press, 1964), vol. 24, pp. 199–215.

63. See chap. 1 in E. Voegelin's, *The New Science of Politics* (Chicago: University of Chicago Press, 1966) on the correlation of social representation and social existence. For a critique see J. G. Gunnell, "The History of Political Philosophy and the Myth of Tradition," *Study v. 2.*

64. S. Toulmin, *An Examination of the Place of Reason in Ethics* (Cambridge: Cambridge University Press, 1968), p. 222.

65. S. Freud, "Thoughts on War and Death," *The Standard Edition of The Complete Psychological Works of Sigmund Freud*, vol. 14, p. 299.

XI / ARTS, LETTERS, AND THE BEAUTIFUL

1. Plato, *Symposium* 210A–212C.

2. Plato, *Republic* X. 596E.

3. *Ibid.,* 597A–598D.

4. W. H. Thorpe, "Ethology and Consciousness," in *Brain*, p. 489.

5. W. H. Thorpe, *Bird Song* (Cambridge: Cambridge University Press, 1961), p. 77 ff. Also R. A. Hinde, ed., *Animal Behavior* (New York: McGraw Hill, 1966), p. 332 ff.

6. *Ibid.,* p. 335 ff.

7. Singing is not limited to birds. There have been reports of whale songs which last from seven to thirty minutes (*Science News*, 97 [1970], p. 555) and of a young dromedary that gave a sound normally observed in newborn dromedaries, as soon as its head appeared from its mother's vulva. Hinde, *op. cit.*, p. 338.

8. Thorpe, "Ethology and Consciousness," *loc. cit.*, p. 487.

9. D. Morris, *The Biology of Art* (London: Methuen, 1966).

10. The happy, frightening, plaintive, or ominous song of the bird is the recurring theme in the poetry of the morning. See the volume, *Eos, an Enquiry into the Themes of Lovers' Meetings and Partings at Dawn in Poetry,* a volume as little known as it is delightful, edited by A. T. Hatto (The Hague: Mouton, 1965). It examines poetry from fifty-nine languages, with samples that span thirty-three centuries. Cf. C. Darwin, *Descent of Man* (New York: Modern Library, n.d.), p. 705. "Mr. Weir has told me of the case of a bullfinch which had been taught to pipe a German Waltz, and who was so good a performer that he cost ten guineas: when the bird was first introduced into a room where other birds were kept and began to sing, all the others, consisting of about twenty linnets and canaries, ranged themselves on the nearest side of their cages and listened with the greatest interest to the new performer."

11. Thorpe, "Ethology and Consciousness," *loc. cit.*, p. 488.

12. E. H. Gombrich, "Moment and Movement in Art," *Journal of the Warburg and Courtland Institutes*, 27 (1963), p. 296. He notes the well-known practice of motion picture advertisements: the scenes purported to be from the movie are usually not selected frames but specially posed tableaux.

13. *Ibid.,* p. 302.

14. *Ibid.,* p. 297.

15. Henri Matisse, "Notes of a Painter," (1908) reprinted in E. Vivas and M. Krieger, eds., *The Problems of Aesthetics* (New York: Holt, Rinehart & Winston, 1953), p. 261.

16. Gombrich, *loc. cit.*, p. 304.

17. "A green parrot is also a green salad *and* a green parrot. He who makes it only a parrot diminishes its reality." Picasso quoted in *The Observer* (July 10, 1960).

18. Gombrich illustrates one technique for the enhancement of the impression of motion upon which we may comment here. In Donatello's *Cantoria* a relief of dancing *putti* is partially hidden by columns. Gombrich attributes the visual enhancement of motion to the ambiguity known as the Poggendorff illusion (a line interrupted by a band tends to lose its identity). This, I believe, is true but incidental. In the *Cantoria* we see symbolism of the unpredictable (the happy, confused group of *putti*), symbolism of being (cyclically repetitive static columns), as well as the Poggendorff illusion. Identical dynamics may be seen in Van Gogh's *Rain*. The orderly countryside and the randomness of the rainlines make one hear the changing rhythm of the afternoon, and make one's senses tingle with the interplay of sensory modalities.

19. A. Ehrenzweig, *The Psycho-analysis of Artistic Vision and Hearing* (New York: Braziller, 1965), p. 189.

20. *Ibid.*, p. 172.

21. The interested reader will enjoy consulting the collection edited by Christian Zervos, entitled *Pablo Picasso* (Réimpression, Cahiers d'Art, 1967), 26 volumes.

22. In: G. Gamow, *Mr. Tompkins in Paperback* (Cambridge: Cambridge University Press, 1969).

23. G. Schaltenbrand, "Cyclic States as Biological Space-Time Fields," *Study*, v. 2.

24. E. Panofsky, *Studies in Iconology* (New York: Harper and Row, 1962), see ch. 3.

25. *Ibid.* Also J. Cohen, "Subjective Time," in *The Voices*, p. 274.

26. Panofsky, *op. cit.*, p. 81.

27. George Kubler, *The Shape of Time* (New Haven: Yale University Press, 1962), p. 83.

28. *Ibid.*, p. 37.

29. Show at summer exhibition at the Whitney and Guggenheim Museums, reported in *Art News* (Summer, 1969), pp. 40–3, by S. Burton, "Time on Their Hands."

30. A tool which suggests the possibility of new forms of art is the computer. In a paper on "Human Robots and Computer Art" (*History Today*, August, 1970, pp. 3–11) John Cohen asked this: "Since the myths of creation were composed men have tried to emulate the gods. Is the twentieth century computer capable of the daemonic urge?" He does not think so. As I see it, a computer, or any available technical display, is as challenging and potentially as great or disappointing as a brush. As an adjunct to man, as a new type of brush, computer art is as promising as its artist. For a visionary and fascinating probing into the effects of science and technology on the sculpture of the twentieth century see J. Burnham: *Beyond Modern Sculpture* (New York: Braziller, 1969), especially chapter 8 on "Robot and Cyborg Art."

31. Plotinus, *Ennead* II. 9. 16.

32. See his sensitive treatise called *On Music,* tr. R. C. Taliaferro, in *The Fathers of the Church* (New York: CIMA, 1947), vol. 4. "Music is the science of mensurating well [modulandi]," p. 172. Augustine's keen interest in music is

likely to have been at the roots of his insistence that we measure time in the mind. See ref. 51, chap. 1. We will also recall that Galileo's readiness to note the isochronism of the pendulum was certainly prepared, even if not explicitly prompted, by his education in music.

33. A. Schopenhauer, *The World as Will and Idea,* tr. R. B. Haldane and J. Kemp (London: Routledge & Kegan Paul, 1950), p. 335.

34. Suzanne Langer, *Feeling and Form* (New York: Scribner, 1953), p. 109.

35. P. Fraisse, *Les Structures Rhythmique* (Louvin Public University, 1956), p. 119. See this also for its valuable bibliography.

36. The musicological remarks are partly based on the following writings by George Rochberg: "Indeterminacy in the New Music," *The Score,* (January, 1960), pp. 9–19; "Duration in Music," *The Modern Composer and His Works,* ed. J. Beckwith and Udo Kasemets (Toronto: University Of Toronto Press, 1961), pp. 56–64; and "The New Image of Music," *Perspectives of New Music,* vol. 2, (1963) no. 1, p. 9.

37. C. K. Ogden and I. A. Richards, *The Meaning of Meaning* (New York: Harcourt Brace, 1938), see chap. 7 on "The Meaning of Beauty."

38. M. Weitz, *Philosophy of the Arts* (Cambridge, Mass.: Harvard University Press, 1950), p. 134 f.

39. Luther wondered once why the Devil should have all the good tunes; Nietzsche asserted that he would not believe in a God that did not dance. From two serious men, these are serious remarks.

40. S. G. F. Brandon, "Time and the Destiny of Man," in *The Voices,* p. 141.

41. Note the similarity of the ecstacy of rock 'n' roll and post-rock 'n' roll dancing to the tribal dances around the fire. Both customs perform group catharsis through ritualized motion of open sexuality. The difference is that, while in its early forms, the efficacy of the dance, as far as we know, was usually directed to a practical goal (for example, to obtain more or less rain), the ecstacy of contemporary dance in its group manifestation is without any specific social focus. Thus, the latter grows into a release of primordial emotions but is stripped of its simple goals by intellectual or commercial censorship; still, the resulting mass screaming, weeping, and "flipping" form an organic continuity in the long history of man's search for timelessness through the ecstacy of the dance. Orpheus, it will be recalled, was slain by females who were maddened by his lyre.

42. L. N. Tolstoy, *Kreutzer Sonata,* ch. XXIII. Quote is from *The Novels and Other Works of Leo Tolstoy,* (New York: Scribner, 1899).

43. Aristotle, *Problems* 920a5. Current understanding suggests that tastes, colors, and smells also bear some relation to dispositions but much less so than does music.

44. Homer, *Iliad* XXIV, lines 613–14. (Everyman's Library Edition, tr. Edward, Earl of Derby.)

45. Aristotle, *Poetics,* 1449a.

46. See the delightful work of Jane Ellen Harrison, *Themis, A Study of the Social Origins of Greek Religion* (Cleveland: World Publishing Co., 1927).

47. J. de Romilly, *Time in Greek Tragedy* (Ithaca: Cornell University Press, 1968), p. 141.

48. Aristotle, *Poetics,* 1449b.

49. *Ibid.,* 1450a.

50. de Romilly, *op. cit.,* p. 11.

51. R. J. Quinones, *The Renaissance Discovery of Time* (Cambridge, Mass.: Harvard University Press, 1972), p. 181.

52. "Lines of extension cannot be maintained. Historical connections vanish as the thread of the present is cut off from past and future. Hamlet says there will be no more marriages and Antony and Cleopatra and Romeo and Juliet are reminders of the tragic truth. The children revolt in *King Lear,* and Hamlet is deprived of his succession (while his mother marries the interloper). . . . The expectations that serve as attainable values in the histories are blighted at their very source in the tragedies. The flowers with which Gertrude had thought to deck Ophelia's marriage bed she strews on her grave. . . . The on-going rhythms of the race and civilization do not provide the needed consolation or triumph. Quite literally Hamlet must endure an 'orphanhood in time,' and find his redemption in the height and depth of the present." *Ibid.,* p. 362.

53. Quinones notes (*ibid.,* p. 365) that tradition honors Shakespeare's birthday and the day of his death both as April 23.

54. Aristotle, *Poetics,* 1449b.

55. The narrowing of the time span from Aristotle's instructions that tragedy should limit itself to "a single circuit of the sun" to Hotspur's cry, "time that takes survey of all the world must have a stop" suggests a historical and collective reevaluation of time, a cultural narrowing of the "Now" from days to hours.

56. T. Ungvári, "Time and the Modern Self, A Change in Dramatic Form," in *Study v. 1,* p. 475.

57. This absence may also be the loss of ability to celebrate—that is, to celebrate anything at all—as Harvey Cox so spiritedly argues in *The Feast of Fools* (Cambridge, Mass.: Harvard University Press, 1969). I believe that the loss of the capacity to celebrate is one symptom of the changing attitudes to time as suggested in the text.

58. F. Dürrenmatt, *Four Plays* (New York: Grove Press, 1965).

59. *Ibid.,* p. 349.

60. Dürrenmatt himself points to such differences between tragedy and comedy, though not in respect to his play, in his essay on "Problems of the Theater," included in his *Four Plays.*

61. Ungvári, *op. cit.,* p. 478.

62. S. Beckett, *Endgame* (New York: Grove Press, 1958).

63. No, to my knowledge, Beckett is not a student of relativity theory. Rather, both modern physics and modern stage reflect identical attitudes to the unresolvable conflicts of time.

64. F. Turner, *Shakespeare and the Nature of Time* (Oxford: Clarendon Press, 1971), p. 100 ff.

65. G. Anders, "Being without Time: On Beckett's Play *Waiting for Godot,*" in *Samuel Beckett,* ed. M. Esslin (Englewood Cliffs: Prentice Hall, 1965), pp. 140–51.

66. The three cruelest things according to an Assyrian saying are to wait for someone who comes not, to try to sleep and sleep not, to try to please and please not.

67. S. Burnshaw, *The Seamless Web* (New York: Braziller, 1970), p. 307.

68. *Ibid.,* p. 308.

69. Herbert Spencer, "On The Origin and Function of Music," *Fraser's Magazine,* October, 1857. Reprinted in: H. Spencer, *Essays on Education and Kindred Subjects* (New York: Dutton, 1963), pp. 310–30.

70. Let three poetic expressions stand here as lonely representatives of time and the edifice of *ars poetica.* One is an expression of anxiety, one of hope, and one of conciliation.

> Goodness me, the clock has struck
> Alackday, and fuck my luck.

"Cinderella at Midnight,"
from Kurt Vonnegut, Jr.,
Slaughterhouse Five (New York:
Dell, 1969).

and in a mystery to be
(when time from time shall set us free)
forgetting me, remember me

"in time of daffodils (who know"
e. e. cummings, 95 *Poems* (New
York: Harcourt, Brace & World,
1958), p. 16.

But the fountain sprang up and the bird sang down
Redeem the time, redeem the dream
The token of the word unheard, unspoken.

"Ash Wednesday" from
Collected Poems 1909–62
by T. S. Eliot

71. Experimental psychology, lagging centuries behind novelists, has made a few cautious steps toward studying the relationship of one variable (learning to postpone) to age, social class, and culture. *Patterning of Time*, p. 244 ff. One test consists of the spontaneous writing of stories, or the completion of given initial plots. The time span of the stories so generated is then tabulated against variables of age, social class, and culture. In a way all novels are such story completions.

72. G. Poulet, *Studies in Human Time,* tr. E. Coleman (New York: Harper and Row, 1956). The American edition contains a brief appendix on time and American writers.

73. The work of Proust, according to Poulet, is one immense "resonance box in which are perceived not only the *times* of an individual existence and the *timeless* traits of a particular spirit but where retrospectively are found also the *times* of French thought, to its origins . . . Like the vast *Summae* which were erected [during the middle ages but subsequently lost], all is simultaneously discovered here on the different levels which are the tiers of times." *Ibid.,* p. 321.

74. G. H. Ford, "Dickens and The Voices of Time," *Nineteenth Century Fiction,* 24 (1970), pp. 428–48.

75. J. H. Raleigh, *Time Place and Idea: Essays on the Novel* (Carbondale: Southern Illinois University Press, 1968), p. 54. See also Jerome Buckley's *The Triumph of Time* (Cambridge, Mass.: Harvard University Press, 1966).

76. "Very deep is the well of the past. Should we not call it bottomless? Bottomless indeed, if—and perhaps only if—the past we mean is the past merely of the life of mankind. . . . For the deeper we sound, the further down into the lower world of the past we probe and press, the more do we find that the earliest foundations of humanity, its history and culture, reveal themselves unfathomable. No matter to what hazardous lengths we let our line they still withdraw again, and further, into the depths, again and further are the right words, for the unresearchable plays a kind of mocking game with our researching ardours; it offers apparent holds and goals. . . ." Thomas Mann, *Joseph and His Brothers,* tr. H. T. Lowe-Porter (New York: Knopf, 1948), p. 3.

77. These are the author's opinions. For a different evaluation, see W. Follett, "Time and Thomas Mann," *Atlantic Monthly,* (June, 1938), pp. 792–94.

78. T. Mann, *The Magic Mountain,* tr. H. T. Lowe-Porter (New York: Knopf, 1952), p. 725.

79. *Ibid.*, p. 541.

80. See also Margaret Church, "Thomas Mann: Time," *The Hopkins Review*, 3 (1950), pp. 20–9.

81. A. Robbe-Grillet, "On Several Obsolete Notions," in, *For A New Novel*, tr. R. Howard (New York: Grove Press, 1965), pp. 25–47.

82. A. Robbe-Grillet, "Time and Description in Fiction Today," in, *For A New Novel*, pp. 143–56.

83. *Ibid.*, p. 155.

84. A. Toffler, *Future Shock* (New York: Bantam Books, 1970), p. 162.

85. The present book does not illustrate this trend.

86. A. Burgess, *The Novel Now* (New York: Norton, 1967), p. 21.

87. The interested reader will profit by inspecting the original work of Father Kircher. Athanasius Kircher, *Ars Magna Lucis et Umbrae*, 1646.

88. This fish is so combatant that it attacks its own image in the mirror. If the image is chopped at high rates the fish keeps on attacking it. If it is chopped at about 30 frames per second or less, the fish ceases its attacks. It is assumed that at the lower rates the fish does not perceive a continuity among the separated frames. See the excellent and interesting paper of G. A. Brecher "Die Entstehung und biologische Bedeutung der subjektiven Zeiteinheit,—des Momentes." *Zeit. f. Vergl. Physiologie*, 18 (1932) pp. 204–43.

89. *Ibid.* See the remark on the dog cinema, p. 224. See also Jakob von Uexküll and G. Kriszat, *Streifzüge durch die Umwelten von Tieren und Menschen* (Frankfurt: S. Fischer, 1970) pp. 33–5 and *passim*.

90. Robbe Grillett, *op. cit.*, p. 152. Cf. Jacques Brunnis, "Every Year in Marienbad," *Renaissance of the Film*, ed. J. Bellone (London: Collier, 1970).

91. Cf. W. A. Luchting, "Hiroshima, Mon Amour, Time and Proust," in *Renaissance of the Film*, pp. 105–26.

92. "The Seventh Seal," in *Four Plays of Ingmar Bergman*, tr. L. Malmstrom and D. Kushner (New York: Simon and Schuster, 1960), p. 106.

93. See plate 95 in C. W. Ceram, *Archeology of the Cinema* (New York: Harcourt Brace & World, 1965).

94. A. Hauser, *The Social History of Art*, tr. S. Godman (New York: Vintage Books, n.d.), vol. 4, p. 246.

INDEXES

AUTHOR INDEX

Italicized numbers refer to the *Notes and References*

SUBJECT INDEX

This index is for guidance only; it does not aspire to completeness.
Italicized numbers refer to the *Notes and References*.
The slash (/) is to be read as *and, versus,* or *or*.